Activating Unreactive Substrates

Edited by
Carsten Bolm and F. Ekkehardt Hahn

Further Reading

Toru, T., Bolm, C. (Eds.)

Organosulfur Chemistry in Asymmetric Synthesis

2008

ISBN: 978-3-527-31854-4

Dupont, J., Pfeffer, M. (Eds.)

Palladacycles

Synthesis, Characterization and Applications

2008

ISBN: 978-3-527-31781-3

Sheldon, R. A., Arends, I., Hanefeld, U.

Green Chemistry and Catalysis

2007

ISBN: 978-3-527-30715-9

Hiersemann, M., Nubbemeyer, U. (Eds.)

The Claisen Rearrangement

Methods and Applications

2007

ISBN: 978-3-527-30825-5

Dyker, G. (Ed.)

Handbook of C-H Transformations

Applications in Organic Synthesis

2 Volumes

2005

ISBN: 978-3-527-31074-6

Knochel, P. (Ed.)

Handbook of Functionalized Organometallics

Applications in Synthesis

2 Volumes

2005

ISBN: 978-3-527-31131-6

Activating Unreactive Substrates

The Role of Secondary Interactions

Edited by
Carsten Bolm and F. Ekkehardt Hahn

WILEY-VCH Verlag GmbH & Co. KGaA

The Editors

Prof. Dr. Carsten Bolm
RWTH Aachen
Institut für Organische Chemie
Landoltweg 1
52056 Aachen
Germany

Prof. Dr. F. Ekkehardt Hahn
Westfälische Wilhelms-Universität Münster
Anorganisch-Chemisches Institut
Corrensstr. 30
48149 Münster
Germany

Library of Congress Card No.: applied for

British Library Cataloguing-in-Publication Data
A catalogue record for this book is available from the British Library.

Bibliographic information published by the Deutsche Nationalbibliothek
The Deutsche Nationalbibliothek lists this publication in the Deutsche Nationalbibliografie; detailed bibliographic data are available on the Internet at http://dnb.d-nb.de.

Typesetting Thomson Digital, Noida, India
Printing betz-druck GmbH, Darmstadt
Binding Litges & Dopf Buchbinderei GmbH, Heppenheim

Printed in the Federal Republic of Germany
Printed on acid-free paper

ISBN: 978-3-527-31823-0

Contents

Activating Unreactive Substrates: The Role of Secondary Interactions.
Edited by Carsten Bolm and F. Ekkehardt Hahn

ISBN: 978-3-527-31823-0

Preface

The DFG Priority Program "Secondary interactions as governing principle for the directed functionalization of less reactive substrates" (SPP 1118) was initiated in 2001 and was funded for a six-year period until 2007. The program originated from two independent initiatives, which initially were focussed on substrate activation in inorganic and organic chemistry. Their merger brought together scientists from different research areas with broad expertise in the field of substrate activation. It was hoped that this approach would generate additional synergies. This concept proved most successful, and the results of the corresponding efforts by various research groups are summarized in the volume presented here.
The use of secondary interactions allows us to program molecular architectures in a way which results in a desired functionality. Such functionality can, for example, be a reactive position at a metal center, and can be determined by the electronic configuration, the orbital symmetry or the spin state of the metal center. Secondary effects caused by the ligands at a metal center are supposed to lead to a specific reactivity. The Priority Program focussed on initiatives to utilize such secondary interactions for the directed functionalization of less reactive substrates. In the chapters of this volume, researchers from various fields of inorganic and organic chemistry describe approaches which range from metal-based catalysts to organocatalysts. The understanding of molecular association processes has formed the basis for gaining insight into molecular orientation and dynamics for a certain reaction profile.
More than ever, substrate activation is becoming a central topic in modern synthetic chemistry. This type of research is alive and well in Germany as well as in the worldwide chemical community. This fascinating interdisciplinary field has matured over recent years, and we have no doubt that the success of the DFG Priority Program 1118 has contributed to this situation.

Aachen and Muenster,
November 2008

Carsten Bolm
F. Ekkehardt Hahn
(Coordinators of SPP 1118)

Activating Unreactive Substrates: The Role of Secondary Interactions.
Edited by Carsten Bolm and F. Ekkehardt Hahn

ISBN: 978-3-527-31823-0

List of Contributors

Nuri Abacilar
Fachbereich Chemie
Philipps-Universität Marburg
Hans-Meerwein Straße
35043 Marburg
Germany

Markus Albrecht
Institut für Organische Chemie
RWTH Aachen University
Landoltweg 1
52072 Aachen
Germany

Perdita Arndt
Leibniz-Institut für Katalyse e. V. an der
Universität Rostock
Albert-Einstein-Str. 29a
18059 Rostock
Germany

Ruediger Beckhaus
Institute of Pure and Applied Chemistry
University of Oldenburg
26111 Oldenburg
Germany

Matthias Beller
Leibniz-Institut für Katalyse e.V. an der
Universität Rostock
Albert-Einstein-Str. 29a
18059 Rostock
Germany

and

Center for Life Science Automation
University of Rostock
Friedrich-Barnewitz-str. 8
18119 Rostock
Germany

Klaus Bergander
Organisch-Chemisches Institut
Westfälische Wilhelms-Universität
Münster
Correnssstrasse 40
48149 Münster
Germany

Bianca Bitterlich
Leibniz-Institut für Katalyse e.V. an der
Universität Rostock
Albert-Einstein-Str. 29a
18059 Rostock
Germany

Activating Unreactive Substrates: The Role of Secondary Interactions.
Edited by Carsten Bolm and F. Ekkehardt Hahn

ISBN: 978-3-527-31823-0

Carsten Bolm
Institut für Organische Chemie
RWTH Aachen University
Landoltweg 1
52056 Aachen
Germany

Jens Bunzen
Kekulé-Institut für Organische Chemie
und Biochemie
Rheinische Friedrich-Wilhelms-
Universität Bonn
Gerhard-Domagk-Str. 1
53121 Bonn
Germany

Vladimir V. Burlakov
Russian Academy of Sciences
A. N. Nesmeyanov Institute of
Organoelement Compounds
Vavilov St. 28
117813 Moscow
Russia

Giuliano C. Clososki
Department Chemie und Biochemie
Ludwig-Maximilians-Universität
München
Butenandtstr. 5-13, Haus F
81377 München
Germany

Peter Comba
Anorganisch-Chemisches Institut
Universität Heidelberg
INF 270
69120 Heidelberg
Germany

Gerhard Erker
Organisch-Chemisches Institut
Westfälische Wilhelms-Universität
Münster
Corrensstrasse 40
48149 Münster
Germany

Chun-An Fan
Kekulé-Institut für Organische Chemie
und Biochemie
Rheinische Friedrich-Wilhelms-
Universität Bonn
Gerhard Domagk Str. 1
53127 Bonn
Germany

Alicja Franke
Department of Chemistry and Pharmacy
Friedrich-Alexander-University of
Erlangen-Nürnberg
Egerlandstr. 1
91058 Erlangen
Germany

Roland Fröhlich
Organisch-Chemisches Institut
Westfälische Wilhelms-Universität
Münster
Corrensstrasse 40
48149 Münster
Germany

Andreas Gansäuer
Kekulé-Institut für Organische Chemie
und Biochemie
Rheinische Friedrich-Wilhelms-
Universität Bonn
Gerhard Domagk Str. 1
53127 Bonn
Germany

Ekatarina Gaoutchenova
Fachbereich Chemie
Philipps-Universität
Hans-Meerwein Straße
35043 Marburg
Germany

Udo Garrelts
Fachbereich Chemie
Philipps-Universität
Hans-Meerwein Straße
35043 Marburg
Germany

Nina Gommermann
Department Chemie und Biochemie
Ludwig-Maximilians-Universität
München
Butenandtstr. 5-13, Haus F
81377 München
Germany

Andreas Grohmann
Institut für Chemie
Technische Unversität Berlin
Straße des 17. Juni 135
10623 Berlin
Germany

F. Ekkehardt Hahn
Institut für Anorganische und
Analytische Chemie
Westfälische Wilhelms-Universität
Münster
Correnssstraße 36
48149 Münster
Germany

Marko Hapke
Leibniz-Institut für Katalyse e. V. an der
Universität Rostock
Albert-Einstein-Str. 29a
18059 Rostock
Germany

Klaus Harms
Fachbereich Chemie
Philipps-Universität
Hans-Meerwein Straße
35043 Marburg
Germany

Jens Hartung
Fachbereich Chemie, Organische
Chemie
Technische Universität Kaiserslautern
Erwin-Schrödinger-Str. Geb. 54
67663 Kaiserslautern
Germany

Jürgen Heck
Department Chemie
Universität Hamburg
Martin-Luther-King-Platz 6
20146 Hamburg
Germany

Christoph Herrmann
Organisch-Chemisches Institut
Westfälische Wilhelms-Universität
Münster
Corrensstrasse 40
48149 Münster
Germany

Natalya Hessenauer-Ilicheva
Department of Chemistry and Pharmacy
Friedrich-Alexander-University of
Erlangen-Nürnberg
Egerlandstr. 1
91058 Erlangen
Germany

Michael Hill
Organisch-Chemisches Institut
Westfälische Wilhelms-Universität
Münster
Corrensstrasse 40
48149 Münster
Germany

Ulrike Jäger-Fiedler
Leibniz-Institut für Katalyse e. V. an der Universität Rostock,
Albert-Einstein-Str. 29a
18059 Rostock
Germany

Joo-Eun Jee
Department of Chemistry and Pharmacy
Friedrich-Alexander-University of Erlangen-Nürnberg
Egerlandstr. 1
91058 Erlangen
Germany

Anja Kaiser
Institut für Organische Chemie
Universität Regensburg
Universitätsstr. 31
93053 Regensburg
Germany

Peter Karbaum
Kekulé-Institut für Organische Chemie und Biochemie
Rheinische Friedrich-Wilhelms-Universität Bonn
Gerhard Domagk Str. 1
53127 Bonn
Germany

Gerald Kehr
Organisch-Chemisches Institut
Westfälische Wilhelms-Universität Münster
Corrensstrasse 40
48149 Münster
Germany

Florian Keller
Kekulé-Institut für Organische Chemie und Biochemie
Rheinische Friedrich-Wilhelms-Universität Bonn
Gerhard Domagk Str. 1
53127 Bonn
Germany

Marion Kerscher
Anorganisch-Chemisches Institut
Universität Heidelberg
INF 270
69120 Heidelberg
Germany

Berthold Kersting
Institut für Anorganische Chemie
Universität Leipzig
Johannisallee 29
04103 Leipzig
Germany

Ulf Kiehne
Kekulé-Institut für Organische Chemie und Biochemie
Rheinische Friedrich-Wilhelms-Universität Bonn
Gerhard-Domagk-Str. 1
53121 Bonn
Germany

Peter Kitaev
Department Chemie
Universität Hamburg
Martin-Luther-King-Platz 6
20146 Hamburg
Germany

Marcus Klahn
Leibniz-Institut für Katalyse e. V. an der Universität Rostock
Albert-Einstein-Str. 29a
18059 Rostock
Germany

Paul Knochel
Department Chemie und Biochemie
Ludwig-Maximilians-Universität
München
Butenandtstr. 5-13, Haus F
81377 München
Germany

Stephan Kohl
Institut für Chemie
Technische Unversität Berlin
Straße des 17. Juni 135
10623 Berlin
Germany

Burkhard König
Institut für Organische Chemie
Universität Regensburg
93040 Regensburg
Germany

Ulrich Kortz
School of Engineering and Science
Jacobs University Bremen
P.O. Box 750 561
28725 Bremen
Germany

Arne Lützen
Kekulé-Institut für Organische Chemie
und Biochemie
Rheinische Friedrich-Wilhelms-
Universität Bonn
Gerhard-Domagk-Str. 1
53121 Bonn
Germany

Claudia May
Fachbereich Chemie
Technische Universität Kaiserslautern
Erwin-Schrödinger-Str. Geb. 54
67663 Kaiserslautern
Germany

Bárbara Menéndez Pérez
Fachbereich Chemie, Organische
Chemie
Technische Universität Kaiserslautern
Erwin-Schrödinger-Str. Geb. 54
67663 Kaiserslautern
Germany

Thomas Nebe
Institut für Anorganische und
Analytische Chemie
Justus Liebig Universität Gießen
Heinrich-Buff-Ring 58
35392 Gießen
Germany

Anna-Katharina Pleier
Fachbereich Chemie
Technische Universität Kaiserslautern
Erwin-Schrödinger-Str. Geb. 54
67663 Kaiserslautern
Germany

Volker Raab
Fachbereich Chemie
Philipps-Universität
Hans-Meerwein Straße
35043 Marburg
Germany

Andreas Reis
Technische Universität Kaiserslautern
Fachbereich Chemie
Erwin-Schrödinger-Str. Geb. 54
67663 Kaiserslautern
Germany

Oliver Reiser
Institut für Organische Chemie
Universität Regensburg
Universitätsstr. 31
93053 Regensburg
Germany

Hongjun Ren
Department Chemie und Biochemie
Ludwig-Maximilians-Universität
München
Butenandtstr. 5-13, Haus F
81377 München
Germany

Christoph J. Rohbogner
Department Chemie und Biochemie
Ludwig-Maximilians-Universität
München
Butenandtstr. 5-13, Haus F
81377 München
Germany

Uwe Rosenthal
Leibniz-Institut für Katalyse e. V. an der
Universität Rostock
Albert-Einstein-Str. 29a
18059 Rostock
Germany

Caroline A. Schall
Institut für Organische Chemie
Universität Regensburg
Universitätsstr. 31
93053 Regensburg
Germany

Siegfried Schindler
Institut für Anorganische und
Analytische Chemie
Justus Liebig Universität Gießen
Heinrich-Buff-Ring 58
35392 Gießen
Germany

Harald Schmaderer
Institut für Organische Chemie
Universität Regensburg
93040 Regensburg
Germany

Anett Schubert
Fachbereich Chemie
Technische Universität Kaiserslautern
Erwin-Schrödinger-Str. Geb. 54
67663 Kaiserslautern
Germany

Dominik Schuch
Fachbereich Chemie, Organische
Chemie
Technische Universität Kaiserslautern
Erwin-Schrödinger-Str. Geb. 54
67663 Kaiserslautern
Germany

Michael Seitz
Institut für Organische Chemie
Universität Regensburg
Universitätsstr. 31
93053 Regensburg
Germany

Anke Spannenberg
Leibniz-Institut für Katalyse e. V. an der
Universität Rostock
Albert-Einstein-Str. 29a
18059 Rostock
Germany

Patrick Spies
Organisch-Chemisches Institut
Westfälische Wilhelms-Universität
Münster
Corrensstrasse 40
48149 Münster
Germany

Anke Spurg
Kekulé-Institut für Organische Chemie
und Biochemie
Rheinische Friedrich-Wilhelms-
Universität Bonn
Gerhard-Domagk-Straße 1
53121 Bonn
Germany

Gerald Stephan
Institut für anorganische Chemie
Christian Albrechts Universität zu Kiel
Otto Hahn Platz 6/7
24098 Kiel
Germany

Yu Sun
Fachbereich Chemie
Technische Universität Kaiserslautern
Erwin-Schrödinger-Str. Geb. 54
67663 Kaiserslautern
Germany

Jörg Sundermeyer
Fachbereich Chemie
Philipps-Universität
Hans-Meerwein Straße
35043 Marburg
Germany

Jiri Svoboda
Institut für Organische Chemie
Universität Regensburg
93040 Regensburg
Germany

Matthias Tamm
Institut für Anorganische und
Analytische Chemie
Technische Universität Braunschweig
Hagenring 30
38106 Braunschweig
Germany

Tobias Thaler
Department Chemie und Biochemie
Ludwig-Maximilians-Universität
München
Butenandtstr. 5-13, Haus F
81377 München
Germany

Werner R. Thiel
Fachbereich Chemie
Technische Universität Kaiserslautern
Erwin-Schrödinger-Str. Geb. 54
67663 Kaiserslautern
Germany

Man Kin Tse
Leibniz-Institut für Katalyse e.V. an der
Universität Rostock,
Albert-Einstein-Str. 29a
18059 Rostock
Germany

and

Center for Life Science Automation,
University of Rostock
Friedrich-Barnewitz-Str. 8
18119 Rostock
Germany

Felix Tuczek
Institut für anorganische Chemie
Christian Albrechts Universität zu Kiel
Otto Hahn Platz 6/7
24098 Kiel
Germany

Rudi van Eldik
Department of Chemistry and Pharmacy
Friedrich-Alexander-University of
Erlangen-Nürnberg
Egerlandstr. 1
91058 Erlangen
Germany

Siegfried R. Waldvogel
Kekulé-Institut für Organische Chemie
und Biochemie
Rheinische Friedrich-Wilhelms-
Universität Bonn
Gerhard-Domagk-Straße 1
53121 Bonn
Germany

Gotthelf Wolmershäuser
Fachbereich Chemie
Technische Universität Kaiserslautern
Erwin-Schrödinger-Str. Geb. 54
67663 Kaiserslautern
Germany

Stefan H. Wunderlich
Department Chemie und Biochemie
Ludwig-Maximilians-Universität
München
Butenandtstr. 5-13, Haus F
81377 München
Germany.

Jing-Yuan Xu (Current address)
Pharmacy College
Tianjin Medical University
300070 Tianjin
China

Dirk Zabel
Fachbereich Chemie
Technische Universität Kaiserslautern
Erwin-Schrödinger-Str. Geb. 54
67663 Kaiserslautern
Germany

Daniela Zeysing
Department Chemie
Universität Hamburg
Martin-Luther-King-Platz 6
20146 Hamburg
Germany

1
Chemistry of Metalated Container Molecules

Berthold Kersting

1.1
Introduction

Since the seminal work of Cram on cyclophanes and resorcinarenes [1] the host–guest chemistry of container molecules has been extensively investigated, and more sophisticated examples with other forms and larger cavities have been reported [2, 3]. Container molecules are of great interest, because their encapsulated guest species often exhibit novel and unusual properties which are not observed in the free or solvated state [4]. They are used today as probes of isolated molecules and of the intrinsic characteristics of the liquid state, and are capable of enantioselective recognition [5], reversible polymerization [6], isolation of reactive species [7], and promoting reactions within their interiors [8].

Several types of closed-shell host molecules with fascinating inner-phase binding and reactivity properties have been described in the literature [9]. These include covalently constructed molecular containers such as the carcerands [10], which encapsulate their guests irreversibly, and the calixarenes or resorcinarenes [11, 12], which have shell holes large enough to permit the exchange of guest molecules [13]. During the last two decades many examples of noncovalent molecular capsules formed by spontaneous aggregation via H-bonds or metal-ligand coordinative bonds have also been described [14, 15].

In recent years, the development of metalated container molecules has become an attractive research goal [16–18] because such compounds allow for an interplay of molecular recognition and transition-metal catalysis [19, 20]. Consequently, a number of research groups are involved in the development of new receptor molecules that create confined environments about active metal coordination sites. In the following, selected types of metalated container molecules and their properties are briefly discussed. Thereafter, our work in this field is reported.

Activating Unreactive Substrates: The Role of Secondary Interactions.
Edited by Carsten Bolm and F. Ekkehardt Hahn

ISBN: 978-3-527-31823-0

1.2
Metalated Container Molecules: A Brief Overview

Metalated container molecules can be viewed as a class of compounds that have one or more active metal coordination sites within or next to a molecular cavity. Several host systems are capable of creating such structures. The majority of these compounds represent mononucleating ligand systems, as for instance calixarenes [21], cyclodextrines [22, 23], and some functionalized tripod ligands [24–26]. Figure 1.1 shows some representative examples. Coordination cages with active coordination sites are comparatively rare [27].

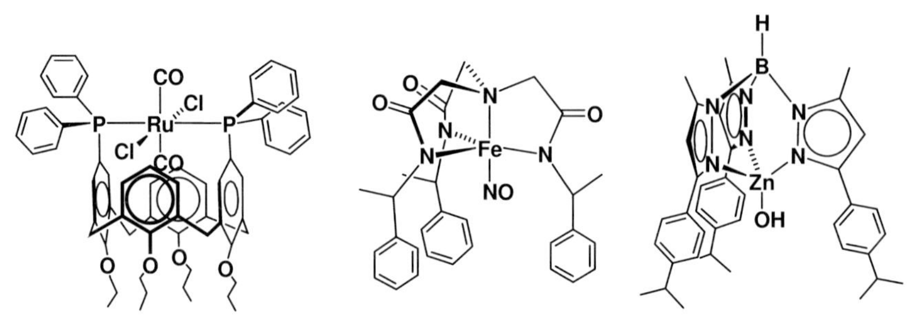

Figure 1.1 Structures of some mononucleating supporting ligands and their complexes [23–25].

Like their organic counterparts, metalated container molecules can not only stabilize reactive species or molecules in uncommon conformations [28] but also act as molecular reaction vessels for encapsulated substrates [29]. Moreover, they offer the advantage of stronger host–guest (substrate) interactions via coordinative bonds, thereby allowing activation and transformation of rather unreactive guest molecules [30, 31]. Complexes of functionalized ligand systems, in which the substituents serve to from a concave surface, have already been shown to exhibit enhanced chemical reactivity when compared with their unmodified analogs [32, 33]. They can be used as catalysts for selective organic transformations [20] and as catalysts for reactions which depend on the reaction medium [34]. Some complexes were also designed to mimic the hydrophobic binding site of metalloproteins [35]. The unusual properties can be traced back to complementary host–guest interactions and the distinct size and form of the binding cavities.

1.3
Metalated Container Molecules of Binucleating Supporting Ligands

Only a few ligand systems are known which impose cage-like structures about polynuclear core structures, and little is known about the chemistry of their corresponding complexes [36, 37]. These observations led us to study binuclear metal complexes with bowl-shaped binding cavities, hoping to modulate their chemical reactivity. In the following, we report the most important findings.

1.3.1
Synthesis

We have been concerned with the development of binucleating supporting ligands for metalated container molecules for now about 6 years [38]. The 24-membered hexaaza-dithiophenolate ligand H_2L^1 and its various derivatives (Scheme 1.1) were found to support such structures (see Section 1.3.2).

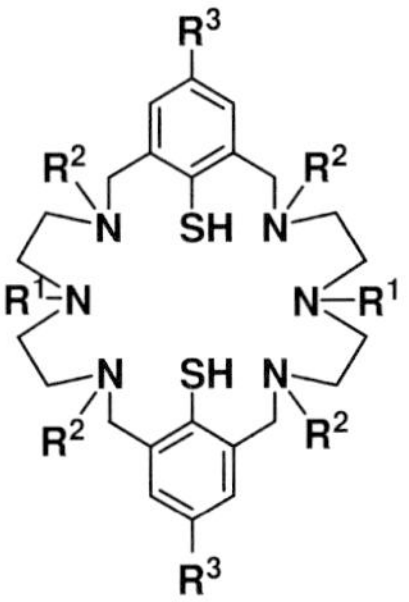

	R^1	R^2	R^3
H_2L^1	H	H	tBu
H_2L^2	CH_3	CH_3	tBu
H_2L^3	CH_3	H	tBu
H_2L^4	CH_3	CH_3	H
H_2L^5	CH_3	CH_3	Ph
H_2L^6	CH_3	CH_3	C_6H_4-tBu
H_2L^7	C_2H_5	C_2H_5	tBu
H_2L^8	C_3H_7	C_3H_7	tBu
H_2L^9	CH_3	C_2H_5	tBu
H_2L^{10}	$(CH_2)_2CN$	$(CH_2)_2CN$	tBu
H_2L^{11}	$(CH_2)_3NH_2$	$(CH_2)_3NH_2$	tBu
H_2L^{12}	$(CH_2)_2OCH_3$	$(CH_2)_2OCH_3$	tBu
H_2L^{13}	$(CH_2)_2OH$	$(CH_2)_2OH$	tBu

Scheme 1.1 Binucleating supporting ligands H_2L.

The new ligands are obtained in good overall yields without the need of metal templates. Scheme 1.2 illustrates the general procedure. Reductive amination of tetraaldehyde **1** [39] with diethylene triamine under medium-dilution conditions affords the macrobicyclic hexaaza-dithioether **2** in excellent yields (>90%) [40]. The aliphatic thioether linkage serves both as a protecting group for the air-sensitive thiophenolate groups and as a template function. It can be readily removed with Na/NH_3 to give the free ligands.

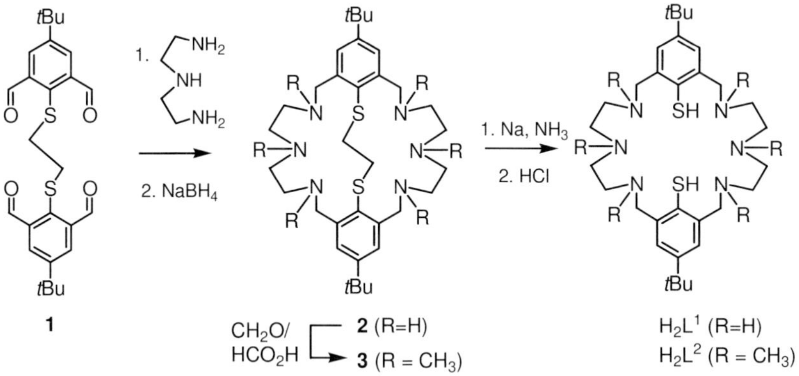

Scheme 1.2 Synthesis of H_2L^1 and H_2L^2.

An attractive feature of the macrobicycle **2** is that all of its six secondary amine functions are readily alkylated without affecting the masked thiolate functions. Thus, reductive methylation of **2** with formaldehyde and formic acid under

Eschweiler–Clarke conditions yields the permethylated derivative **3** in nearly quantitative yield [41] (Scheme 1.2). Likewise *N*-alkyl substituents sterically more demanding than a methyl group (H_2L^7-H_2L^9) can be readily introduced via acylation of **2** with the corresponding acid anhydrides (or acid chlorides) followed by $LiAlH_4$ reduction of the resulting amides [42]. The protocol is also applicable for the synthesis of the derivatives H_2L^4-H_2L^6, which offer differing degrees of encapsulation due to their depth ($R^3 = H$ [43], *t*Bu, Ph, 4-*t*Bu [44]). Moreover, compounds H_2L^{10}-H_2L^{13} containing R^1 and R^2 side arms with terminal CN, OMe, NH_2 and OH groups are also readily accessible [45].

The molecular structures of several of the hexaaza-dithioether intermediates have been determined by X-ray crystallography [43–47]. The macrobicycles adopt highly folded conformations, which are reminiscent of calixarenes [11] and related Schiff-base macrocycles [48, 49].

1.3.2
Coordination Chemistry of Binucleating Supporting Ligands

All ligands are effective dinucleating ligands that support the formation of bioctahedral $[M_2L(L')]^{n+}$ complexes with a range of divalent and trivalent metal ions (M = Mn^{II}, Fe^{II}, Co^{II}, Ni^{II}, Zn^{II} [42, 50], Cd^{II} [51], Co^{III} [52]). Table 1.1 lists a selection of the synthesized complexes and their labels. The macrocycles adopt either a C_{2v}-symmetric 'bowl-shaped' conformation of type B or a C_s-symmetric 'saddle-shaped' conformation of type A (Figure 1.2). The former is observed for multi-atom coligands

Table 1.1 Selected complexes, their labels and structures.[a)]

Complex		Structure	[d(M···M)/Å]	Ref.
4	$[Ni^{II}_2L^1(\mu\text{-Cl})]^+$	A	3.098(2)	[41]
5	$[Co^{II}_2L^1(\mu\text{-Cl})]^+$	n.d.[b)]		[40]
6	$[Ni^{II}_2L^2(\mu\text{-Cl})]^+$	A	3.184(1)	[41]
7	$[Co^{II}_2L^2(\mu\text{-Cl})]^+$	A	3.165(1) [3.194(1)][c)]	[54]
8	$[Ni^{II}_2L^3(\mu\text{-Cl}]^+$	A	3.074(1)	[47]
9	$[Ni^{II}_2L^9(\mu\text{-Cl}]^+$	A	3.2400(4)	[47]
10	$[Mn^{II}_2L^2(\mu\text{-OAc})]^+$	B	3.456(1)	[50]
11	$[Fe^{II}_2L^2(\mu\text{-OAc})]^+$	B	3.421(1)	[50]
12	$[Co^{II}_2L^2(\mu\text{-OAc})]^+$	B	3.448(1)	[50, 52]
13	$[Ni^{II}_2L^2(\mu\text{-OAc})]^+$	B	3.483(1)	[50, 54]
14b	$[Zn^{II}_2L^2(\mu\text{-OAc})]^+$	B	3.427(1)	[54]
15	$[Co^{II}_2L^8(\mu\text{-OAc})]^+$	B	3.482(1)	[42]
16	$[Zn^{II}_2L^8(\mu\text{-OAc})]^+$	B	3.460(1)	[42]
17	$[Ni^{II}_2L^8(\mu\text{-OAc})]^+$	B	3.513(1)	[42]
18	$[Ni^{II}_2L^2(\mu\text{-OH})]^+$	A	3.037(3)	[41]

a) The complexes were isolated as ClO_4^- or BPh_4^- salts.
b) Not determined.
c) There are two independent molecules in the unit cell. Value in square brackets corresponds to the second molecule.

L′, the latter for single-atom bridges [41, 40]. Complexes of type B are amongst the first prototypes for binuclear complexes with confined binding cavities [53]. The two aromatic rings of H_2L form the walls of the binding pockets.

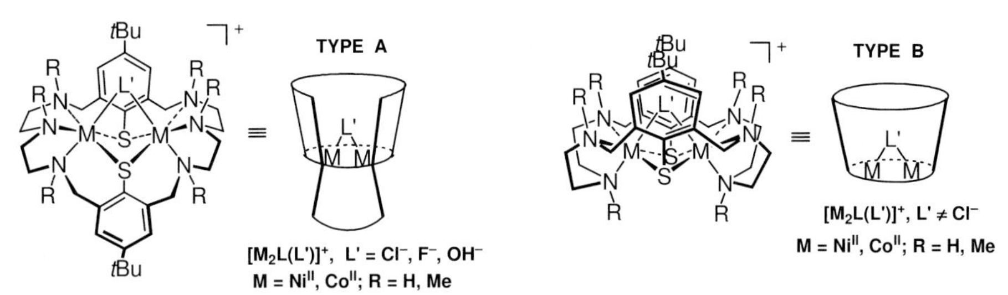

Figure 1.2 Structures of complexes of binucleating supporting ligands H_2L^1 and H_2L^2. The cavity representation used should not be confused with the one used for cyclodextrins [34].

1.3.3
Effects of *N*-alkylation on the Molecular and Electronic Structures of the Complexes

It is well established that the molecular and electronic structures of metal complexes of azamacrocycles are affected upon *N*-alkylation. This is mainly due to two factors: (i) the decrease in the ligand field strength and (ii) the increase in the steric requirements upon going from a secondary to a tertiary amine donor function [55]. In order to examine whether the properties of our complexes are affected by the *N*-alkyl substituents, we have characterized analogous complexes of the two macrocycles H_2L^1 and H_2L^2 by various spectroscopic methods (IR, UV/Vis, EPR, ^{57}Fe-Mössbauer spectroscopy), cyclic voltammetry, temperature-dependent magnetic susceptibility measurements and X-ray crystallography. The most important findings of these investigations are as follows:

- The conversion of the six secondary into tertiary amine donor functions does not change the overall structure of the $[M_2L(L')]$ complexes. That is, the macrocycle conformation remains constant for a given coligand [41]. This offers the opportunity to extend the rims of the binding pocket of the $[M_2L(L')]$ complexes.
- Complexes of the peralkylated macrocycles exhibit longer M-N and shorter M-L′ bond distances than the homologous complexes of the parent compound $(L^1)^{2-}$ [47]. In other words, *N*-functionalization leads to stronger metal–coligand bonding interactions, which in turn results in a stronger polarization of the bridging coligands and thus in higher reactivity of the complexes.
- The *N*-alkylated supporting ligands generate a more hydrophobic cavity about the active coordination site. The NH functions in **4**, for example, are involved in inter- and intra-molecular hydrogen bonding interactions, whereas no such interactions are observed for **6**. This drastically alters the ease of the substitution of the bridging coligands. Thus, while the latter complex reacts readily with NBu_4OH in

acetonitrile to produce the hydroxo-bridged complex **18**, the former complex was found to be unreactive [41].

- The donor strength of H_2L^2 is significantly weaker (by ca. 650 cm^{-1}) than its unmodified derivative as determined by UV/Vis spectroscopy [41]. The different ligand field strengths are also reflected in the redox potentials. In the $[M_2L^2(L')]$ complexes the M^{III} oxidation state is enormously destabilized compared with the M^{II} oxidation state, even in the case of Co [38, 52]. As a consequence of the weak ligand field of $(L^2)^{2-}$, the Mn^{II}, Fe^{II}, Co^{II} and Ni^{II} ions in the acetato-bridged complexes **10–13** adopt high-spin configurations [50].

Overall, the use of the permethylated derivative H_2L^2 in place of H_2L^1 influences many properties of the binuclear $[M_2L(L')]^+$ complexes, including color, molecular and electronic structure, hydrogen bonding interactions, redox potential, complex stability, and ground spin-state. The reactivity is also greatly affected (see Section 1.3.5).

1.3.4
The Ligand Matrix as a Medium

A series of zinc complexes (**14**) bearing different carboxylate coligands were prepared and characterized by NMR spectroscopy in solution (Table 1.2). The most important findings of these investigations are as follows:

- All complexes exist as discrete species in solution.
- The NMR signals of the coligands are shifted downfield from their values in the uncomplexed form [47, 51, 56]. The shift to higher field can be explained by the ring current effect. The coligands are located above the center of the two aryl rings of $(L^2)^{2-}$ in the shielding region [57]. This results in the observed chemical shift change to higher field. It clearly indicates that the ligand matrix functions as a medium.
- Ligand exchange reactions of the zinc complexes are rapid (Scheme 1.3). The equilibrium concentrations of the individual species are attained within the time scale of sample preparation (<30 s).

Table 1.2 Carboxylato-bridged zinc complexes **14a–14f**.

Complex		Coligand	Structure	pK_s	K_{rel}
14a	$[Zn^{II}_2L^2(O_2CH)]^+$	Formate	B	3.75	0.1
14b	$[Zn^{II}_2L^2(O_2CCH_3)]^+$	Acetate	B	4.75	1.0
14c	$[(Zn^{II}_2L^2(O_2CCH_2Cl)]^+$	Chloroacetate	B	2.86	1.1
14d	$[Zn^{II}_2L^2(O_2CCH_2CH_3)]^+$	Propionate	B	4.87	2.7
14e	$[Zn^{II}_2L^2(O_2CC_6H_5)]^+$	Benzoate	B	4.20	5.4
14f	$[Zn^{II}_2L^2(O_2CC_6H_4\text{-}p\text{-}CH_3)]^+$	4-Methyl benzoate	B	4.38	9.0

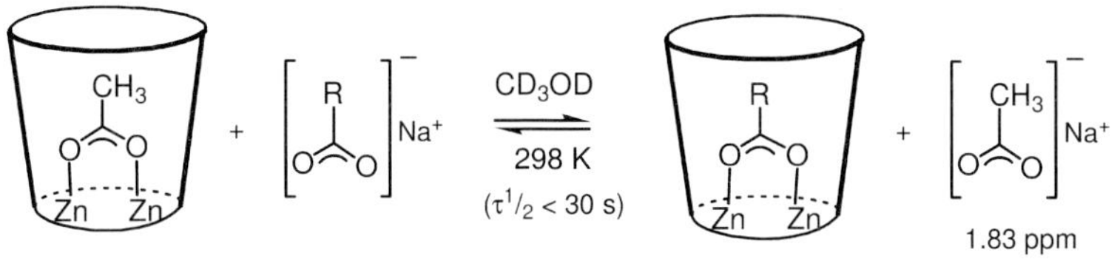

Scheme 1.3 Exchange of carboxylate ligands in dizinc complexes **14**.

In order to probe the hydrophobicity of the binding cavity of the $[Zn_2L^2]^{2+}$ fragment [58], the relative stability constants (K_{rel}) of the carboxylato-bridged zinc complexes were determined [59]. The concentrations of the species were measured by integration of the respective NMR signals and the relative stability constants were determined with the mass action law. Table 1.2 lists the so calculated values. As can be seen, the stability constants differ by about two orders of magnitude and are not correlated with the pK_a values of the acids. Rather, they increase in the order **14a** < **14b** < **14c** < **14d** < **14e** < **14f**. Thus, the larger the organic residue R of the carboxylate anion (RCO_2^-) the larger is the binding constant. The observed trend is indicative of hydrophobic effects (Van der Waals interactions) between the substituents of the carboxylate ion and the macrocyclic ligand. These interactions contribute to the stability of the complexes.

1.3.5
Variation, Coordination Modes and Activation of Coligands

The coordination chemistry of transition-metal complexes with well-defined binding pockets is of interest in many respects. In particular, by adjusting the size and form of the pocket it is often possible to stabilize reactive intermediates or to coordinate coligands in unusual coordination modes. Such assemblies also offer the opportunity to study secondary host–guest interactions between the coordinated coligands and the walls of the host [60]. For systematic investigations it is of importance that various coligands can be accommodated in the binding pocket of the container molecules. Single crystals suitable for X-ray structure determinations are also required such that one can study the effects of the size and form of the binding pocket on the coordination mode of the coligands and *vice versa*.

In this context it is to be noted that a large number of coligands can be readily accommodated in the binding pocket of the $[M^{II}{}_2L^2]^{2+}$ fragments [86]. These include Cl^- [41], OH^- [40], NO_2^-, NO_3^-, N_3^- [61], BH_4^- [62], various carboxylates and alkylcarbonates [43, 52, 63], N_2H_4, pyrazolate, pyridazine [61], $H_2PO_4^-$ [54], SH^- [64] and some biologically relevant molecules such as HCO_3^- [54], $(p\text{-}NO_2C_6H_4O)_2PO_2^-$ and proline [64], to name but a few. Single crystals of X-ray quality were obtained in each case.

More recent examples of metalated container molecules are **20–22** with classical [65, 66] or organometallic metallo coligands (Scheme 1.4) [64]. Such compounds are readily accessible from the labile ClO_4^- complexes **19a,b** [65]. They are stable in a range of organic solvents, and their charged nature gives them good solubility.

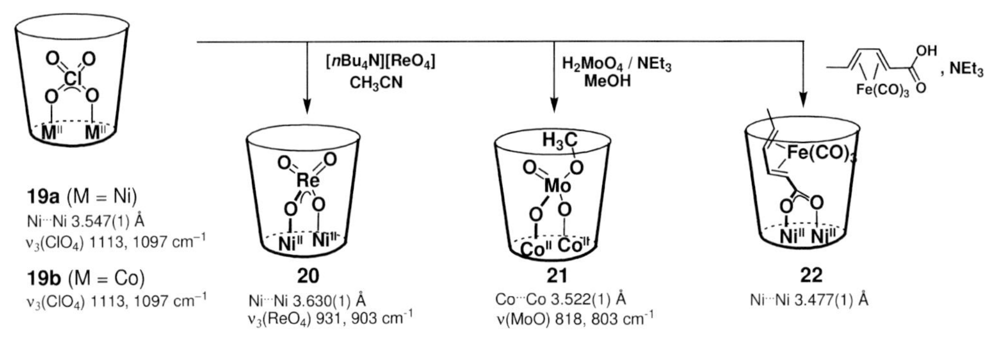

Scheme 1.4 Preparation of complexes **20–22**.

The binding modes of the coligands in the complexes **23–25** merit further consideration (Scheme 1.5). The hydrazine complex **23** provides an example of an unusual conformation of a small inorganic molecule. Free hydrazine exists predominantly in the *gauche* conformation at room temperature (dihedral angle $\tau \sim 100°$) [67]. In **23**, a *cis* (ecliptic) conformation ($\tau = 3.7°$) is observed. This conformation is presumably enforced by repulsive $NH \cdots C_{aryl}$ van der Waals interactions [68]. In **24**, an unusual conformation of an organic molecule is present [43]. The substituents in cyclohexene rings generally assume equatorial positions. In **24**, they are axial. The container molecules can also stabilize resonance extremes. Complex **25**, bearing a *p*-nitrophenolate in the quinoid *aci*-form is an example [69]. This clearly shows that $[M_2L]$ complexes can accommodate guest molecules in unusual coordination modes.

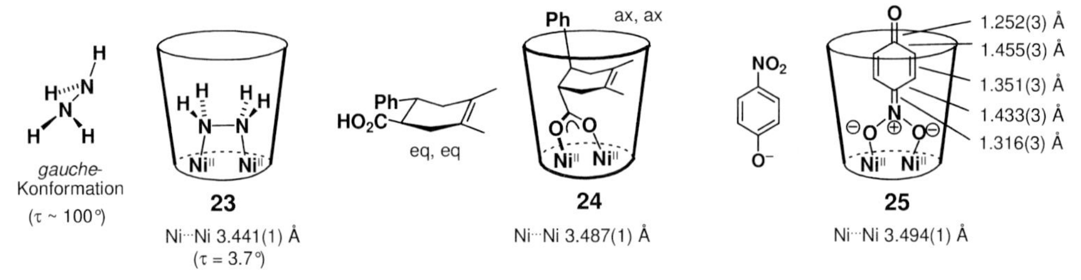

Scheme 1.5 Binding mode of the coligands in **23–25**.

Finally, reactive molecules can also be incorporated in the binding cavities. The BH_4^- complex **26** is an example (Scheme 1.6). In the absence of air and protic reagents this compound is stable for weeks both in the solid state and in solution [62]. This stability is quite remarkable given that nickel(II) complexes of sterically less demanding ligands are readily reduced to nickel boride. The BH_4^- becomes activated upon coordination as indicated by IR by the shifts of the B-H stretching frequencies

(Figure 1.3). The complex reacts rapidly with protic reagents (HCl, H_2O), with electrophiles (CO_2, S_8) and halogenated solvents (CCl_4).

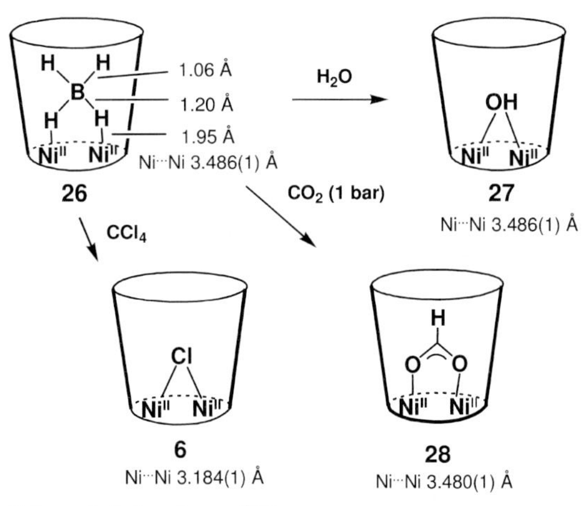

Scheme 1.6 Reactions of **26**.

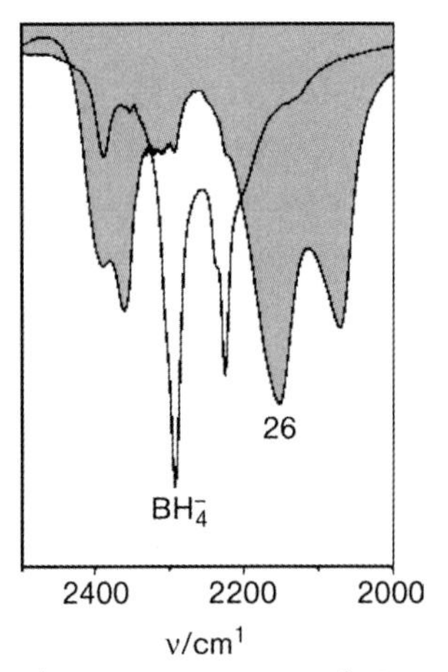

Figure 1.3 IR spectrum of **26**.

In summary, a large number of coligands can be accommodated in the binding pockets of the $[M_2L]^{2+}$ complexes. In each case only one of several possible coordination modes of the coligand is realized. In some cases, the binding pocket confers very unusual coordination modes or conformations on the guest molecules. Stabilization and activation of reactive molecules is observed in other instances. Since the complex integrity is retained in the solution state, the reactivity of these compounds can be examined (see below).

1.3.6
Reactivity of the Complexes

It has already been mentioned that utilization of the permethylated ligand $(L^2)^{2-}$ in place of $(L^1)^{2-}$ drastically alters the ease of substitution reactions of the $[M_2L(Cl)]^+$ complexes (Section 1.3.2). Further studies revealed a remarkable influence of the hydrophobic pocket on the rate and course of several substrate transformations, including for example the fixation of carbon dioxide [54] and the *cis*-bromination of α,β-unsaturated carboxylate ligands [63] (Scheme 1.7).

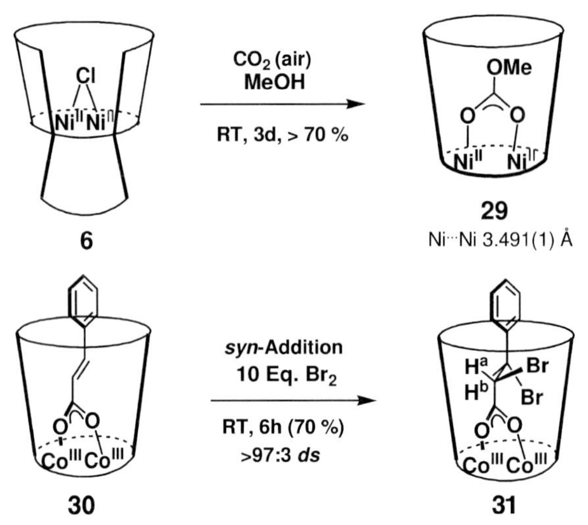

Scheme 1.7 Fixation of CO_2 and *cis*-bromination of α,β-unsaturated carboxylate ligands.

Recently, it has been found that the outcome of some cycloadditions can be altered remarkably when they are performed inside the cavity of cyclodextrines [70], self-assembled molecular capsules [71], or coordination cages [72]. This fact intrigued us greatly and awoke our interest in the Diels–Alder reactivity of the 'calixarene-like' $[M_2L^2(L')]^+$ complexes bearing unsaturated carboxylate coligands L′ [43]. The results are discussed in more detail.

The observation that α,β-unsaturated carboxylate ligands can be readily incorporated led us to study an orienting reaction between **35** and 2,3-dimethylbutadiene (**33**). The free reaction proceeds readily in toluene solution to give the adduct **34** (Scheme 1.8). However, the reaction of **35** with a large excess of **33** did not occur, even when heated at 210 °C for 24 h. The inhibition of the Diels–Alder reaction can be traced to the limited space in the binding pocket of **35**.

The above findings prompted us to study the reaction between a coordinated dienoate ligand and an external alkene. The reaction between sorbinic acid and acrylonitrile was selected (Scheme 1.9). This reaction is rather slow (pseudo-first-order rate constant $k' = 4.8 \times 10^{-6}\,s^{-1}$, $\tau_{1/2} \approx 2$ days) and produces a mixture of the four possible Diels–Alder adducts **36a–d** along with the by-product **38**. The acids

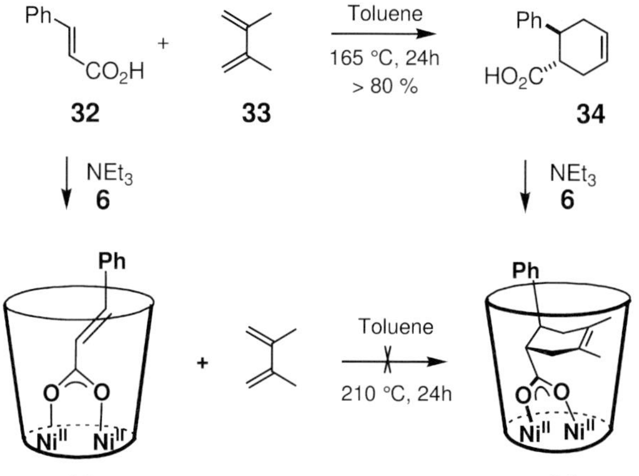

Scheme 1.8 Preparation of **34**, **35** and **24**.

isomerize under the basic reaction conditions to give the α,β-unsaturated derivatives **37a–d**.

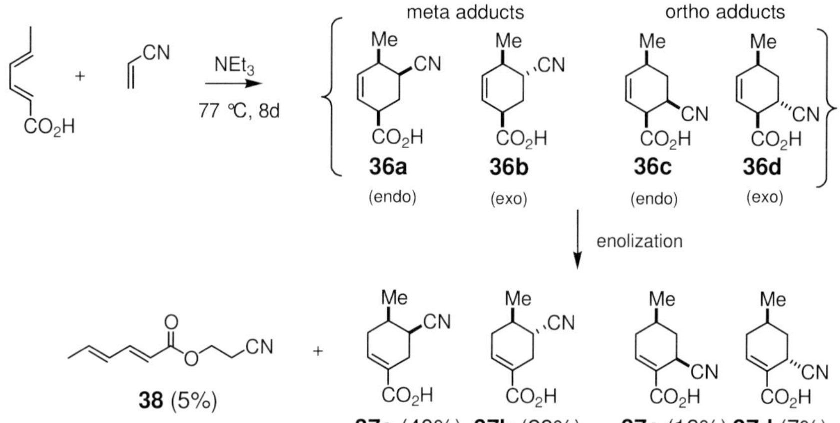

Scheme 1.9 Preparation of **36–38**. Numbers in parentheses refer to yields of isolated products.

Thereafter, the reaction between the coordinated dienoate ligand in **39** and acrylonitrile was examined (Scheme 1.10). Surprisingly, this reaction completes within 56 h (pseudo-first-order rate constant $k' = 1.4 \times 10^{-5}\,s^{-1}$, $\tau_{1/2} \approx 0.5$ days) and affords only two products, **40a** and **40b**, in a ratio of 57:43. The products **36a,b** can be liberated via acid hydrolysis. Thus, in striking contrast to the low regioselectivity observed in the background reaction, the Diels–Alder reaction between the encapsulated dienoate and acrylonitrile proceeds with strict 'meta' regioselectivity. In addition, no by-products are detected.

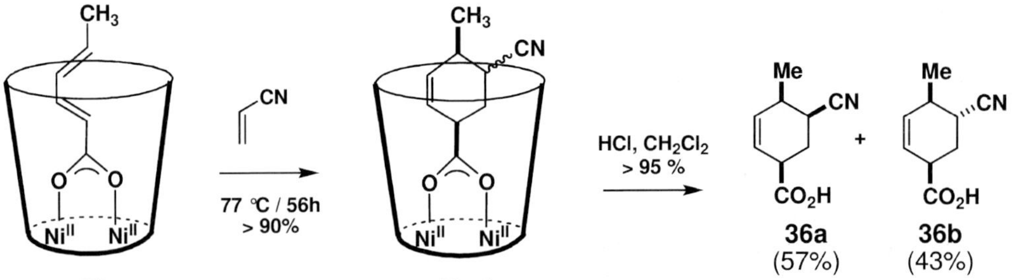

Scheme 1.10 Preparation of complexes **40a,b** and Diels-Alder products **36a,b**.

This and the fact that the Diels–Alder adducts **36a,b** do not isomerize in the binding pocket of the complexes can be attributed to the directing and protecting effect of the binding cavity as schematically represented in Scheme 1.11.

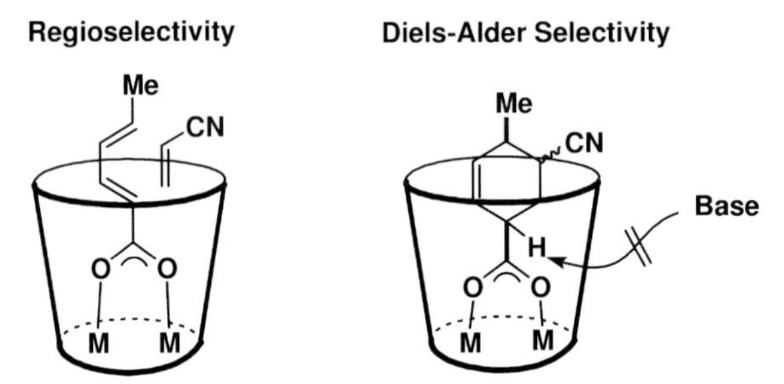

Scheme 1.11 Directing and protecting effect of the binding cavity.

Similar results were obtained for analogous complexes of the de-*tert*-butylated macrocycle $(L^3)^{2-}$ [43]. The *tert*-butyl substituents do not affect the regiochemistry of this particular Diels–Alder reaction, but they clearly increase its rate. The observed trend is indicative of a small stabilization of the transition-state by hydrophobic effects ($\Delta\Delta G^{\ddagger} \approx 3\,\mathrm{kJ/mol}$; $k'_{\mathrm{complex}}/k'_{\mathrm{background}} = \exp(\Delta\Delta G^{\ddagger}/RT)$). This would be consistent with our earlier observation that complexes bearing less polar carboxylate anions have the higher stability constants (see Section 1.3.4). Overall, the reaction between the coordinated sorbinate coligand and acrylonitrile is controlled by the binding cavity of the complexes and is highly regioselective. The new method is currently only applicable to dienes with anchoring carboxylate groups, but expansion of this approach to a general concept for the control of the regioselectivity of Diels-Alder reactions between unsymmetrical dienes and dienophiles appears to be within reach.

1.4 Conclusions

The development, synthesis, and coordination chemistry of novel metal complexes with deep binding cavities have been described. It was demonstrated that the doubly deprotonated forms of the macrocycles $(L)^{2-}$ support the formation of complexes of the type $[M_2L(L')]$ with bowl-shaped, 'calixarene-like' structures. The compounds are amongst the first prototypes for metalated container molecules with open binding cavities. Despite the fact that the active coordination site L' is deeply buried in the center of the molecules, it is readily accessible for a wide range of exogenous guest molecules. A large number of crystal structure determinations have provided insights into the binding modes of the coligands. These sense the size and form of the binding pocket of the $[M_2L]^{2+}$ fragments, as indicated by distinctive binding modes and unusual conformations. The use of the *N*-alkylated macrocycles in place of the unmodified analogs influences many properties of the binuclear complexes, including color, molecular and electronic structure, hydrogen bonding interactions, redox potential, complex stability, and ground spin-state. The reactivity is also greatly affected as shown by the activation and transformation of small molecules such as CO_2, the highly diastereoselective *cis*-bromination of α,β-unsaturated carboxylate coligands, and some regioselective Diels–Alder reactions.

Acknowledgments

Several coworkers named on our joint publications have made major contributions to the chemistry described here. This chemistry would not have reached its current level of broad synthetic utility without their dedication. I am particularly thankful for Prof. H. Vahrenkamp for providing facilities for spectroscopic and X-ray crystallographic measurements. Financial assistance by the Deutsche Forschungsgemeinschaft (SPP 1118, 'Sekundäre Wechselwirkungen'), the University of Freiburg, and the University of Leipzig is gratefully acknowledged.

References

1 Cram, D.J. and Cram, J.M. (1994) *Container Molecules and their Guests*, Royal Society of Chemistry, Cambridge.

2 Oshovsky, G.V., Reinhoudt, D.N. and Verboom, W. (2007) *Angewandte Chemie – International Edition*, **46**, 2366–2393.

3 Rissanen, K. (2005) *Angewandte Chemie – International Edition*, **44**, 3652–3654.

4 Hecht, S. and Fréchet, J.M.J. (2001) *Angewandte Chemie – International Edition*, **40**, 74–91.

5 Rivera, J.M., Martin, T. and Rebek, J., Jr (1998) *Science*, **279**, 1021–1023.

6 Ritter, H., Sadowski, O. and Tepper, E. (2003) *Angewandte Chemie – International Edition*, **42**, 3171–3173.

7 (a) Warmuth, R., Kerdelhué, J.-L. Carrera, S.S., Langenwalter, K.J. and Brown, N. (2002) *Angewandte Chemie – International Edition*, **41**, 96–99. (b) Roach, P. and Warmuth, R. (2003) *Angewandte Chemie – International Edition*, **42**, 3039–3042.

8 (a) Hof, F., Craig, S.L., Nuckolls, C. and Rebek, J., Jr (2002) *Angewandte Chemie – International Edition*, **41**, 1488–1508. (b) Hof, F., Craig, S.L., Nuckolls, C. and Rebek, J., Jr (2005) *Angewandte Chemie – International Edition*, **44**, 2068–2078.

9 (a) Vögtle, F. (1992) *Supramolekulare Chemie*, Teubner, Stuttgart. (b) Lehn, J.M. (1995) *Supramolecular Chemistry*, John

Wiley & Sons, Chichester. (c) Steed, J.W. and Atwood, J.L. (2000) *Supramolecular Chemistry*, John Wiley & Sons, Chichester.

10 Cram, D.J., Tanner, M.E. and Thomas, R. (1991) *Angewandte Chemie – International Edition*, **30**, 1024–1027.

11 Gutsche, C.D. (1998) Calixarenes, Revisited, *Royal Society of Chemistry*, Hertforshire.

12 McKinlay, R.M. and Atwood, J.L. (2007) *Angewandte Chemie – International Edition*, **46**, 2394–2397.

13 Lützen, A. (2005) *Angewandte Chemie – International Edition*, **44**, 1000–1002.

14 Yoshizawa, M., Tamura, M. and Fujita, M. (2006) *Science*, **312**, 251–254.

15 Fujita, M., Tominga, M., Hori, A. and Therrrien, B. (2005) *Accounts of Chemical Research*, **38**, 371–398.

16 Canary, J.W. and Gibb, B.C. (1997) *Progress in Inorganic Chemistry*, **45**, 1–83.

17 Armspach, D. and Matt, D. (1999) *Chemical Communications*, 1073–1074.

18 Cameron, B.R., Loeb, S.J. and Yap, G.P.A. (1997) *Inorganic Chemistry*, **36**, 5498–5504.

19 Molenveld, P., Engbersen, J.F.J. and Reinhoudt, D.N. (2000) *Chemical Society Reviews*, **29**, 75–86.

20 (a) Reetz, M.T. and Waldvogel, S.R. (1997) *Angewandte Chemie (International Edition in English)*, **36**, 865–867. (b) Reetz, M.T. *Catalysis Today*, (1998) **42**, 399.

21 Sénèque, O., Campion, M., Giorgi, M., Mest, Y.L. and Reinaud, O. (2004) *European Journal of Inorganic Chemistry*, 1817–1826.

22 Lejeune, M., Sémeril, D., Jeunesse, C., Matt, D., Peruch, F., Lutz, P.J. and Ricard, L. (2004) *Chemistry - A European Journal*, **10**, 5354–5360.

23 Engeldinger, E., Armspach, D., Matt, D., Jones, P.G. and Welter, R. (2002) *Angewandte Chemie – International Edition*, **41**, 2593–2596.

24 Hammes, B.S., Ramos-Maldonado, D., Yap, G.P.A., Liable-Sands, L., Rheingold, A.L., Young, V.G., Jr and Borovik, A.S. (1997) *Inorganic Chemistry*, **36**, 3210–3211.

25 Vahrenkamp, H. (1999) *Accounts of Chemical Research*, **32**, 589–596.

26 (a) Kitayima, N. and Tolman, W.B. (1995) *Progress in Inorganic Chemistry*, **43**, 419–531. (b) Trofimenko, S. (1999) *Scorpionates: The Coordination Chemistry of Polypyrazolylborate Ligands*, Imperial College Press, London U.K.

27 Marty, M., Clyde-Watson, Z., Twyman, L.J., Nakash, M. and Sanders, J.K.M. (1998) *Chemical Communications*, 2265–2266.

28 Rohde, J.-U., In, J.-H., Lee, M.H., Brennessel, W.W., Bukowski, M.R., Stubna, A., Münck, E., Nam, W. and Que, L., Jr (2003) *Science*, **299**, 1037–1039.

29 Ooi, T., Kondo, Y. and Maruoka, K. (1998) *Angewandte Chemie – International Edition*, **37**, 3039–3041.

30 Yandulov, D.V. and Schrock, R.R. (2003) *Science*, **301**, 76–83.

31 Castro-Rodriguez, I., Nakai, H., Zakharov, L., Rheingold, A.L. and Meyer, K. (2004) *Science*, **305**, 1757–1759.

32 Ruf, M. and Vahrenkamp, H. (1996) *Inorganic Chemistry*, **35**, 6571–6578.

33 Slagt, V.F., Reek, J.N.H., Cramer, P.C.J. and van Leeuwen P.W.N.M. (2001) *Angewandte Chemie – International Edition*, **40**, 4271–4274.

34 (a) Breslow, R. (1995) *Accounts of Chemical Research*, **28**, 146–153. (b) Breslow, R. and Dong, S. *Chem. Rev.*, (1998) **98**, 1997–2011.

35 Seneque, O., Rager M.-N. Giorgi, M. and Reinaud, O. (2001) *Journal of the American Chemical Society*, **123**, 8442–8443.

36 Fontecha, J.B., Goetz, S. and McKee, V. (2002) *Angewandte Chemie – International Edition*, **41**, 4553–4556.

37 Du Bois, J., Mizoguchi, T.J. and Lippard, S.J. (2000) *Coordination Chemistry Reviews*, **200–202**, 443–485.

38 Kersting, B. (2004) *Zeitschrift fur Anorganische und Allgemeine Chemie*, **630**, 765–780.

39 Kersting, B., Steinfeld, G., Fritz, T. and Hausmann, J. (1999) *European Journal of Inorganic Chemistry*, 2167–2172.

40 (a) Klingele, M.H. and Kersting, B. (2001) *Zeitschrift fur Naturforschung*, **56b**, 437–439. (b) Klingele, M.H., Steinfeld, G.

and Kersting, B. (2001) *Zeitschrift fur Naturforschung*, **56b**, 901–907.
41 Kersting, B. and Steinfeld, G. (2001) *Chemical Communications*, 1376–1377.
42 Gressenbuch, M., Lozan, V., Steinfeld, G. and Kersting, B. (2005) *European Journal of Inorganic Chemistry*, 2223–2234.
43 Käss, S., Gregor, T. and Kersting, B. (2006) *Angewandte Chemie – International Edition*, **45**, 101–104.
44 Gregor, T., Weise, C.F., Lozan, V. and Kersting, B. (2007) *Synthesis*, 3706–3712.
45 Gressenbuch, M. (2007) Ph.D. Thesis, Universität Leipzig.
46 Siedle, G. and Kersting, B. (2003) *Zeitschrift fur Anorganische und Allgemeine Chemie*, **629**, 2083–2090.
47 Gressenbuch, M. and Kersting, B. (2007) *European Journal of Inorganic Chemistry*, 90–102.
48 Branscombe, N.D.J., Atkins, A.J., Marin-Becerra A., McInnes, E.J.L., Mabbs, F.E., McMaster, J. and Schröder, M. (2003) *Chemical Communications*, 1098–1099.
49 Brooker, S. (2001) *Coordination Chemistry Reviews*, **222**, 33–56.
50 Journaux, Y., Glaser, T., Steinfeld, G., Lozan, V. and Kersting, B. (2006) *Dalton Transactions*, 1738–1748.
51 Lozan, V. and Kersting, B. (2005) *European Journal of Inorganic Chemistry*, 504–512.
52 Kersting, B. and Steinfeld, G. (2002) *Inorganic Chemistry*, **41**, 1140–1150.
53 Harding, C.J., McKee, V., Nelson, J. and Lu, Q. (1993) *Journal of the Chemical Society, Chemical Communications*, 1768–1770.
54 Kersting, B. (2001) *Angewandte Chemie – International Edition*, **40**, 3988–3990.
55 Barefield, E.K., Freeman, G.M. and Van Derveer D.G. (1986) *Inorganic Chemistry*, **25**, 552–558.
56 Fritz, T., Steinfeld, G., Käss, S. and Kersting, B. (2006) *Dalton Transactions*, 3812–3821.
57 Hausmann, J., Käss, S., Kersting, B., Klod, S. and Kleinpeter, E. (2004) *European Journal of Inorganic Chemistry*, 4402–4411.
58 Smithrud, D.B., Sanford, E.M., Chao, I., Ferguson, S.B., Carcanague, D.R., Evanseck, J.D., Houk, K.N. and Diederich, F. (1990) *Pure and Applied Chemistry*, **62**, 2227–2236.
59 Hausmann, J. (2001) *Diplomarbeit*, Universität Freiburg.
60 Engeldinger, E., Armspach, D., Matt, D., Jones, P.G. and Welter, R. (2002) *Angewandte Chemie – International Edition*, **41**, 2593–2596.
61 Hausmann, J., Lozan, V., Klingele, M.H., Steinfeld, G., Siebert, D., Journaux, Y., Girerd, J.J. and Kersting, B. (2004) *Chemistry - A European Journal*, **10**, 1716–1728.
62 Journaux, Y., Hausmann, J., Lozan, V. and Kersting, B. (2006) *Chemical Communications*, 83–84.
63 Steinfeld, G., Lozan, V. and Kersting, B. (2003) *Angewandte Chemie – International Edition*, **42**, 2261–2263.
64 Lozan, V. and Kersting, B. unpublished results.
65 Lozan, V. and Kersting, B. (2007) *European Journal of Inorganic Chemistry*, 1436–1443.
66 Lozan, V. and Kersting, B. (2006) *Inorganic Chemistry*, **45**, 5630–5634.
67 Hollemann, A.F. and Wiberg, E. (1985) *Lehrbuch der Anorganischen Chemie*, 91–100 edn, Walter de Gruyter, Berlin, pp. 557–559.
68 Meyer, E.A., Castellano, R.K. and Diederich, F. (2003) *Angewandte Chemie – International Edition*, **115**, 1244–1287.
69 Steinfeld, G. and Kersting, B. (2009) *Zeitschrift für Anorganische und Allgemeine Chemie*, in press.
70 Rideout, D.C. and Breslow, R. (1980) *Journal of the American Chemical Society*, **102**, 7816–7817.
71 Kang, J. and Rebek, J., Jr (1997) *Nature*, **385**, 50–52.
72 Kusukawa, T., Yoshizawa, M. and Fujita, M. (2002) *Angewandte Chemie – International Edition*, **41**, 1403–1405.

2
The Chemistry of Superbasic Guanidines

Jörg Sundermeyer, Volker Raab, Ekatarina Gaoutchenova, Udo Garrelts, Nuri Abacilar, and Klaus Harms

The current report is devoted to recent developments in the chemistry of multidentate guanidines as receptors or ligands for protons or metal cations. Whereas amines, imines and azaaromatic donors have been traditionally used as building blocks for the construction of multidentate ligands, up to the year 2000 no such systematic investigations were known for multidentate bis- and tris- or higher guanidines.

Inspired by the reports of Simchen [1] and Barton [2] on the synthesis and the highly basic and less nucleophilic character of sterically hindered peralkyl monoguanidine bases, we embarked on a research program dedicated to the development of such multifunctional guanidines. This investigation was funded by the DFG priority program 1118 focussing on the activation of unreactive substrates via secondary ligand substrate interactions. The present account discusses a selection of our results in the context of the work of other groups.

2.1
Properties of the Guanidine Functionality

The most prominent feature of a guanidine base is its pronounced proton acceptor character. Peralkyl guanidines, in particular, are among the strongest purely organic neutral bases known. Because of excellent stabilization of the positive charge in their resonance-stabilized guanidinium cations, the corresponding proton acceptors are several orders of magnitude superior in their basicity to tertiary amines: for example, the pentamethyl guanidinium ion $[H\text{-}NMe\text{-}C(NMe_2)_2]^+$ has a pK_{BH^+} (MeCN) of 25.00. It is more than six orders of magnitude less acidic than the 1,2,2,6,6-pentamethylpiperidinium cation with a pK_{BH^+} (MeCN) of 18.62 [3]. For this reason, guanidines appear to exist almost exclusively in their protonated form in aqueous media, and accurate pK measurements are typically conducted in less acidic solvents such as MeCN, DMSO or THF.

Activating Unreactive Substrates: The Role of Secondary Interactions.
Edited by Carsten Bolm and F. Ekkehardt Hahn

ISBN: 978-3-527-31823-0

It was suggested that superbases are those proton acceptors that show a higher gas-phase basicity (GB) or proton affinity (PA) – calculated for 298 K applying the TΔS term – than 1,1,3,3-tetramethylguanidine (TMG) with a GB of 982.8 kJ mol^{-1}, a PA of 1017.0 kJ mol^{-1} [4], and a pK_{BH^+} (MeCN) of 23.3 [3]. Because of solvation effects, which depend on the shape, polarity, polarizability and hydrogen bonding properties of both solvent molecules and bases, GB or PA values do not exactly correlate with the basicity scale. We suggest that the borderline for superbasicity – if it has to be defined arbitrarily – should be a calculated GB of 1000 kJ mol^{-1} and a pK_{BH^+} (MeCN) of 25.0 determined via spectrophotometric [5] and/or nmr titration methods [6] with reference bases. The measured pK_{BH^+} (MeCN) of pentamethylguanidine (PMG) is 25 [3]. This is slightly higher than the reported pK_{BH^+} (MeCN), 24.23, of the amidine reference base DBU. However within the limits of error of the calculations, PMG and DBU have the same GB of 1000 ± 3 kJ mol^{-1} [4]. Our own preparative investigations were accompanied by computational studies by Maksic *et al.* who used an isodensity polarized continuum model for MeCN ($\varepsilon = 36.64$) to calculate pK_{BH^+} (MeCN) values from absolute PA values. Calculated data are in perfect accord with experimental data [7].

The guanidine moiety is incorporated in many biological complexes, for example in the form of guanine, the DNA nucleobase matching as hydrogen bond receptor with cytosine, or in the form of the amino acid arginine that serves as a water solubility mediating functional group or as a carboxylate-specific hydrogen bond donor in proteins and other natural products.

Coordination chemists are interested in the fact that guanidines are neutral ligands with a basicity in between the basicity of tertiary amines and anionic amido ligands [NR_2]. The question arises whether the high basicity correlates with a strong π donor and perhaps even a weak π donor character of these neutral ligands toward various metal cations. In fact, it was suggested that in hydrophobic domains of cytochrome C enzymes the guanidine functionality of arginine may bind as neutral *N*-donor to metal cations: this was proven for $[PtCl(trpy)]^+$ as an artificial marker [8].

2.2 Design of Superbasic Proton Sponges

Since the discovery of the unusual basicity of 1,8-diaminonaphthalene (DMAN) by Alder *et al.* [9], the so-called 'proton sponges' received continuous interest from a number of research groups [10]. A proton sponge can be envisaged as the perfect chelating pincer ligand for the smallest electrophile known, the proton. A typical feature of a proton sponge is the steric demand of the chelate functionalities, which inhibits very fast intermolecular proton exchange and also an orientation of the nitrogen lone pairs allowing the uptake of one proton in an intramolecular [N–H...N] hydrogen bond (IHB). An increase of basicity is achieved by the destabilization of the base as a consequence of strong repulsion of unshared electron pairs and by relief from this steric strain upon protonation. Three general concepts to raise the thermodynamic basicity or proton affinity have been followed: one is to replace the naphthalene skeleton by other aromatic spacers, another is to promote basicity by the substituents at the spacer ('buttressing effect'), and the obvious third one, which has been followed in our research, is to replace the tertiary amino functionality by intrinsically superbasic donor functionalities.

Scheme 2.1 displays the evolution of proton pincers of increasing basicity in our laboratory.

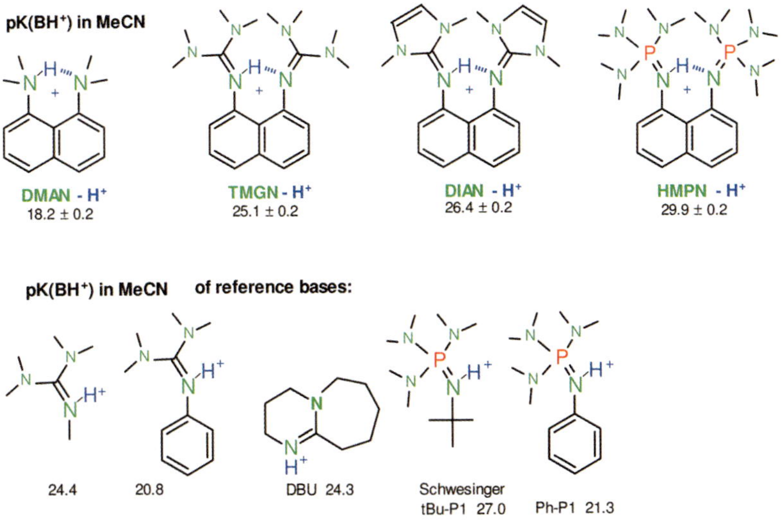

pK(BH$^+$) in MeCN

DMAN - H$^+$ 18.2 ± 0.2

TMGN - H$^+$ 25.1 ± 0.2

DIAN - H$^+$ 26.4 ± 0.2

HMPN - H$^+$ 29.9 ± 0.2

pK(BH$^+$) in MeCN of reference bases:

24.4

20.8

DBU 24.3

Schwesinger tBu-P1 27.0

Ph-P1 21.3

Scheme 2.1 DMAN [9] and superbasic proton sponges TMGN [6], DIAN [11, 12], HMPN [13], their experimental $pK_{BH}{}^{+}$ (MeCN) values at 25°C and selected reference bases.

TMGN is a crystalline compound, relatively easy to synthesize even in large quantities by activating tetramethyl urea with phosgene or oxalylchloride and coupling the intermediate *N,N,N′,N′*-tetramethyl-chlorformamidinium chloride with 1,8-diaminonaphthalene [6, 14]. TMGN has become commercially available from FLUKA. It has a thermodynamic basicity nearly seven orders of magnitude higher than classical DMAN. Furthermore, proton self-exchange nmr studies reveal that, because of its sterically less hindered nature compared to DMAN, it has a much higher kinetic basicity [6]. This proton activity could be an advantage if proton sponges are not only used as proton scavengers but as catalytic proton mediators.

The basicity of guanidine-based proton sponges can be enhanced by incorporating nitrogen lone pairs of the CN_3 unit into planar aromatic 1,3-dialkyl-1,3-imidazole-2-ylidenamino functionalities as realized in DIAN (Scheme 2.3). This strategy leads to a strong polarization of the exocyclic C–N bond into a zwitterionic bonding situation with a quasi-anionic exocyclic nitrogen atom [15]. DIAN has an experimental $pK_{BH}{}^{+}$(MeCN) of 26.4. We find that the corresponding saturated imidazoline DMEGN [11] derivative is about three orders of magnitude less basic than its quasi aromatic counterpart DIAN.

Up to now the most basic proton sponge is the bis-phosphazene proton pincer 1,8-bis(hexamethylphosphazenyl)naphthalene HMPN with an experimental and calculated $pK_{BH}{}^{+}$(MeCN) of about 30 [13]. HMPN represents a combination of Schwesinger's monodentate phosphazene base concept [16], and Alder's chelating proton sponge principle. The thermodynamic basicity of all new proton pincers was measured by nmr titration experiments with corresponding neutral organic super bases of known $pK > 25$ in CD_3CN, while their kinetic basicity was measured by dynamic nmr proton exchange measurements. Their nucleophilic versus basic character was studied in competitive reactions with ethyl iodide and their hydrolytic stability in reactions with aqueous DCl and NaOD.

All these proton sponges were structurally characterized by XRD analyses of their base form and their protonated form. We always observe an asymmetric intramolecular hydrogen bridge in the solid state. However, in solution, both pincer arms are chemically equivalent at all temperatures on the nmr time scale. It is this asymmetric IHB, a *secondary intramolecular interaction of the proton* with the second *N*-donor that is causing the dramatic increase of both proton affinity and basicity. In this respect it is interesting to compare the basicity of the chelating naphthalene-based pincer ligands with the basicity of the monodentate phenyl-substituted proton acceptors cut in half. The secondary chelate effect increases basicity of guanidines by about 4 orders and of phosphazenes by nearly 9 orders of magnitude! Because of the secondary intramolecular interaction, HMPN becomes about two orders of magnitude more basic than monodentate Schwesinger base *tert*-Bu-P_1.

2.3
Some Perspectives in Proton Sponge Chemistry

TMGN, despite its narrow bite angle, can be used as a constrained chelate ligand to metal cations such as Be^{2+} and even to larger coinage and platinum metals [12]. We believe that a combination of the strong donor capability, steric strain and demand will lead to unusual properties of corresponding complexes.

An ongoing study is the use of TMGN and related chelating guanidines as ligands to copper(I). Cu(I) halides form nicely crystalline complexes and were tested for their catalytic ability to promote copper-catalyzed Goldberg–Buchwald type amidations of aryl halides [17]:

I
+
O
N
H
Me
5mol% CuI, toluene
10mol% diamine ligand
vs 10mol% bis-guanidine
K_3PO_4, 110 °C, 24h
O
N
Me

One of the best ligands for this conversion is the secondary amine N,N′-dimethylethylene diamine (DMEDA) with a reported yield (GC) of 67%. Tertiary amine TMEDA did not yield any product [17]. Following exactly the Buchwald protocol but replacing the bis-amines by bis-guanidines, we obtained isolated yields of 30% with the proton-sponge complex [Cu(TMGN)I] (see Figure 2.1) and 14% isolated yield for the aliphatic bis-guanidine [Cu(TMGEn)I] (TMGEn = 1,2-tetramethylguanidinoethane). Even though these are very preliminary results, we learn that guanidines, and in particular the proton sponge TMGN, are promising catalyst ligands. Further studies with *NH*-functional guanidines are ongoing. It is important to note that the performance of this catalytic reaction strongly depends on the quality and commercial source of the heterogeneous inorganic base K_3PO_4. We could not fully reproduce the yields reported for DMED; isolated yields tend to be lower. Therefore it is expected that optimization of the guanidine protocol will lead to higher conversions. Furthermore, we learned that the proton sponge TMGN could not replace the inorganic base as proton scavenger.

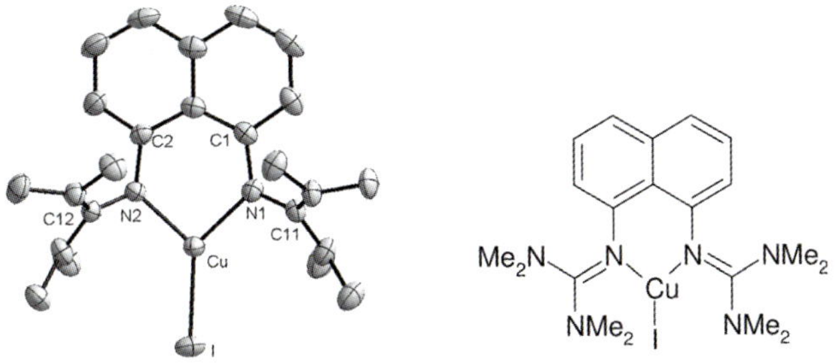

Figure 2.1 Molecular structure of proton-sponge catalyst complex [(TMGN)CuI] (ORTEP of 50% ellipsoids). Selected bond distances Cu-N1 197.7(8), Cu-N2 197.1(2), Cu-I 245.7(1), N2-C2 141.0(3), N2-C12 133.5(4), N1-C1 140.8(4), N1-C11 133.2(1) pm.

Another interesting application of TMGN is its use in the heterolytic cleavage of dihydrogen. This interesting reaction has been realized by frustrated Lewis acid – Lewis base pairs that for steric reasons cannot quench their desire to form a direct donor–acceptor bond. To date, the heterolytic cleavage of the H–H bond [18, 19], the strongest homonuclear sigma bond known, was realized with alkyl phosphanes as bases. We discovered that, because of their intrinsic proton affinity, our nitrogen bases with guanidine or 1,3-dialkyl-1,3-imidazole-2-ylidenamino functionalities, TMGN and DIAN, quantitatively cleave dihydrogen in the presence of tris(pentafluorphenyl)borane (BCF) as hydride acceptor.

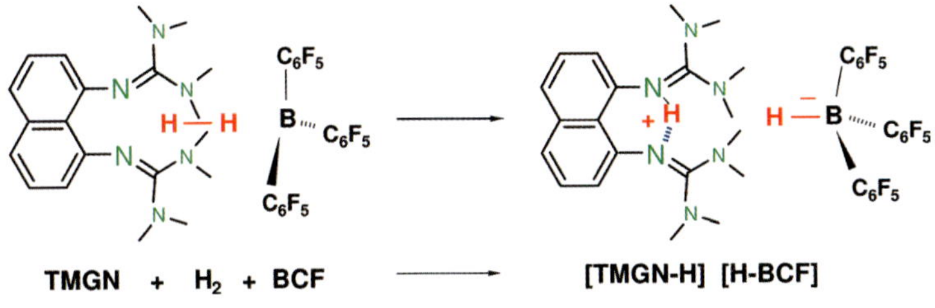

TMGN + H_2 + **BCF** ⟶ **[TMGN-H] [H-BCF]**

The product [TMGN-H] [H-BCF] is a colorless ionic solid with all characteristic spectroscopic features of the protonated pincer ligand and of the corresponding hydridoborate: ESI-MS (MeCN): m/z positive mode 355 [TMGN-H], negative mode 513 [H-BCF]; ^1H = NMR (400 MHz, CD_2Cl_2, 25 °C): $\delta = 3.60$ (q br, $^1J = 92.3$ Hz, $^{11}B-H$) ppm; $^{11}B\{^1H\}$ NMR (400 MHz, CD_2Cl_2, 25 °C): $\delta = -26.2$ ppm; CHN analysis calcd. for $C_{38}H_{32}N_6F_{15}B$: C 52.55, H 3.71, N 9.68; found C 51.95; H 3.97; N 9.65.

2.4
Multidentate Superbasic Guanidine Ligands as Receptors for Metal Cations

In the year 2000 we started to investigate and develop the unknown coordination chemistry of bidentate, tridentate and tetradentate guanidine ligands [20, 21] (Scheme 2.2). At the same time chelating guanidine ligands have been independently developed by the group of G. Henkel [22]. Some of the new ligands of interest are displayed in Scheme 2.2.

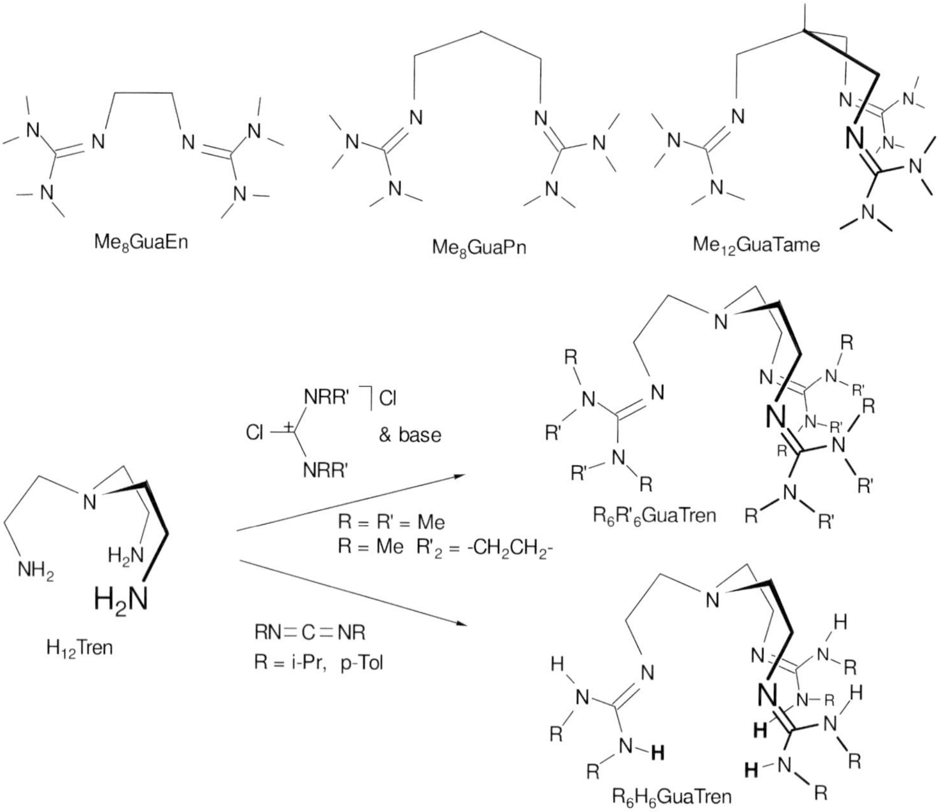

Scheme 2.2 Selected bidentate, tridentate and tetradentate guanidine ligands developed in this research program. Tame = 1,1,1-Tris(aminomethyl)ethane, Tren = tris(2-aminoethyl)amine.

This research program focussed on the guanidine coordination chemistry of copper and zinc. Two working concepts were followed. It was proposed that superbasic tripodal and sterically demanding tren-derived guanidine ligands when coordinated to copper(I) would create a very electron-rich metal center perfectly embedded in a molecular pocket. The principle of substrate activation would be *electron transfer* from copper(I) to small electron-poor molecules (NO, O_2, PhIO, RN_3

etc.) that are coordinatively activated and partially reduced at the reactive metal site. The activation principle for corresponding zinc complexes would be *proton transfer* from a coordinatively activated, slightly proton acidic substrate XH to one of the superbasic pendant guanidine arms. It was observed that coordinated or dangling guanidine functionalities in our tripodal complexes can – because of their high proton affinity – be involved in protolysis reactions with water and some alcohols. Both concepts are displayed in Scheme 2.3:

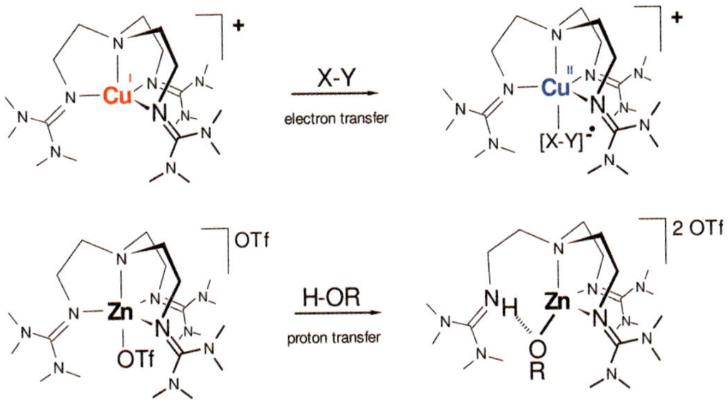

Scheme 2.3 Activation principles: electron transfer mediated by copper versus proton transfer mediated by zinc.

The principle of stabilization of intermediates may involve secondary ligand–ligand IHB interactions. Therefore two classes of guanidine ligands, peralkylated ones and *NH*-functionalized ones, were investigated. The latter are in the position to stabilize complex activated substrates displaying lone pairs by IHB interactions: these are intramolecular hydrogen bonds between a metal-coordinated substrate and the *NH*-functional guanidine ligand (Scheme 2.4).

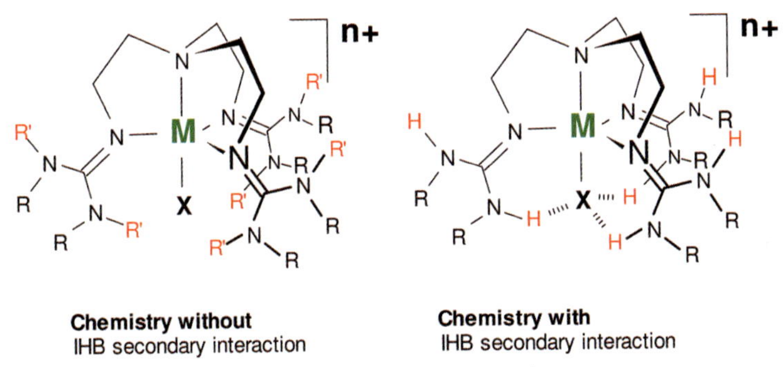

Scheme 2.4 Intramolecular hydrogen bond interactions between coordinated substrate X and guanidines.

In order to screen our chemistry for the existence of a secondary ligand effect, we are looking for differences in the reactivity patterns of *N*-peralkylated and *NH*-functionalized guanidine ligand complexes.

2.5 The Chemistry of Guanidine Copper Complexes

Peralkylguanidines are strong sp^2 nitrogen donors that typically display shorter dative bonds to a metal cation than corresponding tertiary amines as sp^3 donor ligands. Because of their donor capability and perfect charge delocalization they stabilize cationic and dicationic complexes and in some cases even copper(III) complexes or complex intermediates. On the other hand guanidines have a softer character compared to amine ligands. In this respect, it is a great advantage that copper(I) guanidine complexes do not tend to disproportionate into copper and copper(II). We have synthesized and fully characterized a series of cationic copper(I) and copper(II) complexes with tren derivatives of guanidine ligands [23]. The following three copper (I) complexes were selected for further reactivity studies (TMG_3tren code represents a tetramethylguanidyl-tren ligand):

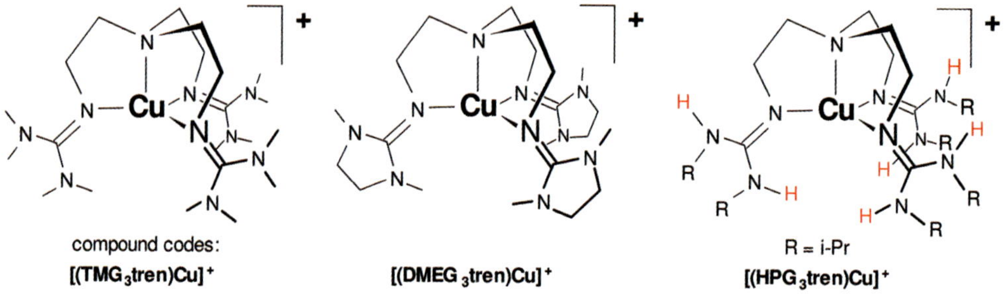

compound codes:
$[(TMG_3tren)Cu]^+$ $[(DMEG_3tren)Cu]^+$ $[(HPG_3tren)Cu]^+$
R = i-Pr

We discovered that $[(TMG_3tren)Cu]Cl$ is catalytically active in the oxidative carbonylation of methanol to dimethyl carbonate and water, the so-called ENICHEM process [24]:

$$2MeOH + 0.5\,O_2 + CO - \text{cat.} \rightarrow (MeO)_2CO + H_2O$$

$[(TMG_3tren)Cu]Cl$ reacts with dioxygen in the presence of methanol to give copper (II) methoxy species (or masked coordinated methoxy radicals), which are oxidatively transferred to CO to yield dimethylcarbonate under recovery of the copper(I) equivalents so that the cycle may start again. For this reason, the reaction of $[(TMG_3tren)Cu]^+$ with dioxygen and CO has been studied in more detail. Within the collaboration network of this DFG priority program, we invited S. Schindler to obtain more detailed insight into the equilibrium with O_2 by applying low-temperature stopped-flow techniques established in his group. These measurements proved what our nmr experiments suggested: a fully reversible association of dioxygen at our copper(I) guanidine complexes. At temperatures below $-40\,^{\circ}C$ the bright green colour of a superoxo species with its characteristic charge transfer band at 410 nm was developing. At this stage a low-temperature proton nmr sample of the green solution did show very broad bands in the range -5 to $+24$ ppm. It was proven by spectra of O_2-saturated solutions of a corresponding zinc complex and by comparison with spectra of authentic copper(II) complexes of this ligand that these signals had to be assigned to a paramagnetic copper–oxygen species. Surprisingly, when the sample was evaporated or warmed to room temperature, the nmr signals of the diamagnetic colorless Cu(I) species and its original UV-Vis spectrum were obtained again without any indication of side products. In collaboration with the Schindler group, we had discovered a fully reversible dioxygen uptake at a copper(I) complex. The key to success is the steric demand of our guanidine ligand that efficiently inhibits the attack of a second Cu(I) equivalent at the coordinated dioxygen molecule; hence its further irreversible reduction to peroxide.

The question arose whether dioxygen is bonded to copper in an *end-on* or *side-on* fashion. DFTcalculations of M. Holthausen clearly showed a preference for an *end-on* bonding mode while low-temperature resonance Raman data of a mixture of $^{16}O-^{18}O$ isotopomers of the dioxygen complex obtained in collaboration with Schindler and Schneider at Erlangen suggested a *side-on* mode on the basis of three ν(O–O) bands at 1117, 1059, and 1089 cm^{-1}. Combined spectroscopic and theoretical evidence including computed harmonic frequencies of the isotopomers (M. Reiher) finally allowed us to propose the existence of a persistent *end-on* copper superoxo complex [25]. It took more than three years of testing different conditions until E. Gaoutchenova and K. Harms obtained a high-precision low-temperature XRD analysis on dark-green single crystals of [(TMG$_3$tren)Cu(η^1-O$_2$)]SbF$_6$ that had been grown by a promising anion and solvent combination discovered by our collaboration partner Ch. Würtele at Gießen [26]. Figure 2.2 displays the molecular

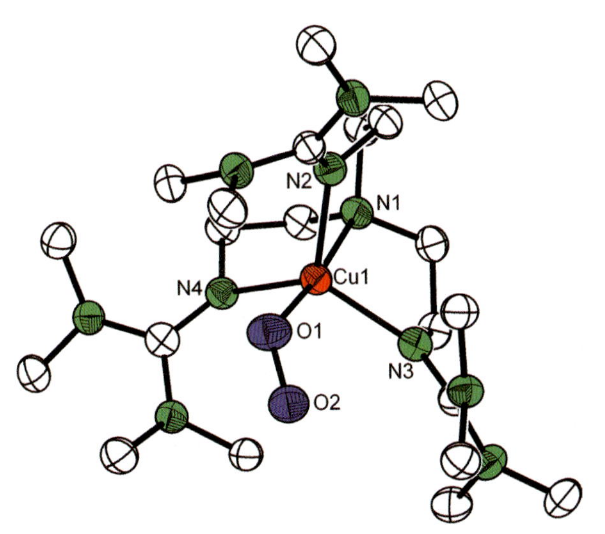

Figure 2.2 Molecular structure of the first synthetic *end-on* copper dioxygen complex. P $2_1/c$ (Z = 4), 100 K; selected bonding distances and angles: Cu1-O1 192.7(5) pm, Cu1-O1 192.7(5) pm, Cu1-O1 192.7(5) pm, Cu1-O1-O2 123.5(2)°.

structure of this reference compound, which has only been obtained by a kind of guanidine-based ‘picket-fence’ approach.

One of the remaining challenges in copper oxygen chemistry was the structural, electronic and magnetic characterization of the first product $[Cu^{II}(\eta^1\text{-}O_2^-)]$ of the dioxygen reduction cascade (Scheme 2.5):

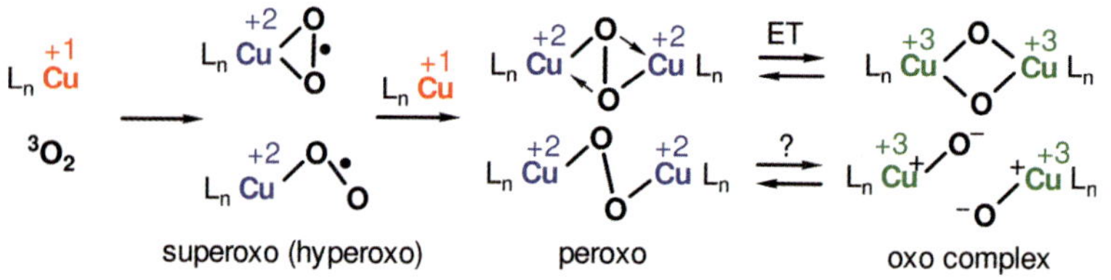

Scheme 2.5 The copper oxygen reduction cascade.

Triplet oxygen is reduced by one copper(I) equivalent to a mononuclear superoxo intermediate $[Cu^{II}(\eta^1\text{-}O_2^-)]$ or $[Cu^{II}(\eta^2\text{-}O_2^-)]$. Typically this is further reduced by a second Cu(I) equivalent to a bridging peroxo copper(II) core $[Cu^{II}(\mu\text{-}\eta^1{:}\eta^1\text{-}O_2)Cu^{II}]$ or $[Cu^{II}(\mu\text{-}\eta^2{:}\eta^2\text{-}O_2)Cu^{II}]$ depending on the steric bulkiness of the ligands L. If the ligands L are very strong donors that stabilize Cu(III), a valence equilibrium between the $[Cu^{II}(\mu\text{-}\eta^2{:}\eta^2\text{-}O_2)Cu^{II}]$ and $[Cu^{III}(\mu\text{-}O)_2Cu^{III}]$ cores has been observed [27]. With our sterically demanding tripodal ligands, no bridging peroxo ligands are conceivable unless one of the ligand arms dissociates into a dangling mode. However independent work of Henkel and Herres [28] as well as Tamm and Tolman [29] shows that with sterically less demanding bidentate chelating guanidine ligands the Cu-O_2 reduction cascade ends up in the valence equilibrium between $[Cu^{II}(\mu\text{-}\eta^2{:}\eta^2\text{-}O_2)Cu^{II}]$ and $[Cu^{III}(\mu\text{-}O)_2Cu^{III}]$ cores.

While these results and collaboration within the DFG priority program SPP 1118 closed the gap of structural evidence of an *end-on* superoxo intermediate, its magnetic properties and electronic structure remained an open question for some time. The frozen green solution of our superoxo complex turned out to be epr silent in the X-band. Furthermore, we were not able to transfer the green crystals into a SQUID capillary at temperatures below −40 °C. However, we were able to employ the Evans method for the measurement of the magnetic susceptibility in acetone at 190 K [30, 31]. The obtained average magnetic moment μ_{eff} of $3.02 + 0.25\,\mu_B$ turned out to be within the error of the spin-only value of $2.83\,\mu_B$ of a ground-state triplet species. This explains the nmr results, with large paramagnetic shifts, that were obtained. The message is that the spins in this *end-on* superoxo complex are ferromagnetically coupled, whereas they are antiferromagnetically coupled in all *side-on* superoxo complexes known so far. Our first structurally characterized reference complex $[Cu^{II}(\eta^1\text{-}O_2^-)]$ is epr silent in the X-band – probably because of an open-shell singlet electronic structure – whereas the first structurally characterized $[Cu^{II}(\eta^2\text{-}O_2^-)]$ reference complex of Kitajima *et al.* is epr silent because of the antiferromagnetic coupling of the spins at copper and oxygen [32].

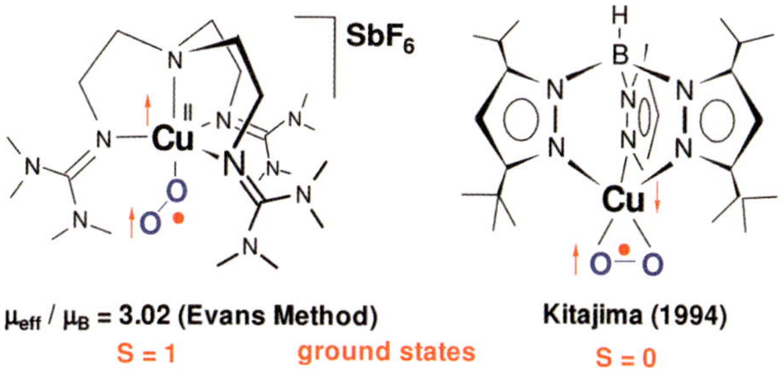

SbF_6

μ_{eff} / μ_B = 3.02 (Evans Method) Kitajima (1994)

S = 1 ground states S = 0

At this stage, a highly sophisticated study of the equilibrium isotope effect upon O_2 binding and electronic structure of the product was initiated by J. Roth [30], while *K. Karlin* offered his expertise in well-designed reactivity studies on the superoxo complex. Reactivity is relevant as *end-on* copper complexes play a role in nature's enzymatic CH hydroxylation reactions. At the same time, when we proposed the existence of this *end-on* superoxo complex, Prigge *et al.* presented an unusual *end-on* Cu-O_2 complex observed in an XRD structure of pepdidylglycine-α-hydroxylating monooxygenase [33] refined to 185 pm. This important finding supports the view that an *end-on* superoxo mono copper site is the precatalytic species in the α-hydroxylation of C–H bonds by this enzyme. The question arose whether our structural model might serve as a biomimetic functional model as well. With this in mind the group of Karlin investigated the reactions of $[(TMG_3tren)Cu(\eta^1\text{-}O_2)]^+$ with substituted phenols. There was an ongoing debate about whether the superoxo or perhaps other oxygen species might be the most potent oxidants in this system. A most relevant insight of this study is that a hydroperoxo complex that is formed via hydrogen abstraction of the precatalytic (resting) superoxo species at a phenol substrate will lead to a more potent oxidizing hydroperoxo species $[(TMG_3tren)Cu(\eta^1\text{-}OOH)]^+$ [34]. Hydrogen transfer is the trigger for a complex cascade of reactions including the hydroxylation of ligand C–H bonds and the oxidation of the phenolic substrate (Scheme 2.6).

The metal-mediated combination of superoxide and NO is considered to play a role in nitric oxide biochemistry. A reaction of our superoxo complex with NO offers the fascinating opportunity to study the formation and decomposition of peroxonitrite in the coordination sphere of copper (Scheme 2.7). It was demonstrated by ESI mass spectroscopy that copper-mediated coupling of isotopically labeled superoxo ligands with NO leads to the formation of $[(TMG_3tren)Cu(\mu^1\text{-}{}^{16}O\text{-}{}^{16}O\text{-}NO)]$ $B(C_6F_5)_4$ and $[(TMG_3tren)Cu(\eta^1\text{-}{}^{18}O\text{-}{}^{18}O\text{-}NO)]$ $B(C_6F_5)_4$ [35]. The borate anion allows monitoring of this reaction in solution at low temperature. DFT calculations support the η^1 coordination mode of the peroxonitrite ion. Most interestingly, peroxonitrite does not rearrange to nitrate but decomposes to dioxygen and an O-nitrito complex. The latter, which has been structurally characterized, is the product of the reaction of the parent cation $[(TMG_3tren)Cu]^+$ with excess of NO gas (Scheme 2.6) [31].

To date, mononuclear copper(III)-oxo species $[LCu(\eta^1\text{-}O)]^+$ or copper(III)-imido species $[LCu(\eta^1\text{-}NR)]^+$ (Scheme 2.9) have never been isolated and structurally characterized. Therefore we follow the concept of generating these intermediates within the sterically demanding ligand regime of these guanidine tripod ligands. As, for steric reasons, bridging peroxo species with a κ^4-bonded ligand TMG_3tren have

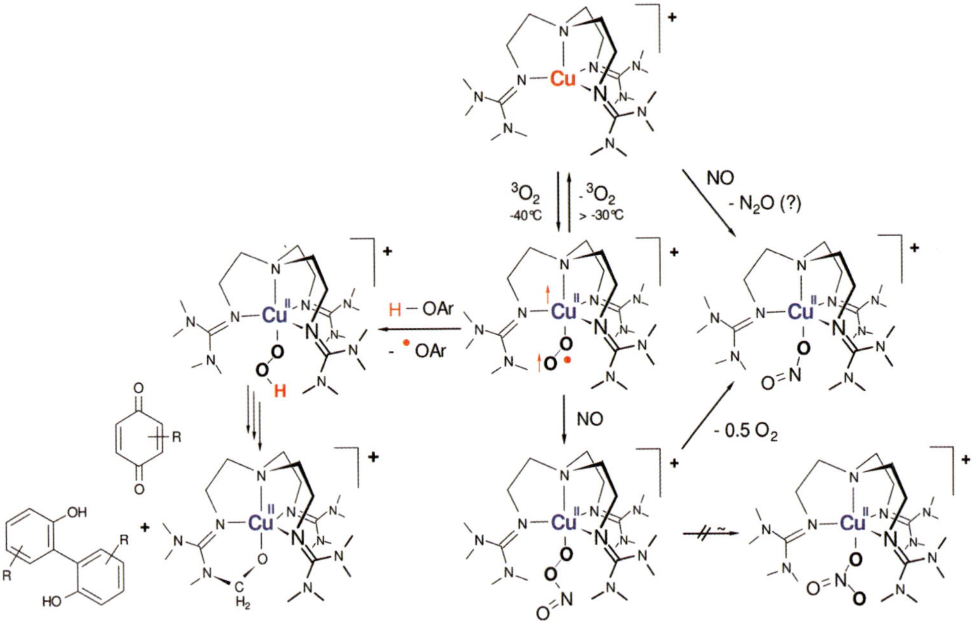

Scheme 2.6 Modeling copper-monooxygenase C-H hydroxylation and peroxonitrite formation and decomposition.

never been detected, different synthetic routes for their synthesis, namely the reactions of $[(TMG_3tren)Cu]^+$, $[(DMEG_3tren)Cu]^+$ and $[(HPG_3tren)Cu]^+$ with TsN_3, $[PhI(NTs)]_x$ and $[PhIO]_x$ in MeCN, were chosen.

Quantitative conversion of the copper(I) complex was only achieved, when 1.5 mole equivalents of the oxidant reacted with peralkyl guanidine ligand complexes. Under these conditions a rather selective oxidation takes place, the products being the reduced oxidant (0.5 mole) and Cu(II) complexes with α-hydroxylated or α-aminated guanidine ligands (Scheme 2.7).

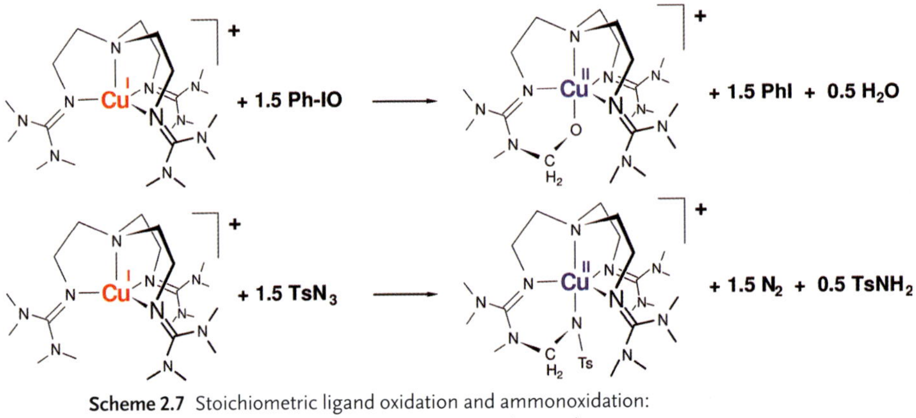

Scheme 2.7 Stoichiometric ligand oxidation and ammonoxidation: two different substrates – one and the same mechanism!

Unfortunately, $[PhIO]_x$, because of its polymeric nature, is quite insoluble and TsN_3 quite unreactive at low temperatures. Therefore the reactions had to proceed at relatively high temperatures in the region of 0 °C, a temperature range that has not so far allowed the spectroscopic characterization of the proposed intermediates $[LCu(\eta^1\text{-}O)]^+$ or $[LCu(\eta^1\text{-}NTs)]^+$. Instead, a more rapid subsequent step – believed to be an α-hydrogen abstraction at the ligand – is observed. The solutions immediately change from colorless to green. Crystallization using many different ligand–anion–solvent combinations ultimately gave non-disordered single crystals suitable for an XRD analysis. The results and plausible pathways of product formation are shown in Schemes 2.8 and 2.9.

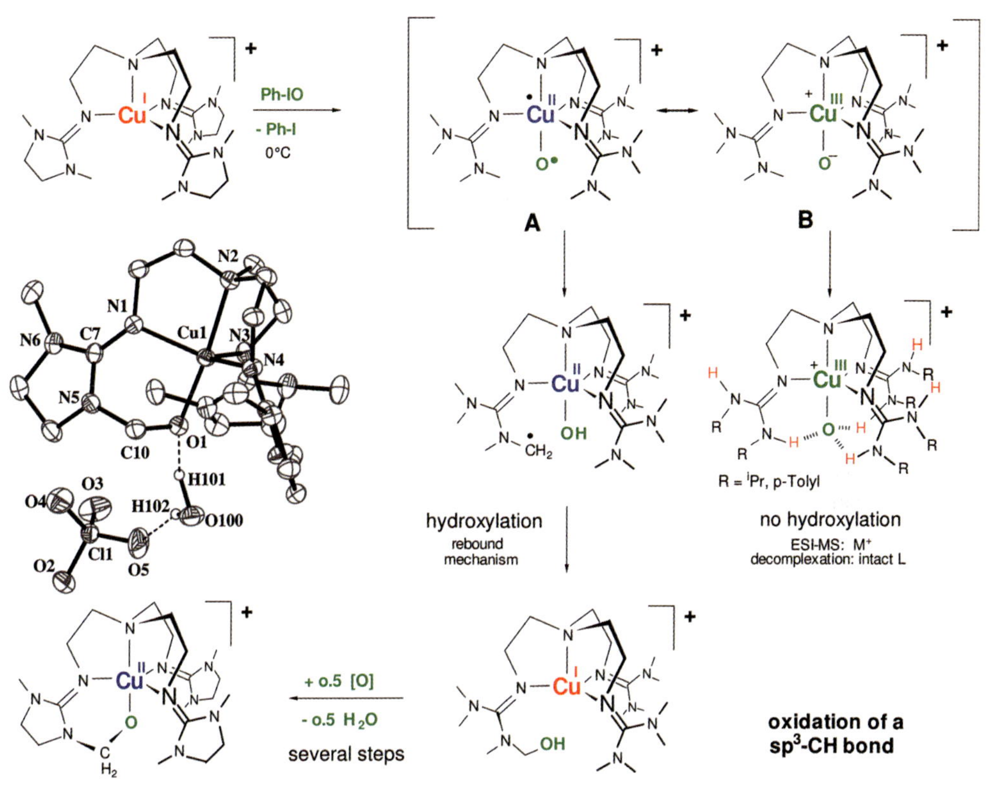

Scheme 2.8 Formation and degradation pathways of a proposed mononuclear oxo species $[(DMEG_3tren)Cu(O)]^+$. Molecular structure of $[(DMEG_3trenO)Cu]ClO_4{\cdot}H_2O$.

It is plausible that the oxidants, solvated $PhIO(MeCN)_n$ or TsN_3, are activated via electron transfer from Cu(I). This might proceed either via an inner-sphere mechanism and a Cu(I) precoordinated oxidant or via an outer-sphere mechanism. The result of a single-electron transfer would be the formation of oxidant radical anions coordinated to Cu(II) and elimination of stable leaving groups N_2 or PhI. Two

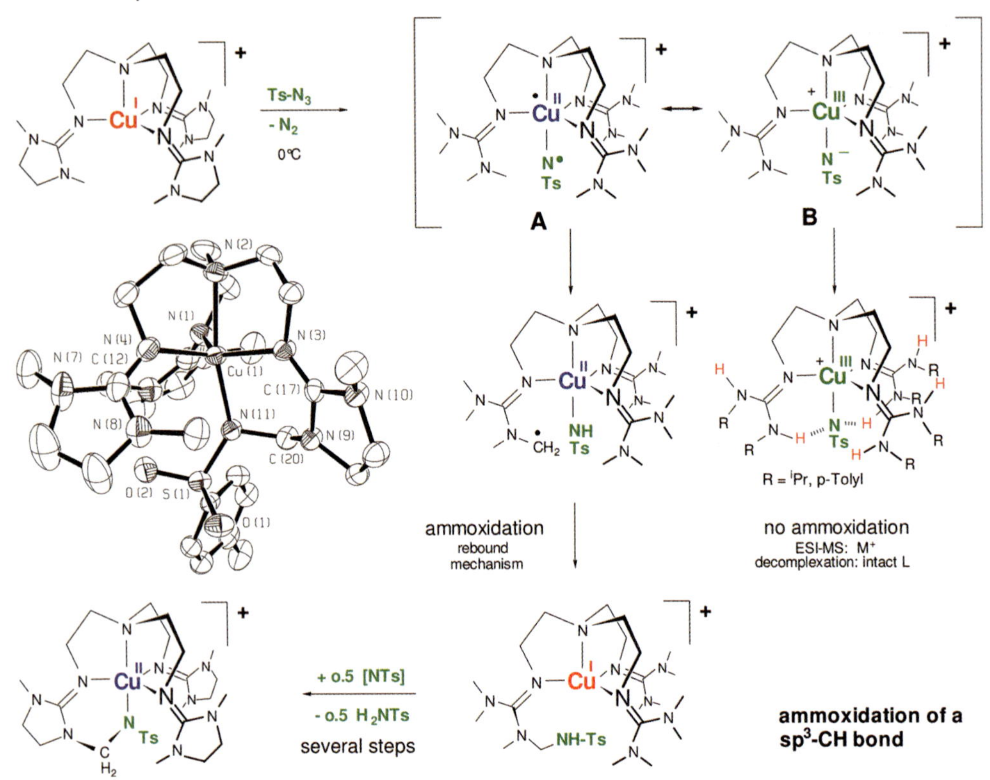

Scheme 2.9 Formation and degradation pathways of a proposed mononuclear oxo species $[(DMEG_3tren)Cu(NTs)]^+$. Molecular structure of $[(DMEG_3trenNTs)Cu]ClO_4$.

electronic Cu structures are plausible for the activated oxidant complex, a radical anion coordinated at Cu(II) (form A) or a fully reduced closed-shell anion $[O]^{2-}$ or $[NTs]^{2-}$ coordinated at a diamagnetic Cu(III) site (form B). In form A, the spins at d^9 Cu and the coordinated radical could be either parallel or antiparallel. Extensive DFT calculations by M. Holthausen clearly favor form A with an $S = 1$ ground state. For $[(TMG_3tren)Cu(O)]^+$, state A is about 22 kcal lower in energy than the $S = 0$ state B [36]. Further scanning of the potential energy surface reveals that the coordinated O radical (triplet T) selectively and exothermically abstracts a hydrogen atom of the C–H bond in α-position to the ligand nitrogen atom with formation of a Cu(II) hydroxo species and a C radical (T state). Recombination of the C radical and the d^9 [Cu–OH] radical is a spin-forbidden process, as the product is a singlet d^{10} Cu(I) complex with hydroxylated ligand. This change in the triplet to singlet hyperface may be envisaged as a copper-analog rebound mechanism, and is hampered by a relatively high activation barrier. Formally, a hydrogen atom (or a proton and an electron) has to be abstracted from this singlet species in order to get to the final product, a Cu(II) alkoxo complex. This electron and proton balance explains why another half equivalent of the oxidant is needed and why water (or $TsNH_2$) is formed on the way to the final products. It is worth mentioning that only half of the yield of $[(DMEG_3trenO)Cu]ClO_4{\cdot}H_2O$ crystallizes; the other half does not. Water, the by-product (0.5 mol%) is needed to form a 1 : 1 hydrogen bond network in the crystalline state.

We furthermore observed an interesting effect, which is believed to be the result of secondary IHB ligand interactions: the same oxidation reactions with the *NH*-functional guanidine complexes turned out to be quantitative after addition of one equivalent of the oxidant. The green reaction mixture does contain at least two paramagnetic species that could not be fully separated. The ESI-MS (MeCN) displays (next to other species) a monocationic species with a molecular ion consistent with a complex $[(HPG_3tren)Cu(O)]^+$. By decomplexation with cyanide or hydrogen sulfide it was proven that no ligand hydroxylation occurred. In both cases (oxidation with $[PhIO]_x$ and TsN_3), the intact guanidine ligand was recovered. An explanation would be that the singlet form B, with a more polar Cu–O bond, might be stabilized by three intramolecular hydrogen bonds to three oxygen lone pairs. A similar intramolecular hydrogen network has been structurally characterized in the zinc chemistry of our ligands. This fascinating part of the research project is still under investigation.

2.6 The Chemistry of Guanidine Zinc Complexes

Zinc is the most prominent metal in enzymes with hydrolase activity. Important biological functions of zinc enzymes include CO_2 hydration and HCO_3^- dehydration as well as hydrolytic cleavage of peptides and organophosphates [37]. Secondary IHB interactions of zinc-coordinated substrates to proton donors of the enzyme such as the guanidinium group of arginine or the imidazolium group of histidine are important parts of an H-bond network that can serve as a proton relay. Numerous biomimetic carbonato and hydrogencarbonato complexes of zinc were structurally characterized [38]. We became interested in investigating the performance of *NH*-functional guanidine ligands in the stabilization of mono- and dinuclear complexes with Zn–OH, Zn–OCOOH, Zn–OCOO–Zn and Zn–O–Zn functionalities. One of the challenges we were focussing on in our zinc project is the stabilization of molecular ZnO in a mononuclear neutral complex [LZnO]. This molecule would have a highly reactive, polar Zn–O bond. The task of a ligand is to block the intermolecular Zn–O attractive forces by intramolecular secondary ligand–ligand interactions that would have to be stronger than the intermolecular forces. To date, no such molecule has been isolated and characterized by its structure and reactivity, although a number of species very close to this target were synthesized. A selection of these are reported here [39].

The idea was to use the R_6H_6GuaTren type ligands in order to stabilize highly basic and nucleophilic functional groups coordinated at the zinc site via intramolecular 'partial protonation' by the guanidine *NH*-donors. The principle is demonstrated by the following two examples:

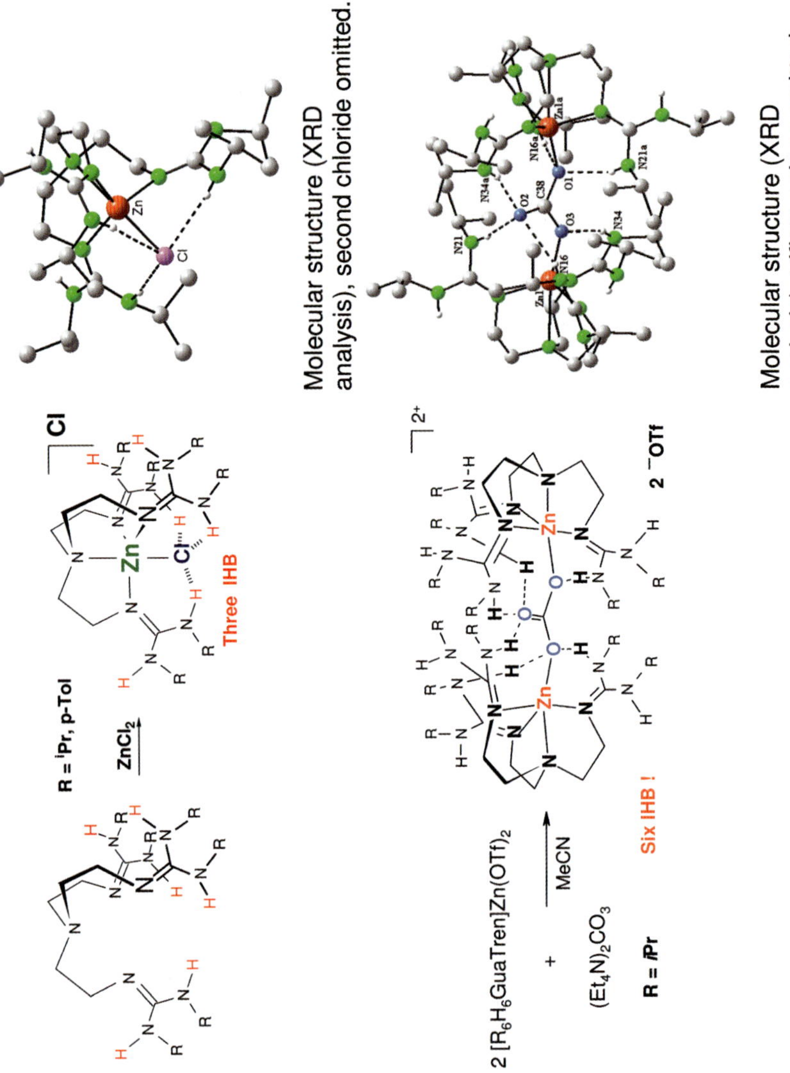

R = iPr, p-Tol
ZnCl2
Three IHB
Cl
Zn
Molecular structure (XRD analysis), second chloride omitted.
2 [R6H6GuaTren]Zn(OTf)2
+
(Et4N)2CO3
MeCN
R = iPr
Six IHB !
2+
2 −OTf
Molecular structure (XRD analysis), triflate anions omitted.

The ligands R_6H_6GuaTren convert anhydrous zinc chloride into cationic complexes, with one chloro ligand and one chloride in the outer coordination sphere. A remarkable structural feature are three IHB of the C_3 symmetrical ligand to the coordinated chloro ligand with its three electron lone pairs, even though the ligand constraints do not allow a perfect alignment with plausible sp^3 orbitals at chlorine.

We tested numerous synthons for nucleophilic replacement of the chloro ligand by oxo, sulfido or carbonato ligands, without any success. Either the reagents were not reactive – such as $(Et_4N)_2CO_3$ in MeCN – or they led to decomplexation and precipitation of ZnO or ZnS. The reactions became much more selective with zinc triflate as electrophile. This strategy allowed us to isolate and fully characterize a remarkable dinuclear zinc complex with altogether six IHB of the bridging carbonato ligand to the guanidine NH proton donor functionalities. Interestingly the electrophilic zinc precursor complex crystallizes in tetrahedral coordination, which is accomplished by a proton shift within the ligand. After axial coordination of an anion, the proton and corresponding ligand arm shifts back into the usual position. It is worth mentioning that the corresponding Cu(I) complex, because of its d^{10} configuration and absence of ligand field effects, does not show this tautomeric ligand equilibrium.

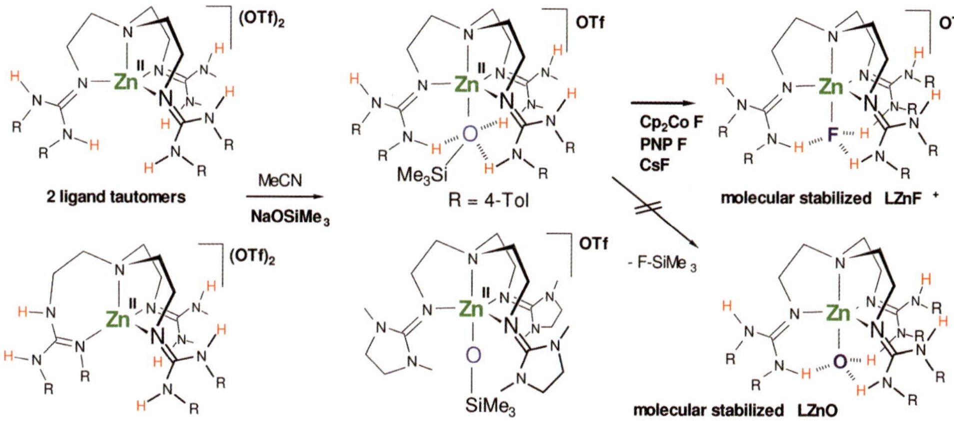

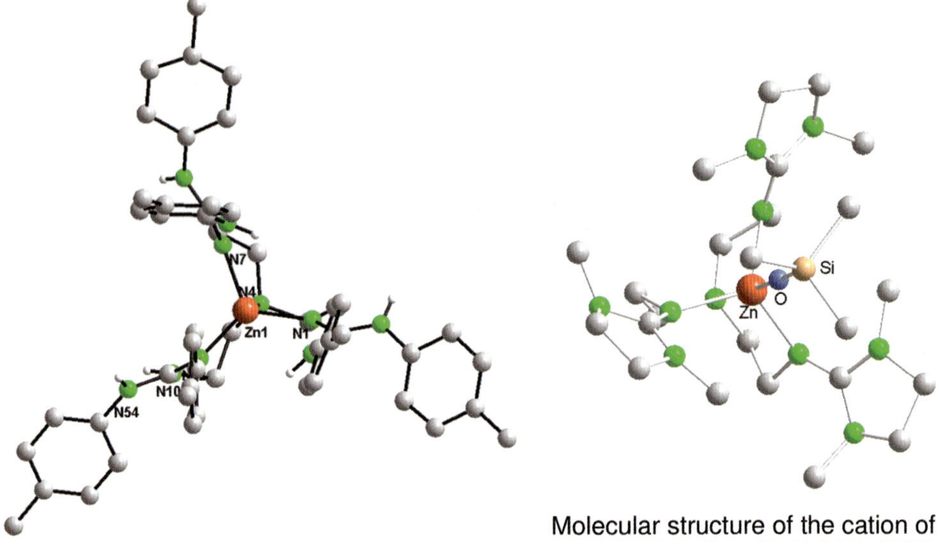

Molecular structure of the tetrahedral tautomer of [(p-Tol_6H_6GuaTren)Zn]$(OTf)_2$.

Molecular structure of the cation of [($DMEG_3$GuaTren)Zn($OSiMe_3$)]OTf Zn-O-Si 176.1°; av. N....O 329.8 pm; Si-O 157.4(2) pm.

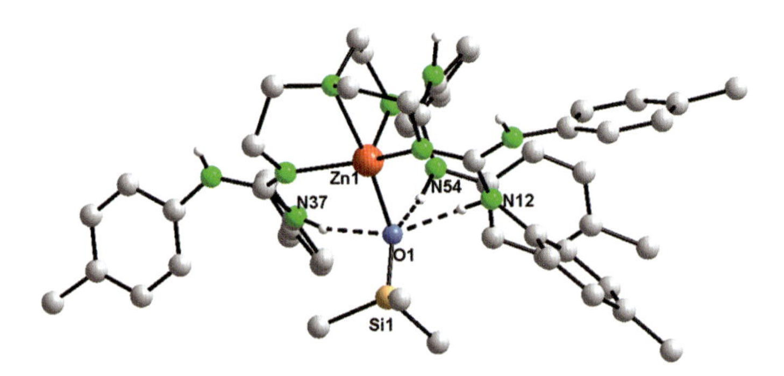

Molecular structure of the cation of [(p-Tol_6H_6GuaTren)Zn($OSiMe_3$)]OTf Zn-O-Si 131.3°; av. N....O 287.4 pm; Si-O 163.1(3) pm.

Unfortunately, straight reactions of the guanidine zinc triflates with synthons for OH^-, SH^-, O^{2-}, S^{2-} again were not as successful as the reactions with halides, pseudohalides, carbonate and trimethylsilanolate. We discovered a remarkable difference in the coordination of the $[Me_3SiO]^-$ ligand depending on the ability (or lack of it) of the guanidine ligand to involve the oxygen lone pair in intramolecular hydrogen bonds. The three IHB lead to a considerable lengthening of the O–Si bond and to a bending of the Zn-O-Si axis. In [(*p*-Tol_6H_6GuaTren)Zn($OSiMe_3$)]$^+$ the silyl group tends to be less strongly bonded to the oxygen atom. We hoped to attack this labilized O–Si bond with fluoride at silicon in order to induce a desilylation which would lead to the desired IHB stabilized molecular [LZnO] complex or a dimer of it. After numerous of attempts with different fluoride donors, we had to accept that fluoride replaces the whole siloxide anion. The obtained complex [(*p*-Tol_6H_6GuaTren) Zn(F)]OTf displays a beautiful quartet at $\delta_F = -125.9$ ppm ($^1J_{HF} = 45$ Hz) for the fluoro ligand due to three strong intramolecular and chemically equivalent hydrogen bonds to the *NH*-functional guanidine ligand. The corresponding proton resonances appear as a sharp doublet at $\delta_H = 9.81$ ppm ($^1J_{HF} = 45$ Hz). What is chemistry without persistent dream reactions?

2.7
Conclusions

The development of multidentate guanidine chemistry enabled valuable advances to be made in research areas that prior to this work had been dominated by amines, imines and azaaromatic building blocks. Guanidines serve as perfect pincer ligands for a proton in the most basic and synthetically easy accessible proton sponges known to date. Since their first systematic introduction into coordination chemistry in 2000, they have become more and more popular. Because of their superior donor character and steric demand, they made targets long looked for available such as the *end-on* superoxo species [(TMG$_3$Tren)Cu(η^1-O_2)]$^+$ or [(DMEG$_3$Tren)Cu(η^1-O_2)]$^+$. These will serve as reference structures and as valuable precursors for many new reactivity studies on coordinated superoxide. According to the aim of this priority research program *NH*-functional guanidine ligands offer the opportunity to stabilize or influence other coordinated ligands via intramolecular ligand–ligand hydrogen bonds. Without the financial support of the DFG, SPP 1118, this development would not have been possible.

Acknowledgments

We are grateful for financial support by the Deutsche Forschungsgemeinschaft, SPP 1118, and thank the collaborating partners named in the article for their valuable contributions.

References

1 Wieland, G. and Simchen, G. (1985) *Justus Liebigs Annalen der Chemie*, 2178.
2 Barton, D.H.R., Elliott, J.D. and Géro, S.D. (1982) *Journal of the Chemical Society-Perkin Transactions*, 2085.
3 Schwesinger, R. (1990) *Nachrichten Aus Chemie Technik Und Laboratorium*, **38**, 1214.
4 Raczynska, E.D., Maria, P.-C., Gal, J.-F. and Decouzon, M. (1994) *Journal of Physical Organic Chemistry*, **7**, 725.

5 Kaljurand, I., Rodima, T., Leito, I., Koppel, I. and Schwesinger, R. (2000) *The Journal of Organic Chemistry*, **65**, 6202.

6 Raab, V., Kipke, J., Gschwind, R.M. and Sundermeyer, J. (2002) *Chemistry - A European Journal*, **8**, 1682.

7 Kovačević, B. and Maksić, Z.B. (2002) *Chemistry - A European Journal*, **8**, 1694.

8 (a) Ratilla, E.M.A. and Kostic, N.M. (1988) *Journal of the American Chemical Society*, **110**, 4427; (b) Ratilla, E.M.A., Scott, B.K., Moxness, M.S. and Kostic, N.M. (1990) *Inorganic Chemistry*, **29**, 918.

9 Alder, R.W., Bowman, P.S., Steele, W.R.S. and Winterman, D.R. (1968) *Journal of the Chemical Society. Chemical Communications*, 723.

10 (a) Staab, H.A. and Saupe, T. (1988) *Angewandte Chemie*, **100**, 895; (b) Alder, R.W. (1989) *Chemical Reviews*, **89**, 1215; (c) Pozharskii, A.F. (1998) *Russian Chemical Reviews*, **67**, 1.

11 Raab, V., Harms, K., Sundermeyer, J., Kovacevic, B. and Maksic, Z.B. (2003) *The Journal of Organic Chemistry*, **68**, 8790.

12 Abacilar, N. (2008/09) unpublished results, Dissertation Marburg projected.

13 Raab, V., Gaoutchenova, K., Merkoulov, A., Harms, K., Sundermeyer, J., Kovačević, B. and Maksić, Z.B. (2005) *Journal of the American Chemical Society*, **127**, 15738.

14 Raab, V. and Sundermeyer, J. DE 101 43566.

15 Kuhn, N., Grathwohl, M., Steinmann, M. and Henkel, G. (1998) *Zeitschrift fur Naturforschung*, **53 b**, 997.

16 (a) Schwesinger, R. and Schlemper, H. (1987) *Angewandte Chemie*, **99**, 1212; *Angewandte Chemie (International Edition in English)*, (1987) **26**, 1167; (b) Schwesinger, R., Hasenfratz, C., Schlemper, H., Walz, L., Peters, E..-M., Peters, K.an0d von Schnering, H.G. (1993) *Angewandte Chemie (International Edition in English)*, **32**, 1361.

17 Klapars, A., Huang, X. and Buchwald, S.L. (2002) *Journal of the American Chemical Society*, **124**, 7421.

18 Welch, G.C., Juan, R.R.S., Masuda, J.D. and Stephan, D.W. (2006) *Science*, **314**, 1124.

19 Spies, P., Erker, G., Kehr, G., Bergander, K., Fröhlich, R., Grimme, S. and Stephan, D.W. (2007) *Chemical Communications*, 5072.

20 Wittmann, H., Schorm, A. and Sundermeyer, J. (2000) *Zeitschrift Fur Anorganische Und Allgemeine Chemie*, **626**, 1583.

21 Wittmann, H., Raab, V., Schorm, A., Plackmeyer, J. and Sundermeyer, J. (2001) *European Journal of Inorganic Chemistry*, 1937.

22 Pohl, S., Harmjanz, M., Schneider, J., Saak, W. and Henkel, G. (2000) *Journal of The Chemical Society-Dalton Transactions*, 3473.

23 Raab, V., Kipke, J. and Sundermeyer, J. (2001) *Inorganic Chemistry*, **40**, 6964.

24 Raab, V., Merz, M. and Sundermeyer, J. (2001) *Journal of Molecular Catalysis A-Chemical*, **175**, 51.

25 Schatz, M., Raab, V., Foxon, S.P., Brehm, G., Schneider, S., Reiher, M., Holthausen, M.C., Sundermeyer, J. and Schindler, S. (2004) *Angewandte Chemie*, **116**, 4460; *Angewandte Chemie (International Edition in English)*, (2004) **43**, 4360.

26 Würtele, Ch., Gaoutchenova, E.V., Harms, K., Holthausen, M.C., Sundermeyer, J. and Schindler, S. (2006) *Angewandte Chemie*, **118**, 3951; *Angewandte Chemie – International Edition*, (2006) **45**, 3867.

27 Halfen, J.A., Mahapatra, S., Wilkinson, E.C., Kaderli, S., Young, V.G., Jr., Que, L., Jr., Zuberbühler, A. and Tolman, W.B. (1996) *Science*, **271**, 1397.

28 (a) Herres, S., Heuwing, A., Flörke, U., Schneider, J. and Henkel, G. (2005) *Inorganica Chimica Acta*, **358**, 1089; (b) Herres-Pawlis, S., Flörke, U. and Henkel, G. (2005) *European Journal of Inorganic Chemistry*, 3815.

29 Petrovic, D., Hill, L.M.R., Jones, P.G., Tolman, W.B. and Tamm, M. (2008) *Dalton Transactions*, 887.

30 Lanci, M.P., Smirnov, V.V., Cramer, C.J., Gaoutchenova, E.V., Sundermeyer, J. and Roth, J. (2007) *Journal of the American Chemical Society*, **129**, 14697.

31 Gaoutchenova, E.V. (2006) Dissertation Marburg.

32 Fujisawa, K., Tanaka, M., Moro-oka, Y. and Kitajima, N. (1994) *Journal of the American Chemical Society*, **116**, 12079.

33 Amzel, L.M. and Prigge, S.T. (2004) *Science*, **304**, 864.

34 Maiti, D., Lee, D.-H., Gaoutchenova, E.V., Würtele, C., Holthausen, M.C., Narducci Sarjeant, A.A., Sundermeyer, J., Schindler, S. and Karlin, K.D. (2008) *Angewandte Chemie*, **120**, 88.

35 Maiti, D., Lee, D.-H., Narducci Sarjeant, A.A., Pau, M.Y., Solomon, E.I., Gaoutchenova, E.V., Sundermeyer, J. and Karlin, K.D. (2008) *Journal of the American Chemical Society*, **130**, 6700.

36 Holthausen, M. and Sundermeyer, J.manuscript in preparation.

37 Rombach, M., Maurer, C., Weis, K., Keller, E. and Vahrenkamp, H. (1999) *Chemistry - A European Journal*, **5**, 1013.

38 (a) Erras-Hanauer, H., Mao, Z..-W., Liehr, G., Clark, T. and van Eldik, R. (2003) *European Journal of Inorganic Chemistry*, **8**, 1562; (b) Dussart, Y., Harding, C., Dalgaard, P., McKenzie, C., Kadirvelraj, R., McKee, V. and Nelson, J. (2002) *Journal of The Chemical Society-Dalton Transactions*, **8**, 1704; (c) Dietrich, J., Heinemann, F.W., Schrodt, A. and Schindler, S. (1999) *Inorganica Chimica Acta*, **288**, 206; (d) Meyer, F. and Rutsch, P. (1998) *Chemical Communications*, 1037.

39 Garrelts, U. and Sundermeyer, J. (2008) manuscript in preparation, U. Garrelts, Dissertation Marburg projected.

3
Iron Complexes and Dioxygen Activation

Thomas Nebe, Jing-Yuan Xu, and Siegfried Schindler

3.1
Introduction

Dioxygen activation at transition metal sites is important for the understanding of catalytic and selective oxidations of organic substrates using dioxygen (air) as the oxidant. While it is easy to completely oxidize (burn) unreactive hydrocarbons such as methane to carbon dioxide and water, it is quite a challenge to partially oxidize hydrocarbons selectively to alcohols, for example methanol. In contrast, the iron enzyme (soluble) [1] methane monooxygenase (MMO) easily accomplishes this reaction at ambient conditions [2]. Therefore, investigations of non-heme iron enzymes as well as their model compounds have gained a lot of interest during the last decades [3–5]. Iron compounds in general play an important role in biological electron transfer reactions, most likely because of a combination of their ready availability and their easy changes of oxidation states. Understanding the mechanisms of dioxygen activation of iron-containing enzymes (oxidases and oxygenases), therefore, is important for understanding (and suppressing) free radical pathways of oxidative damage in biological systems as well as for the synthesis of new selective oxidation catalysts. In that regard, dioxygen and, to a lesser extent, hydrogen peroxide are ideal oxidants for the chemical industry because they are readily available and environmentally clean. For a better understanding of dioxygen activation at the iron centers and the subsequent reactions it is necessary to investigate the mechanisms of these reactions by detailed kinetic measurements [5, 6]. If possible, important intermediates should be characterized through their structural, electronic, and spectroscopic properties [3–5]. In a recent review on this topic we have summarized the results obtained from a large number of relevant mechanistic studies during recent years [5].

A general reaction sequence for the reaction of a mononuclear iron(II) complex with dioxygen is presented in Scheme 3.1 [3–5]. These reactions can be easily extended to dinuclear or polynuclear complexes as well, but reaction rates in these reactions can be quite different because of the pre-organization of some of the

Activating Unreactive Substrates: The Role of Secondary Interactions.
Edited by Carsten Bolm and F. Ekkehardt Hahn

ISBN: 978-3-527-31823-0

complexes, thus making their formation much more facile. Furthermore, additional bridging groups can be present and inter- and intramolecular reactions are possible.

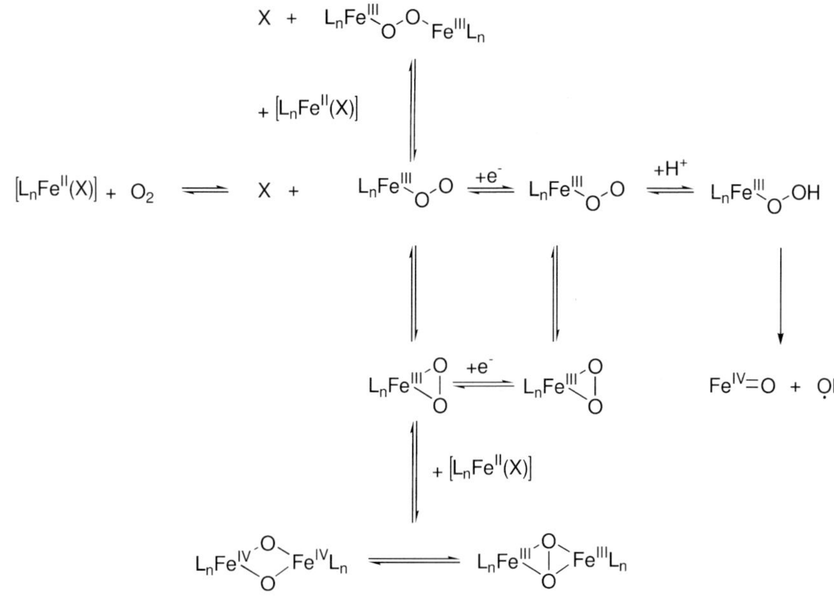

Scheme 3.1

The mechanisms of iron dioxygen-activating enzymes depend on the number of metal centers. The dinuclear systems, such as MMO, that utilizes a Fe_2^{IV}/Fe_2^{III} redox couple for a formally two-electron oxidation of methane, take advantage of both iron centers in catalysis [2]. In addition to electron and charge delocalization over two irons, which stabilize high-valent intermediates, the second metal ion can also be important for substrate coordination at the dinuclear center.

While observable iron oxygen intermediates can be generated from iron(II) complexes and dioxygen according to Scheme 3.1, it is furthermore possible to obtain these species from reactions with superoxide, hydrogen peroxide, alkyl hydroperoxides, peroxoacids, iodosobenzene, etc. [3–5, 7].

3.2 Dinuclear Iron Peroxo Complexes

In contrast to analogous studies on model compounds for copper proteins, only a few of the 'oxygen adduct' iron complexes shown in Scheme 3.1 could be structurally characterized [3–5, 7]. Dinuclear iron peroxo complexes shown in Figure 3.1 could be analyzed in great detail and were studied independently mainly by the research groups of Que and Lippard [8–12].

Our kinetic studies in cooperation with the Lippard group, using low-temperature and high-pressure stopped-flow techniques, allowed us to perform a detailed

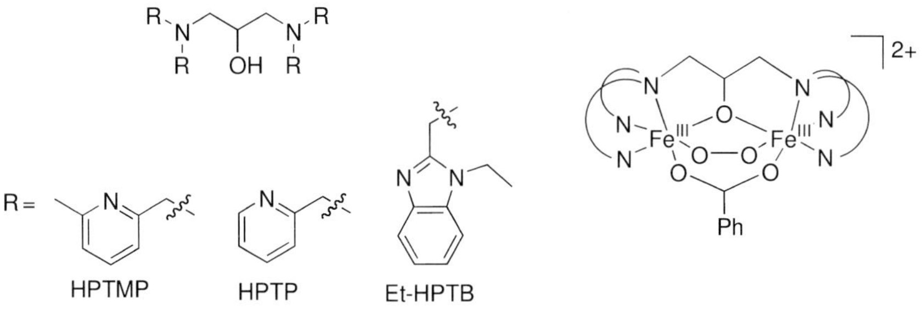

Figure 3.1 Dinuclear iron-peroxo complexes with the ligands HTPMP, HPTP and Et-HPTB.

mechanistic analysis of the binding of dioxygen to the iron(II) compounds to form the dinuclear peroxo complexes depicted in Figure 3.1 [13].

Furthermore, in cooperation with the Krebs group, we could show that the same complexes can be prepared much more easily from the relevant iron(III) complexes and hydrogen peroxide [14]. These reactions could be investigated in protic solvents at ambient temperatures, and it turned out that the product complexes were much more persistent under these conditions (hours in contrast to seconds). This surprising effect could be explained by our detailed kinetic analysis, and the proposed mechanism is shown in Scheme 3.2 (S and T are additonal ligands such as nitrate or solvent molecules, e.g. acetonitrile). Formation of the peroxo complexes is irreversible and does not depend on the way of its formation. Therefore, differences between the reaction pathways of the iron(II) complexes with dioxygen and those of the iron(III) complexes with hydrogen peroxide could not be the reason for the different stabilities. Instead, we have proposed that the final products of the decomposition reactions of the peroxo complexes are responsible for this effect. These products are tetranuclear oxo-bridged iron(III) complexes that can react again with hydrogen peroxide to reform the blue peroxo compounds. This reaction is reversible, and therefore peroxo complexes prepared from an excess of hydrogen peroxide and iron(III)

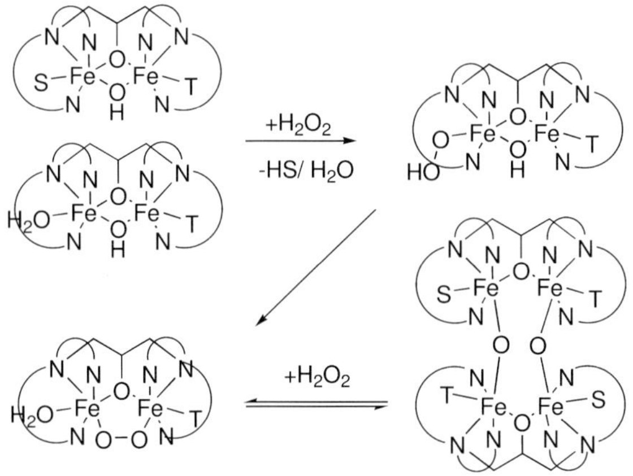

Scheme 3.2

complexes are stable for hours in contrast to those prepared from iron(II) complexes and dioxygen. If no hydrogen peroxide is present the peroxo complexes decompose very fast.

Preliminary studies also demonstrated that the formed peroxo complexes could be used to oxidize alkanes, even methane, to alkyl hydroperoxides as a major product if certain amino acids needed were added [15]. However, this reaction in general would need further optimization.

3.3 Tripodal Tetradentate Ligands and Derivatives

3.3.1 Tmpa

Tetradentate tripodal ligands have attracted a great deal of interest during recent years [16]. Especially tris-((2-pyridyl)methyl)amine (= tmpa, also abbreviated as tpa in the literature, Figure 3.2) has been used successfully in the bioinorganic chemistry of copper [17–19] and to quite a large extent in experimental studies to model iron enzymes [3–5].

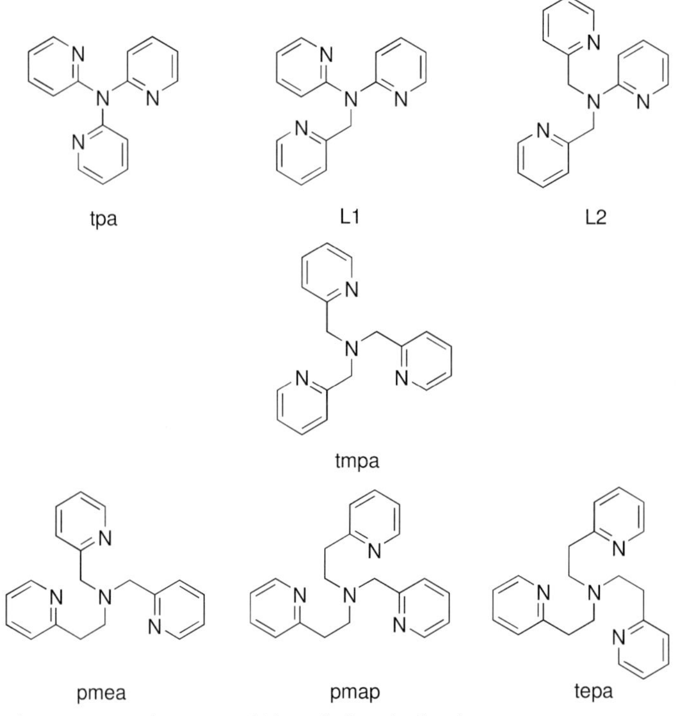

Figure 3.2 Tetradentate pyridyl ligands described in the text.

Que and coworkers showed that the iron(III) complex of tmpa is one of the most efficient biomimetic model compounds for intradiol-cleaving catechol dioxygenases [3, 20]. Catechol dioxygenases are mononuclear non-heme iron enzymes that catalyze the oxidative cleavage of catechol derivatives (by insertion of both dioxygen atoms into the substrate), a key step in the degradation of aromatic compounds.

Que and coworkers postulated a mechanism for the reaction of [Fe(tmpa)(dbc)]B$(C_6H_5)_4$ (dbc = 3,5-di-*tert*-butylcatecholate anion) with dioxygen (Figure 3.3). However, a quantitative kinetic study of this important reaction was lacking [3, 20]. We had hoped to observe one of the postulated reactive intermediates using low-temperature stopped-flow methods (an approach that has been used by us and others successfully in copper dioxygen chemistry) [21], but we could not detect such a species spectroscopically [22].

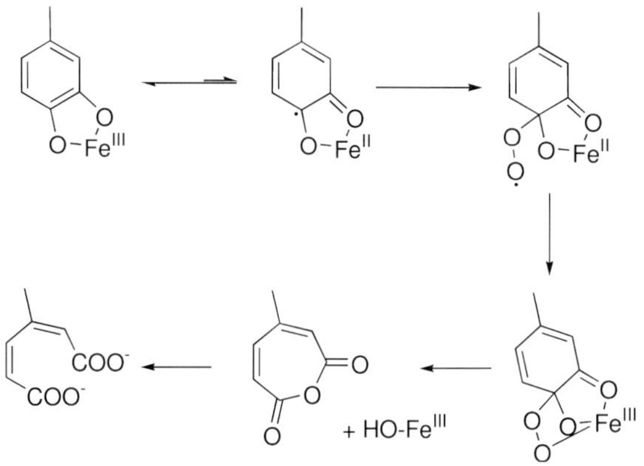

Figure 3.3 Postulated mechanism for the reaction of [Fe(tmpa)(dbc)]B$(C_6H_5)_4$ with dioxygen.

Temperature-dependence studies allowed the calculation of the activation parameter from the Eyring equation as $\Delta H^* = 23 \pm 1\,\mathrm{kJ\,mol^{-1}}$ and $\Delta S^* = -199 \pm 4\,\mathrm{J\,mol^{-1}\,K^{-1}}$ [22]. These data are in line with related reactions in copper chemistry and indicate that the attack of dioxygen on the catecholate complexes is the rate-determining step. Our kinetic data therefore do not contradict the postulated reaction mechanism; however, they also do not provide additional evidence for the occurrence of the described reaction steps. At which site this attack takes place cannot be determined in a kinetic study without the observation of intermediates. Modifications of chelate ring sizes could affect the reaction rates and/or stabilize reactive intermediates (as has been observed previously in our studies on copper complexes) [23, 24]. Therefore we tested iron(III) complexes with the whole ligand series shown in Figure 3.2 together with different catecholates. We were able to structurally characterize several iron complexes with these ligands, but rates could not be enhanced nor could reaction intermediates be trapped [22]. However, it should be pointed out, that L1 and L2 (Figure 3.2) have been synthesized and characterized by us for the first time and have been used in copper coordination chemistry previously [24]. These ligands might become quite useful in coordination chemistry and showed interesting properties for supporting the formation of coordination polymers.

3.3.2
Uns-penp

In our previous studies on the dioxygen activation with copper complexes we have been quite successful with our systematic approach to substituting pyridyl arms in tmpa through aliphatic amine arms, thus obtaining uns-penp, apme and the commercially available tren ligand (Figure 3.4) [17, 21, 25]. Uns-penp (*N′*,*N′*-bis[(2-pyridyl)methyl]ethylenediamine) is a very interesting ligand that can be easily modified (see below), and different research groups have only recently started to use it in different areas of coordination chemistry. (Unfortunately most of the authors do not refer to the original synthesis by Mandel *et al.*) [26, 27]. We have been applying this ligand and its derivatives previously in copper coordination chemistry and have improved its preparation [25]. Interestingly, so far only an iron(II) complex of uns-penp has been described in the literature that has been investigated in regard to its spin cross-over properties [28]. Furthermore, very recently an iron(II) complex of a derivative of uns-penp was described [29]. However, so far no iron(III) complexes have been described in the literature. We were able to easily synthesize iron(III) uns-penp complexes and structurally characterize them [30]. However, again the catechol 1,2-dioxygenase activity of this complex system was lower than with tmpa and therefore was not investigated further.

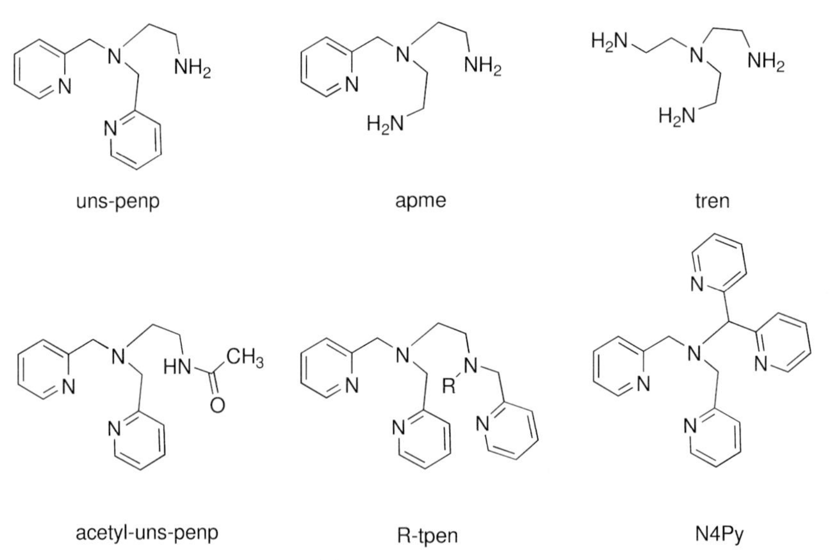

Figure 3.4 The ligand uns-penp and derivatives.

During the synthesis of uns-penp, its acetyl amide derivative, acetyl-uns-penp (*N*-acetyl-*N′*,*N′*-bis[(2-pyridyl)methyl]ethylenediamine, Figure 3.4), was prepared, an interesting compound/ligand itself [27, 30]. In the solid state it forms a dimer that could be structurally characterized and our efforts to synthesize the mononuclear deprotonated iron(III) complex [Fe(acetyl-uns-penp)]$(ClO_4)_2$ did lead to the formation of the dinuclear complex $[Fe_2\ (acetyl\text{-}uns\text{-}penp)_2O](ClO_4)_2 \times H_2O$ (Figure 3.5) [30].

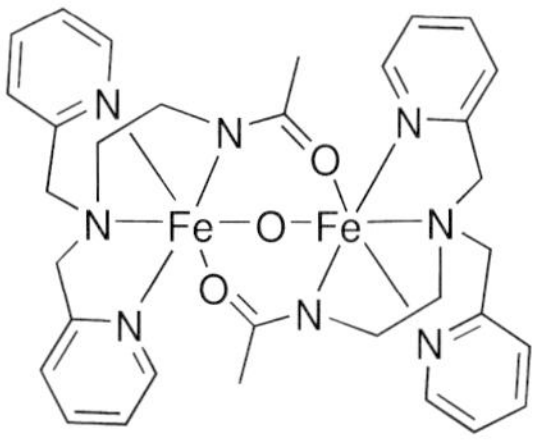

Figure 3.5 Dimeric solid state structure of the complex $[Fe_2(acetyl\text{-}uns\text{-}penp)_2O](ClO_4)_2 \times H_2O$.

It is more than interesting that the iron complex of the related ligand $PaPy_3H$ described by Mascharak and coworkers is structurally completely different from our compound [31]. In their complex the oxo-bridge assembles two mononuclear amide complexes (with coordinated deprotonated amide nitrogen atoms while the oxygen atoms are not coordinated) to form the dimer (as one would expect). In contrast, in our complex, the ligand acetyl-uns-penp shows an unusual pentadentate coordination mode displaying deprotonated carboxamido NCO bridging groups.

In contrast to our hope, replacing one pyridyl moiety of the tmpa ligand with a deprotonated carboxamide function again did not improve the oxidation of catecholates compared with the tmpa system. However, reactions of iron(III) salts with tetrachlorocatechol (tcc = tetrachlorocatecholate anion) and triethylamine in solution allowed the isolation and structural characterization of the complex $[(Fe(acetyl\text{-}uns\text{-}penp)(tcc))_2O]$ (Figure 3.6) [30].

The most striking feature of this complex is its secondary interactions, the intramolecular hydrogen bonds between the carboxamide nitrogen N(4) and O(2) of the catecholate (the donor–acceptor distance has a typical value of 2.755(3) Å and the distance between Fe(1) and N(4) is 3.692 Å). This intramolecular hydrogen bond between one arm of the tripodal ligand and a coordinated substrate molecule makes this compound an excellent model for the second substrate deprotonation step in the reaction cycle of intradiol cleaving catechol dioxygenases.

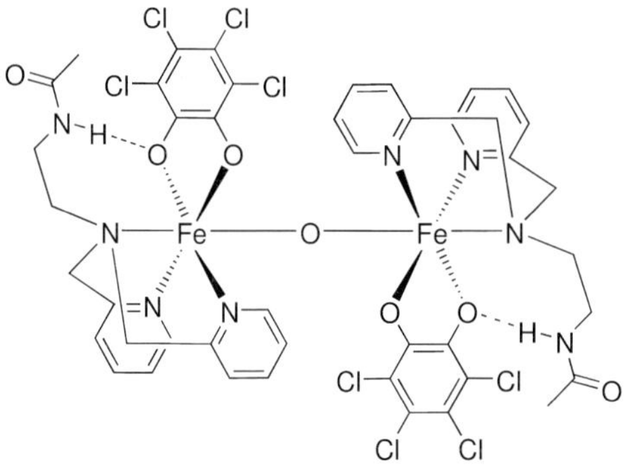

Figure 3.6 Solid state structure of the complex $[(Fe(acetyl\text{-}uns\text{-}penp)(tcc))_2O]$.

3.4
Mononuclear Iron Peroxo Complexes

Efforts by us and other research groups so far have not been successful in structurally characterizing an end-on superoxo iron complex, the first 'adduct dioxygen' compound formed during the reaction of an iron(II) complex with dioxgyen (Scheme 3.1). Only Que and coworkers have reported the spectroscopic characterization of such a species in a dinuclear system [32]. In contrast, in cooperation with the research groups of Sundermeyer and Holthausen, we were able to crystallize an end-on copper superoxo complex and report its molecular structure previously [33]. However, it is important to note at this point that it is not always easy to formulate such a transition metal 'dioxygen adduct' complex correctly as a superoxo compound. The real electron density distribution, for example in a copper complex, might not show complete oxidation of the Cu(I) ion to Cu(II) or it might be more correct to speak of a Cu(III) peroxo species [34].

Previous investigations and further recent studies on our reported copper superoxo complex as well as on related systems indicate that most likely a hydroperoxo complex is the reactive species. From our results, reported by us recently, on a end-on cobalt superoxide complex, we believe that this reaction in general presents a possible pathway for the oxidation of transition metal complexes with dioxygen [35]. The hydroperoxo complexes are formed as depicted for iron(II) complexes in Scheme 3.1. However, hydroperoxo transition metal complexes again can undergo subsequent reactions and form high-valent mononuclear oxo complexes (Fe(IV) = O in Scheme 3.1), which are discussed for iron compounds in more detail below [36, 37].

Instead of oxidizing the low-valency complexes with dioxygen, hydroperoxo complexes can be obtained more easily from reactions with hydrogen peroxide. Again, in copper chemistry such an end-on hydroperoxo complex could be structurally characterized, using a ligand (bppa, Figure 3.7) that stabilized this species through secondary interactions via hydrogen bonding [38]. In iron chemistry this has not yet been achieved, although Borovik and coworkers have obtained, with a deprotonated amide ligand (H_6**1**, Figure 3.7) related to bppa, a monomeric Fe(III) = O complex also stabilized through hydrogen bonding [39].

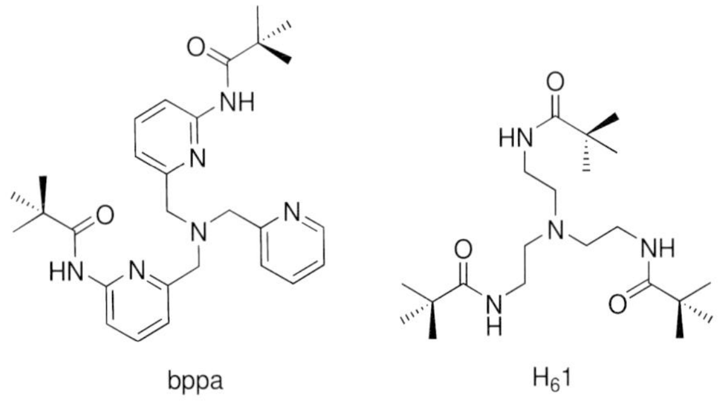

Figure 3.7 The ligands bppa and H_6**1**.

As discussed above, uns-penp is an interesting ligand that can be easily modified at the aliphatic amine arm. Introduction of a further pyridyl group together with different other groups R can be accomplished by facile reductive amination leading to a series of ligands, abbreviated as R-tpen (Figure 3.4). This method avoids the use of the unpleasant picolyl chloride applied in the original syntheses described [40, 41].

In cooperation with the McKenzie group we investigated the reaction of the iron(II) complex [Fe(Bz-tpen)Cl]ClO_4 (Figure 3.4, R = Bz) with hydrogen peroxide using stopped-flow techniques [42]. Time-resolved UV-Vis spectra for the formation of [Fe(Bz-tpen)OOH]$(ClO_4)_2$ are shown in Figure 3.8.

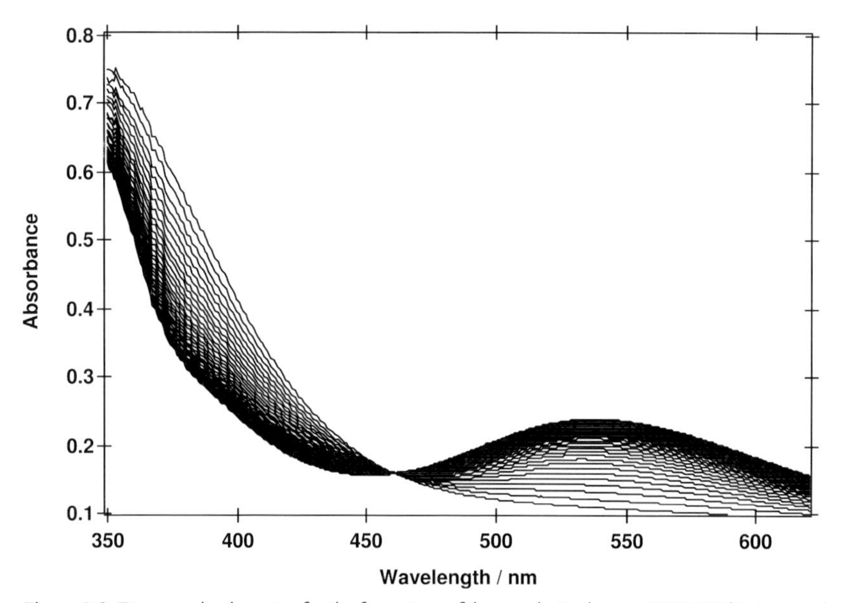

Figure 3.8 Time resolved spectra for the formation of the purple [Fe(bz-tpen)OOH]$(ClO_4)_2$ complex.

The mechanism proposed for this reaction is presented in Scheme 3.3. Immediately after the fast oxidation of Fe(II) to Fe(III) the hydroperoxo complex starts to

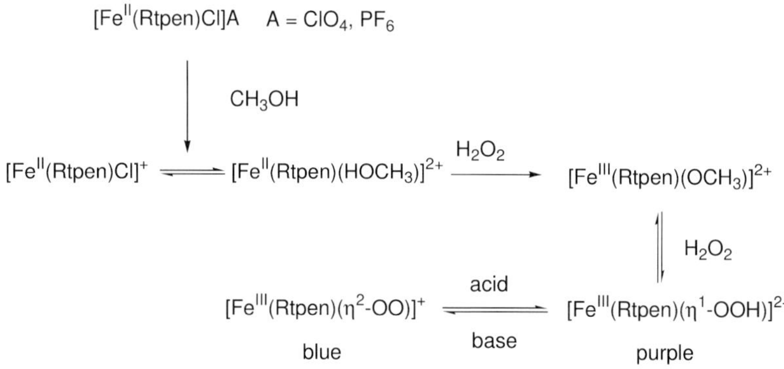

$[Fe^{II}(Rtpen)Cl]A$ A = ClO_4, PF_6

↓ CH_3OH

$[Fe^{II}(Rtpen)Cl]^+ \rightleftharpoons [Fe^{II}(Rtpen)(HOCH_3)]^{2+} \xrightarrow{H_2O_2} [Fe^{III}(Rtpen)(OCH_3)]^{2+}$

$\updownarrow H_2O_2$

$[Fe^{III}(Rtpen)(\eta^2\text{-}OO)]^+ \underset{base}{\overset{acid}{\rightleftharpoons}} [Fe^{III}(Rtpen)(\eta^1\text{-}OOH)]^{2+}$

blue   purple

Scheme 3.3

form. Deprotonation with base leads to a side-on peroxo iron complex. Unfortunately, so far all our numerous efforts to obtain crystals of one of these species for structural characterization have been unsuccessful.

In one attempt to stabilize either the hydroperoxo or the peroxo complex through secondary interactions, the ligand R-tpen was used with R = etOH to create a situation suitable for an intramolecular H-bonding interaction as shown in Figure 3.9. Unfortunately, no significant stabilization was attained, and indeed this complex even decomposed faster than the other derivatives of this complex [42].

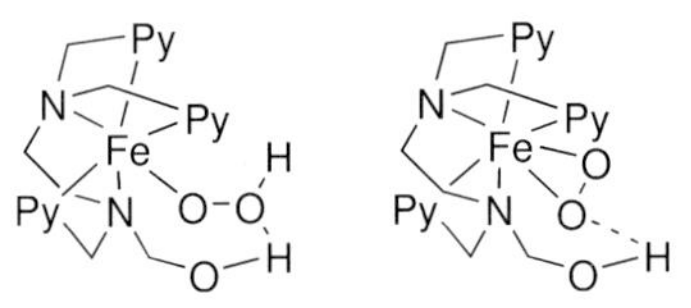

Figure 3.9 Attempt to stabilize either the hydroperoxo or peroxo intermediate by secondary H-bonding interaction in the complexes $[Fe(etOHtpen)OOH]^{2+}$/$[Fe(etOHtpen)OO]^{+}$.

From our kinetic studies, a negative activation entropy ($\Delta S^{*} = -72 \pm 8\,J\,mol^{-1}\,K^{-1}$) was obtained for the formation of $[Fe(II)(bz\text{-}tpen)OOH]ClO_4$, suggesting an associative mechanism. However, this was in contrast to a series of related reactions which have been investigated in the past and which accounted for a pure interchange mechanism [43]. Due to the fact that ΔS^{*} values usually have a very large error we reinvestigated the oxidation of $[Fe(II)Cl(bz\text{-}tpen)]ClO_4$ using high pressure techniques (in cooperation with the van Eldik group) to obtain more reliable activation volumes, ΔV^{*}, and indeed the results clearly indicate that here too the reaction follows an interchange substitution mechanism and not an associative mechanism (as the ΔS^{*} values had suggested) [44].

3.5 Mononuclear Iron Oxo Species

Fully characterized high-valent (oxidation states V and VI) iron complexes have been recently described [45, 46]. However, in bioinorganic chemistry iron(IV) oxo species probably play a more important role as active species in most selective substrate oxidations [47–49]. Excellent work by Que and coworkers allowed for the first time the structural characterization of a mononuclear Fe(IV) = O complex, using a simple macrocycle, methylated cyclam, as a ligand [7, 50]. From the full spectroscopic description of this system it was easy to spectroscopically detect the occurrence of this species as an intermediate with a series of ligands in solution and, quite surprisingly, even in water [7, 51–53]. During these studies it could be furthermore demonstrated that iron complexes with different ligands, for example Bz-tpen as well as the related N4py (Figure 3.4), can form iron(IV) oxo compounds and that these species are able to oxidize different substrates, even alkanes such as cyclohexane [36, 54].

Most recently Que and coworkers demonstrated that using the ligand acetyl-uns-penp (Figure 3.4), introduced by us into iron chemistry (see above), leads to quite interesting oxidation reactions of the relevant iron complexes using an oxidant and a substrate [55].

3.6
Work in Progress

Que and coworkers have demonstrated in the past that tmpa and its derivatives are quite useful in the synthesis and characterization of diamond core oxo complexes that seem to play an important role in the active site of MMO [3, 56]. Due to the fact that in here a dinuclear complex is formed from two monomers, the application of a preorganized dinucleating ligand would favor this reaction. Therefore, we recently synthesized bridged derivatives of the uns-penp ligands (Figure 3.10) and are currently investigating the reactivity of their iron complexes [57].

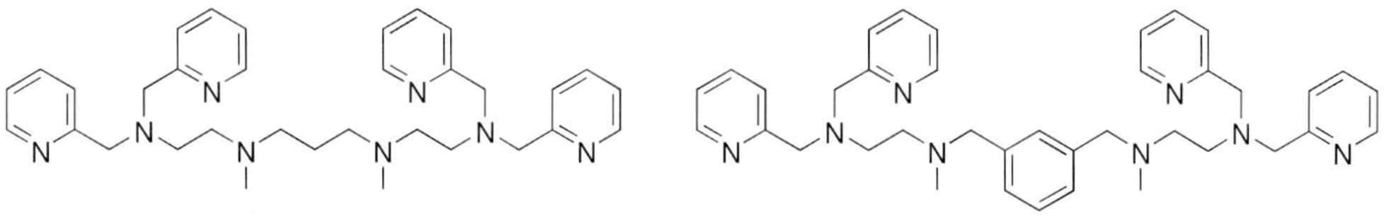

L^1 L^2

Figure 3.10 Bridged derivatives of the ligand uns-penp.

3.7
Conclusions

Today, oxidation reactions using iron complexes and dioxygen or hydrogen peroxide are much better understood than they were in the past and we believe that this will allow the development of simple iron complexes as oxidation catalysts in the near future. Additionally, great interest has developed in the last years in the investigation of the possible application of all kinds of iron catalysts in organic chemistry. The advantage can be seen in the usually low-cost materials (compared with platinum, palladium, etc.), and typically iron complexes are less toxic and/or more environmentally friendly. Furthermore, some of the compounds investigated could be interesting for medical applications, for example for the development of artificial superoxide dismutases [58].

Acknowledgments

We appreciate the collaboration with the research groups mentioned in this chapter and we would like to thank the DFG very much for financial support of our work.

References

1 Balasubramanian, R. and Rosenzweig, A.C. (2007) *Accounts of Chemical Research*, **40**, 573.

2 Murray, L.J. and Lippard, S.J. (2007) *Accounts of Chemical Research*, **40**, 466.

3 Costas, M., Mehn, M.P., Jensen, M.P. and Que L., Jr. (2004) *Chemical Reviews*, **104**, 939.

4 Tshuva, E.Y. and Lippard, S.J. (2004) *Chemical Reviews*, **104**, 987.

5 Kryatov, S.V., Rybak-Akimova, E.V. and Schindler, S. (2005) *Chemical Reviews*, **105**, 2175.

6 Korendovych, I.V., Kryatov, S.V. and Rybak-Akimova, E.V. (2007) *Accounts of Chemical Research*, **40**, 510.

7 Que, L., Jr. (2007) *Accounts of Chemical Research*, **40**, 493.

8 Hayashi, Y., Suzuki, M., Uehara, A., Mizutani, Y. and Kitagawa, T. (1992) *Chemistry Letters*, 91.

9 Menage, S., Brennan, B.A., Juarez-Garcia, C., Münck, E. and Que, L., Jr. (1990) *Journal of the American Chemical Society*, **112**, 6423.

10 Brennan, B., Chen, Q., Juarez-Garcia, C., True, A., O'Connor, C. and Que, L., Jr. (1991) *Inorganic Chemistry*, **30**, 1937.

11 Dong, Y., Ménage, S., Brennan, B.A., Elgren, T.E., Jang, H.G., Pearce, L.L. and Que, L.J. (1993) *Journal of the American Chemical Society*, **115**, 1851.

12 Feig, A.L. and Lippard, S.J. (1994) *Journal of the American Chemical Society*, **116**, 8410.

13 Feig, A.L., Becker, M., Schindler, S., van Eldik, R. and Lippard, S.J. (1996) *Inorganic Chemistry*, **35**, 2590.

14 Westerheide, L., Müller, F.K., Than, R., Krebs, B., Dietrich, J. and Schindler, S. (2001) *Inorganic Chemistry*, **40**, 1951.

15 Nizova, G.V., Krebs, B., Süss-Fink, G., Schindler, S., Westerheide, L., Gonzalez, L.G. and Shul'pin, G.B. (2002) *Tetrahedron*, **58**, 9231.

16 Blackman, A.G. (2005) *Polyhedron*, **24**, 1.

17 Schindler, S. *European Journal of Inorganic Chemistry*, (2000) 2311.

18 Mirica, L.M., Ottenwaelder, X. and Stack, T.D.P. (2004) *Chemical Reviews*, **104**, 1013.

19 Lewis, E.A. and Tolman, W.B. (2004) *Chemical Reviews*, **104**, 1047.

20 Jang, H.G., Cox, D.D. and Que, L., Jr. (1991) *Journal of the American Chemical Society*, **113**, 9200.

21 Weitzer, M., Schatz, M., Hampel, F., Heinemann, F.W. and Schindler, S. (2002) *Dalton Transactions*, 686.

22 Merkel, M., Pascaly, M., Krebs, B., Astner, J., Foxon, S.P. and Schindler, S. (2005) *Inorganic Chemistry*, **44**, 7582.

23 Schatz, M., Becker, M., Thaler, F., Hampel, F., Schindler, S., Jacobson, R.R., Tyeklar, Z., Murthy, N.N., Ghosh, P., Chen, Q., Zubieta, J. and Karlin, K.D. (2001) *Inorganic Chemistry*, **40**, 2312.

24 Foxon, S.P., Walter, O. and Schindler, S. (2002) *European Journal of Inorganic Chemistry*, 111.

25 Schatz, M., Leibold, M., Foxon, S.P., Weitzer, M., Heinemann, F.W., Hampel, F., Walter, O. and Schindler, S. (2003) *Dalton Transactions*, 1480.

26 Mandel, J. and Douglas, B. (1989) *Inorganica Chimica Acta*, **155**, 55.

27 Mandel, J., Maricondi, C. and Douglas, B. (1988) *Inorganic Chemistry*, **27**, 2990.

28 Matouzenko, G.S., Bousseksou, A., Lecocq, S., van Koningsbruggen, P.J., Perrin, M., Kahn, O. and Collet, A. (1997) *Inorganic Chemistry*, **36**, 2975.

29 Davies, C.J., Fawcett, J., Shutt, R. and Solan, G.A. (2005) *Dalton Transactions*, 2630.

30 Xu, J.-Y., Astner, J., Walter, O., Heinemann, F.W., Schindler, S., Merkel, M. and Krebs, B. (2006) *European Journal of Inorganic Chemistry*, 1601.

31 Patra, A.K., Afshar, R.K., Rowland, J.M., Olmstead, M.M. and Mascharak, P.K. (2003) *Angewandte Chemie – International Edition*, **42**, 4517.

32 Shan, X. and Que, L., Jr. (2005) *PNAS*, **102**, 5340.

33 Würtele, C., Gaoutchenova, E., Harms, K., Holthausen, M.C., Sundermeyer, J. and Schindler, S. (2006) *Angewandte Chemie*, **118**, 3951.

34 Cramer, C.J. and Tolman, W.B. (2007) *Accounts of Chemical Research*, **40**, 601.

35 Müller, J., Würtele, C., Walter, O. and Schindler, S. (2007) *Angewandte Chemie*, in press.

36 Bukowski, M.R., Comba, P., Lienke, A., Limberg, C., Lopez de Laorden, C., Mas-Ballesté, R., Merz, M. and Que, L., Jr. (2006) *Angewandte Chemie*, **118**, 3524.

37 Rosenthal, J. and Nocera, D.G. (2007) *Accounts of Chemical Research*, **40**.

38 Wada, A., Harata, M., Hasegawa, K., Jitsukawa, K., Masuda, H., Mukai, M., Kitagawa, T. and Einaga, H. (1998) *Angewandte Chemie*, **110**, 874.

39 MacBeth, C.E., Golombek, A.P., Young, V.G.J., Yang, C., Kuczera, K., Hendrich, M.P. and Borovik, A.S. (2000) *Science*, **289**, 938.

40 Jensen, K.B., McKenzie, C.J., Nielsen, L.P., Zacho Pedersen, J. and Svendsen, H.M. (1999) *Chemical Communications*, 1313.

41 Bernal, I., Jensen, I.M., Jensen, K.B., McKenzie, C.J., Toftlund, H. and Tuchagues, J.-P. (1995) *Dalton Transactions*, 3667.

42 Hazell, A., McKenzie, C.J., Nielsen, L.P., Schindler, S. and Weitzer, M. (2002) *Journal of the Chemical Society, Dalton Transactions*, 310.

43 Than, R., Schrodt, A., Westerheide, L., van Eldik, R. and Krebs, B. (1999) *European Journal of Inorganic Chemistry*, 1537.

44 Nebe, T., van Eldik, R. and Schindler, S.manuscript in preparation.

45 Berry, J.F., Bill, E., Bothe, E., DeBeer, S., Mienert, B., Neese, F. and Wieghardt, K. (2006) *Science*, **312**, 1937.

46 de Oliveira, F.T., Chanda, A., Banerjee, D., Shan, X., Mondal, S., Que, L., Jr., Bominaar, E., Münck, E. and Collins, T.J. (2007) *Science*, **315**, 835.

47 Krebs, C., Fujimori, D.G., Walsh, C.T. and Bollinger, J.M., Jr. (2007) *Accounts of Chemical Research*, **40**, 484.

48 Shaik, S., Hirao, H. and Kumar, D. (2007) *Accounts of Chemical Research*, **40**, 532.

49 Nam, W. (2007) *Accounts of Chemical Research*, **40**, 522.

50 Rohde, J.-U., In, J.-H., Lim, M.H., Brennessel, W.W., Bukowski, M.R., Stubna, A., Münck, E., Nam, W. and Que, L., Jr. (2003) *Science*, **299**, 1037.

51 Bukowski, M.R., Comba, P., Limberg, C., Merz, M., Que, L., Jr. and Wistuba, T. (2004) *Angewandte Chemie*, **116**, 1303.

52 Pestovsky, O., Stoian, S., Bominaar, E.L., Shan, X., Münck, E., Que L., Jr. and Bakac, A. (2005) *Angewandte Chemie*, **117**, 7031.

53 Bautz, J., Bukowski, M.R., Kerscher, M., Stubna, A., Comba, P., Lienke, A., Münck, E. and Que, L., Jr. (2006) *Angewandte Chemie*, **118**, 5810.

54 Kaizer, J., Klinker, E.J., Oh, N.Y., Rohde, J.-U., Song, W.J., Stubna, A., Kim, J., Münck, E., Nam, W. and Que, L., Jr. (2004) *Journal of the American Chemical Society*, **126**, 472.

55 Que, L., Jr. *Presentation ICBIC Vienna*, 2007.

56 Hsu, H.-F., Dong, Y., Shu, L., Young, V.G., Jr. and Que, L., Jr. (1999) *Journal of the American Chemical Society*, **121**, 5230.

57 Würtele, C., (2005) *Diplomarbeit*, Universität Gießen.

58 Kovacs, J.A. and Brines, L.M. (2007) *Accounts of Chemical Research*, **40**, 501.

4
Tuning of Structures and Properties of Bispidine Complexes

Peter Comba and Marion Kerscher

4.1
Introduction

Structures and electronics of transition metal complexes are interrelated and determine their thermodynamic properties and reactivities. Metal–ligand bond distances and angles involving the metal centers are generally softer than those of the carbohydrate backbone of the ligands, where torsions around single bonds are the only soft modes. Therefore, structures and properties of transition metal complexes are largely enforced by the ligand system, and the rigid ligands obviously are of importance. It follows that a careful design and choice of the ligand enables the preparative coordination chemist to obtain complexes with the desired properties [1].

Bispidine-derived ligands (3,7-diazabicyclo[3.3.1]nonane derivatives, see Scheme 4.1) are very rigid and, with respect to the two tertiary amine groups, highly preorganized [2]. The only flexibility in the ligand backbone of the tetradentate ligands (L^{1-4}) is that associated with the torsional angles which involve the two trans-disposed pyridine donors. Usually, distances from the metal center to the in-plane amine N3 are shorter than those to the axial amine N7, and the two cis-disposed sites for substrate coordination (trans to N3 and trans to N7) are therefore sterically as well as electronically different. In the pentadentate ligands (L^{5-11}), either of these two sites is blocked, usually by a third pyridine-derived donor. One of the more interesting features in bispidine coordination chemistry is that, while the ligands are very rigid, the coordination geometry is elastic [2, 3]. This is not a contradiction and has been observed with other non-cyclic rigid ligands such as phenanthroline. The rigid bispidine cavity is open on one side, and the metal center may find various minimum energy positions in the bowl-shaped cavity, where the position of the pyridine groups may adjust to the geometry [4]. The result is a potential energy surface with steep edges, a flat bottom part and with various shallow minima. This landscape may be tuned by subtle changes in the ligand backbone, primarily with respect to the substituents at N3, N7 and the two or three aromatic donor groups [2, 3].

Here, we report on possibilities to tune the structural and electronic properties of bispidine-derived ligands and their complexes. All examples discussed involve

Activating Unreactive Substrates: The Role of Secondary Interactions.
Edited by Carsten Bolm and F. Ekkehardt Hahn

ISBN: 978-3-527-31823-0

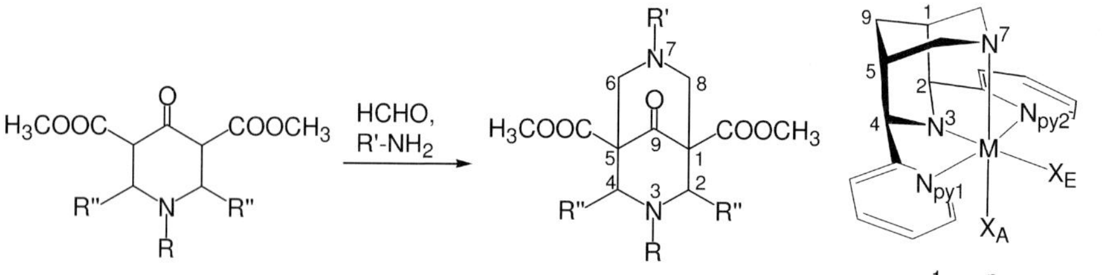

H_3COOC … $COOCH_3$, R'' N R'' , R —($HCHO$, R'-NH_2)→ L^n (R', N^7, 6, 8, O, 9, H_3COOC, $COOCH_3$, 5, 1, 4, 2, 3, R'', N, R'', R)

$[M(L^1)X_2]^{n+}$ (9, 1, N^7, 5, 2, 4, N^3, N_{py2}, M, X_E, N_{py1}, X_A)

subsituents at 1,3,5,7,9 omitted

ligand	R	R'	R''
L^1	CH_3	CH_3	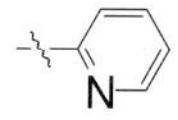 N
L^2	CH_3	CH_3	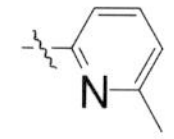 N
L^3	CH_3	CH_3	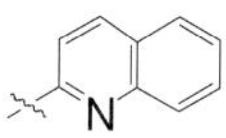 N
L^4	CH_3	CH_3	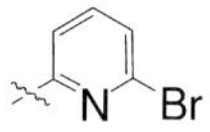 N Br
L^5	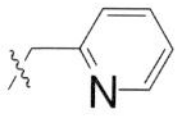 N	CH_3	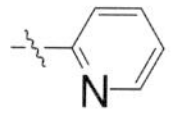 N
L^6	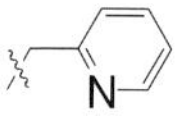 N	C_2H_5	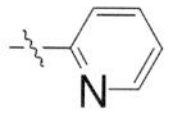 N
L^7	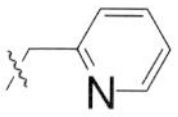 N	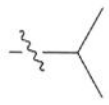	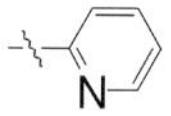 N
L^8	CH_3	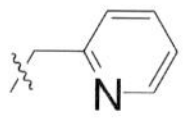 N	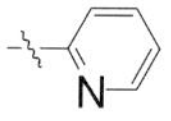 N
L^9	CH_3	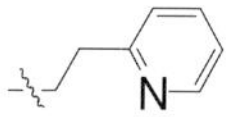 N	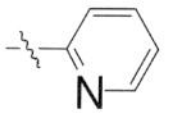 N
L^{10}	CH_3	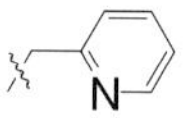 N	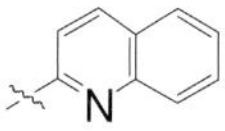 N
L^{11}	CH_3	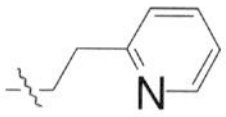 N	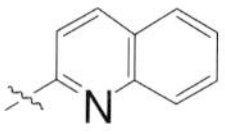 N
L^{12}	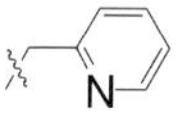 N	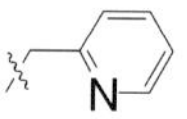 N	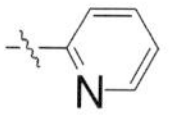 N
L^{13}	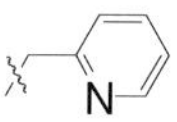 N	C_2H_4—L^5	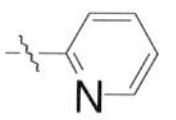 N

Scheme 4.1 Bispidine ligands.

Jahn–Teller-type effects, and the systems discussed involve copper(II), iron(IV) and cobalt(III) complexes of the ligands shown in Scheme 4.1.

4.2 Jahn–Teller Isomerism with Copper(II) Bispidines

Vibronic coupling between the two components of the vibrational mode of, for example, symmetry (in O_h) and the ^2E electronic wave function (in O_h) of hexacoordinate copper(II) complexes leads to tetragonal (Q_θ) or rhombic (Q_ε) distortions, and generally the resulting coordination geometries are elongated octahedral or square pyramidal. In general, there are three possible isomeric structures with the elongation along each of the three coordinate axes. This feature is described by the 'Mexican hat' potential energy surface with a warped rim which describes three distinct minima (the three 'isomeric' elongations, $d_{x^2-y^2}$ ground state), the corresponding maxima (the three 'isomeric' compressions, d_{z^2} ground state) and linear combinations involving these Q_θ distortions and the Q_ε modes in between [5–7]. Depending on the relative energies of the three minima and the related barriers (the compressed geometries), there is stabilization of a particular elongated (or in rare cases compressed) structure or fluxionality (dynamic Jahn–Teller effects). The third possibility, the isolation and structural and/or spectroscopic characterization of two or three isomeric structures related to the minima on the 'Mexican hat' potential energy surface was first reported for copper(II) bispidine complexes [8]. Some of the examples, where X-ray data are available, are presented in Figure 4.1.

With the tetradentate ligand L^2 all three structural forms with elongations along Cu-N7 (coligand trans to N3), Cu-N3 (coligand trans to N7) and py1-Cu-py2 (bidentate coligand) were isolated and structurally as well as spectroscopically characterized [9, 10]. Based on a conventional force field analysis, combined with spectroscopic and structural data as well as with ligand field and DFT calculations, it was concluded that the stabilization of one of the three minima is the result of a subtle balance between steric and electronic influences exerted by the coligands [10]. With the pentadentate ligand L^5 and $NCCH_3$ as coligand trans to N3 two minima were observed (elongation along py1-Cu-py2 or N7-Cu-py3); with other coligands (Cl^-, OH_2) only the latter structure was stabilized [8]. A preliminary analysis indicates that the isomer with the elongated py1-Cu-py2 axis is slightly more stable than that with a long Cu-N7 bond, but is destabilized by bulky substituents at N7 or relatively bulky coligands such as Cl^- or H_2O. Therefore, it appears that the tuning of the structures in this case is based on steric effects. All structures with the isomeric pentadentate bispidine L^8 have an elongated Cu-N7 bond, except for the trinuclear hexacyanoferrate-derivative with two trans-disposed $\{Cu(L^8)\}$ end groups [2]. This was attributed to the fact that the strong Cu-NC bond cannot be on the Jahn–Teller elongation axis, which therefore is switched to the py1-Cu-py2 direction. That is, in this example the tuning of the structure appears to be the result of electronic effects, and this interpretation is supported by 'electronically doped' force field calculations (LFMM) [11].

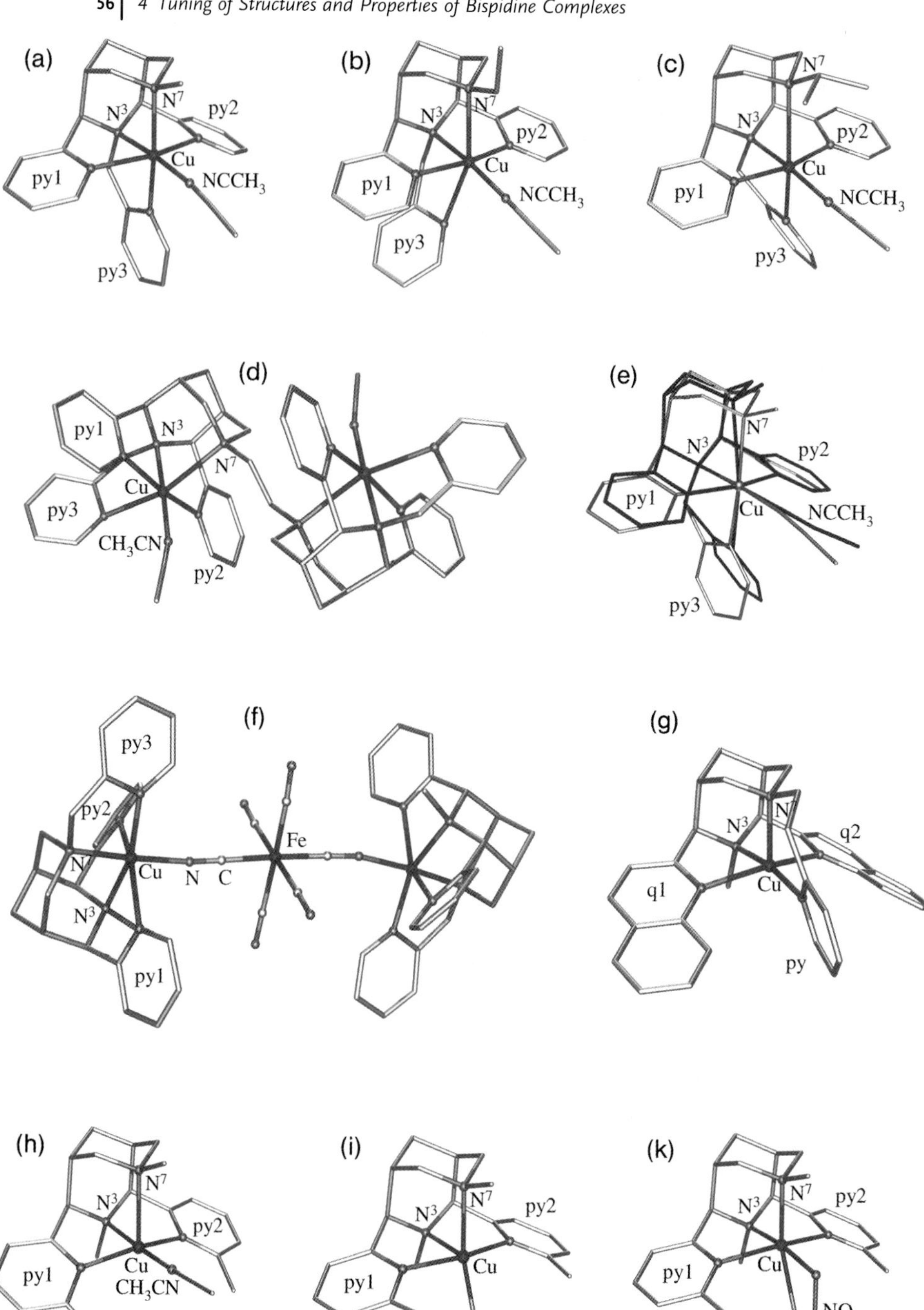
(a)
N^3
N^7
py2
py1
Cu
NCCH$_3$
py3
(b)
N^3
N^7
py2
py1
Cu
NCCH$_3$
py3
(c)
N^7
N^3
py2
py1
Cu
NCCH$_3$
py3
(d)
py1
N^3
N^7
py3
Cu
CH$_3$CN
py2
(e)
N^3
N^7
py2
py1
Cu
NCCH$_3$
py3
(f)
py3
py2
Fe
N
Cu
N
C
N^3
py1
(g)
N^3
N
q2
q1
Cu
py
(h)
N^3
N^7
py2
py1
Cu
CH$_3$CN
(i)
N^3
N^7
py2
py1
Cu
Cl
(k)
N^3
N^7
py2
py1
Cu
NO$_3$

Force field calculations with a ligand-field-derived energy term (LFMM) [12–16] have been shown to be able to accurately predict the geometries and relative energies of Jahn–Teller-distorted copper(II) systems. Consequently, the whole series of our structurally characterized copper(II) bispidine complexes has been thoroughly studied with this model. Presented in Table 4.1 is a selection of these results, where for each complex all relevant minima were optimized and, based on the relative energies, the correct structures were predicted [11].

Table 4.1 Comparison of experimental (plain) and calculated (italic) structural parameters of copper(II) complexes $[Cu(L^n)X]^{2+}$ with various bispidine ligands.

Distances	$L^{1a)}$	$L^{2a)}$	$L^{3a)}$	$L^{5a)}$	$L^{8a)}$	$L^{10a)}$	$L^{11b)}$	$L^{12c)}$
Cu-N3	2.04	**2.15**	2.15	2.07	2.04	2.09	1.95	2.09
Cu-N7	**2.27**	**2.12**	2.13	**2.48**	**2.37**	2.11	2.10	2.05
Cu-ar1	2.02	2.06	2.00	2.01	2.03	**2.34**	**2.31**	**2.28**
Cu-ar2	2.02	2.06	2.02	1.98	2.03	**2.61**	**2.27**	**2.57**
Cu-ar3	–	–	–	**2.54**	2.02	2.02	1.93	2.01
Cu-X	2.23	2.22	2.27	2.25	**2.72**	2.31	–	2.03
Angles [°]								
N3-Cu-N7	85	87	86	80	83	86	89	86
ar1-Cu-ar2	158	164	166	165	160	151	154	149

a) X = Cl.
b) 5-coordinate complex (X not available).
c) X = py (as part of the ligand).

The interesting and obvious question is, whether and how a particular geometry may be enforced by modifications of the backbone. This has been studied by DFT, LFMM and conventional force field calculations [10, 11]. The more interesting question is how the properties of the complexes depend on the ligand-enforced structures [2, 10, 17–19]. An interesting observation is the fact that because of the α-substituent at the pyridine groups (6-Me-pyridine vs pyridine (L^2 vs L^1)), there is a large difference in Cu-Cl bond stability, the {CuL} stability constant, the redox potential of the corresponding chloro complexes and the copper-catalyzed aziridination reactivity (see Table 4.2) [2, 18–20].

◀

Figure 4.1 Plots of the X-ray structures of Cu^{II} bispidine complexes with various directions of the Jahn–Teller elongation (hydrogen atoms and substituents at C1, C5 and C9 omitted); (a) $[Cu(L^5)(NCCH_3)]^{2+}$, (b) $[Cu(L^6)(NCCH_3)]^{2+}$, (c) $[Cu(L^7)(NCCH_3)]^{2+}$, (d) $[Cu(L^{13})(NCCH_3)]^{2+}$, (e) overlay plot of the chromophores of (a) and (d), (f) $\{Fe(CN)_6[Cu(L^1)]_2\}^+$, (g) $[Cu(L^{11})]^{2+}$, (h) $[Cu(L^2)(NCCH_3)]^{2+}$, (i) $[Cu(L^2)(Cl)]^+$, (k) $[Cu(L^2)(NO_3)]^+$.

Table 4.2 Thermodynamic properties and efficiency in catalytic aziridination of copper(II) bispidine complexes.

Ligand	Log K $[Cu(L)]^{2+}$	K $[\{CuL\}(Cl)]^{+}$	$E_{1/2}$ [mV vs $Ag/AgNO_3$]	Aziridination[a)]
L^1	16.6	20	−417	38% in 7.0 h
L^2	9.6	2.0	−98	35% in 2.0 h
L^4	–	–	+53	94% in 0.5 h

a) 5 mol% catalyst, 22-fold excess of styrene vs PhINTs, rt.

Also of interest is the fact that, within the group of tetra-, penta- and hexadentate bispidine ligands L^1, L^2, L^5, L^8, L^{12} with tertiary amine and pyridine donors and based on a constant backbone, there is an interesting variation of stability constants, which rather reflects the structural preferences than the type and number of donor groups. In addition, there is a non-Irving–Williams-type behavior of the complex stabilities as a function of the metal center. Both observations are related to the structures enforced by the ligands (see Table 4.3) [21].

Table 4.3 Complex stabilities (log K_{ML}) of the metal ion bispidine complexes in aqueous solution, $\mu = 0.1$ M (KCl).

Log K (ML^n)	L^1	L^2	L^5	L^8	L^{12}
Cu(I)[a)]	5.6	5.5	6.3	5.7	5.0
	9.5	8.9	8.8	9.6	7.9
Cu(II)[b)]	16.6	9.6	15.7	18.3	16.3
	12.5	7.1	14.5	15.3	16.2
Co(II)[c)]	5.5	–	13.7	6.2	7.3
Ni(II)	–	7.5	9.5	6.1	5.0
Zn(II)	11.4	–	13.6	8.3	9.2
Li(I)[d)]	2.7	3.9	3.7	–	3.7

a) By NMR titration, in MeCN.

b) Calculated in H_2O $\left(\ln\frac{K_{CuI}}{K_{CuII}} = \frac{E^\circ \cdot n \cdot F}{RT}\right)$.

c) Calculated in CH_3CN.

d) By UV-Vis titration.

While, for copper(II) this may also be related to Jahn–Teller-type distortions, for other metal ions, specifically for zinc(II), this obviously is not the case. It was concluded that some of the ligands (L^1 and L^8) enforce a structure which is preferred by the Jahn–Teller-labile copper(II) ion, i.e. these ligands have a high degree of complementarity for copper(II). This leads to increased stabilities for the copper(II) complexes and to destabilization of some of the other complexes, and also to relatively small copper(II) stabilities with some of the ligands (L^2 vs L^1, L^5 vs L^8, L^{12}). In conclusion, a careful choice of bispidine ligand enables us to enforce a particular structure and carefully tune the thermodynamic properties and reactivities [2, 18–21].

4.3
Stabilization of High-spin Ferryl Complexes

There is an increasing interest in mononuclear non-heme high-valent iron-oxo species, largely bcause of their importance in biological oxygen activation and the hydroxylation and hydroperoxidation of saturated and unsaturated C-H bonds [22–24]. A large body of experimental and computational data, primarily from the last decade and including structural and spectroscopic data as well as mechanistic work, helps to refine the models for the proposed pathways of dioxygen activation and oxygen transfer to substrates [23, 25].

One of the more intriguing remaining questions derives from the observation that the known and thoroughly studied enzymes have a high-spin oxo-iron(IV) center ($S = 2$), while all fully characterized model complexes have an intermediate-spin ($S = 1$) electronic ground state; in other examples oxo-iron(V) sites have been proposed as the catalytically active species [26–29]. Based on spectroscopic studies and the analyses of the cyclooctene-derived oxidation products as well as extensive DFT calculations, we have proposed a high-spin iron(IV) active catalyst together with a novel low-spin iron(IV) dihydroxo complex (see Figure 4.2) [30, 31]. However, so far we have not been able to trap and fully characterize the elusive high-spin ferryl complex.

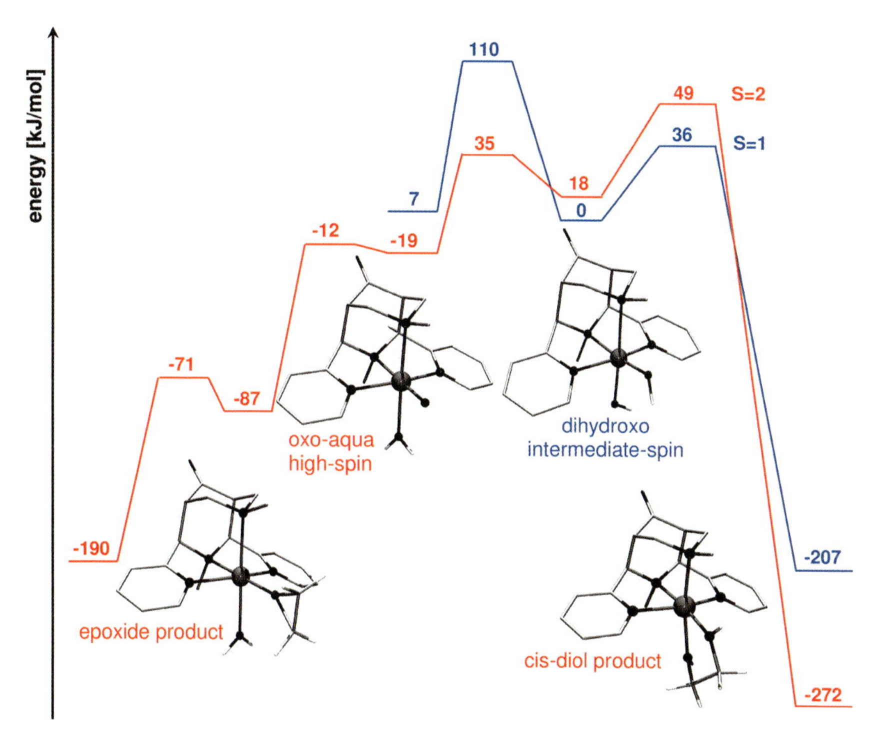

Figure 4.2 Computed pathway for the [Fe(L[1])]-catalyzed epoxidation and *cis*-dihydroxylation of ethylene.

Because of the enforced *cis*-octahedral coordination geometry, the ligand rigidity and the resulting flat potential energy surface for the bispidine ligands are ideal conditions for tuning the spin state of the ferryl complexes [32]. The main design principles, based on a DFT-based analysis, are that (i) independently of the spin state, substrates are strongly bound in the plane of the pyridine donors, that is trans to N3, and there are relatively weak interactions trans to N7, (ii) independently of the spin state, there is a strong trans influence exerted by the oxo group, and (iii) in contrast to the intermediate-spin species, the high-spin ferryl complex is Jahn–Teller labile, and the Jahn–Teller axis needs to be in the plane perpendicular to the iron-oxo bond. The data listed in Table 4.4 indicate that there is a good chance, based on the design principles listed above and the computed data in Table 4.4, of designing a bispidine-based ligand for a high-spin ferryl complex.

Table 4.4 Bond lengths, Mulliken spin densities, absolute and relative energies and triplet-quartet splittings for the bispidine $Fe^{IV}=O$ complexes. Relative energies are given with respect to the most stable spin state of a particular complex, regardless of geometry. Triplet-quintet splittings are given for each isomer.

			Bond lengths [Å]					E_{rel} [kJ mol^{-1}]	$\Delta E_{T\rightarrow Q}$ [kJ mol^{-1}]
	Fe=O	S	Fe-O	Fe-N7	Fe-N3	Fe-Npy	Fe-X		
6-coordinate $[(X)(L^{1,2})Fe^{IV}=O]^{2+}$									
$[(OH_2)(L^1)Fe^{IV}=O]^{2+}$	*trans*-N3	1	1.62	2.10	2.10	1.99	2.07	+7	−7
		2	1.62	2.28	2.08	2.03	2.34	0.0	
	trans-N7	1	1.61	2.27	1.97	2.00	2.08	+22	+9
		2	1.61	2.23	2.17	2.09	2.18	+32	
$[(NCCH_3)(L^1)Fe^{IV}=O]^{2+}$	*trans*-N3	1	1.62	2.14	2.10	1.99	1.99	0.0	+4
		2	1.61	2.32	2.10	2.04	2.30	+4	
	trans-N7	1	1.62	2.31	1.99	2.00	1.99	+13	+22
		2	1.61	2.26	2.18	2.11	2.17	+36	
$[(pyridine)(L^1)Fe^{IV}=O]^{2+}$	*trans*-N3	1	1.62	2.20	2.12	1.99	2.09	0.0	+3
		2	1.61	2.36	2.11	2.06	2.31	+3	
	trans-N7	1	1.61	2.45	2.02	2.02	2.07	+16	+12
		2	1.61	2.29	2.18	2.19	2.15	+28	
$[(OH)(L^1)Fe^{IV}=O]^{+}$	*trans*-N3	1	1.62	2.28	2.12	1.98	1.85	+19	−18
		2	1.62	2.33	2.19	2.21	1.81	+1	
	trans-N7	1	1.62	2.07	2.36	1.98	1.85	+23	−23
		2	1.61	2.30	2.21	2.22	1.80	0.0	
$[(NCCH_3)(L^2)Fe^{IV}=O]^{2+}$	*trans*-N3	1	1.62	2.14	2.09	2.05	1.99	+7	−7
		2	1.61	2.26	2.08	2.12	2.30	0.0	
6-coordinate $[(L^{5.8})Fe^{IV}=O]^{2+}$									
$[(L^5)Fe^{IV}=O]^{2+}$	*trans*-N3	1	1.62	2.14	2.07	1.99	1.98	0.0	+27
		2	1.61	2.24	2.09	2.13	2.09	+27	
$[(L^8)Fe^{IV}=O]^{2+}$	*trans*-N7	1	1.62	2.23	2.00	2.00	1.99	+10	+23
		2	1.61	2.21	2.19	2.19	2.06	+32	

4.4 Jahn–Teller-distorted Cobalt(III) Complexes

Cobalt(III) hexa- and pentaamine complexes are the prototypes of low-spin octahedral complexes with a strong ligand field and octahedral coordination geometry. However, based on structural data of the hexacoordinate cobalt(II) and zinc(II) and the pentacoordinate copper(II) complexes of the pentadentate bispidine L^9 it was not with surprise but with excitement that a pentacoordinate cobalt(III) complex was isolated and structurally characterized (see Figure 4.3a and Table 4.5) [33]. Interestingly, and as expected, the corresponding hexacoordinate 'precursor' could also be isolated and structurally characterized (see Figure 4.3b). A force-field- and DFT-based analysis confirmed that the main reason for the observed distortion is a strong repulsion of the axial donor (trans to N7) exerted by the pendent pyridine group trans to N3. This is the result of the ethylene bridge which, in contrast to the usual methylene bridge (ligand L^8), leads to a rather large N7-Co-py3 angle (97.8° vs 86.1°). The consequence is a relatively long bond to the monodentate coligand trans to N7 and, as a consequence, a relatively low energy d_{z2} orbital and a small energy gap between the homo (d_{xy}) and the lumo (d_{z^2}). The consequence is a low-energy excited high-spin state ($S = 2$), which is Jahn–Teller active and leads to a further elongation of the Co^{III}-coligand bond and eventually to loss of the monodentate ligand trans to N7, i.e. to the observed high-spin pentacoordinate complex (see Figure 4.3).

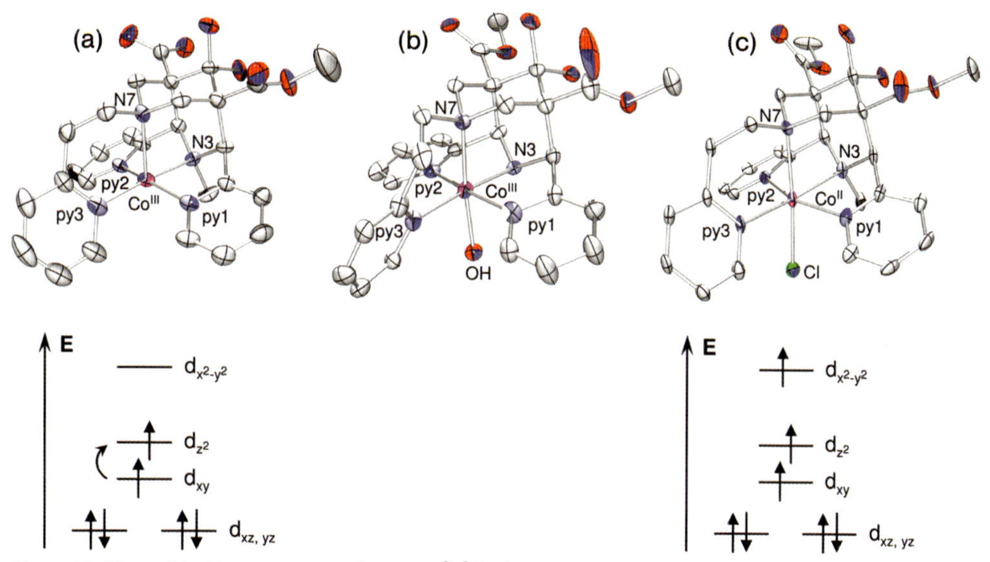

Figure 4.3 Plots of the X-ray structures of (a) $[Co(L^9)]^{3+}$, (b) $[Co(L^{9'})(OH)]^+$ and (c) $[Co(L^9)(Cl)]^+$ including the electronic occupation of the d-orbitals of the metal center for complexes (a) and (c).

This model explains why L^9 is able to enforce a square pyramidal structure with cobalt(III), while the corresponding cobalt(II) and zinc(II) complexes are expected and observed to lead to stable hexacoordinate complexes (see Table 4.5).

Table 4.5 Experimental metal-donor bond distances for transition metal complexes of the pentadentate ligand L^9.

Distances [Å]	Cu^{2+} (5-coord)	Co^{3+} (5-coord)	Co^{3+} (6-coord)	Co^{2+} (6-coord)	Zn^{2+} (6-coord)
M-N3 [Å]	2.01	1.97	1.91	2.19	2.22
M-N7	2.24	2.15	2.03	2.26	2.25
M-py1	2.04	1.99	1.95	2.14	2.13
M-py2	2.02	1.94	1.93	2.15	2.14
M-py3	1.98	1.98	2.00	2.17	2.09
M-X	–	–	1.95 (OH)	2.45 (Cl)	2.21 (OH_2)

4.5 Conclusions

The three examples discussed here indicate that subtle structural changes of the bispidine ligands may lead to strikingly different coordination geometries, electronic ground states and molecular properties (redox potentials, reactivities and stabilities). The first example discussed above involves the ligands L^1 and L^2, which differ only in the α-methyl substituents at the pyridine groups and lead to Cu^{II} complexes which have a difference in redox potentials of around 300 mV (energy difference of around 30 $kJmol^{-1}$). The second example is based on the tetradentate bispidine ligand L^1, which is assumed to stabilize with iron(IV) the elusive high-spin electronic configuration. The basis is a subtle balance between trans-influences and Jahn–Teller distortions, which are highly directional because of the ligand rigidity and preorganization. The third example shows that a simple extension of the chelate ring size may lead to a predictable but extremely rare situation for classical cobalt(III) complexes, where an easily populated excited state leads to a Jahn–Teller distortion and eventually to ligand dissociation and a stable pentacoordinate cobalt(III) complex. The basis for all three examples is a very rigid ligand backbone, a rather low ligand field due to a slightly too large and very rigid ligand cavity, and consequently a flat potential energy surface with various shallow minima.

Acknowledgments

Generous financial support by the German Science Foundation (DFG) is gratefully acknowledged.

References

1 Comba, P. (1999) *Coordination Chemistry Reviews*, **182**, 343.
2 Comba, P., Kerscher, M. and Schiek, W. (2007) *Progress in Inorganic Chemistry*, **55**, 613.
3 Comba, P. and Schiek, W. (2003) *Coordination Chemistry Reviews*, **238–239**, 21–29.
4 Comba, P., Kerscher, M., Merz, M., Müller, V., Pritzkow, H., Remenyi, R., Schiek, W. and Xiong, Y. (2002) *Chemistry - A European Journal*, **8**, 5750.
5 Bersuker, I.B. (2001) *Chemical Reviews*, **101**, 1067–1114.
6 Hitchman, M.A. (1994) *Comments on Inorganic Chemistry*, **15**, 197–254.
7 Figgis, B.N. and Hitchman, M.A. (2000) *Ligand Field Theory and its Applications*, Wiley-VCH, Weinheim, New York, p. 354.
8 Comba, P., Hauser, A., Kerscher, M.adn Pritzkow, H. (2003) *Angewandte Chemie – International Edition*, **42**, 4536–4540.
9 Comba, P., Lopez de Laorden, C. and Pritzkow, H. (2005) *Helvetica Chimica Acta*, **88**, 647–664.
10 Comba, P., Martin, B., Prikhod'ko, A. Pritzkow, H. and Rohwer, H. (2005) *Comptes Rendus Chimie*, **6**, 1506.
11 Benz, A., Comba, P., Deeth, R.J., Kerscher, M., Seibold, B. and Wadepohl, H. (2008) *Inorganic Chemistry*, **47**, 9518.
12 Burton, V.J., Deeth, R.J., Kemp, C.M. and Gilbert, P.J. (1995) *Journal of the American Chemical Society*, **117**, 8407–8415.
13 Deeth, R.J. (2001) *Coordination Chemistry Reviews*, **212**, 11–34.
14 Deeth, R.J., Fey, N. and Williams-Hubbard, B.J. (2005) *Journal of Computational Chemistry*, **26**, 123.
15 Deeth, R.J. and Hearnshaw, L.J.A. (2005) *Journal of The Chemical Society-Dalton Transactions*, 3638.
16 Deeth, R.J. and Hearnshaw, L.J.A. (2006) *Journal of The Chemical Society-Dalton Transactions*, 1092.
17 Comba, P. and Lienke, A. (2001) *Inorganic Chemistry*, **40**, 5206.
18 Comba, P., Merz, M. and Pritzkow, H. (2003) *European Journal of Inorganic Chemistry*, 1711.
19 Comba, P. and Kerscher, M. (2004) *Crystal Engineering*, **6**, 197–211.
20 Comba, P., Lang, C., Lopez de Laorden, C. Muruganantham, A., Rajaraman, G., Wadepohl, H. and Zajaczkowski, M. (2007) *Chemistry - A European Journal*, submitted.
21 Born, K., Comba, P., Ferrari, R., Kuwata, S., Lawrance, G.A. and Wadepohl, H. (2007) *Inorganic Chemistry*, **46**, 458.
22 Meunier, B., de Visser, S.P. and Shaik, S. (2004) *Chemical Reviews*, **104**, 3947.
23 Costas, M., Mehn, M.P., Jensen, M.P. and Que, L. Jr. (2004) *Chemical Reviews*, **104**, 939.
24 Nam, W. (2007) *Accounts of Chemical Research*, **40**, 522–531.
25 Chen, K., Costas, M. and Que, L. Jr. (2002) *Journal of The Chemical Society-Dalton Transactions*, 672.
26 Bassan, A., Blomberg, M.R.A., Siegbahn, P.E.M. and Que, L. Jr. (2005) *Angewandte Chemie*, **117**, 2999.
27 Bassan, A., Borowski, T., Lundberg, M. and Siegbahn, P.E.M., (2006) in *Concepts and Models in Bioinorganic Chemistry* (eds H.-B. Kraatz and N. Metzler-Nolte), Wiley-VCH, p. 63.
28 Jackson, T.A. and Que, L., Jr., (2006) in *Concepts and Models in Bioinorganic Chemistry* (eds H.-B. Kraatz and N. Metzler-Nolte), Wiley-VCH, p. 259.
29 Quinonero, D., Morokuma, K., Musaev, D.G., Morokuma, K., Mas-Balleste, R. and Que, L., J. (2005) *Journal of the American Chemical Society*, **127**, 6548.
30 Comba, P., Rajaraman, G. and Rohwer, H. (2007) *Inorganic Chemistry*, **46**, 3826.
31 Bautz, J., Comba, P., Lopez de Laorden, C. Menzel, M. and Rajaraman, G. (2007) *Angewandte Chemie – International Edition*, **46**, 8067.
32 Anastasi, A., Comba, P., McGrady, J., Lienke, A. and Rohwer, H. (2007) *Inorganic Chemistry*, **46**, 6420.
33 Comba, P., Kerscher, M., Lawrance, G.A., Martin, B., Wadepohl, H. and Wunderlich, S., (2008) *Angewandte Chemie – International Edition*, **47**, 4742.

5
Novel Phosphorus and Nitrogen Donor Ligands Bearing Secondary Functionalities for Applications in Homogeneous Catalysis

Anna-Katharina Pleier, Yu Sun, Anett Schubert, Dirk Zabel, Claudia May, Andreas Reis, Gotthelf Wolmershäuser, and Werner R. Thiel

5.1
Introduction

During the last two decades, decisive improvements in homogeneous catalysis in terms of activity and selectivity have been achieved. This makes the present situation different from that in earlier years, where the portfolio of catalytic reactions had to be filled up. Such a situation is quite typical for the development of innovation in scientific domains, when the increase in basic knowledge is transferred into practical applications over the years.

While the elementary steps of most catalytic reactions have become well understood, more and more subtle effects can now be worked out. This includes on one hand the adaptation of known catalytic reactions to special solvent systems like aqueous/organic or organic/fluorinated two-phase compositions and supercritical CO_2 and ionic liquids. Such solvents had only rarely been used in catalysis before [1–3]. On the other hand, further effects are due to electronic structuring and steric shaping of the ligand sphere, an approach which in some cases can clearly be seen to be copied from nature. In biological systems a series of enzymes may, for example, use the same co-factor, although the catalyzed reactions may differ fundamentally from a mechanistic point of view. Using a similar approach for the construction of (small) ligands (let us call it biomimetic ligand design), one may try to realize a certain ligand geometry not only by the stepwise formation of covalent bonds, but also by the implementation of secondary intra- and intermolecular interactions. This may finally lead to the realization of exceptional ligand geometries or to substrate recognition (Scheme 5.1).

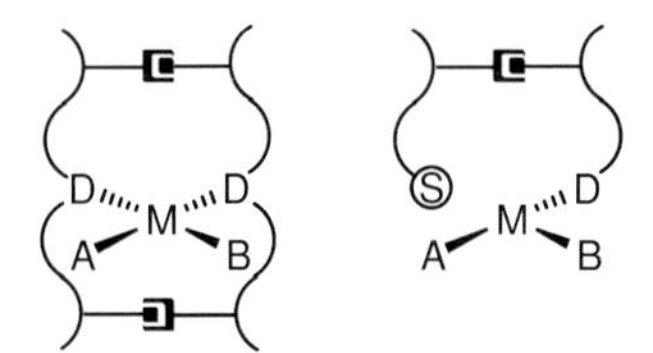

Scheme 5.1 Ligand shaping (left) and molecular recognition of a substrate (right) by reversible interactions.

Activating Unreactive Substrates: The Role of Secondary Interactions.
Edited by Carsten Bolm and F. Ekkehardt Hahn

ISBN: 978-3-527-31823-0

Such an approach, which unifies homogeneous catalysis with biochemical and supramolecular aspects of structure formation, is in our opinion a novel and promising new concept for ligand design. It clearly differs from the search for model compounds for the active sites in metallo enzymes. The structures of metallo proteins are determined by a multitude of weak chemical interactions [4, 5], which are difficult to be quantified completely by the modern methods of structure determination (NMR, X-ray). Additionally, the cost of synthesis of more and more complex ligands will increase and will become a critical factor, at least for industrially relevant systems. We are thus not looking for new model complexes for metallo proteins but are trying to implement new functionalities in known ligand systems, which allow us to transfer some of the principles realized by nature in enzymes to homogeneous catalysis. One thing we are interested in is the increase of rigidity of a given ligand backbone by reversible interactions (e.g. coordinative or hydrogen bonds) or irreversible but self-organizing covalent interactions. The desired ligands should therefore be accessible in just a few steps. Additionally, these ligands may possess extra binding sites either for substrates or reagents (Scheme 5.1), which will influence the activity or selectivity of the catalyst without interacting directly with the active site.

5.2 Phosphine Ligands

Substitution in the ortho positions of the phenyl rings of triphenylphosphine first of all sterically affects a catalytically active site. This is expressed by an increase of Tolman's cone angle, which gives a hint of the shielding of the metal site by the ligand. Bulky substituents in the ortho position stabilize low coordination numbers which, for example, favors the reductive elimination in catalytic reactions. Secondly, ortho substituents in aryl phosphines are in a position which allows them to interfere in the coordination sphere. There are prominent examples of so-called semi-labile ligands using this concept to provide free coordination sites. Additionally, *ortho*-methyl substituents can even undergo deprotonation and lead to *P*,*C*-chelation, which again can be favorable for the performance of a catalytic system. While ortho substitution thus has a direct effect on the metal site, meta and para substitution is mainly used to tune the electronic properties of the electron-donating site or to adapt a given ligand to a special solvent system. This can, for example, be done by the introduction of fluorinated alkyl chains, sulfonate groups or polythers. In contrast to this, we tried to implement moieties for intra- and intermolecular interactions to reach self-organization of the ligand backbone as well as selective substrate recognition.

5.2.1 Cooperative Effects for Ligand Self-organization

We have achieved ligand self-organization by the implementation of a modular design principle which allows the rapid and variable generation of mono- and bidentate phosphine ligands. In this construction kit, we take 3-formylphenyl(diphenyl) phosphine (**1**) as the building block which is responsible for the coordination to the metal center and provides a reactive aldehyde group for binding a linker via a stable imine group. This strategy allows the introduction of almost any functional group which is capable of performing a linkage, for example, via hydrogen bridges, disulfide moieties or the coordination of hard Lewis acids. The precursor **1** is available in excellent yields in a four-step reaction sequence in quantities up to 50 g (Scheme 5.2) following a modified published procedure [6]. The functionalization in the meta position was assigned to be beneficial by molecular modeling studies.

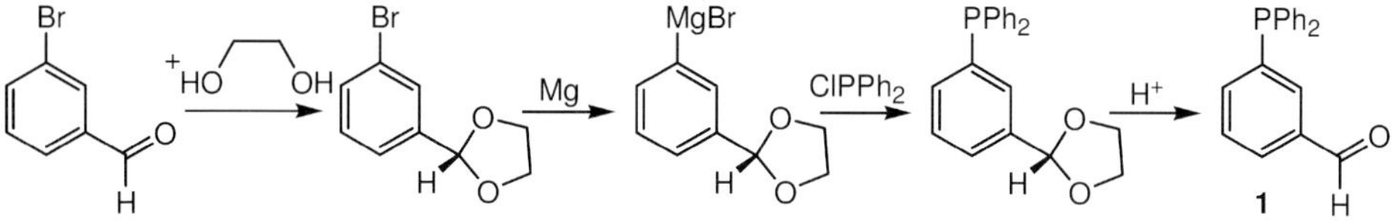

Br, + HO OH, O, H, Br, O, O, H, Mg, MgBr, O, O, H, ClPPh2, PPh2, O, O, H, H+, PPh2, O, 1, H

Scheme 5.2 Synthesis of 3-formylphenyl(diphenyl) phosphine (**1**).

As the second component for ligand construction we used substituted anilines, which are accessible by standard reactions of organic chemistry. The synthesis of these compounds allows us to attach the functionalities for secondary interactions either directly or via a short CH_2 linker. It is either possible to build up these molecules starting from 3-nitrobenzyl chloride or via a nucleophilic aromatic substitution with 4-fluoronitrobenzene. In both cases the remaining nitro group can be reduced to the desired amino function.

Following these strategies we first used polyether chains of different lengths to link two PPh_3 ligands as shown in Scheme 5.3. Glycols like ethylene or tetraethylene glycol are α, ω-substituted with 3-nitrobenzyl units when treated with 3-nitrobenzyl chloride in a molar ratio of 1 : 2. The reaction requires deprotonation of the OH group with NaH in THF or strongly basic conditions (NaOH) in a biphasic system (CH_2Cl_2/H_2O) with (NBu_4)Cl as a phase-transfer catalyst. The resulting di(nitrophenyl) compounds **2** and **3** are smoothly reduced by 10% Pd-C/$NH_4(HCO_2)$ in dry MeOH to give the corresponding anilines **4** and **5** in excellent yields.

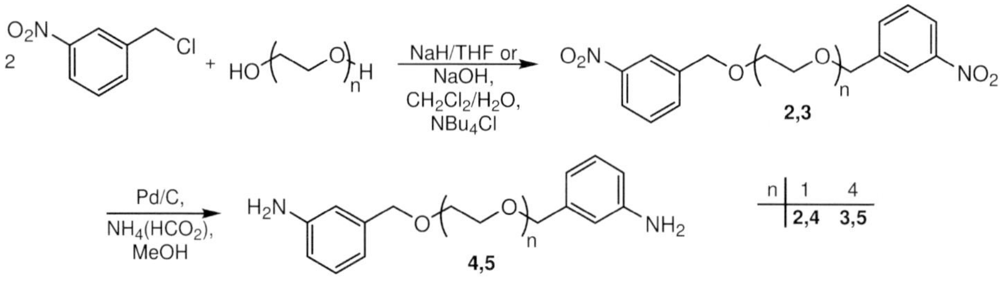

O2N, Cl, 2, + HO(~O)nH, NaH/THF or NaOH, CH2Cl2/H2O, NBu4Cl, O2N, O, O, n, NO2, 2,3

Pd/C, NH4(HCO2), MeOH, H2N, O, O, n, NH2, 4,5

n	1	4
	2,4	3,5

Scheme 5.3 Synthesis of the polyether linker units **3** and **5**.

For a reversible coordinative linkage, the pyrazolyl pyridine-substituted aniline **7** was synthesized. It can be obtained by a nucleophilic aromatic substitution of 4-fluoronitrobenzene with pyrazolyl pyridine leading to the nitro derivative **6**

(Scheme 5.4), which could be reduced to the aniline **7** under the same conditions as described above.

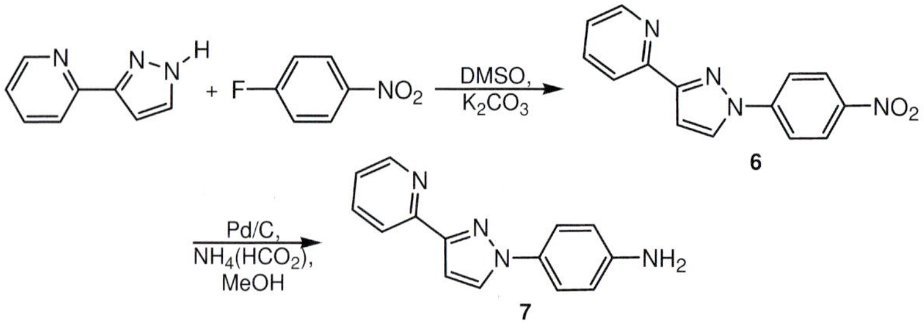

DMSO, K_2CO_3

6

Pd/C, $NH_4(HCO_2)$, MeOH

7

Scheme 5.4 Synthesis of the pyrazolylpyridine linker unit **7**.

From this reaction sequence two derivatives could be structurally characterized: 4-(3-(2-pyridyl)pyrazol-1-yl)nitrobenzene (**6**) reacts with $Mo(CO)_4(pip)_2$ (pip = piperidine) to give the corresponding tetracarbonyl complex (Figure 5.1, top) [7], one of the rare examples of a low-valent organometallic species bearing an oxidizing nitro group in the ligand system. The amine **7** could be derivatized and crystallized as the amide of decanoic acid (Figure 5.1, bottom).

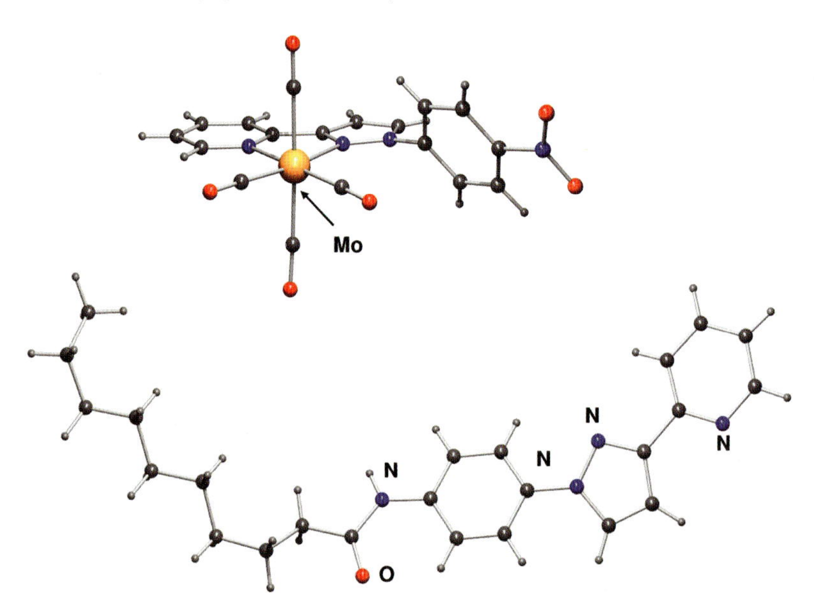

Figure 5.1 Structurally characterized derivatives of **6** and **7**.

Condensation of 3-formylphenyl(diphenyl) phosphine (**1**) with the aromatic amines **4**, **5** and **7** gave the corresponding phosphine-substituted Schiff bases **8**, **9** and **10**, respectively (Scheme 5.5) [8]. The reactions were carried out under reflux conditions in dry EtOH [9]. After the crystallization from MeOH (except for **9**, which is an oil), the Schiff bases were obtained in almost quantitative yields.

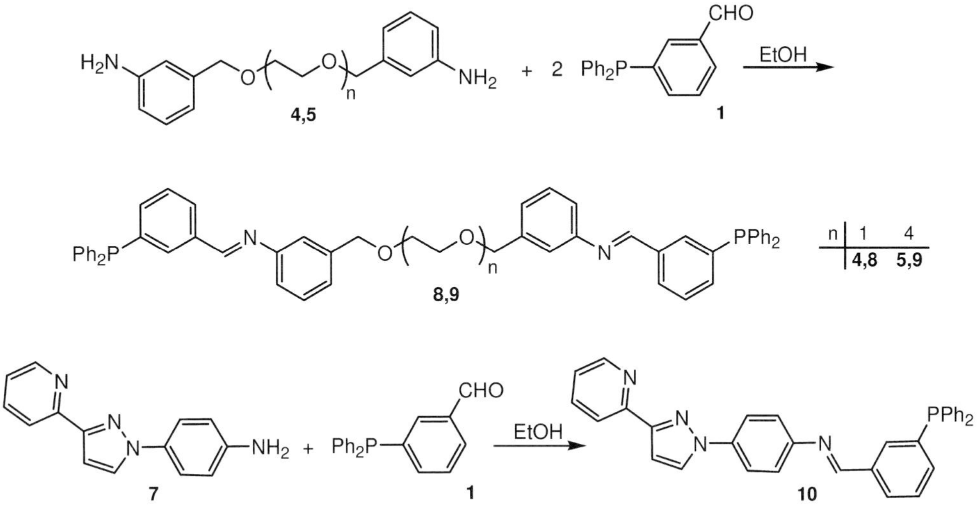

Scheme 5.5 Synthesis of the Schiff-base functionalized phosphines **8–10**.

Treatment of **8–10** with (CO)AuCl gave the corresponding gold [10] complexes **11–13** (Scheme 5.6).

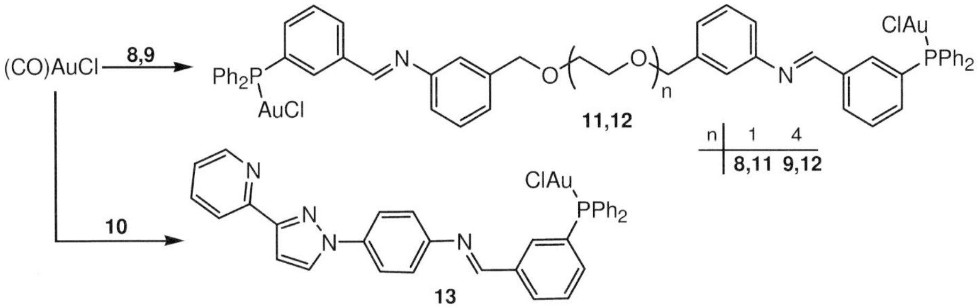

Scheme 5.6 Gold complexes derived from functionalized triphenyl phosphine ligands.

The rhodium complexes **14** and **15** bearing the chelating bisphosphine ligands **8** and **9** were prepared by slowly dropping diluted solutions of $[(CO)_2Rh(\mu\text{-}Cl)]_2$ and the appropriate ligand simultaneously into large volumes of benzene to prevent the formation of polymeric aggregates (Scheme 5.7) [8]. Complex **16** is accessible in more concentrated solution from ligand **10**.

The complexes **11–16** were characterized by means of NMR spectroscopy (^{1}H, ^{13}C, ^{31}P), infrared spectroscopy ($\nu_{C=N}$: 1620–1626 cm^{-1}) and mass spectroscopy (ESI-TOF). A single CO absorption at about 1970 cm^{-1} proves that the phosphine ligands coordinate the rhodium center in trans configuration [11].

Interestingly, the ^{31}P NMR spectra of compounds **14** and **15** show two doublets with chemical shifts of 28.2 and 28.3 ppm (ratio: 3 : 1) and $^1J_{RhP}$ of 127.1 and 126.8 Hz for **14** and with chemical shifts of 28.7 and 28.6 ppm (ratio: 4 : 1) and $^1J_{RhP}$ of 128.4 and 127.6 Hz for **15**. Molecular mechanics calculations suggest two conformers

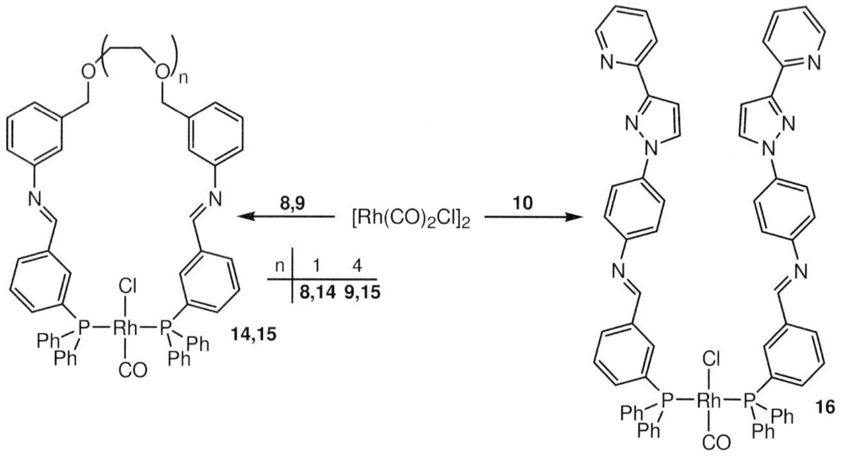

Scheme 5.7 Rhodium complexes derived from functionalized triphenyl phosphine ligands.

differing in the relative orientation of the ligand backbone (Scheme 5.8). Because of the polyether linkage, the rotations around the two critical P-C_{Ph} bonds are hindered on the time scale of ^{31}P NMR, thus allowing the observation of two conformers with a syn or an anti orientation of the imine groups.

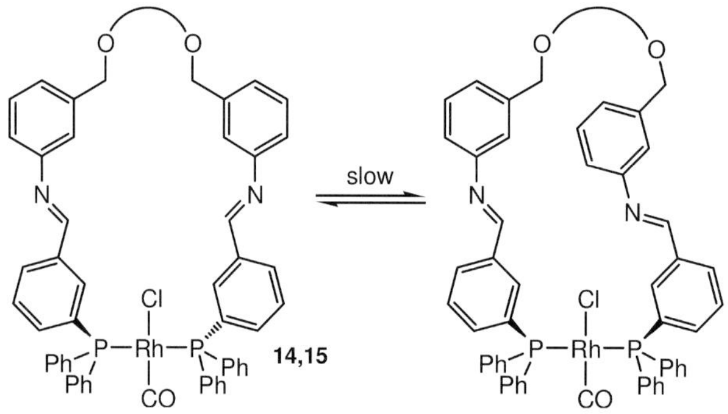

Scheme 5.8 Anti (left) and syn (right) conformer of the rhodium complexes bearing polyether linked phosphine ligands [12].

In compound **16** free rotation around the critical P-C_{Ph} bonds is possible and fast with respect to the time scale of ^{31}P NMR spectroscopy preventing the formation of conformers. Therefore only one doublet is observed in the ^{31}P NMR spectrum of **16** ($\delta = 28.4$ ppm, $^{1}J_{RhP} = 127.8$ Hz).

Addition of KPF_6 to solutions of **14** does not affect its ^{31}P resonances. In contrast, the signals of the syn and the anti conformer of complex **15** (with the longer polyether chain) start to coalesce already after the addition of 1 equiv. of KPF_6. This means that the exchange between the conformers is accelerated in the presence of the Lewis acid K^+. For complex **13** the exchange between the syn and anti conformer is rapid since the donor sites are not covalently connected. Here, the addition of $Zn(OTf)_2$ leads to a splitting of the ^{31}P NMR resonance into two signals, which we again assign to belong to the syn and the anti conformer.

In collaboration with the group of M. Beller at the IfoK in Rostock, the rhodium complexes **14–16** were tested as catalysts for hydroaminomethylation reactions [13]. The reactions of 1-pentene and styrene with piperidine (Scheme 5.9) were investigated. Selected examples of our study are summarized in Tables 5.1 and 5.2.

R = $CH_3CH_2CH_2$, C_6H_5

Scheme 5.9 Catalytic hydroaminomethylation of pentene and styrene with piperidine.

In general, the conversions and selectivities obtained by the rhodium(I) complexes **14–16** are comparable with the standard systems $Rh(CO)_2(acac)$ and $Rh(CO)_2(acac)$ + PPh_3 as long as identical CO pressures are applied. Simple $Rh(CO)_2(acac)$ gives almost 100% selectivity for the formation of the amine with 1-pentene, while the addition of triphenylphosphine reduces the selectivity to about 70–90%. The addition of the Lewis acids KPF_6 or $Zn(OTf)_2$ to standard systems makes only slight differences in conversion or selectivity. However, an interesting effect was observed for compound **11** in the hydroaminomethylation of 1-pentene: the amine selectivity

Table 5.1 Results of the Rh-catalyzed hydroaminomethylation of 1-pentene with piperidine.

Catalyst	Lewis acid	P_{CO} [bar]	Conversion [%]	Selectivity [%]		N : iso	
				Amine	Enamine	Amine	Enamine
14	–	10	93	66	31	89 : 11	71 : 29
14	–	5	88	80	8	90 : 10	17 : 83
15	–	10	93	57	37	88 : 12	78 : 22
16	–	10	92	60	33	87 : 13	64 : 36
14	KPF_6	5	95	94	6	87 : 13	0 : 100
15	KPF_6	10	91	54	36	83 : 17	64 : 36
16	$Zn(OTf)_2$	10	92	76	20	89 : 11	40 : 60
$Rh(CO)_2(acac) + PPh_3$	–	5	90	84	15	93 : 7	49 : 51
$Rh(CO)_2(acac) + PPh_3$	KPF_6	5	96	93	6	92 : 8	0 : 100
$Rh(CO)_2(acac) + PPh_3$	$Zn(OTf)_2$	5	85	70	24	90 : 10	45 : 55
$Rh(CO)_2(acac)$	–	5	95	96	–	60 : 40	–
$Rh(CO)_2(acac)$	KPF_6	5	98	98	–	62 : 38	–
$Rh(CO)_2(acac)$	$Zn(OTf)_2$	5	97	95	–	58 : 42	–

Reaction conditions: 1-pentene (10 mmol), piperidine (10 mmol), solvent: THF (30 mL). Catalyst 0.1 mol-%, Lewis acid: 0.4 mol-%, P_{H2} 50 bar, T 95 °C, t 12 h.

Table 5.2 Results of the Rh-catalyzed hydroaminomethylation of styrene with piperidine.

Catalyst	Lewis acid	P_{CO} [bar]	Conversion [%]	Selectivity [%]		N : iso	
				Amine	Enamine	Amine	Enamine
14	–	5	90	35	63	67 : 33	63 : 37
14	KPF_6	5	94	43	53	65 : 35	59 : 41
$Rh(CO)_2(acac) + PPh_3$	–	5	92	51	45	41 : 59	65 : 35

Reaction conditions: styrene (10 mmol), piperidine (10 mmol), solvent: THF (30 ml). Catalyst 0.1 mol-%, Lewis acid: 0.4 mol-%, P_{H2} 50 bar, T 95 °C, t 12 h.

rises from 80% to 94% after the addition of a fourfold molar excess of KPF_6. Parallel to this, the iso enamine is found as the only isomer. This is also observed for the catalytic system $Rh(CO)_2(acac) + PPh_3$.

With these investigations we could prove that secondary interactions between functionalized phosphine ligands and the Lewis acids are not only detectable by means of ^{31}P NMR spectroscopy but are of relevance for catalysis. The work on systems which carry binding motives for certain substrates in the ligand backbone is still going on.

5.2.2
Phosphines with Pyrazole and Pyrimidine Substituents

A second route to functionalized phosphines was opened up by the introduction of pyrazole or pyrimidine groups into triaryl phosphines. To our surprise such systems were unknown until our investigations. The monoortho-functionalized derivatives are structurally closely related to diphenyl(2-(2-pyridyl)phenyl) phosphine, which has been used as a ligand in different catalytic transformations [14]. Additionally, multiple functionalization of PPh_3 with pyrazole or pyrimidine groups, which is as simple as mono functionalization, opens up a novel access to water-soluble ligands, because of the basic (pyrazole, pyrimidine) and protic (pyrazole) properties of the heterocycles or by further derivatization. Such ligands therefore may also find application in aqueous-phase organometallic catalysis [15].

Pyrazoles or pyrimidines are obtained by reacting 1,3-diketones with either hydrazine or reagents bearing a $NH{=}C(R){-}NH_2$ unit. Up to now, only a few arylphosphines with β-diketone side chains have been described, probably because of difficulties related to the synthesis [16]. We have been working on the synthesis of arylpyrazoles for some time and have realized an access using 1-aryl-3-dimethylamino-2-propen-1-ones as 'masked' aromatic 1,3-diketones [17]. For the application of this route in phosphine chemistry, an acetyl group attached to one or more of the phenyl rings of PPh_3 is required which is transformed to an aminopropenone chain by reaction with dimethylformamide dimethylacetal (DMFDMA, Scheme 5.10) [18]. The aminopropenone-functionalized phosphines are obtained in good yields; only the threefold ortho substituted derivative **17aα** could not be obtained, probably for steric reasons [19]. We are now working on the shortening of the synthesis.

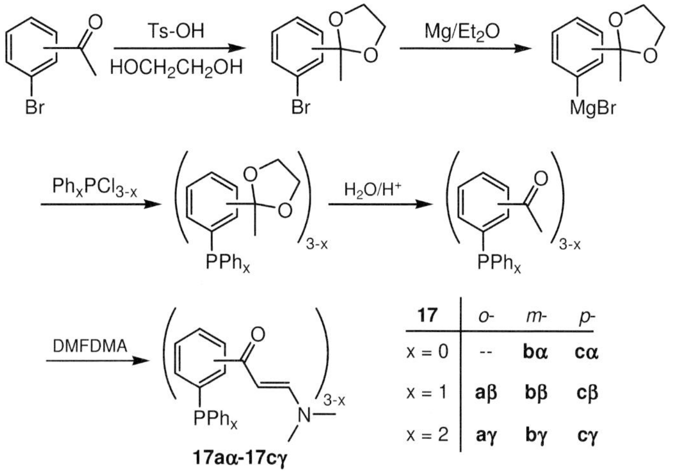

17	*o-*	*m-*	*p-*
x = 0	--	**bα**	**cα**
x = 1	**aβ**	**bβ**	**cβ**
x = 2	**aγ**	**bγ**	**cγ**

Scheme 5.10 Synthesis of aminopropenone functionalized arylphosphines.

Ring closure with excess of hydrazine generates the pyrazolyl-substituted derivatives **18aα-cγ** as colorless solids in excellent yields (Scheme 5.11). The threefold substituted derivatives **18bα** and **18cα** are already soluble in diluted acids and thus represent a novel class of water-soluble phosphine ligands.

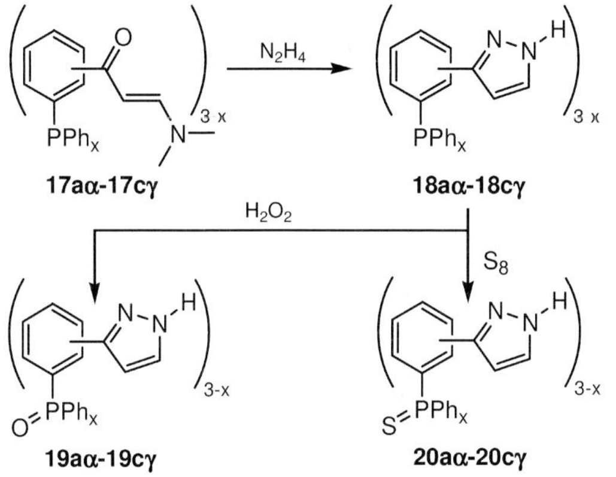

17-20	*o-*	*m-*	*p-*
x = 0	--	**bα**	**cα**
x = 1	**aβ**	**bβ**	**cβ**
x = 2	**aγ**	**bγ**	**cγ**

Scheme 5.11 Synthesis of pyrazole functionalized arylphosphines and derived phosphine oxides and sulfides.

As an example, the solid-state structure of the phosphine oxide **19bγ**, monofunctionalized in the *meta* position, was determined by single-crystal X-ray structure analysis (Figure 5.2). Hydrogen bonds dominate the solid-state structure: the proton of the N–H function of the pyrazole undergoes hydrogen bonding to the phosphine oxide group of a neighboring molecule, which results in the formation of planar zigzag chains parallel to the *a*-axis of the unit cell.

Reaction of the aminopropenones **17aα-cγ** with an excess of guanidinium carbonate in EtOH and aqueous KOH at reflux temperature for several hours resulted in

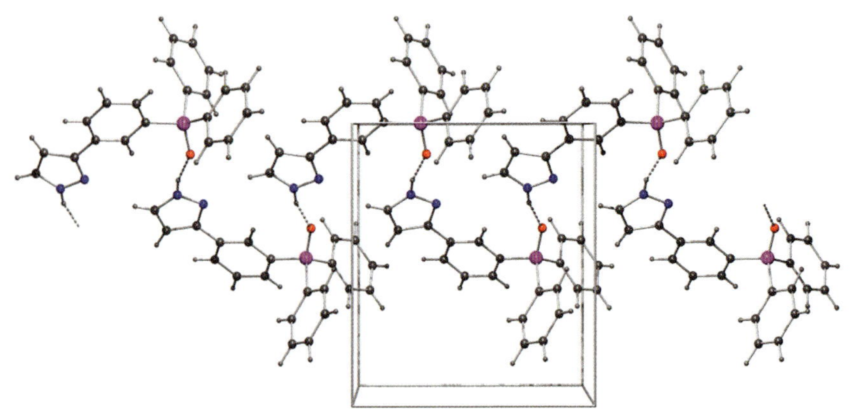

Figure 5.2 Solid state structure of phosphine oxide **19bγ** [20].

the formation of the corresponding 4-(2-amino)pyrimidinyl substituted derivatives **18aα-cγ** (Scheme 5.12) in moderate yields.

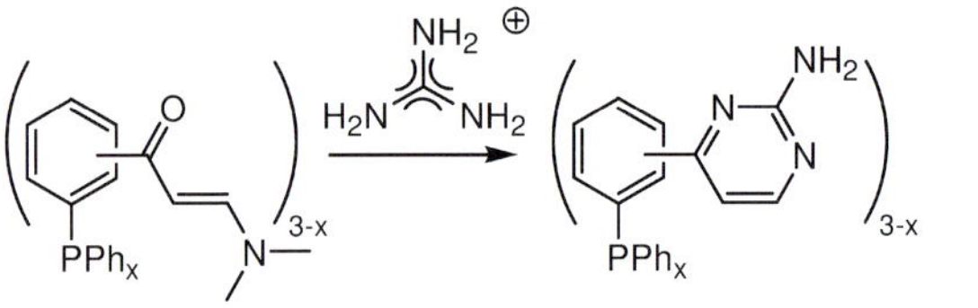

17aα-17cγ → **21aα-21cγ**

17,21	*o-*	*m-*	*p-*
x = 0	--	**bα**	**cα**
x = 1	**aβ**	**bβ**	**cβ**
x = 2	**aγ**	**bγ**	**cγ**

Scheme 5.12 Synthesis of pyrimidine functionalized arylphosphines.

The introduction of 3(5)-pyrazolyl and 4-(2-amino)pyrimidinyl rings only slightly influences the electronic situation of the phosphorus center. The ^{31}P NMR resonances of compounds **18** and **21** are observed in the range of −4.4 to −10.5 ppm, which is typical for triarylphosphines.

For the evaluation of the performance of these compounds in catalysis we synthesized palladium complexes of the mono-functionalized ortho- and meta-substituted systems **18aγ** and **18bγ** as well as the threefold meta-substituted ligand **18bα** [21]. Reaction of **18aγ** with $(PhCN)_2PdCl_2$ gave solely complex **22aγ** (Scheme 5.13), even when an excess of the ligand is applied.

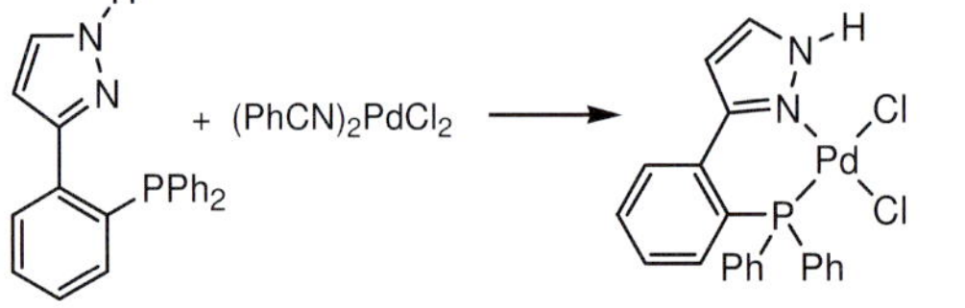

18aγ + $(PhCN)_2PdCl_2$ → **22aγ**

Scheme 5.13 Synthesis of a palladium complex bearing a P,N chelate ligand.

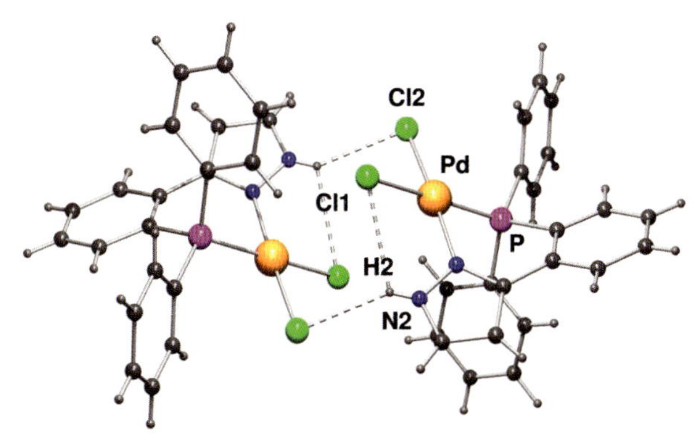

Figure 5.3 Solid state structure of the palladium complex **22** [20].

Here the ligand coordinates as a chelating *P,N* donor, which was confirmed by X-ray structure analysis (Figure 5.3). **22a**γ crystallizes by forming dimers via intermolecular hydrogen bonding between the N–H proton of one molecule and one of the chloro ligands of a neighboring complex. Additionally, an intramolecular hydrogen bond is found.

Ligands like **18b**γ and **18b**α with substituents in the meta position are not capable of undergoing chelation. Therefore, two of these donors coordinate in cis geometry to complete the square-planar coordination around the $PdCl_2$ fragment (Scheme 5.14).

2 **18bα,18bγ** + $(PhCN)_2PdCl_2$ → **22bα,22bγ**

18,22	R
bα	C_6H_4-$C_3H_3N_2$
bγ	C_6H_5

Scheme 5.14 Synthesis of a palladium complexes with meta-functionalized arylphosphines.

The palladium-catalyzed Heck coupling is a versatile tool for the formation of C–C bonds in organic synthesis [22]. Catalyst development has led to a variety of different ligand systems, among which phosphines giving palladacycles [23], bulky and electron-rich monodentate phosphines [24], and carbon donors like *N*-heterocyclic carbenes (NHCs) [25] should be mentioned here because of the high activities of the derived palladium catalysts. Bidentate *P,N*-ligands have not generally been applied for the palladium-catalyzed Heck coupling. The direct bonding of an *N*-heterocycle to the phosphorus center will decrease the electron-density at the phosphorus center, which is unfavorable to the oxidative addition step in the catalytic cycle. Most *P,N*-ligands which have been reported up to now to be applied in the Heck coupling have an *N*-heterocycle attached to the ortho position of a phenyl ring. This was realized for oxazolines [26], imino groups [27], imidazolines [28] or benzoxazines [29]. Some of these functional groups allow us to introduce chirality into the ligand system and thus have been synthesized for the asymmetric Heck coupling.

We chose the palladium-catalyzed Heck coupling of bromobenzene with styrene as the test reaction for the activity of palladium catalysts derived from **18aγ**, **18bγ** and **18bα**. Since Pd(0) is generally accepted to be the starting active component in the catalytic cycle, $Pd_2(dba)_3$ was first applied as a precatalyst in the presence of four equivalents of the corresponding phosphine ligand (Table 5.3). After 9 h at 130 °C, the best result (43% yield of Heck products) was obtained with **18bα**. The other ligands led to rapid formation of palladium black.

Table 5.3 Heck coupling of bromobenzene with styrene catalyzed by $Pd_2(dba)_3$ in the presence of triarylphosphines.[a)]

Catalyst	Conv.% of bromobenzene	Yield% of *trans*-stilbene	Yield% of 1,1-diphenylethylene
$Pd_2(dba)_3 + 4\ PPh_3$	71.6[b)]	54.6	8.9
$Pd_2(dba)_3 + 4$ **18bα**	50.3	37.4	5.7
$Pd_2(dba)_3 + 4$ **18bγ**	4.0[c)]	2.2	0.4
$Pd_2(dba)_3 + 4$ **18aγ**	traces[c)]	traces	traces

a) Reaction conditions: bromobenzene (5 mmol), styrene (7.5 mmol), NaOAc (7.5 mmol), $Pd_2(dba)_3$ (1.5 mol%), phosphine ligand (6 mol%) (all relative to bromobenzene), DMAc (12.5 mL, 130 °C), 9 h.
b) Reaction time: 6 h.
c) Rapid formation of palladium black was observed.

To prevent the rapid formation of palladium black, $PdCl_2(phosphine)_2$ type complexes were synthesized with the functionalized phosphines and directly used as catalysts in the Heck coupling (Table 5.4). For all palladium catalysts tested in this series we found improved conversions and selectivities compared to the $Pd_2(dba)_3$/phosphine systems. Even after 24 h at a reaction temperature of 130 °C no formation of palladium black was observed. In-situ generated catalysts using $Pd^{II}(OAc)_2$ gave comparable results. The chelating ligand **18aγ** led to an excellent regio isomer ratio.

The performance $Pd^{II}(OAc)_2$/**18aγ** (1 : 1 ratio) was additionally tested in the Heck coupling of styrene with chlorobenzene and 4-chloroacetophenone. It is well known

Table 5.4 Heck coupling of bromobenzene with styrene by $PdCl_2(L)_2$ and $PdCl_2(L)$ type complexes.[a)]

Catalyst	Conv.% of bromobenzene	Yield% of *trans*-stilbene	Yield% of 1,1-diphenylethylene
$PdCl_2(PPh_3)_2$	95.6	81.3	10.0
$PdCl_2$(**18bα**)$_2$	90.8[b)]	70.9	10.2
$PdCl_2$(**18bγ**)$_2$	97.2	78.7	11.7
$PdCl_2$(**18aγ**)	74.4	65.4	5.6

a) Reaction conditions: analogous to Table 5.3, except that the Pd loading was 3 mol% (based on bromobenzene) and reaction time was 24 h.
b) Pd loading was 2 mol% (based on bromobenzene).

that iodo salts can accelerate the Heck reaction when non-activated aryl halides are used as substrates [30]. Without or with only 20% of $n\mathrm{Bu_4NI}$, very low activity was found for 4-chloroacetophenone. When one equivalent of $n\mathrm{Bu_4NI}$ was added, an overall yield of 49% of Heck-type coupling products could be obtained. For the catalyst system $\mathrm{Pd^{II}(OAc)_2/2\ PPh_3}$, the addition of 1 equiv. of $n\mathrm{Bu_4NI}$ did not enhance the reaction at all but resulted in less than 1% yield of the Heck products. The Heck coupling of chlorobenzene with styrene could not be accelerated in the presence of one equivalent of $n\mathrm{Bu_4NI}$ using $\mathrm{Pd^{II}(OAc)_2}$/**18a**γ.

5.3 Nitrogen Donor Ligands Without Phosphorus Sites

Pyrazoles are ideal to prove how slight changes in the electronic properties of a donor ligand will influence the performance of a given catalytically active system. We were able to show this for two examples: molybdenum-catalyzed olefin epoxidation [31] and copper-catalyzed radical polymerization by atom transfer (ATRP) of styrene [32]. If one takes 2,2′-bipyridine and 2,2′ : 6′,2″-terpyridine as the fundamental structures for bi- and tridentate ligands, the formal exchange of one or two pyridine ligands will lead to 2-(3(5)-pyrazolyl)pyridine, 3(5),3′(5′)-dipyrazolyl and 2,6-di(3(5)pyrazolyl)pyridine (Scheme 5.15). These ligands do not differ essentially in their coordinating properties from the mother systems but allow a series of modifications in the ligand backbone to fine-tune the electronic and steric situation, which is not possible with the ligands purely based on pyridine fragments.

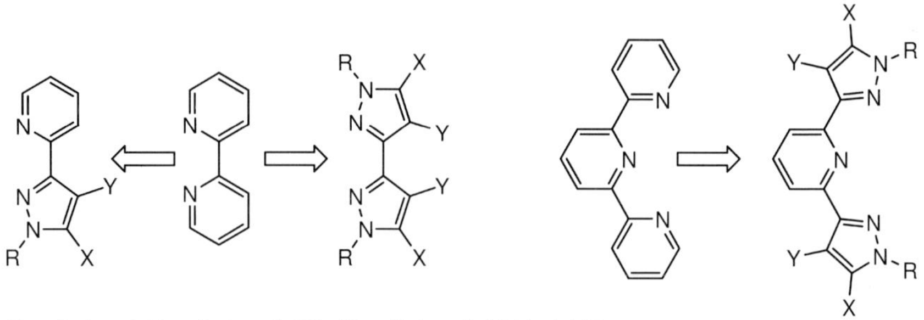

R = alkyl, aryl; X = alkyl, aryl, CF_3; Y = alkyl, aryl, Cl, Br, I, NO_2, ...

Scheme 5.15 Ligand development for the fine-tuning of coordination properties.

We have been working with 2-(3(5)-pyrazolyl)pyridines for a couple of years. In the present project we focused on 2,6-di(3(5)pyrazolyl)pyridines (**23**) and 2,6-di(pyrimidin-4-yl)pyridines (**24**) as lead structures for the implementation of motifs for secondary interactions. It has to be mentioned here that a series of tridentate *N*,*N′*,*N″* donors have been described in the literature up to now: 2,6-di(imino)pyridines [33], 2,6-di(oxazol-2-yl)pyridines (pybox) [34], 2,6-di(imidazolinyl)pyridines [35] and 2,6-di(pyrazol-1-yl)pyridines [36]. Partially, these systems are used for catalytic reactions.

2,6-Di(3(5)pyrazolyl)pyridines (**24**) as well as 2,6-di(pyrimidin-4-yl)pyridines (**25**) are synthesized starting from versatile pyridine-2,6-dicarboxylic acid (Scheme 5.16) [37].

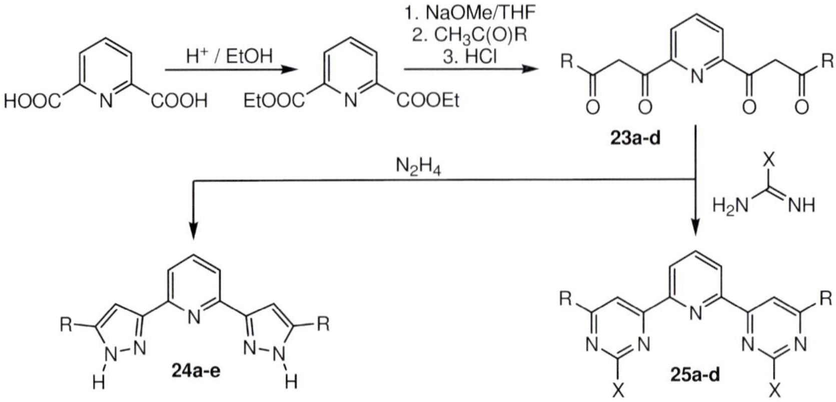

23,24	a	b	c	d	e
R =	Et	*n*-Pr	*n*-Bu	*t*-Bu	Ph

25: X = NH_2, OH, C_6H_5

Scheme 5.16 Synthesis of 2,6-di(3(5)pyrazolyl)pyridines (**24**) and 2,6-di(pyrimidin-4-yl)pyridines (**25**) starting from pyridine-2,6-dicarboxylic acid.

The intermediate tetraketone **23d** could be obtained in crystalline form by crystallization from EtOH. Figure 5.4 shows the molecular structure of this compound as obtained from X-ray diffraction data.

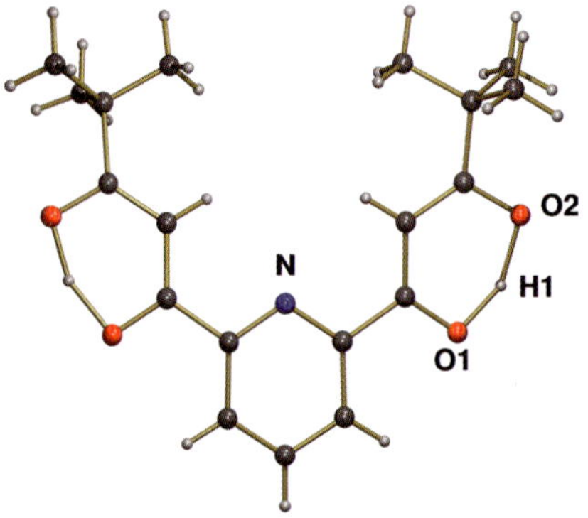

Figure 5.4 Molecular structure of the tetraketone **23d** [20].

The pyrazole-derived compounds could be further derivatized by either bromination in the 4-position of the pyrazole ring or by nucleophilic substitution reactions with deprotonated NH groups (Scheme 5.17).

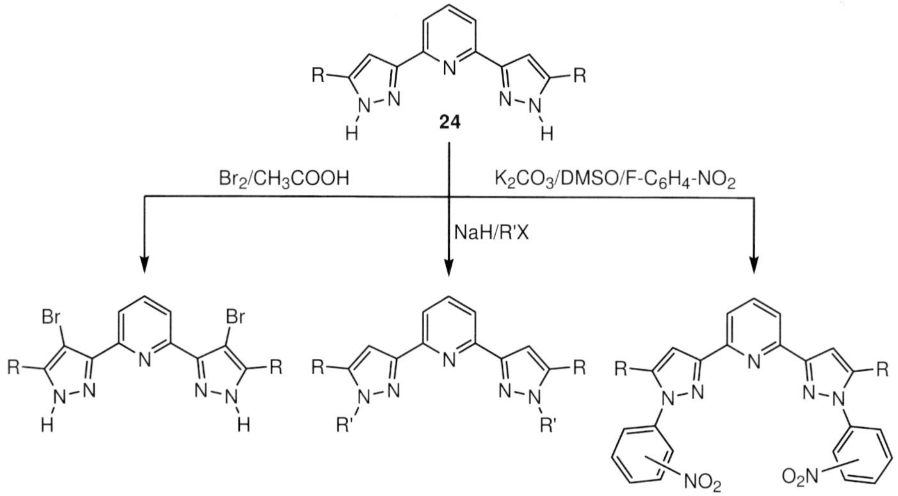

Scheme 5.17 Derivatization of 2,6-di(3(5)pyrazolyl)pyridines (**24**).

Such molecules have then been taken as starting materials for the implementation of functionalities for substrate recognition, as known from 'host/guest chemistry', and substrate activation.

Substrate recognition was initially studied by means of molecular modeling. As expected, it turned out that thiourea or urea moieties would be ideal for binding carbonyl functions as, for example, realized in *N*-formyl-1-amino-2-oxopent-4-ene. This led to the design of ligand system **26** derived from tridentate 2,6-di(3(5) pyrazolyl)pyridine as shown in Figure 5.5 [37b]. Catalytic experiments applying such ligands in hydrogenation reactions are planned. The corresponding carboxylic acid amides can be obtained by following a similar reaction sequence.

As a second structural motif, polyether-bridged tridentate ligands were synthesized. These systems should undergo polyether binding to hard Lewis acids as was the case for the phosphine ligands described above [8]. The difference here is that, owing to the molecular structure of the ligand, the Lewis acid will be transferred right into the coordination sphere of a catalytically active transition metal bound to the *N*, N',N' chelating moiety. Here it can interact with additional ligands such as halogenido anions and thus open up a free coordination site at the active center (Scheme 5.18).

To achieve such ligands, nucleophilic aromatic substitution was applied [37, 38], which gave the corresponding twofold *meta*-nitrophenyl-substituted 2,6-di(3(5)pyrazolyl)pyridine in almost quantitative yields. Reduction of the nitro group with Pd-C/ NH_4HCO_2 and treatment of the derived amine **27** with 3,6-dioxaoctanoic diaciddichloride, 3,6,9-trioxaundecanoic diaciddichloride and 3,6,9,12-tetraoxatetradecan

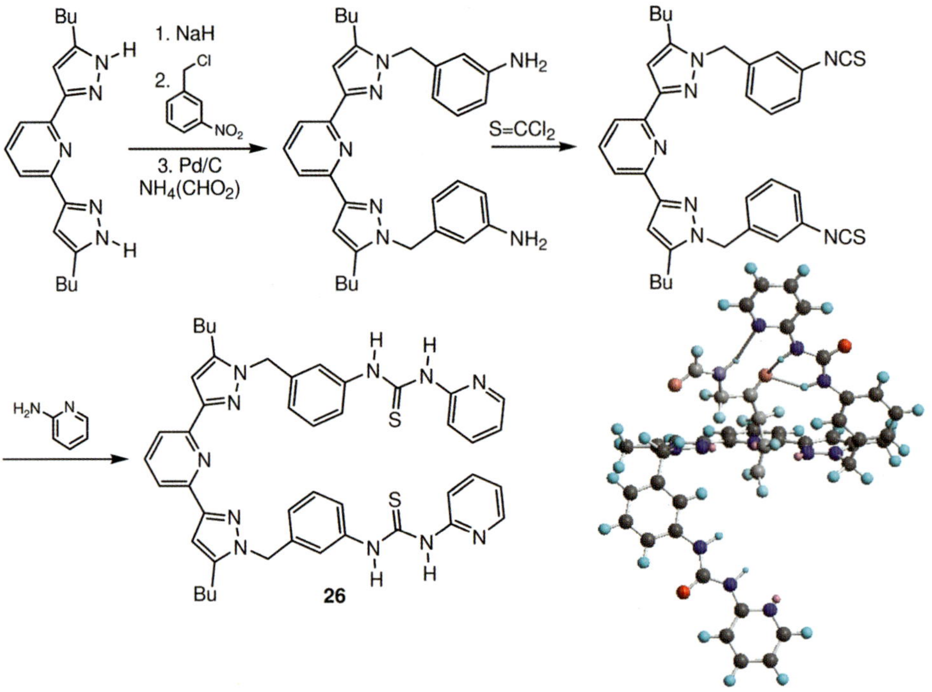

Figure 5.5 Synthesis of the twofold thiourea functionalized 2,6-di(3(5)pyrazolyl)pyridine **26** and the recognition of *N*-formyl-1-amino-2-oxopent-4-ene with the C_2 symmetrically folded ligands.

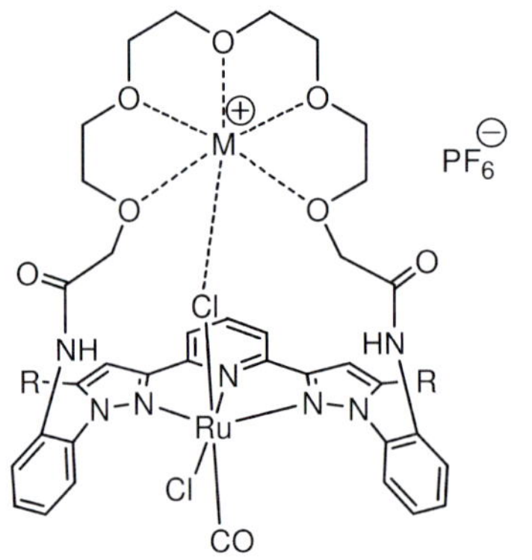

Scheme 5.18 Activation of a ruthenium complex by Lewis-acid interaction.

diaciddichloride, respectively, gave the polyether linked systems **28a-c** in moderate to good yields (Scheme 5.19) [37b].

As an example, Figure 5.6 shows the ^{1}H NMR spectrum of compound **28a**. Three singlets with a typical shift in the range of aliphatic CH groups are assigned to Pz-CH_3

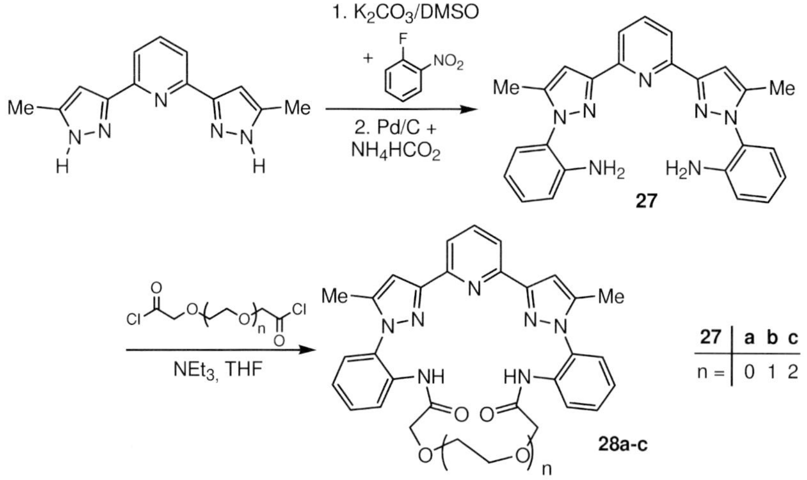

Scheme 5.19 Synthesis of polyether linked tridentate ligands.

(2.29 ppm), $O(CH_2)_2$ (3.22 ppm) and $COCH_2O$ (3.85 ppm). The CH proton of the pyrazoles is observed at 6.93 ppm and the amide proton at 9.32 ppm.

A series of model complexes were synthesized to evaluate the general performance of the obtained tridentate *N,N′,N′* ligands for coordination chemistry and catalysis. Reacting substituted 2,6-di(pyrazol-3-yl)- **24** and 2,6-di(pyrimidin-4-yl)pyridines **25** in dry methanol with $FeCl_2$ and $CoCl_2$, respectively, resulted in the formation of the corresponding pentacoordinate complexes $(N,N',N')MCl_2$**29M** and **30M**

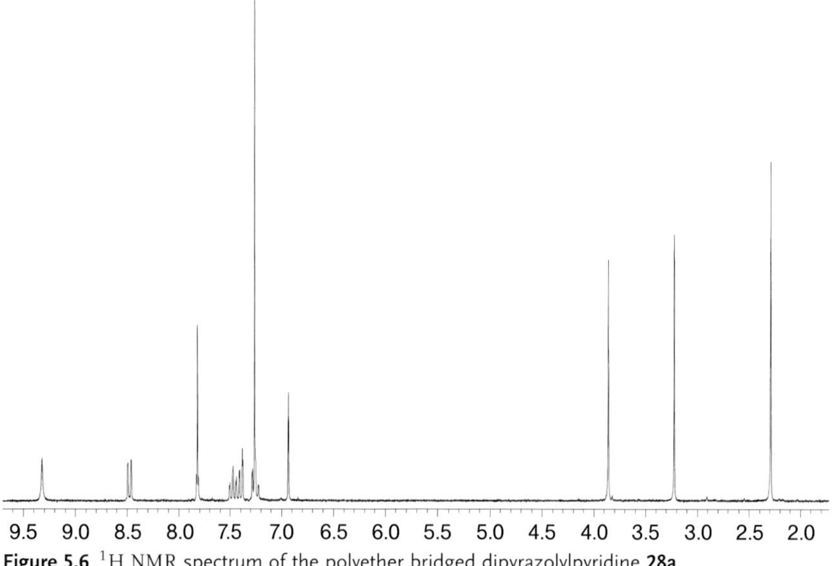

Figure 5.6 ^{1}H NMR spectrum of the polyether bridged dipyrazolylpyridine **28a**.

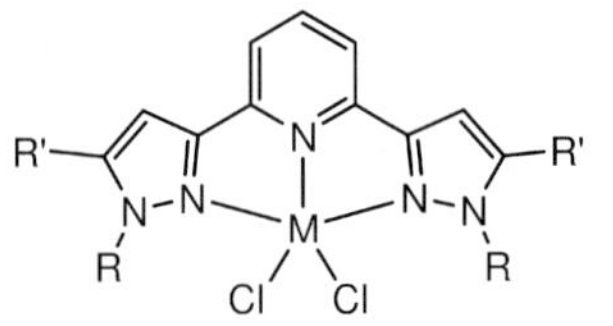

29	R	R'
a	*p*-$C_6H_4NO_2$	^nBu
b	*o*-$C_6H_4NO_2$	^nBu
c	$CH_2C_6H_5$	^nBu
d	^nBu	^nBu

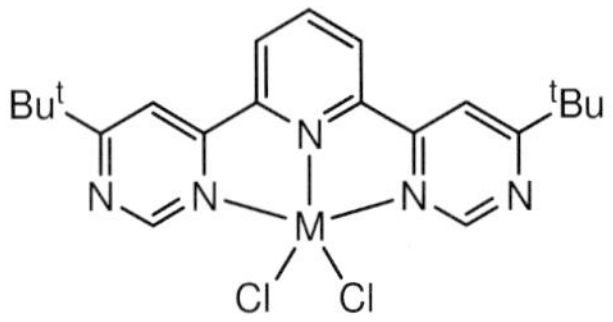

29Ma-d

30M

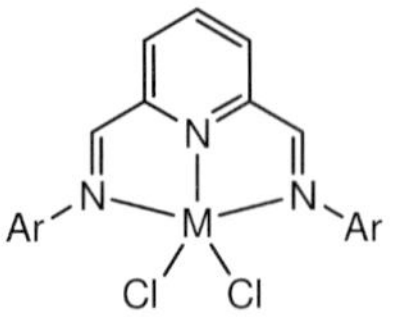

Scheme 5.20 Complexes of the type $(N,N',N')MCl_2$ (M = Fe, Co) and isostructural 2,6-diiminopyridine compounds (bottom right).

(M = Fe, Co) in almost quantitative yields (Scheme 5.20). These compounds are isostructural to 2,6-diiminopyridine complexes, first described by Brookhart and Gibson, which are highly active catalysts for ethylene polymerization [39].

These complexes were investigated for activity in ethylene polymerization with MAO as an activator. Only the iron and cobalt complexes **28Ma** showed activity. It has to be mentioned here that oligomerization of ethylene was not investigated in these experiments. Table 5.5 summarizes the results of these polymerization experiments, which were carried out in cooperation with Dr. L. Jones from BASELL.

Table 5.5 Ethylene polymerization with **29Fea** and **29Coa**.

Catalyst	Amount of catalyst [mg]	Al/m	*P*(ethylene) [atm]	Yield of PE [g]	Activity [kg_{pe} $mmol^{-1}$ h^{-1}]
29Fea	13.7	500	1	0.1	0.02
	10	1000	10	12.7	0.44
29Coa	19	500	1	0	0
	10	1000	10	16.6	0.58

The molecular mass distributions of the polymers obtained at an ethylene pressure of 10 atm were determined by means of gel permeation chromatography. The results are listed in Table 5.6.

Table 5.6 Material properties of the polyethylenes obtained at 10 atm.

catalyst	mp [°C]	M_w [g/mol]	M_n [g/mol]	*Q*
29Fea	134	524177	155089	3.38
29Coa	135	485026	194428	2.49

If iron(II) or cobalt(II) perchlorates are used for coordination to substituted 2,6-di(pyrazol-3-yl)-**24** and 2,6-di(pyrimidin-4-yl)pyridines **25** instead of the chlorides, homoleptic complexes of the type $[(N,N',N')_2M](ClO_4)_2$ (**31M**, M = Fe, Co) are obtained. The complex [2,6-di(6-*tert*-butylpyrimidin-4-yl)pyridine)iron]diperchlorate could be obtained by slow diffusion of diethylether into a concentrated solution of the complex in acetonitrile as dark violet crystals, suitable for X-ray diffraction (Figure 5.7).

Reacting one equivalent of $RuCl_3{\cdot}H_2O_x$ with the tridentate N,N',N'' ligands gives the six-coordinate Ru(III) complexes **32a,b** and **33** in almost quantitative yields, and these, when treated with disodium 2,6-pyridinedicarboxylate (pybox) in aqueous solution are transformed to the corresponding Ru(pybox) complexes **34a,b** and **35** according to Ref. [40] (Scheme 5.21). From **34b**, crystals suitable for single crystal X-ray diffraction could be obtained. Figure 5.8 shows the structure of this compound in the solid state. The bond lengths and angles are in agreement with those of (pybox)Ru complexes bearing 2,6-di(oxazol-2-yl)pyridine and 2,6-di(oxazinyl)pyridine ligands [41].

The ruthenium pybox complexes **34a,b** and **35** described above were tested for catalytic activity in olefin epoxidation with hydrogen peroxide as the oxidizing agent, in analogy to procedures published by Beller *et al.* [41, 42]. With *tert*-amylalcohol as the solvent, aryl-substituted olefins are transferred to the corresponding epoxides in quantitative yields; alkyl-substituted olefins are not epoxidized at all. Our experiments can rule out one of the possible routes for the oxygen transfer: it was proposed that one of the outer pyridyl rings of terpy is oxidized intermedially to an *N*-oxide when terpy is used as the N,N',N' ligand. This route is impossible with pyrazole

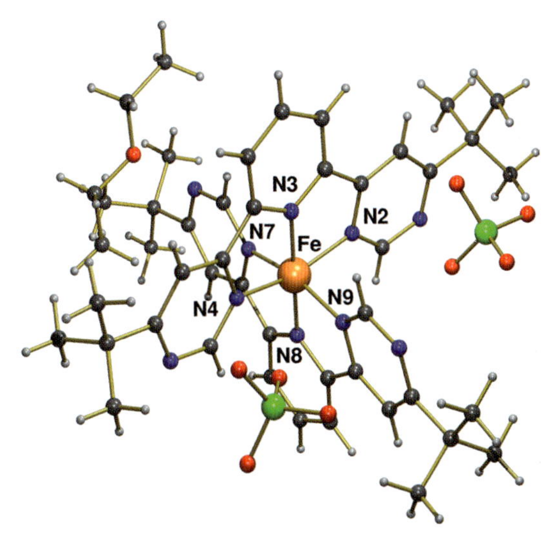

Figure 5.7 Molecular structure of the (2,6-di(6-*tert*-butylpyrimidin-4-yl)pyridine)iron(II) diperchlorate complex, the unit cell contains one additional molecule of diethylether. Characteristic bond lengths [Å] and angles [°]: Fe–N2 1.975(2), Fe–N3 1.8823(19), Fe–N4 1.979(2), Fe–N7 1.972(2), Fe–N8 1.8729(19), Fe–N9 1.959(2), N2–Fe1–N3 81.29(8), N2–Fe1–N4 161.51(8), N3–Fe1–N4 80.32(8), N7–Fe1–N8 81.02(8), N7–Fe1–N9 161.91(8), N8–Fe1–N9 80.89(8).

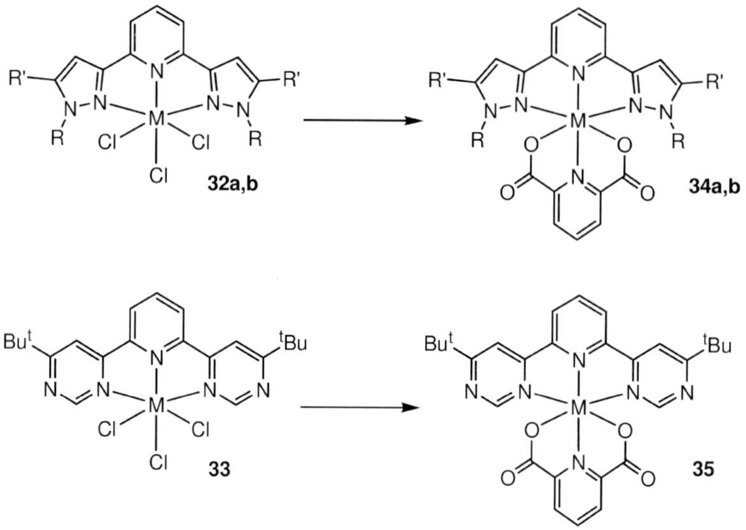

32,34	R	R'
a	CH_3	^{n}Bu
b	CH_3	^{t}Bu

Scheme 5.21 Synthesis of the pybox complexes **34a,b** and **35**.

instead of pyridine, since pyrazole *N*-oxide will not form under the mild conditions of this reaction. Additionally the ruthenium complexes **34a,b** and **35** are highly active catalysts for the oxidation of hydroquinones to quinones. The reactions were carried out according to a protocol published by Iwasa *et al.* [43].

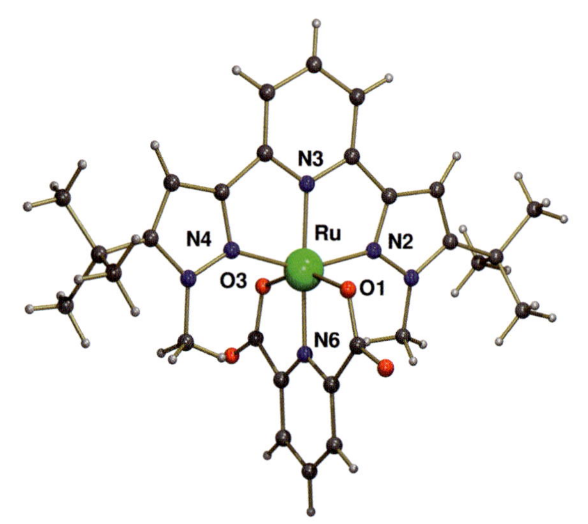

Figure 5.8 Molecular structure of the (pybox)Ru complex **34a**. Characteristic bond lengths [Å] and angles [°]: Ru1–N2 2.0550 (14), Ru1–N3 1.9931(14), Ru1–N4 2.0788(14), Ru1–N6 1.9614 (14), Ru1–O1 2.0971(14), Ru1–O3 2.1169(12), N2–Ru1–N3 77.67 (6), N2–Ru1–N4 155.03(6), N3–Ru1–N4 77.37(6), O1–Ru1–N6 78.93(5), O3–Ru1–N6 78.33(5), O1–Ru1–O3 157.23(5).

5.4 Conclusion

Motivated by the idea of implementing certain features of enzymes as a strategy for ligand design, we started to establish a 'construction kit' style of assembling different binding motifs and linkers for the generation of ligands which should be capable of undergoing secondary interactions for the binding of small molecules, ions and substrates. Additionally, we have been able to obtain and investigate a series of novel pyrazole-based chelating ligands, with either phosphorus or nitrogen donors, as ligands in different catalytic reactions.

Acknowledgments

We thank the *Deutsche Forschungsgemeinschaft* (SPP 1118, projects Th 550/6-1, Th 550/7-1, Th 550/7-2 and Th 550/10-1).

References

1 Cornils, B. and Herrmann, W.A. (eds) (1998) *Aqueous-Phase Organometallic Catalysis*, Wiley-VCH, Weinheim.

2 Horváth, I.T. (1998) *Accounts of Chemical Research*, **31**, 641.

3 Kainz, S., Koch, D. and Leitner, W. (1998) *Selective Reactions of Metal-Activated Molecules* (eds H. Werner and W. Schreier), Vieweg, Wiesbaden.

4 Lippard, S.J. and Berg, J.M. (1994) *Principles of Bioinorganic Chemistry*, University Science Books, Mill Valley.

5 Kaim, W. and Schwerderski, B. (1995) *Bioanorganische Chemie*, Teubner, Stuttgart.

6 Schiemenz, G.P., Kaack, H. and Hermann, J. (1973) *Annalen Der Chemie-Justus Liebig*, **9**, 1480.

7 Sun, Y., Herdtweck, E., Thiel, W.R., unpublished results.

8 Sun, Y., Ahmed, M., Jackstell, R., Beller, M. and Thiel, W.R. (2004) *Organometallics*, **23**, 5260.

9 Lam, F. and Chan, K.S. (1995) *Tetrahedron Letters*, **36**, 919.

10 Sun, Y., Herdtweck, E., Thiel, W.R., unpublished results.

11 There are just a few cis configured complexes of the Type (P-P)Rh(CO)Cl, which can be obtained solely with chelate ligands possessing a short linkage between the donor sites. These complexes show CO stretching vibrations at about 2010 cm^{-1}: (a) Sanger, A.R. (1977) *Journal of The Chemical Society-Dalton Transactions*, 120. (b) Dyer, G., Wharf, R.M. and Hill, W.E. (1987) *Inorganica Chimica Acta*, **133**, 137. (c) Thurner, C.L., Barz, M., Spiegler, M. and Thiel, W.R. (1997) *Journal of Organometallic Chemistry*, **541**, 39.

12 MM calculations were carried out with the MM2 module implemented in the program package CHEM3D (version 4.0). Since the parameters for the description of the binding situation at the rhodium center were not known, the structure of the fragment *trans*-$P_2RhCl(CO)$ was generated from X-ray data. The structure of this fragment was then preserved during the energy optimization.

13 Review: Eilbracht, P., Bärfacker, L., Buss, C., Hollmann, C., Kitsos-Rzychon, B.E., Kranemann, C.L., Rische, T., Roggenbuck,

R. and Schmid, A. (1999) *Chemical Reviews*, **99**, 3329.

14 (a) Takacs, J.M., Schröder, S.D., Han, J., Gifford, M., Jiang, X., Saleh, T., Vayalakkada, S. and Yap, A.H. (2003) *Organic Letters*, **5**, 3595. (b) Gsponer, A. and Consiglio, G. (2003) *Helvetica Chimica Acta*, **86**, 2170. (c) Consiglio, G. and Gsponer, A. (2002) *Polymeric Materials Science and Engineering*, **87**, 78.

15 Cornils, B. and Herrmann, W.A. (eds) (1998) *Aqueous-Phase Organometallic Catalysis*, Wiley-VCH, Weinheim, Germany.

16 (a) Rauchfuss, T.B., Wilson, S.R. and Wrobleski, D.A. (1981) *Journal of the American Chemical Society*, **103**, 6769. (b) Wrobleski, D.A., Wilson, S.R. and Rauchfuss, T.B. (1982) *Inorganic Chemistry*, **21**, 2114. (c) Wrobleski, D.A. and Rauchfuss, T.B. (1982) *Journal of the American Chemical Society*, **104**, 2314–2316. (d) Wrobleski, D.A., Rauchfuss, T.B., Rheingold, A.L. and Lewis, K.A. (1984) *Inorganic Chemistry*, **23**, 3124. (e) Wrobleski, D.A., Day, C.S., Goodman, B.A. and Rauchfuss, T.B. (1984) *Journal of the American Chemical Society*, **106**, 5464.

17 (a) Pleier, A.-K., Glas, H., Grosche, M., Sirsch, P. and Thiel, W.R. (2001) *Synthesis*, 55. (b) Pleier, A.-K., Herdtweck, E., Mason, S.A. and Thiel, W.R. (2003) *European Journal of Organic Chemistry*, 499.

18 (a) Schiemenz, G.P. (1966) *Chemische Berichte*, **99**, 504. (b) Schiemenz, G.P. (1966) *Chemische Berichte*, **99**, 514. (c) Schiemenz, G.P. and Kaack, H.J. (1973) *Annalen Der Chemie-Justus Liebig*, **9**, 1480. (d) Schiemenz, G.P. and Kaack, H.J. (1973) *Annalen Der Chemie-Justus Liebig*, **9**, 1494.

19 (a) Sun, Y., Hienzsch, A., Grasser, J., Herdtweck, E. and Thiel, W.R. (2006) *Journal of Organometallic Chemistry*, **691**, 291. (b) Reis, A., Sun, Y., Wolmershäuser, G. and Thiel, W.R., (2007) *European Journal of Organic Chemistry*, 777.

20 Spek, A.L. (2000) PLATON, A Multipurpose Crystallographic Tool, Utrecht University, Utrecht, The Netherlands.

21 Sun, Y. and Thiel, W.R., (2006) *Inorganica Chimica Acta*, **359**, 4807.

22 A recent excellent review on the Heck-coupling: Beletskaya, I.P. and Cheprakov, A.V. (2000) *Chemical Reviews*, **100**, 3009.

23 (a) Herrmann, W.A., Brossmer, C., Reisinger, C.-P., Riermeier, T.H., Öfele, K. and Beller, M. (1997) *Chemistry - A European Journal*, **3**, 1357. (b) Herrmann, W.A., Böhm, V.P.W. and Reisinger, C.-P. (1999) *Journal of Organometallic Chemistry*, **576**, 23.

24 Zapf, A. and Beller, M. (2005) *Chemical Communications*, 431. (and references of Fu, Buchwald, Hartwig and Beller's work cited therein).

25 (a) Herrmann, W.A., Elison, M., Fischer, J., Köcher, C. and Artus, G.R.J. (1995) *Angewandte Chemie – International Edition*, **34**, 2371. (b) Herrmann, W.A. (2002) *Angewandte Chemie – International Edition*, **41**, 1291.

26 (a) Loiseleur, O., Hayashi, M., Keenan, M., Schmees, N. and Pfaltz, A. (1999) *Journal of Organometallic Chemistry*, **576**, 16. (b) Loiseleur, O., Hayashi, M., Schmees, N. and Pfaltz, A. (1997) *Synthesis*, 1338. (c) Kiely, D. and Guiry, P.J. (2003) *Journal of Organometallic Chemistry*, **687**, 545. (d) Hou, X., Dong, D. and Yuan, K. (2004) *Tetrahedron-Asymmetry*, **15**, 2189.

27 (a) Catsoulacos, D.P., Steele, B.R., Heropoulos, G.A., Micha-Screttas, M. and Screttas, C.G. (2003) *Tetrahedron Letters*, **44**, 4575. (b) Schultz, T., Schmees, N. and Pfaltz, A. (2004) *Applied Organometallic Chemistry*, **18**, 595.

28 Busacca, C., Grossbach, D., So, R.C., O'Brien, E.M. and Spinelli, E.M. (2003) *Organic Letters*, **5**, 595.

29 Kündig, E.P. and Meier, P. (1999) *Helvetica Chimica Acta*, **82**, 1360.

30 Examples of $n\mathrm{Bu_4NI}$ as an additive in C-C cross-coupling reactions: Piber, M., Jensen, A.E., Rottländer, M. and Knochel, P. (1999) *Organic Letters*, **1**, 1323. (b) Jensen, A.E. and Knochel, P. (2002) *The Journal of Organic Chemistry*, **67**, 79.

31 (a) Thiel, W.R. and Eppinger, J. (1997) *Chemistry - A European Journal*, **3**, 696. (b) Hroch, A., Gemmecker, G. and Thiel, W.R. (2000) *European Journal of Inorganic Chemistry*, **3**, 1107. (c) Thiel, W.R., Barz, M., Glas, H. and Pleier, A.-K. (2000) '*Olefin Epoxidation Catalyzed by Molybdenum Peroxo Complexes: A Mechanistic Study*' in '*Peroxide Chemistry: Mechanistic and Preparative Aspects of Oxygen Transfer*' (ed W. Adam), Wiley-VCH, p. S.433

32 Schubert, A. Pleier, A.-K. Sun, Y. Thiel, W. R.publication in preparation.

33 (a) Britovsek, G.J.P., Gibson, V.C., Kimberley, B.S., Maddox, P.J., McTavish, S.J., Solan, G.A., White, A.J.P. and Williams, D.J. (1998) *Chemical Communications*, 849. (b) Britovsek, G.J.P., Bruce, M., Gibson, V.C., Kimberley, B.S., Maddox, P.J., Mastroianni, S., McTavish, S.J., Redshaw, C., Solan, G.A., Strömberg, S., White, A.J.P. and Williams, D.J. (1999) *Journal of the American Chemical Society*, **121**, 8728. (c) Bianchini, C., Mantovani, G., Meli, A., Migliacci, F., Zanobini, F., Laschi, F. and Sommazzi, A. (2003) *European Journal of Inorganic Chemistry*, 1620. (d) Kleigrewe, N., Steffen, W., Blömker, T., Kehr, G., Fröhlich, R., Wibbeling, B., Erker, G., Wasilke, J.-C., Wu, G. and Bazan, G.C. (2005) *Journal of the American Chemical Society*, **127**, 13955. (e) Pelascini, F., Wesolek, M., Peruch, F. and Lutz, P.J. (2006) *European Journal of Inorganic Chemistry*, 4309–4316.

34 (a) Tse, M.K., Bhor, S., Klawonn, M., Anilkumar, G., Jiao, H., Döbler, C., Spannenberg, A., Mägerlein, W., Hugl, H. and Beller, M. (2006) *Chemistry - A European Journal*, **12**, 1855. (b) Tse, M.K., Bhor, S., Klawonn, M., Anilkurnar, G., Jiao, H., Döbler, C., Spannenberg, A., Mägerlein, W., Hugl, H. and Beller, M. (2006) *Chemistry - A European Journal*, **12**, 1875.

35 Bhor, S., Anilkumar, G., Tse, M.K., Klawonn, M., Döbler, C., Bitterlich, B., Grotevendt, A. and Beller, M. (2005) *Organic Letters*, **7**, 3393.

36 (a) Jameson, D.L. and Goldsby, K.A. (1990) *The Journal of Organic Chemistry*, **55**, 4992. (b) Sun, X., Yu, Z., Wu, S. and Xiao, W.-J. (2005) *Organometallics*, **24**, 2959.

37 (a) Schubert, A. (2003) diploma thesis, Technische Universität Chemnitz. (b) Zabel, D. (2007) PhD thesis, Technische Universität Kaiserslautern.

38 Rößler, K., Kluge, T., Schubert, A., Sun, Y., Herdtweck, E. and Thiel, W.R. (2004) *Zeitschrift fur Naturforschung*, **59b** 1253.

39 (a) Small, B.L., Brookhart, M. and Bennet, A.M.A. (1998) *Journal of the American Chemical Society*, **120**, 4049. (b) Small, B.L. and Broohart, M. (1998) *Journal of the American Chemical Society*, **120**, 7143. (c) Small, B.L. and Broohart, M. (1999) *Macromolecules*, **32**, 2120. (d) Dias, E.L., Brookhart, M. and White, P.S. (2000) *Organometallics*, **19**, 4995. (e) Glentsmith, G.K.B., Gibson, V.C., Hitcock, P.B., Kimberley, B.S. and Rees, C.W. (2002) *Chemical Communications*, 1498. (f) Gibson, V.C., Tellmann, K.P., Humphries, M.J. and Wass, D.F. (2002) *Chemical Communications*, 2316.

40 Couchman, S.M., Dominguez-Vera, J.M., Jeffery, J.C., McKee, C.A., Nevitt, S., Pohlmann, M., White, C.M. and Ward, M.D. (1998) *Polyhedron*, **17**, 3541.

41 Tse, M.K., Bhor, S., Klawonn, M., Anilkumar, G., Jiao, H., Döbler, C., Spannenberg, A., Mägerlein, W., Hugl, H. and Beller, M. (2006) *Chemistry - A European Journal*, **12**, 1855.

42 (a) Tse, M.K., Bhor, S., Klawonn, M., Anilkumar, G., Jiao, H., Döbler, C., Spannenberg, A., Mägerlein, W., Hugl, H. and Beller, M. (2006) *Chemistry - A European Journal*, **12**, 1875. (b) Bhor, S., Anilkumar, G., Tsc, M.K., Klawonn, M., Döbler, C., Bitterlich, B., Grotevendt, A. and Beller, M. (2005) *Organic Letters*, **7**, 3393. (c) Tse, M.K., Klawonn, M., Bhor, S., Döbler, C., Anilkumar, G., Mägerlein, W., Hugl, H. and Beller, M. (2005) *Organic Letters*, **7**, 987. (d) Tse, M.K., Döbler, C., Bhor, S., Klawonn, M., Mägerlein, W., Hugl, H. and Beller, M. (2004) *Angewandte*

Chemie – International Edition, **43**, 5255. (e) Tse, M.K., Bhor, S., Klawonn, M., Döbler, C. and Beller, M. (2003) *Tetrahedron Letters*, **44**, 7479–7483.

43 Iwasa, S., Fakhruddin, A., Widago, H.S. and Hishiyama, H. (2005) *Advanced Synthesis and Catalysis*, **347**, 517.

6
Square-Pyramidal Coordinated Phosphine Iron Fragments: A Tale of the Unexpected

Andreas Grohmann and Stephan Kohl

6.1
Introduction

The *coordination octahedron* is one of the fundamental concepts first introduced by Alfred Werner when he laid the foundations of modern coordination chemistry some 115 years ago. It was only with Werner's concept of metal centers being surrounded by ligands in a well-defined geometry that previously baffling observations, such as the seemingly unsystematic exchange of bonding partners or different manifestations of isomerism, finally yielded to a firmly grounded understanding, and predictability. Werner-type coordination compounds contain what has since come to be called 'classical ligands', having nitrogen, oxygen, and halogen donors. Depending on the type of donor, the ligands are monodentate or can be chelating, with ammonia (ammine) or ethylenediamine (en) as examples of well-established bonding partners even as early as in Werner's days [1].

It is against this background that we designed a very special type of chelate ligand which, having C_{2v} symmetry, would occupy five coordination positions in a coordination octahedron and thereby provide the metal ion with a square-pyramidal 'coordination cap': All possible reactivity is focused on the remaining sixth coordination site. We began by studying complexes of the NN_4 ligand **1**, which is the closest possible chelating analog of the pentaammine donor set $(NH_3)_5$. Of the range of possible transition metal ions, Fe (in both its +II and +III oxidation state) has so far yielded the most extensive new chemistry.

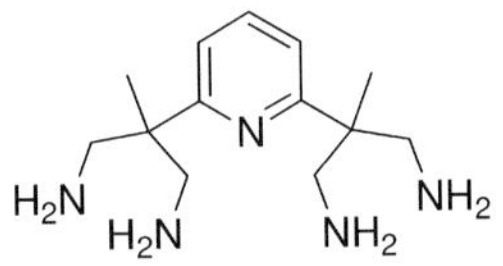

H_2N H_2N N NH_2 NH_2

1

An inventory of the previous literature had shown that whenever iron, ferrous or ferric, is present in a nitrogen-dominated coordination environment, the ligands

Activating Unreactive Substrates: The Role of Secondary Interactions.
Edited by Carsten Bolm and F. Ekkehardt Hahn

ISBN: 978-3-527-31823-0

usually contain imine donors or they are macrocyclic (e.g. tris(phenanthroline)iron or haemoglobin, or sarcophagine [2], cyclam [3], and triazacyclononane complexes [4]). These ligands either have π^* orbitals available for back-donation of electron density from the metal, or the stability of complexes is ensured through the operation of the cryptate or macrocyclic effects.

It is therefore reasonable to expect iron complexes with a large number of *monodentate saturated* nitrogen donor ligands to be rare, as is indeed the case: the only example for which there is reliable evidence is the hexaammine complex $[Fe(NH_3)_6]^{2+}$ [5]. Pentaammine complexes of iron have been studied theoretically [6], but not reproducibly prepared [7]. In stark contrast to iron, such complexes are numerous for its heavier homologs, i.e., ruthenium and osmium. Here, the $(NH_3)_5M$ modules (M = Ru, Os) often contain extremely interesting ligands in the sixth coordination position: $[(NH_3)_5Ru(N_2)]^{2+}$ is the prototypical dinitrogen complex [8], $[(NH_3)_5Ru(pyrazine)Ru(NH_3)_5]^{5+}$, the Creutz–Taube ion, has been studied extensively for its mixed valence and electron exchange characteristics [9], and the high-yield preparation of organometallic $[(NH_3)_5Os(\eta^2\text{-}C_2H_4)]^{2+}$ attests to the fact that the pentaammine-coordinated metal ion binds π-bonding ligands such as olefins, acetylenes and arenes with particular ease [10].

Ammonia is a pure σ-donor ligand; hence the hypothetical $[(NH_3)_5Fe]^{n+}$ fragment is expected to be electron-rich, which should give rise to unusual chemistry at the remaining coordination site. While $[(NH_3)_5Fe]^{n+}$ is inaccessible under normal conditions, we reasoned that $[(\mathbf{1})Fe]^{n+}$, with similar electronic properties, should be amenable to experimental study. This is indeed the case, with very interesting results: ligand **1** readily forms complexes with both iron(II) and iron(III), allowing the stabilization of previously unknown bridging motifs, such as Fe^{II}-(OH)-Fe^{II} [11]. The ferrous κ-*N*-nitro complex shows unprecedented nitrite reductase-like reactivity [12], and the related aqua complex can be oxidized cleanly with aerobic oxygen to give the diferric μ-oxo complex without oxidative degradation of the ligand (such as unwanted imine formation) [13]. When combined with a suitable monodentate *N*-donor ligand in the sixth coordination position, iron(II) complexes of **1** undergo reversible spin transitions (between the low- and high-spin states, $S = 0$ and 2, respectively) in response to a change in temperature [11]. Our most recent finding concerns the iron(III) azido complex $[(\mathbf{1})Fe(N_3)]^{2+}$, which, upon collision-induced loss of dinitrogen in the mass spectrometer, is transformed into the iron(V) nitrido intermediate $[(\mathbf{1})FeN]^{2+}$. This highly reactive species then undergoes further transformations resulting in C—H and N—H bond activation within the 'coordination cap' [14].

These results provide the context in which we extended our studies of square-pyramidal coordinating ligands, to include polyphosphines having an NP_4 donor set. This part of our work was supported by the DFG initiative set up to study the role of secondary interactions in directing transformations within the coordination environment of a metal center (Schwerpunktprogramm 1118). More details of the chemistry outlined in this account may be found in the pertinent publications [15–18].

6.2 Polyphosphine Ligands with Three and Four Coordinating Arms

Our first objective was to produce a square-pyramidal coordinating NP_4 ligand with a sterically demanding periphery, which would shield the sixth coordination position in an octahedron to the extent that access would be possible for only the smallest ligands, such as hydride. To this effect, we synthesized the phenyl-substituted tetraphosphine **2**.

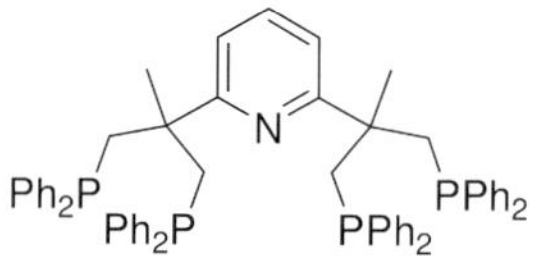

N
Ph_2P Ph_2P PPh_2 PPh_2

2

We tested complex formation of **2** with a variety of transition metal ions, including iron [15]. Whereas well-defined mononuclear products can be isolated with, *inter alia*, Ni^{2+}, Co^{2+} and Ru^{2+}, the reactions with iron salts do not proceed cleanly, and intractable product mixtures are obtained. Complexes such as $[(\mathbf{2})Co](BF_4)_2$ and $[(\mathbf{2})RuCl]BF_4$ proved inert with respect to reaction with small molecules such as dioxygen or dihydrogen, or ligand exchange. We therefore deferred the investigation of this ligand and turned to the methyl-substituted analog **3**, expecting its considerably reduced steric bulk to translate into enhanced complex reactivity. A closely related ligand is **4**, in which one dimethylphosphinyl substitutent has been removed to give a C_s-symmetrical tripodal tetradentate phosphine.

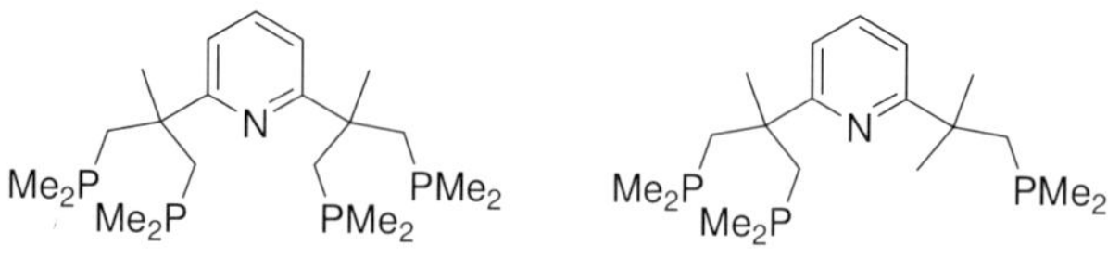

N
Me_2P Me_2P PMe_2 PMe_2

3

N
Me_2P Me_2P PMe_2

4

The polyphosphines **3** and **4** (systematic names: 2,6-bis(2-methyl-1,3-bis(dimethylphosphino)propan-2-yl)pyridine [*py*(PMe_2)$_4$] and 2-(2-methyl-1,3-bis(dimethylphosphino)-propan-2-yl)-6-(2-methyl-1-(dimethylphosphino)propan-2-yl)pyridine [*py*(PMe_2)$_3$], respectively) may be prepared in good yields (>80%) from the reaction of the corresponding polybromide and the required equivalents of $LiPMe_2$ [19] in ether. The synthetic protocol for **4** is summarized in Scheme 6.1. The preparation of **3** is largely analogous, starting from 2,6-diethylpyridine and proceeding via the tosylated tetraalcohol. The intermediate preparation of the respective polybromides is essential for successful syntheses, as the mesyl or tosyl esters do not react cleanly with lithium phosphide.

Before we proceed to summarize the unexpected bond activation chemistry we have so far recorded for ligands **3** and **4**, let us briefly delineate the 'history' of C−H and C−P bond activation observed with tertiary phosphines. This class of ligands plays a prominent role in well-defined molecular catalysts for application in homogeneous solution. Cleavage, or activation, of bonds within the catalyst backbone

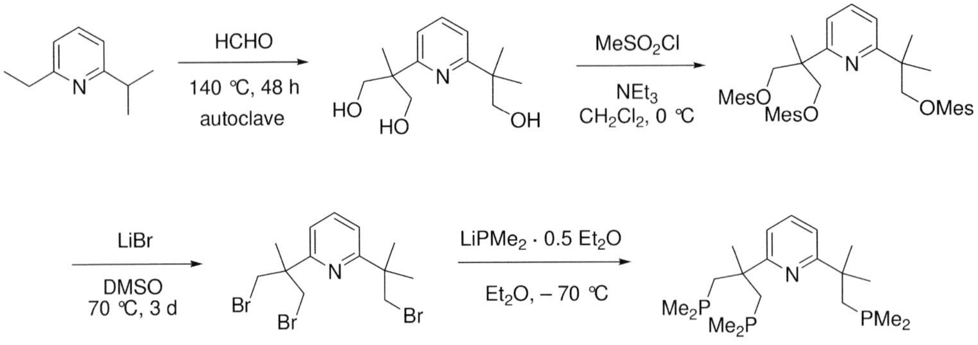

HCHO
140 °C, 48 h
autoclave
$MeSO_2Cl$
NEt_3
CH_2Cl_2, 0 °C
LiBr
DMSO
70 °C, 3 d
$LiPMe_2 \cdot 0.5\ Et_2O$
Et_2O, – 70 °C
4

Scheme 6.1

under reaction conditions is not uncommon. It was reported first for C–H bonds ('cyclometalation') and subsequently also for C–P bonds [20]. In the latter case, the observed reactivity follows the order $P–C_{sp} > P–C_{sp2} > P–C_{sp3}$, with phosphinoacetylenes being the most reactive. The next in line, arylphosphine ligands, have the largest number of reported C–P bond activations (which may be a reflection of their preferred use in catalysis). An external stimulus is usually required, such as the addition of acid [21]. In a few and only recently observed cases C–P bond cleavage in such ligands can be induced by the solvent, the latter acting as a nucleophile [22]. There appear to be no previous examples of solvent-induced C–P bond activation in alkylphosphines. The reverse reaction, i.e. the specific remaking under different conditions of a C–P bond that had previously been broken, is without precedent.

6.3 C–P Bond Activation and Agostic Interactions in Iron Complexes of Polypodal Phosphine Ligands

When stoichiometric amounts of the tetraphosphine **3** and $Fe(BF_4)_2 \cdot 6\,H_2O$ are reacted in methanol at room temperature, a red microcrystalline material (**5**) forms within a few minutes. The first indication that an unexpected reaction had taken place came from the well-resolved ^1H NMR spectrum ([D$_6$]DMSO, r.t.; 200 MHz) of this product, which has two striking features: the pyridine protons are separated into three signals (ABC, t/d/d), and a singlet corresponding to 3 protons is observed at high field ($\delta = -3.7$ ppm). In the ^{31}P NMR spectrum there are four signals, one at very low field ($\delta = 174.5$ ppm relative to 85% H_3PO_4) and a set of the three others at 50.3 ... 17.0 ppm. Whereas the original ligand has C_{2v} symmetry, the apparent loss of symmetry (C_1) served as a first indication that cleavage of the ligand might have occurred.

Structural characterization of the product (Figure 6.1) showed it to have what is now a *tri*podal *tetra*dentate ligand coordinated to iron, in addition to an unusual monodentate phosphinite ligand (dimethylphosphinous acid methyl ester,

$P(OMe)Me_2$, a constitutional isomer of trimethylphosphine oxide). An agostic interaction with iron of one of the C–H bonds in what now is a CH_3 group (formerly P-bonded CH_2 in the ligand backbone) completes the coordination sphere. Full structural and spectroscopic data of this complex were reported (see the references cited in the Introduction), and only a few features will be singled out here. The Fe–P bond lengths are in the range 2.1720(6) to 2.2710(6) Å, the bond trans to the agostic interaction being the shortest. This attests to the weak π-acceptor quality of the coordinated C–H bond, giving rise to a polarization of the appropriate iron d orbital towards the diametrically opposite P atom, with concomitant shortening of the bond. In **5**, the ^1H NMR signal of the methyl group in close proximity to the central iron ion is expected to be strongly shifted, as is indeed observed (rotation of this group, which is fast on the NMR time scale, causes the 3 protons to show as a broadened singlet). When the complexation is carried out in [D4]methanol under otherwise identical conditions, the deuterium label appears in the phosphinite ligand Me_2POCD_3 and a monodeuterated methyl group, CH_2D.

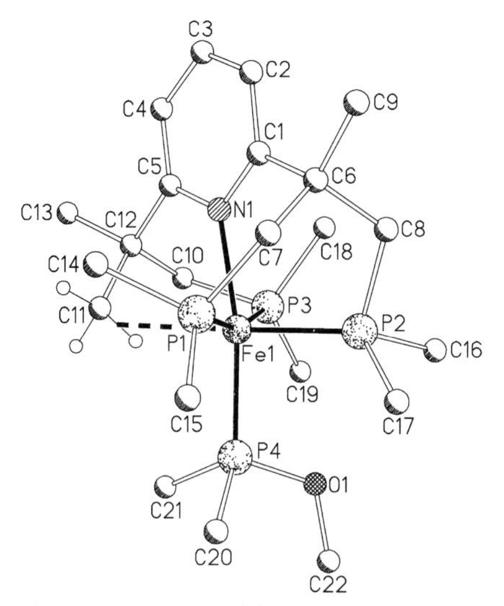

Figure 6.1 Structure of the dication in **5** · 0.5 MeOH (tetrafluoroborate salt). Hydrogen atoms are shown only for the methyl group in agostic interaction with the metal center.

We were intrigued to observe a different outcome for the complexation reaction when it is performed with the perchlorate salt, i.e. $Fe(ClO_4)_2 \cdot 6\,H_2O$, under otherwise identical conditions. The solution turns orange in the course of the reaction but does not deposit a precipitate. Careful addition of ether causes the deposition of an orange microcrystalline solid (**6**) which is quite unstable with respect to perchlorate-induced decomposition: the material explodes, particularly when dry, if excessive pressure is exerted during handling, for example with a spatula. In the 200 MHz ^1H NMR spectrum of **6** ([D4]methanol, r.t.), there are no signals at $\delta < 0$ ppm, which in **5** are diagnostic of an agostic interaction. Similarly to the spectrum of **5**, the signals of the pyridine protons suggest an ABC system, indicating the loss of C_{2v} symmetry. In the ^{31}P NMR spectrum there are four signals, with one (178.6 ppm) again clearly separated from the rest (43.5 ... 18.4 ppm). Elemental analysis data are compatible with an AB salt, containing a *mono*cation. Conclusive evidence comes from the X-ray crystal structure (Figure 6.2): a dimethylphosphinyl group has again been cleaved, but the carbanionic residue of the ligand is now coordinated to the iron ion, giving a monocationic complex. Whereas methoxide-induced cleavage of a dimethylphosphinyl group has again produced a monodentate phosphinite ligand, the fate of the methanol proton is as yet undetermined (it likely produces 1 equiv. of $HClO_4$). Similarly to **5**, the geometry at iron is again distorted octahedral (see Table 6.1). While the Fe–N bond lengths in **5** and **6** are almost the same, the Fe–P bond length patterns in **5** and **6** are in stark contrast (Table 6.1): The shortest such bond now connects the phosphinite ligand in the axial position (accentuating the weak π acceptor quality of the pyridine ring), while all three equatorial bonds are significantly longer but of overall similar length. Carbanion coordination does not induce a significant lengthening of the bond in the trans position (no trans influence on Fe–P2); quite on the contrary, this bond is the shortest of the three equatorial Fe–P bonds, the diametrically opposite Fe–C unit being only a weak π acceptor. The electron density provided to the metal by the carbanionic ligand seems to be funneled into the bond to the phosphine ligand in the trans position via increased π back donation. From the combined structural and spectroscopic information (carbanionic donor plus a monoanion), we assign a +II oxidation state to the central iron ion. Remarkably, **6** dissolves in water without decompositon, even in the presence of aerobic oxygen.

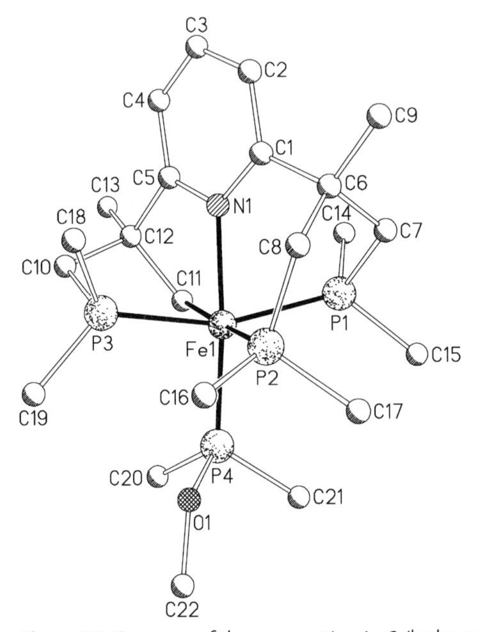

Figure 6.2 Structure of the monocation in **6** (hydrogen atoms omitted for clarity).

Table 6.1 Selected bond distances [Å] and angles [°] in **5**, **6**, **9**, and **10** (standard deviations in parentheses). Values in italics are parameters involving the 'agostic' methyl carbon atom.

	5 · 0.5 MeOH	**6**	**9** · H_2O	**10** · 2 MeOH
Fe1—N1	2.058(2)	2.065(2)	2.023(1)	2.143(2)
Fe1—P1	2.2681(6)	2.2432(7)	2.2766(5)	2.2629(6)
Fe1—P2	2.1720(6)	2.2105(7)	2.3127(5)	2.2394(6)
Fe1—P3	2.2710(6)	2.2388(7)	2.2945(5)	2.2482(6)
Fe1—P4	2.2041(6)	2.1659(7)	2.2353(6)	2.2460(6)
Fe1 ··· C11 (**5**); Fe1–C11(**6**, **9**)	*2.643(2)*	2.068(2)	2.079(2)	–
Fe1—C22	–	–	–	1.739(2)
N1-Fe1-P4	168.97(5)	170.81(5)	160.55(4)	–
N1-Fe-C22	–	–	–	177.55(8)
Fe1-C22-O1	–	–	–	177.5(2)
P1-Fe1-P3	163.49(2)	160.69(3)	163.47(2)	159.30(2)
P2-Fe1-C11	*172.11(5)*	175.93(7)	175.91(6)	–
P2-Fe1-P4	91.74(2)	91.46(2)	97.71(2)	165.53(2)

A careful study of preparation conditions revealed that both **5** and **6** may be obtained in 'different states of protonation' depending on the reaction temperature, as outlined in Scheme 6.2 (the isolated yields are uniformly between 80 and 90%): the nature of the acid HX which is produced in the course of phosphinite formation, and the solubility of the respective salt, appear to determine which product is formed at any one temperature. Conversely, for one and the same counterion, the protonation is temperature-dependent: when reacting $Fe(BF_4)_2 \cdot 6\ H_2O$ with **1** at $-50\,°C$, the only

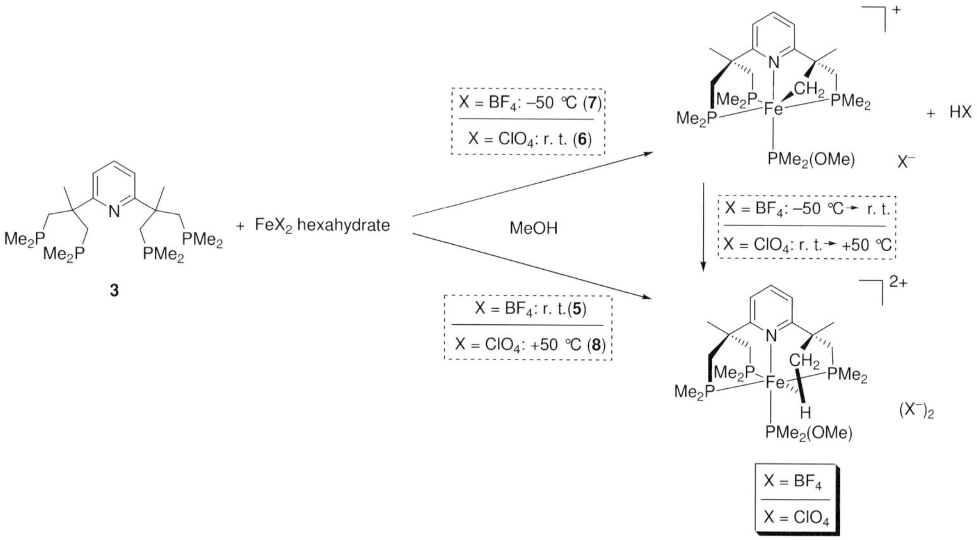

Scheme 6.2

isolable product (upon removal of the solvent and washing with ether; microcrystalline red solid, 75% yield) is the tetrafluoroborate salt **7**, containing the carbanionic ligand (Scheme 6.2). Once isolated at $-50\,°C$, this salt is stable even at room temperature; if, however, it is not isolated, and the reaction mixture allowed to warm to room temperature, the only detectable product is compound **5**, which has the agostic methyl group. Similar observations are made in the case of $X = ClO_4$, the only difference being that while the 'carbanionic' complex **6** forms at room temperature, preparation of the 'agostic' complex **8** requires heating (to $+50\,°C$, Scheme 6.2).

Attempts to reverse the protonation reaction, aiming to re-form the Fe–C bonded complex by addition of LiOMe to the 'agostic complexes' **5** and **8**, have so far been unsuccessful (room temperature). However, deprotonation of the agostic methyl group is facile upon oxidation: a careful quantitative study revealed that when reacting two equivalents of **5** with half an equivalent of dioxygen (Eq. (1)), the products are two equivalents of the *carbanionic* iron(III) complex **9** and, remarkably, one equivalent of *water*.

2 $[(Me_2P)_2\text{-}Fe^{II}(CH_2\text{-}H)(PMe_2)\,PMe_2(OMe)]^{2+}(BF_4^-)_2$ **5** + 0.5 O_2 $\xrightarrow{MeOH}$ 2 $[(Me_2P)_2\text{-}Fe^{III}(CH_2)(PMe_2)\,PMe_2(OMe)]^{2+}(BF_4^-)_2$ **9** + H_2O (1)

Compound **9** · H_2O forms as a crystalline green precipitate in 95% isolated yield. It is soluble in bulk water without decomposition. In the solid state structure of the dication (bis(tetrafluoroborate) hydrate salt, Figure 6.3 and Table 6.1), the iron-

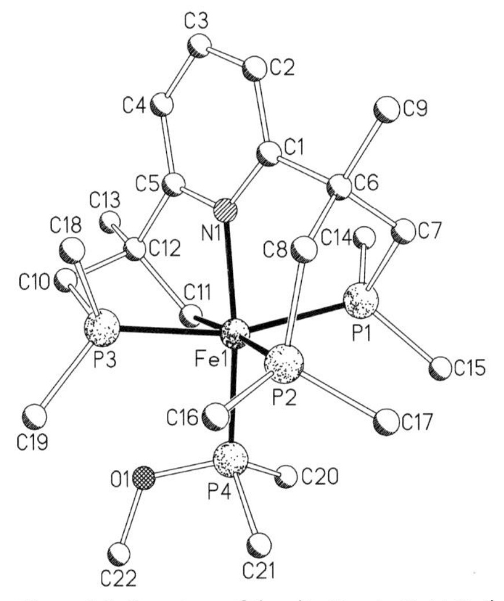

Figure 6.3 Structure of the dication in **9** · H_2O (hydrogen atoms omitted for clarity).

pyridine nitrogen bond length is shorter than in **5** or **6**, as expected for a more highly oxidized metal center and a ligand of predominant σ donor character. By contrast, all iron-phosphorus bonds are *longer* than in the iron(II) complexes, reflecting strongly decreased π back donation from iron(III). A detailed structural comparison of the complexes having $Fe^{II}\cdots(H_3CR)$, Fe^{II}–CH_2R, and Fe^{III}–CH_2R moieties (**5**, **6**, and **9**) is given in the relevant publication (see references cited in the Introduction).

Both the $\{Fe^{II}\cdots(H_3CR)\}$- and the $\{Fe^{II}$–$CH_2R\}$-containing compounds (tetrafluoroborate salts, complexes **5** and **7**) were reacted with carbon monoxide in an autoclave in order to study possible displacement reactions ($p_{CO} = 10.5$ bar, methanol solution, 80 °C, 20 h). While the outcome for **5** does not suggest a clean reaction, we were surprised to find for **7**, upon workup of the resultant yellow solution, the isolated product to be the iron(II) carbonyl complex of the intact tetraphosphine ligand **3**, $[(\mathbf{3})Fe(CO)](BF_4)_2$ (**10**). This material is obtained as a yellow (single-crystalline) solid. While the isolated yield is still low (12% based on **7**), NMR spectroscopic analyses (^{1}H, ^{31}P) of the mother liquor confirm the latter to contain a major amount of **10** in addition to other as yet unidentified compounds (^{31}P NMR). The NMR spectroscopic data of **10** (^{1}H, ^{31}P, ^{13}C) suggest C_{2v} symmetry for the diamagnetic cation. The carbonyl ligand gives rise to a prominent band in the solid-state IR spectrum (KBr; $\tilde{\nu} = 1969\ cm^{-1}$). In the mass spectrum (ESI), the molecular ion M^{2+} is detected at $m/z = 258$.

In the solid state, the dication is distorted from octahedral geometry by displacement of diametrically opposite pairs of P donors from the equatorial plane towards a tetrahedral P_4 arrangement (Figure 6.4, Table 6.1). The nitrogen-iron-carbonyl moiety is linear (N1-Fe1-C22 177.55(8)°, Fe1-C22-O1 177.5(2)°).

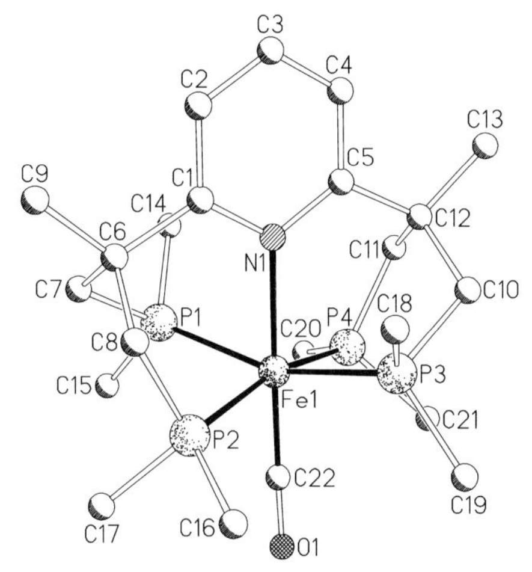

Figure 6.4 Structure of the dication in **10**·2 MeOH (hydrogen atoms omitted for clarity).

The reason for the observed reactivity of **7**, resulting in the re-formation under different conditions of a bond that was split when **3** was first reacted with iron(II), is as yet unclear. We are currently addressing the question whether the process is *intra*molecular (and thereby entails breaking of the P–O bond in the phosphinite ligand which, as such, should have considerable strength) or *inter*molecular (the required PMe_2 unit being provided by another equivalent of **3**, thereby explaining the presence of additional signals in the ^{31}P NMR spectrum of the mother liquor and the moderate yield). We expect the reaction pattern of a 'mixed' complex such as **11** to shed light on the reaction mechanism, as the carbonyl complex of a mixed 'py $(PMe_2)_3(PEt_2)$' ligand should be formed in high yield if the process is *intra*molecular.

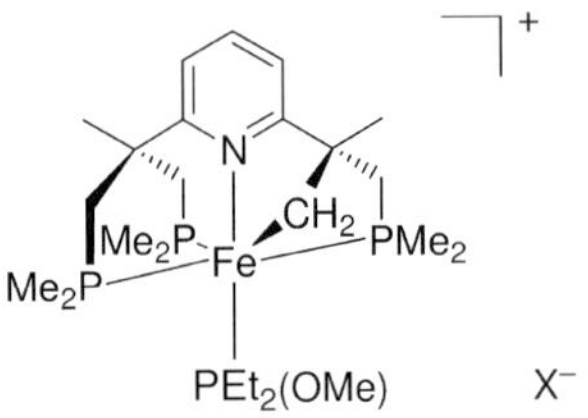

$PEt_2(OMe)$ X^-

11

We envisaged the independent synthesis of **11** by complexation of the tripodal tetradentate phosphine **4**, reaction of the complex with diethyl(methyl phosphinite), and subsequent deprotonation. The first part of this two-step sequence yielded, however, a mixture of products, as shown in Scheme 6.3.

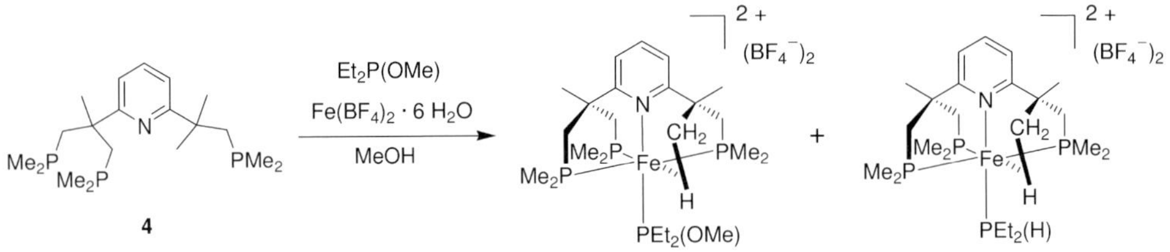

$Et_2P(OMe)$, $Fe(BF_4)_2 \cdot 6\,H_2O$, MeOH

4 → **12** ($PEt_2(OMe)$, $2+$, $(BF_4^-)_2$) + **13** ($PEt_2(H)$, $2+$, $(BF_4^-)_2$)

Scheme 6.3

The simultaneous formation of the diethylphosphine complex **13** (varying amounts depending on reaction stoichiometry; NMR, single crystal X-ray studies) was entirely unexpected. The target complex **12** is obtained pure only if $Fe(BF_4)_2$ hexahydrate is used in substoichiometric amount (10%) [23]. We are still working to clarify the role of the solvent in the transformation of the phosphinite ligand and to find a sufficiently gentle method for the deprotonation of **12**, avoiding decomposition of the phosphinite ligand.

For the observed bond activation in the tetrapodal ligand **3** upon complexation with iron, the choice of solvent is critical. Both methanol and ethanol effect ready cleavage of a dimethylphosphinyl group. The reaction of an anhydrous iron precursor such as $Fe(SO_3CF_3)_2 \cdot 2\,CH_3CN$ with **3** in an aprotic solvent (acetonitrile) proceeds with complexation of the intact ligand. In THF, P–C bond activation is still observed if the iron salt is a hydrate, as in $Fe(BF_4)_2 \cdot 6\,H_2O$, where hydrate water will act as the cleaving nucleophile, to produce metal-coordinated dimethylphosphinous acid. The latter is a tautomer of dimethylphosphine oxide, which is the thermodynamically stable form. The reaction of **3** with anhydrous ferrous bromide, $FeBr_2$, in methanol again produces a dimethylphosphinite ligand, but the complex now contains *ferric* iron coordinated by a carbanionic residual chelate ligand, implicating H^+ as the oxidizing agent under these conditions. Scheme 6.4 gives an overview of these reactions.

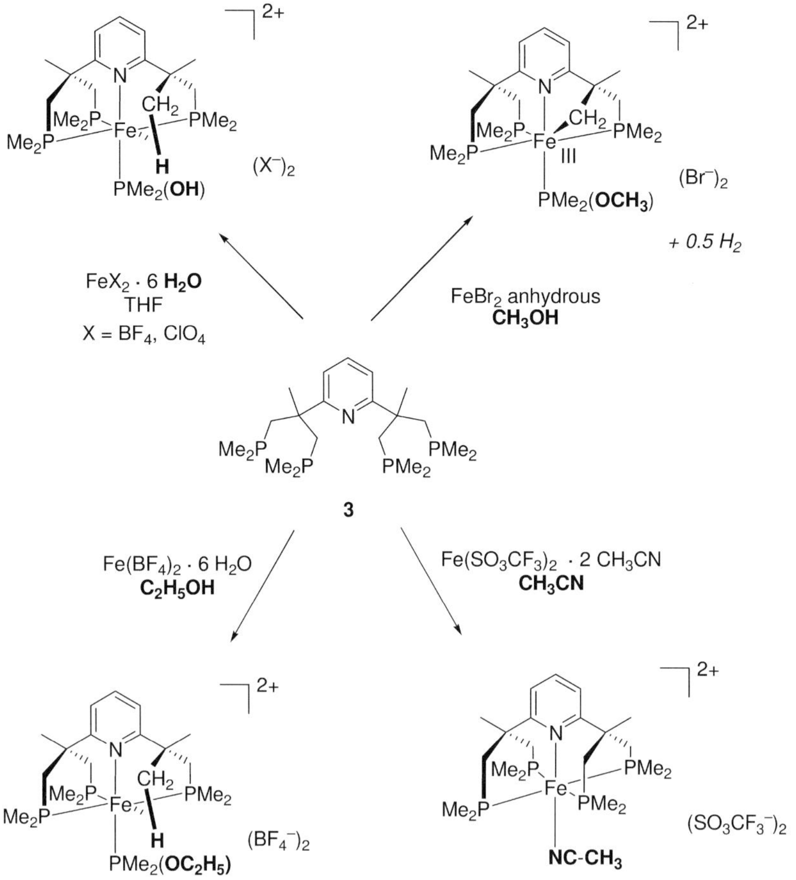

$FeX_2 \cdot 6\ H_2O$, THF, X = BF_4, ClO_4

$FeBr_2$ anhydrous, CH_3OH

$Fe(BF_4)_2 \cdot 6\ H_2O$, C_2H_5OH

$Fe(SO_3CF_3)_2 \cdot 2\ CH_3CN$, CH_3CN

Scheme 6.4

6.4
Mechanistic Considerations

The ligand cleavage observed upon reaction of **3** with iron(II) may be rationalized as follows. The square-pyramidal capping ligand provides $5 \times 2 = 10$ valence electrons through its NP_4 donor set, so that complexation to Fe^{II} (d^6) yields a 16 valence electron, and hence coordinatively unsaturated, complex. This coordinative unsaturation is likely removed by coordination of the Lewis-basic solvent MeOH, thereby significantly increasing the latter's acidity and priming it for nucleophilic attack. Other nucleophiles induce ligand cleavage in the same way: in tetrahydrofuran solution, solvate water from $FeX_2{\cdot}6\ H_2O$ produces the monodentate ligand P(OH) Me_2 exclusively. On the other hand, when acetonitrile is used as the (non-nucleophilic) solvent in conjunction with *anhydrous* iron(II) salts, such as iron(II) triflate, the product obtained is the regular complex $[(\mathbf{3})FeNCMe]^{2+}$, in which the ligand functions as a tetrapodal pentadentate 'coordination cap'.

As regards P—C bond breaking, a plausible sequence of events may thus be as follows. Intermediate coordination of methanol forms the complex $[(\mathbf{3})FeHOMe]^{2+}$, and intramolecular nucleophilic attack by methoxide then removes an adjoining PMe_2 substituent, producing a phosphinite ligand and a metal-coordinated *alkyl* residue. The nature of the acid (HX) formed in the process (Scheme 6.2) then controls the rate of protonation and crystallization. With tetrafluoroborate, the protonated product will precipitate at room temperature, whereas, with perchlorate, protonation/precipitation is not observed at room temperature. The overall process bears a distant resemblance to what has been termed a platinum-alkoxide/phosphorus-aryl methathesis, as reported by van Leeuwen, Orpen *et al.* [24] P—C bond activation in *aryl* phosphines (M = Ru) has, however, been postulated to require the initial formation of an unsaturated precursor (16 VE), which undergoes insertion of the metal ion into a phosphorus-carbon bond to give a saturated (18 VE) system which is then liable to attack by an incoming nucleophile [22]. As an alternative mechanism to the one described above, this process may be operative in our case also (Eq. (2)). It is noteworthy that, depending on what resonance forms are drawn, this entails formal involvement of either an iron(IV) or a phosphenium (R_2P^+) ion [25]. A particularly striking aspect in our case is the specificity with which only that P—C bond is cleaved which links the dimethylphosphinyl unit to the rest of the ligand; metal insertion into a P—Me bond is not observed.

$$\underset{16\ VE}{[Fe^{II}]\text{–}PMe_2R'} \rightleftharpoons \underset{18\ VE}{\left\{ R'\text{–}[Fe]\text{=}PMe_2 \leftrightarrow (R'^-)[Fe^{II}](^+PMe_2) \leftrightarrow (R'^-)[Fe^{IV}](^-PMe_2) \right\}} \xrightarrow{MeO^-} \underset{18\ VE}{R'\text{–}[Fe]\text{–}PMe_2(OMe)} \qquad (2)$$

6.5 Conclusion

Our work delineates some unusual reactivity patterns in iron(II) complexes of highly symmetrical aliphatic polyphosphines. In methanol, the tetrapodal pentadentate phosphine ligand $C_5H_3N[CMe(CH_2PMe_2)_2]_2$ (**3**), upon complexation to ferrous iron, undergoes selective cleavage of a methylene carbon-phosphorus bond, with concomitant formation of a phosphinite ligand, Me_2POMe. Given the relative inertness of tertiary alkylphosphines in other contexts (e.g. such compounds do not undergo alkali metal-induced cleavage [26], in contrast to arylphosphines or mixed aryl/alkylphosphines), the observed reactivity is even more remarkable. The reaction product is an organometallic complex, with a direct Fe–C bond or an *agostic* Fe–H–C interaction, depending on the reaction temperature and the nature of the counterion. The product containing the Fe–C bond reacts, in methanol under CO pressure, in such a way as to reform the intact ligand **3**, coordinated to iron(II), with a carbonyl ligand at the sixth coordination site. This remaking of a phosphorus-carbon bond that had previously and under different conditions been selectively cleaved is a particularly intriguing finding, and work is under way to elucidate the mechanism of this reaction.

References

1 Gade, L.H. (2002) *Chemie in Unserer Zeit*, **36**, 168.

2 Martin, L.L., Martin, R.L. and Sargeson, A.M. (1994) *Polyhedron*, **13**, 1969.

3 Berry, J.F., Bill, E., Bothe, E., Neese, F. and Wieghardt, K. (2006) *Journal of the American Chemical Society*, **128**, 13515.

4 Song, Y.-F., Berry, J., Bill, E., Bothe, E., Weyhermüller, T. and Wieghardt, K. (2007) *Inorganic Chemistry*, **46**, 2208.

5 Behrens, H. and Wakamatsu, H. (1963) *Zeitschrift fur Anorganische und Allgemeine Chemie*, **320**, 30.

6 (a) Marynick, D.S. and Kirkpatrick, C.M. (1983) *The Journal of Physical Chemistry*, **87**, 3273; (b) Kai, E., Misawa, T. and Nishimoto, K. (1980) *Bulletin of the Chemical Society of Japan*, **53**, 2481.

7 Mosbæk, H. and Poulsen, K.G. (1969) *Chemical Communications*, 479.

8 Senoff, C.V. (1990) *Journal of Chemical Education*, **67**, 368.

9 Bolvin, H. (2007) *Inorganic Chemistry*, **46**, 417.

10 Smith, P.L., Chordia, M.D. and Harman, W.D. (2001) *Tetrahedron*, **57**, 8203.

11 Pitarch López, J., Kämpf, H., Grunert, M., Gütlich, P., Heinemann, F.W., Prakash, R. and Grohmann, A. (2006) *Chemical Communications*, 1718.

12 Pitarch López, J., Heinemann, F.W., Prakash, R., Hess, B.A., Horner, O., Jeandey, C., Oddou, J.-L., Latour, J.-M. and Grohmann, A. (2002) *Chemistry - A European Journal*, **8**, 5709.

13 Pitarch López, J., Heinemann, F.W., Grohmann, A., Horner, O., Latour, J.-M. and Ramachandraiah, G. (2004) *Inorganic Chemistry Communications*, **7**, 773.

14 Schlangen, M., Neugebauer, J., Reiher, M., Schröder, D., Pitarch López J., Haryono, M., Heinemann, F.W., Grohmann, A. and Schwarz, H. (2008) *Journal of the American Chemical Society*, **130**, 4285.

15 Zimmermann, C., Heinemann, F.W. and Grohmann, A., (2005) *European Journal of Inorganic Chemistry*, 3506.

16 Kohl, S.W., Heinemann, F.W., Hummert, M., Weißhoff, H. and Grohmann, A. (2006) *European Journal of Inorganic Chemistry*, 3901.

17 Kohl, S.W., Heinemann, F.W., Hummert, M., Bauer, W. and Grohmann, A. (2006) *Chemistry - A European Journal*, **12**, 4313.

18 Kohl, S.W., Heinemann, F.W., Hummert, M., Bauer, W. and Grohmann, A. (2006) *Dalton Transactions*, 5583.

19 Karsch, H.H. and Appelt, A. (1983) *Zeitschrift fur Naturforschung*, **38b**, 1399.

20 Garrou, P.E. (1985) *Chemical Reviews*, **85**, 171.

21 Geldbach, T. and Pregosin, P.S. (2002) *European Journal of Inorganic Chemistry*, 1907.

22 Caballero, A., Jalón, F.A., Manzano, B.R., Espino, G., Pérez-Manrique, M.,

Mucientes, A., Poblete, F.J. and Maestro, M. (2004) *Organometallics*, **23**, 5694.

23 Gentschow, S.A. (2007) *Diplomarbeit*, Technische Universität, Berlin.

24 van Leeuwen, P.W.N.M., Roobeek, C.F. and Orpen, A.G. (1990) *Organometallics*, **9**, 2179.

25 Cowley, A.H. and Kemp, R.A. (1985) *Chemical Reviews*, **85**, 367.

26 'Phosphorus Ligands', Downing, J.H. and Smith, M.B. (2003) *Comprehensive Coordination Chemistry II*, vol. **1**, (eds J. A. McCleverty and T. J. Meyer), Elsevier Pergamon, Amsterdam, p. 264.

7
Regioselective Catalytic Activity of Complexes with NH, NR-substituted Heterocyclic Carbene Ligands

Siegfried R. Waldvogel, Anke Spurg, and F. Ekkehardt Hahn

7.1
Introduction

The unique stability and electronic properties of *N*-heterocyclic carbenes (NHCs) has led to a valuable and multi-purpose ligand system [1]. NHCs are employed in a broad spectrum of applications in modern synthetic organic chemistry (Scheme 7.1). In particular, transition metal complexes with NHC ligands have been shown to be valuable catalysts for various applications [2] such as olefin metathesis [3]. The substituents on the nitrogen centers are usually sterically demanding aryl, for example mesityl, or alkyl moieties, respectively. In many cases the *N*-heterocyclic carbenes are stable as free species, whereas the NH substituted congeners are immediately tautomerized to the imidazole derivatives. Therefore, NR,NH-stabilized heterocyclic carbenes have to be stabilized in metal complexes. Since the spatial effects of the *N,N'*-substituents are normally exploited in the catalytic system, *N*-heterocyclic carbene complexes exhibiting NH functions were largely disregarded in catalytic applications [2c]. However, the proximity of the N–H group as a potential hydrogen bonding donor and the unusual rigidity of the whole NHC-metal complex gives access to secondary interactions close to the catalytic center.

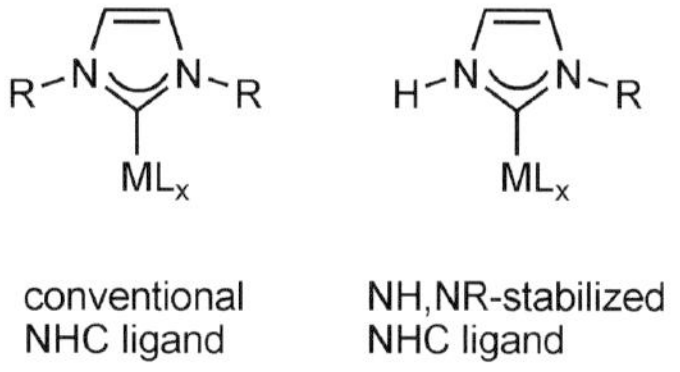

R–N N–R, ML_x — conventional NHC ligand

H–N N–R, ML_x — NH,NR-stabilized NHC ligand

Scheme 7.1 Complexes with NR,NR- and NH,NR-stabilized NHC ligands.

7.2
Concept of Regioselective Substrate Activation

The high substrate selectivity of proteins is normally caused by an equilibrium of molecular recognition preceding the actual enzymatic transformation. Adapting

Activating Unreactive Substrates: The Role of Secondary Interactions.
Edited by Carsten Bolm and F. Ekkehardt Hahn

ISBN: 978-3-527-31823-0

this approach to molecular chemistry, various attempts have been reported to use supramolecular interactions for the control of selectivity in catalytic processes [4]. The combination of a receptor unit with a catalytically active metal complex as found in I (Figure 7.1) enables such substrate recognition. This concept is based on the pre-complexation of the desired substrate out of a collection of potential substrate molecules. The substrate-catalyst-complex II is based on a two-point interaction. Aside from the supramolecular interaction of the substrate with the recognition unit, the reactive group of the substrate also coordinates to the metal center. The catalytic transformation proceeds from this situation giving complex III. The conversion of the reactive metal-coordinated group of the substrate changes this group and prevents its further coordination to the metal center leading to complex IV with a one-point interaction. Complex IV, held together only by the weak supramolecular interaction, allows the introduction of a new substrate molecule capable of forming a two-point interaction. This situation effectively prevents product inhibition of the catalyst.

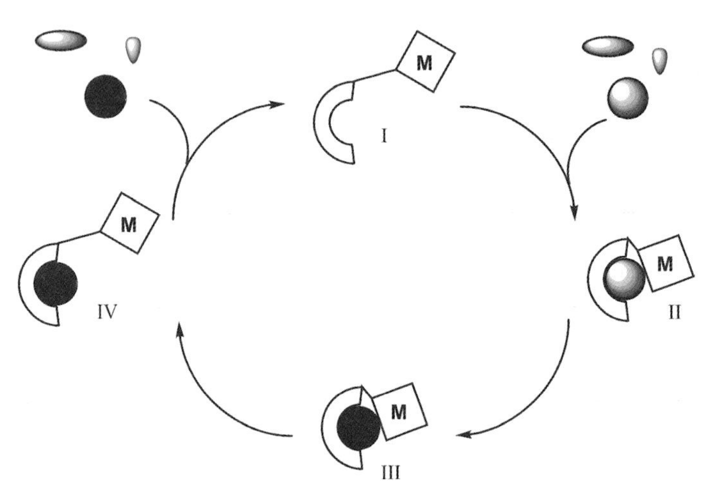

Figure 7.1 Concept of the substrate-selective supramolecular metal catalysis with preceding substrate recognition equilibrium (M = catalytically active metal center).

Selected diphosphine ligands which are able to coordinate to catalytically active metal centers have been used for substrate recognition. Complexes with the diphosphine ligand derived from glycouril developed by Nolte *et al.* (Figure 7.2, left) recognize and preferably transform substrates possessing a catechol unit [5]. However, the association constants of the substrate and the reaction product are high, which prevents the exchange to such an extent that catalysis without recognition becomes an important side reaction.

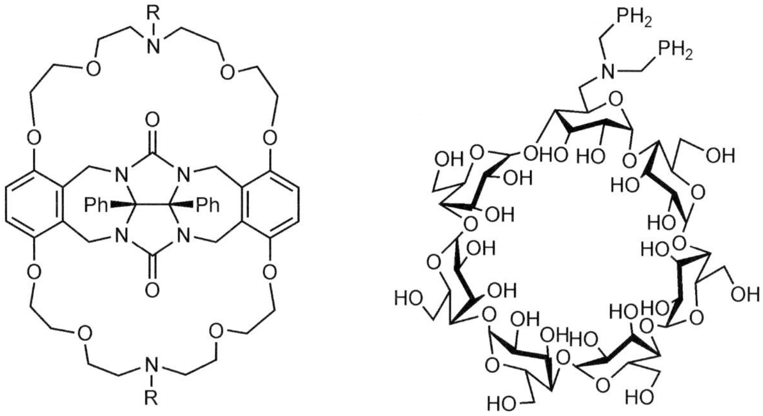

$R = CH_2C_6H_4OP(OPh)_2$

Figure 7.2 Diphosphine ligands which are combined with a supramolecular receptor (left according to Nolte *et al.* [5], right according to Reetz *et al.* [6]).

Reetz and coworkers substituted β-cyclodextrins with diphosphine ligands (Figure 7.2, right) [6]. The Rh^I complex with this ligand was used successfully in a substrate-selective catalytic hydrogenation reaction. While the exchange of substrate and reaction product proceeds rapidly, the large heterocycles generated upon metal coordination cause undesired side reaction particularly at high substrate concentrations.

Breit and coworkers developed a related approach [7]. The reaction of allylic alcohols with 2-diphenyl phosphonic acid yields substrate-functionalized phosphines (Figure 7.3). Such substrates coordinate to Rh^I. During catalysis, phosphine coordination to the metal center guarantees control of the selectivity in the subsequent hydroformylation reaction. Breit and coworkers demonstrated the synthetic value of this approach with a number of catalytic transformations.

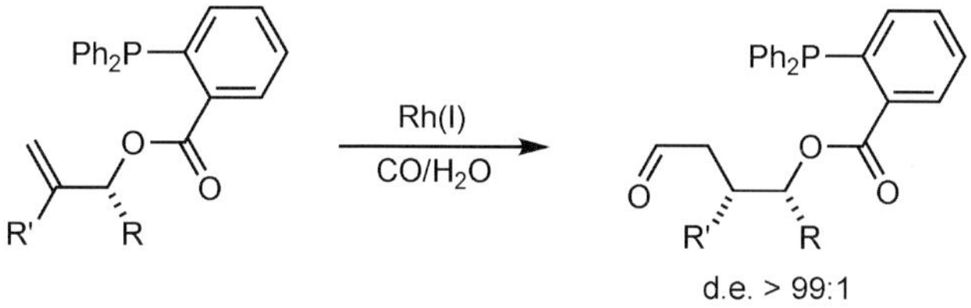

d.e. > 99:1

Figure 7.3 2-Diphenylphosphino benzoic acid in a reagent-controlled hydroformylation reaction.

It would be of special interest if a synthetic strategy could be developed which avoids the introduction of the phosphine auxiliary. The selectivity-determining interaction of the substrate with the metal complex should then be of supramolecular nature. Such a situation can be achieved by coordination of the substrate to the reaction center with both the reactive group and an additional donor, as was observed during the Rh^I-catalyzed hydrogenation of *N*-acetamido cinnamic acids (Figure 7.4) [8]. The application of this concept to other substrates, however, has been problematic.

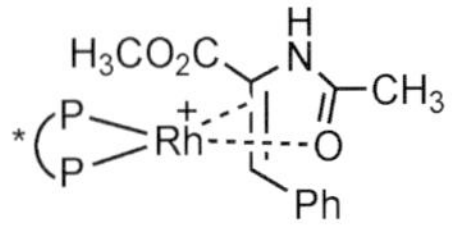

Figure 7.4 Transition state for the enantioselective hydrogenation of *N*-acetamido cinnamic acids with Rh^I catalysts.

The concept of selectivity control using supramolecular interactions in metal-catalyzed reactions has been demonstrated in some promising studies. However, up to now no simple system combining high transformation rates and substrate concentrations is known. Particularly the introduction of a small recognition unit for common molecular groups in substrates located in close proximity to a catalytically active metal center would constitute a useful combination of the concepts described above.

Complexes with NR,NR-substituted *N*-heterocyclic carbene ligands are important catalysts for various transformations [2]. However, much less is known about complexes with NH,NR-substituted carbene ligands, which thus have been disregarded for catalytic applications [2c]. We became interested in complexes with NH, NR-substituted carbene ligands because the NH function could act as a hydrogen-bonding donor. In a catalytically active complex bearing an NH,NR-functionalized carbene ligand the NH group could lead to recognition and temporary binding of a substrate with a hydrogen-bonding acceptor moiety like a carbonyl compound. The NH group of an NH,NR-stabilized carbene ligand can thus act as a supramolecular recognition site. Its close proximity to the metal center is unusual. In most described cases the supramolecular host moieties are connected to the metal center by a flexible spacer causing entropic problems [5, 6]. Secondary interactions of the NH group with selected substrates should be suitable to direct metal-mediated conversions. In the Rh^I-catalyzed hydrogenation reaction of a suitably carbonyl-functionalized olefin, for example, the substrate will interact initially by multiple contacts while some attractive interactions will disappear after the catalytic transformation at the metal center (Scheme 7.2), which generates an interesting motif for directed transition metal catalysis

7.3
Synthesis of Complexes with NH,NR-stabilized NHC Ligands

The majority of complexes with NHC ligands have been obtained using cyclic azolium precursors [1, 2]. However, aminocarbenes with an NH group can be obtained when protic nucleophiles like alcohols or primary or secondary amines attack a coordinated isocyanide. Unintentionally and most likely unrecognized, this method was first applied as early as 1915 when Tschugajeff and Skanawy-Grigorjewa

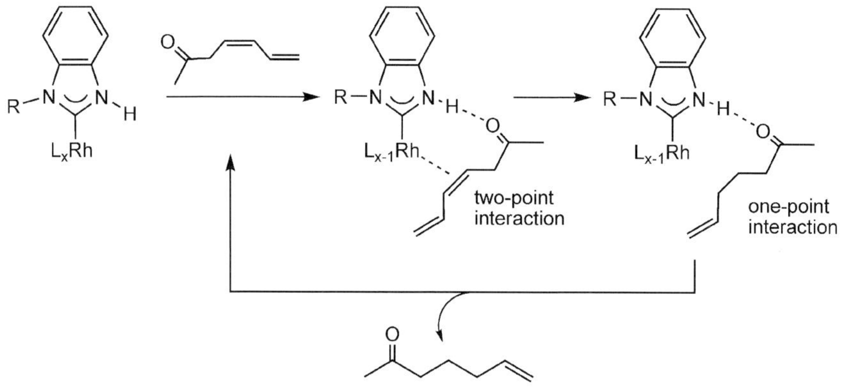

Scheme 7.2 Concept of two-point interaction/recognition of substrates with complexes bearing NH,NR-stabilized *N*-heterocyclic carbene ligands.

treated tetrakis(methyl isocyanide)platinum(II) with hydrazine and isolated a red salt [9] which was very likely the first carbene complex to be synthesized in pure form, predating Fischer's seminal carbene synthesis from metal carbonyls [10] by 49 years. Tschugajeff's red salt was unambiguously established as a carbene complex in 1970 [11a,b], and a number of derivatives, partially forming higher associates via hydrogen bonds, have been described recently [11c–f].

While the nucleophilic attack of a base at a coordinated isocyanide leads to acyclic heterocarbenes, the use of functionalized isocyanides gives access to cyclic heterocarbenes via an intramolecular 1,2-addition to the C≡N bond. Various groups have developed methods for the preparation of such nucleophile-functionalized isocyanide ligands at selected metal centers [12]. Much easier than the template-controlled generation of the functionalized isocyanides at a metal center is their direct use. Fehlhammer et al. used 2-hydroxyalkyl isocyanides which contain both the nucleophile and the isocyanide in the same molecule. Upon activation of the isocyanide via coordination to an electron-poor metal center, this type of ligand spontaneously cyclizes to give a coordinated oxazolidin-2-ylidene [13]. The nucleophile and the isocyanide in 2-hydroxyphenyl isocyanides are not only connected within the same molecule but they are also oriented in one plane for the intramolecular nucleophilic attack. This orientation as well as the aromaticity of the benzoxazolin-2-ylidene obtained after the intramolecular nucleophilic attack favor the cyclization reaction.

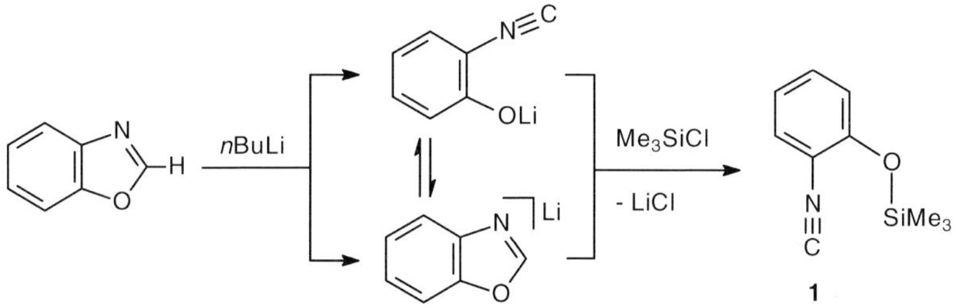

Scheme 7.3 Synthesis of trimethylsiloxyphenyl isocyanide (**1**).

While 2-hydroxyalkyl isocyanides are normally stable, 2-hydroxyphenyl isocyanide is not, but cyclizes to give benzoxazole [14]. Benzoxazole is opened by *n*BuLi, and subsequent reaction with trimethylsilyl chloride yields the 2-trimethylsiloxyphenyl isocyanide **1** (Scheme 7.3) [15].

The 2-trimethylsiloxyphenyl isocyanide **1** coordinates to Lewis-acidic metal centers to give isocyanide complexes of type **2**. Upon hydrolysis of the O-$SiMe_3$ bond with KF the ligand cyclizes to yield complexes of type **3** with an NH,O-stabilized benzoxazolin-2-ylidene ligand (Scheme 7.4) [16]. The heterocycle in complexes of type **3** is easily *N*-alkylated giving complexes of type **4**. NH,O- and NR,O-stabilized benzoxazolin-2-ylidene ligands have been generated at various metal centers like W [17, 18], Cr [18], Pd, Pt [19], B [20], Fe [21] and Re [22]. The intramolecular nucleophilic attack of the hydroxyl function at the isocyanide is influenced by the degree of $d \rightarrow \pi^*$ backbonding from the metal center to the isocyanide. A strong backbonding prevents the cyclization reaction and the 2-hydroxyphenyl isocyanide ligand is stabilized at the metal center. The IR spectra of complexes of type **2** and the force constants for the N$\equiv$C bonds calculated from these allow a prediction of the reaction after Si—OMe_3 bond cleavage [23]. An immediate cyclization was observed in the Re^{V} complex, while 2-hydroxyphenyl isocyanide is stabilized at an Re^{III} center [22]. An Fe^{II} complex containing both cyclized benzoxazolin-2-ylidene and 2-hydroxyphenyl isocyanide ligands is also known [24]. The reactivity of β-functionalized isocyanides has been reviewed [25].

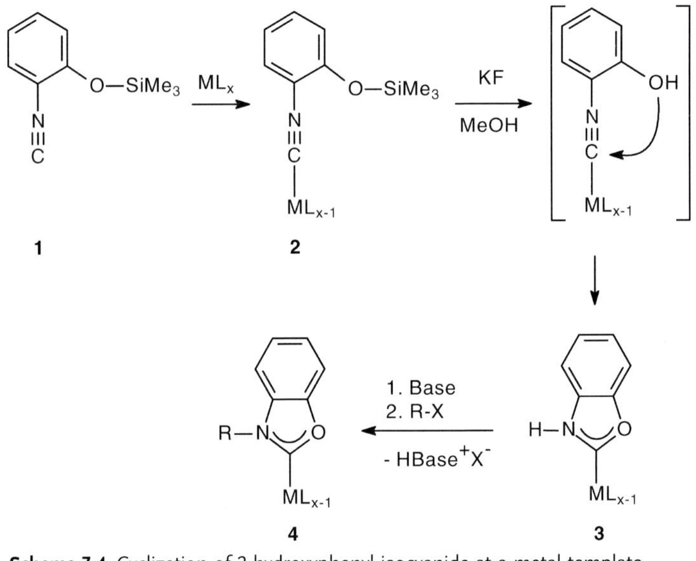

Scheme 7.4 Cyclization of 2-hydroxyphenyl isocyanide at a metal template.

The cyclization of β-functionalized isocyanides can also be employed for the preparation of cyclic diaminocarbenes. The preparation of complexes with benzannulated *N*-heterocyclic carbenes required a synthon for 2-aminophenyl isocyanide which is unstable toward cyclization to benzimidazole. It has been demonstrated that

2-azidophenyl isocyanide (**5**) is a suitable synthon. It coordinates to various metal centers to give complexes of type **6** (Scheme 7.5) [26]. The amine function for the intramolecular cyclization reaction is generated through a Staudinger reaction [27]. For this purpose complexes of type **6** are reacted with a tertiary phosphine to give the iminophosphorane complexs of type **7**. The N=P bond in complexes of type **7** can be hydrolyzed to yield triphenylphosphine oxide and the complexes with the elusive 2-aminophenyl isocyanide ligand. As described for coordinated 2-hydroxyphenyl isocyanide the 2-aminophenyl isocyanide immediately cyclizes to yield the complexes with a benzannulated NH,NH-stabilized NHC ligand (**8**). The alkylation of the nitrogen atoms of the heterocycle in **8** to yield **9** is straightforward. Complex **9** contains a benzannulated *N*-heterocyclic carbene ligand of the type which is also accessible in the free and uncoordinated state by reductive desulfurization of benzimidazolin-2-thiones [28] or deprotonation of benzimidazolium salts. The NH,NH-stabilized carbene ligand in **8**, however, cannot be prepared by any of these methods. A similar cyclization of 2-(azidomethyl)phenyl isocyanide has been described by Michelin *et al.* [29].

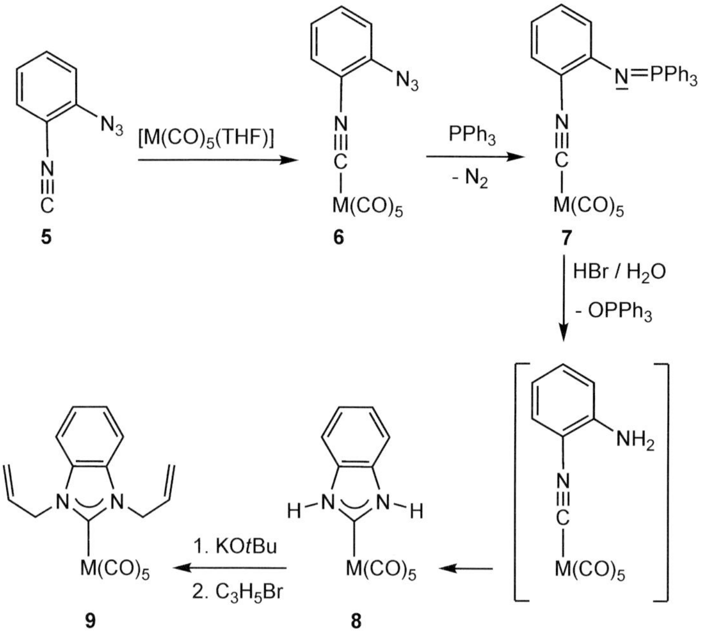

Scheme 7.5 Template synthesis of benzannulated NH,NH- and NR,NR-stabilized NHC ligands.

NH,NH-Stabilized benzannulated NHC ligands can also be generated by reduction of the nitro group in coordinated 2-nitrophenyl isocyanide (Scheme 7.6) [30]. The reduction of the nitro group with Sn/HCl yields complexes of type **8** while the Raney-Nickel/hydrazine reduction proceeds only to the hydroxylamine derivative, which also cyclizes to give the complex with an NH,NOH-stabilized carbene ligand **10**. Alkylation of the NH and OH functions is again non-problematic [30].

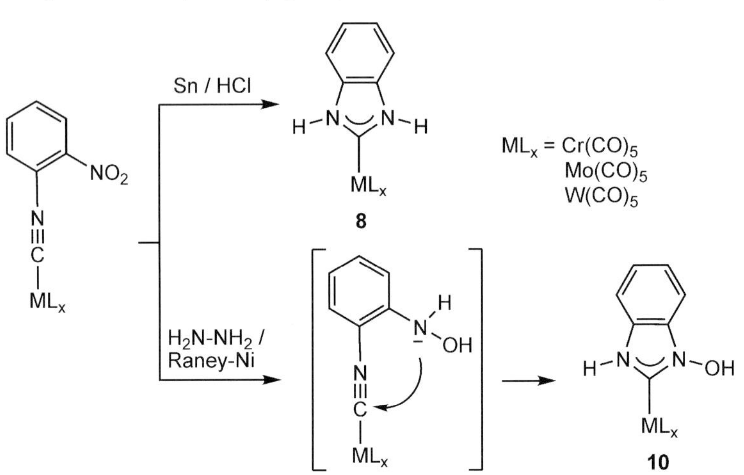

Scheme 7.6 Reduction and cyclization of coordinated 2-nitrophenyl isocyanide.

The easily accessible 2-azidoethyl isocyanide reacts in analogy to 2-azidophenyl isocyanide. Complex **11** with an NH,NH-stabilized imidazolidin-2-ylidene ligand was obtained in this reaction (Scheme 7.7) [31]. The same complex had been obtained previously by Lin *et al.* via the reaction of an amine-phosphinimine with $[W(CO)_6]$ [12i–j].

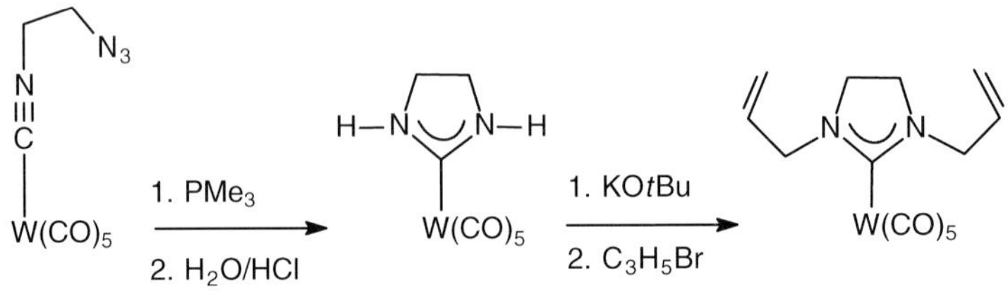

Scheme 7.7 Cyclization of 2-azidoethyl isocyanide at a metal template.

The template-controlled cyclization reactions of 2-azidoethyl and 2-azidophenyl isocyanide constitute an alternative method for the preparation of NHC complexes. This method is advantageous when azolium salt precursors are unavailable for the synthesis of NHC complexes by classical methods. In addition, the intermediate complexes with NH,NH-stabilized carbene ligands are important precursors for the template-controlled preparation of cyclic polycarbene ligands [32]. Most important for applications within the concept of selective substrate activation presented in Scheme 7.2 is the possibility to generate complexes with NH,NR-stabilized carbene ligands by monoalkylation of an NH,NH-stabilized carbene ligand (Scheme 7.8).

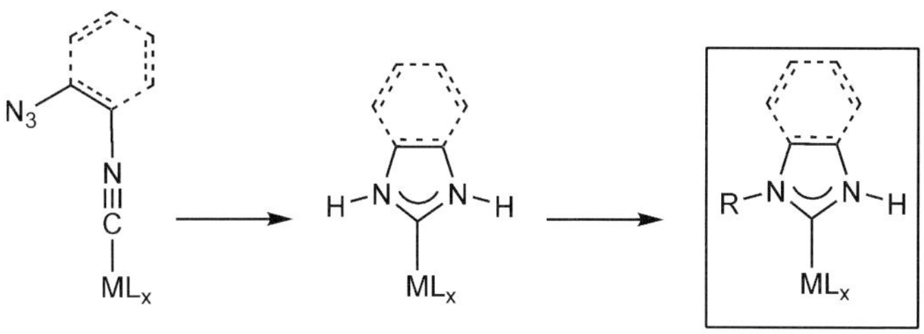

Scheme 7.8 Template synthesis of complexes with NH,NR-stabilized carbene ligands.

While the template synthesis of complexes with NH,NR-stabilized NHC ligands constitutes a general method for the preparation of such derivatives, the preparation of the 2-azido substituted isocyanides is time consuming and tedious. Therefore, we have developed an alternative method starting from monoalkylated benzimidazoles. *N*-Ethylbenzimidazole can be deprotonated with *n*BuLi to give the lithium salt **12**. This salt reacts with $[W(CO)_5(THF)]$ under C^2-metalation of the heterocycle. Subsequent treatment with HCl/diethyl ether directly produces complex **13** with an NH, NR-stabilized NHC ligand (Scheme 7.9) [33]. This constitutes an alternative and time-saving approach to complexes with NH,NR-substituted *N*-heterocyclic carbene ligands compared to the intramolecular cyclization of coordinated isocyanide ligands followed by *N*-alkylation.

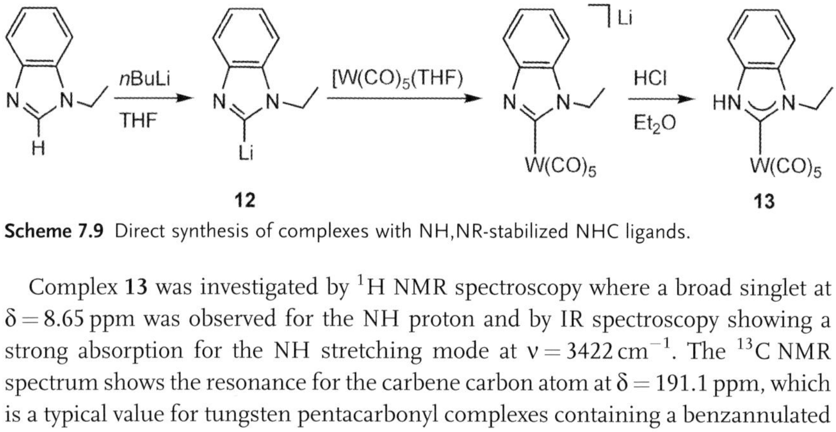

Scheme 7.9 Direct synthesis of complexes with NH,NR-stabilized NHC ligands.

Complex **13** was investigated by ^1H NMR spectroscopy where a broad singlet at $\delta = 8.65$ ppm was observed for the NH proton and by IR spectroscopy showing a strong absorption for the NH stretching mode at $\nu = 3422\ cm^{-1}$. The ^{13}C NMR spectrum shows the resonance for the carbene carbon atom at $\delta = 191.1$ ppm, which is a typical value for tungsten pentacarbonyl complexes containing a benzannulated *N*-heterocyclic carbene [26, 30]. The molecular structure of **13** was determined by X-ray diffraction confirming the connectivity within the complex. Bond distances and angles in **13** show no unusual features and are within the range observed previously for tungsten pentacarbonyl complexes with saturated [31] or benzannulated [26, 28, 30] *N*-heterocyclic carbene ligands.

In order to manipulate a catalytic reaction a weak secondary interaction is highly desired since not only the formation of a reactive intermediate is required but also a quick exchange of the final product. Very little is known about the hydrogen bonding capabilities of complexes with *N*-heterocyclic carbenes exhibiting free NH functions. Hydrogen bonding between the NH group of an NH,O-stabilized carbene ligand at an Re^V center and triphenylphosphine oxide has been described [22]. Because of the successful deprotonation/alkylation sequences on the NH groups of NH,NH-stabilized carbene ligands similar properties to carboxamides were anticipated. Therefore, we became interested in the hydrogen-bonding properties of complex **13**.

For studying secondary interactions a whole range of analytical tools are established in supramolecular chemistry [34]. Complex **13** is yellow in color [33]. Secondary interactions by hydrogen bonding will alter the electronic nature of the carbene ligand and should consequently be observed in the UV/Vis spectra. It is noteworthy, that the UV/Vis titration can be carried out at ambient conditions, but prolonged exposure of **13** to ambient conditions leads to some decomposition. The intense absorbance of **13** allows the titration with DMPU at low concentration, e.g. $<10^{-5}$ M. However, no change in the spectral data was found. Most probably high dilution impedes the observation of hydrogen-bound species indicating a binding rate <100 M^{-1}. Mass spectrometric investigations support this assumption. No aggregates of **13** and DMPU were found employing electro spray ionisation techniques. ^{1}H NMR techniques allowed a detailed insight into the hydrogen bonding involving the NH proton of the *N*-heterocyclic carbene. First, self aggregation of the carbene complex **13** as well as the formation of higher aggregates could be excluded by dilution experiments. Since **13** involves both a hydrogen bond donor (NH) and acceptor (CO), self assembly to supramolecular aggregates seemed possible as indicated in Scheme 7.10. The investigated concentration range covered several orders of magnitude. Dilution experiments were performed in CD_2Cl_2. No changes in the ^{1}H NMR chemical shifts were observed in the range of 10^{-5} to 5×10^{-1} M.

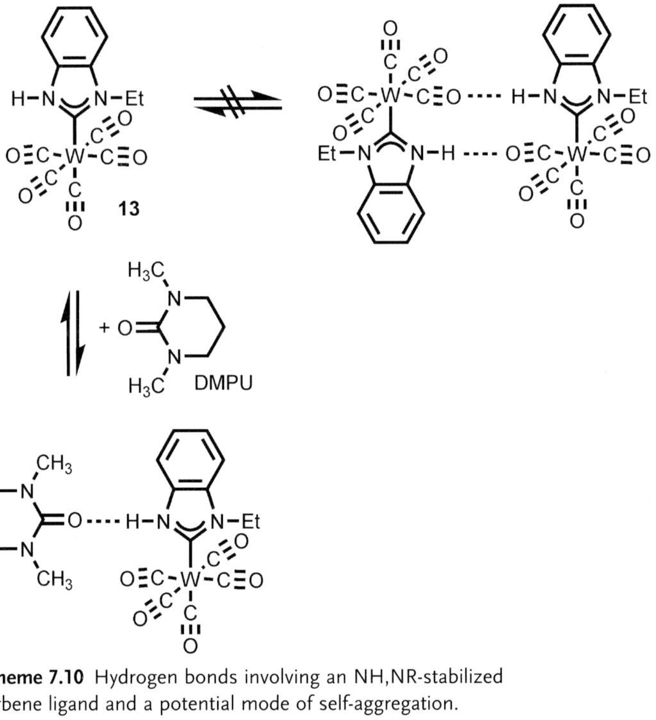

Scheme 7.10 Hydrogen bonds involving an NH,NR-stabilized carbene ligand and a potential mode of self-aggregation.

Complex **13** was titrated with the strong hydrogen bond acceptor DMPU (Figure 7.5) as a model for carbonyl compounds in $CDCl_3$, and the reaction was monitored by ^1H NMR spectroscopy. The NMR titration was carried out at 27 °C in CD_2Cl_2 and demonstrated a significant downfield shift of the resonance for the NH proton upon addition of DMPU while the chemical shifts for all other protons remain essentially unchanged. Only broadening of the NH signal was observed after addition of more than one equivalent of DMPU (up to 5 equivalents). Furthermore, a Job-pot of the titration of **13** with DMPU indicated a clear 1:1 ratio in the interaction.

The chemical shifts were fitted using standard tools [35]. Multiple experiments revealed a binding constant of about $40 \pm 10\,M^{-1}$ for the single hydrogen bond which corresponds to a ΔG_{ass} of 2.2 kcal mol^{-1}. The hydrogen bonding is supported by chlorinated solvents and THF, whereas protic media like alcohols impede the supramolecular interaction. Therefore, titration in $[O_7]$-*N,N*-dimethylformamide yields no effect at all.

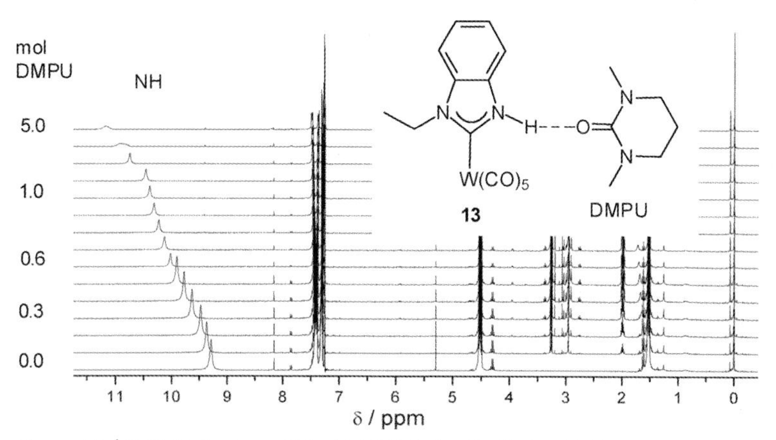

Figure 7.5 ^1H NMR titration of **13** with DMPU (in CD_2Cl_2, $T = 27$ °C).

After demonstrating the formation of an N−H···O hydrogen bond between **13** and DMPU we tried to generate NH,NR-stabilized NHC ligands at catalytically active metal centers. The Rh^I complex **14** was prepared from $[RhCl(coe)_2]_2$, *N*-methylbenzimidazole and PPh_3 (Scheme 7.11). Two isomers of **14** were observed, which appear to interconvert in solution. The major isomer with a C^2−Rh bond (**14a**) tautomerizes to the minor isomer with an N^3−Rh bond (**14b**). Complex **15**, which is similar to **14a** but appears not to undergo the tautomerism has been isolated by Bergman et al. from the reaction of $[RhCl(coe)_2]_2$, *N*-methylbenzimidazole and PCy_3 (Scheme 7.10) [36].

At this time it appears that the tautomerism observed for complex **14** is a common feature of NHC complexes of redox-variable metals bearing NH,NR-stabilized cyclic

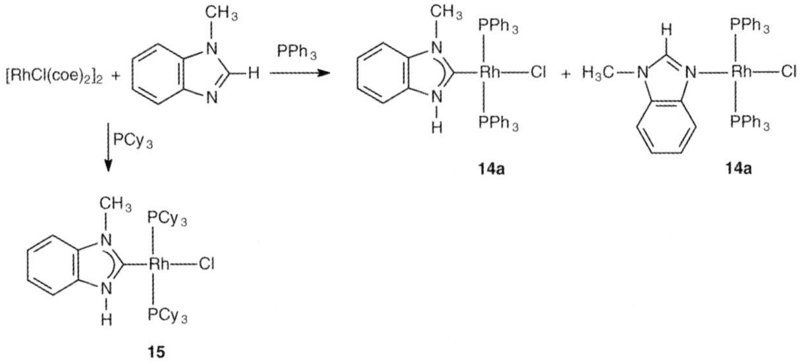

Scheme 7.11 Synthesis of complexes **14** and **15** and tautomers **14a/14b**.

diaminocarbene ligands. Ruiz and Perandones observed a base-promoted tautomerization of *N*-coordinated imidazole ligands at Mn^{I} (Scheme 7.12) [37]. Reaction of the cationic Mn^{I} imidazole complex **16** with a base rapidly produces the neutral, somewhat unstable C^2 deprotonated species **17**. Protonation of the isomer mixture of **17** leads to the NHC complex **18**, which is a tautomer of **16**. If possible, the Mn^{I} moiety apparently prefers *C*-binding rather than *N*-binding. This tendency is reversed when an even softer metal center becomes available. Consequently, deprotonation of **18** in the presence of $[AuCl(PPh_3)]$ yields the dinuclear complex with the imidazole nitrogen atom coordinating to Mn^{I} and the C^2 carbene center coordinating to the softer Ag^{I} center [37]. Similar tautomeric rearrangements have recently been discussed [38].

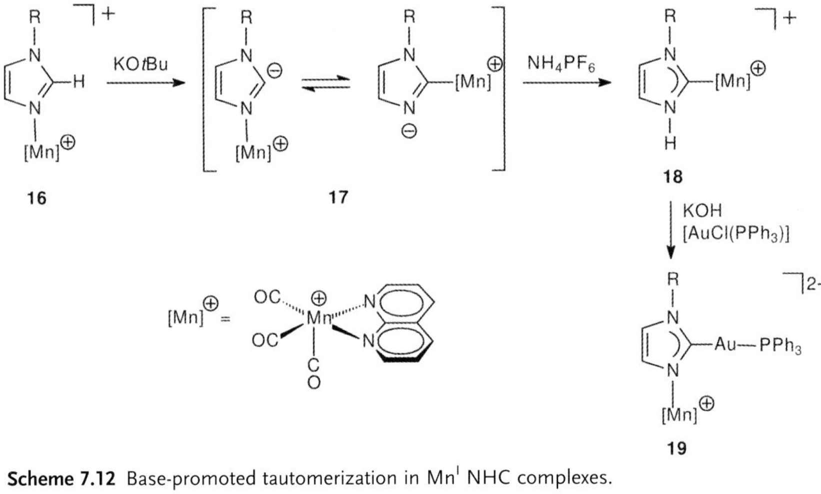

Scheme 7.12 Base-promoted tautomerization in Mn^{I} NHC complexes.

7.4
Preparation of Substrates for Catalytic Experiments

The hydrogen bonding from NH,NR-stabilized carbene ligands to substrates that contain ubiquitous functionalities like carbonyl groups can be used to demonstrate the secondary effect in catalysis. This type of substrate recognition could, for example, lead to the preferred conversion of only one out of a collection of substrates in a metal-catalyzed reaction. We have become interested in substrate-selective catalytic hydrogenation reactions. A single hydrogen bond makes an energy difference of about 2–4 kcal mol^{-1}. It could lead to a two-point interaction, as in **20**, and additional ligand/substrate combinations like the three-point interactions in **21** and **22** (Figure 7.6).

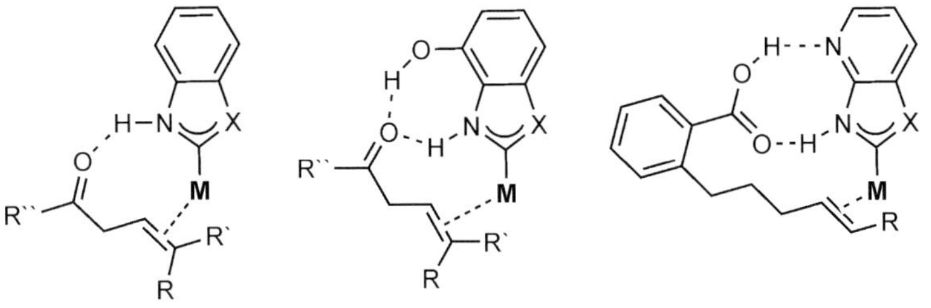

M = Rh, Ir, Pd
X = O, N-alkyl

Figure 7.6 Possible multiple-point interactions for substrate activation.

To demonstrate the substrate selectivity in catalytic hydrogenations, some selected substrates were prepared in multi-gram quantities. For the scenario involving intermediates of type **20**, olefins with carbonyl groups at an appropriate distance from the olefinic group were prepared (Figure 7.7 and Scheme 7.13). A competitive hydrogenation involving a hydrocarbon like **23** and the corresponding oxygenated substrates (**24**–**26**) would provide the proof of the principle. The catalysis was monitored by GC and GC/MS techniques. For valid analytical data the substrates have to be in the same boiling range and should be easily separated by gas chromatography. Of course the hydrogenated products as well as the by-products obtained by isomerization have to be clearly identified and determined. The hydrocarbon which also has the role as non-functionalized reference is 1-dodecene. The substrates for the competitive experiments are depicted in Figure 7.7.

While **23** is commercially available, **24**–**26** had to be prepared by standard procedures. Vinylacetic acid and 4-pentanoic acid were esterificated *in situ* by addition of thionyl chloride in the presence of 1-pentanol or 1-butanol, respectively. It is noteworthy that this protocol yields significant amounts of dipentyl- or dibutylether, respectively. These by-products were removed by fractional distillation. The analytical data for **24** and **25** correspond to literature data [39, 40]. From 5-hexenoic acid the known ethyl ester was made by alkylation of the carboxylate [41]. Tailor-made substrates for the three-point interaction shown in **22** (Figure 7.6) were also prepared. Terminal olefins like **27** are accessible from *ortho*-toluic acid by *ortho*-directed metalation and subsequent installation of the alkene moiety by alkylation of the carbanion (Scheme 7.13) [42].

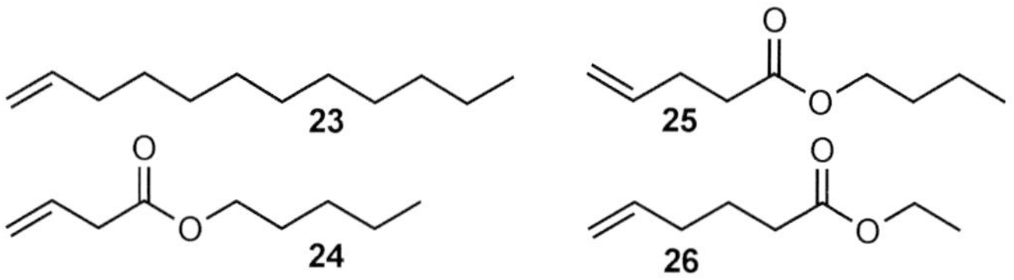

Figure 7.7 Substrates for substrate selective hydrogenation experiments.

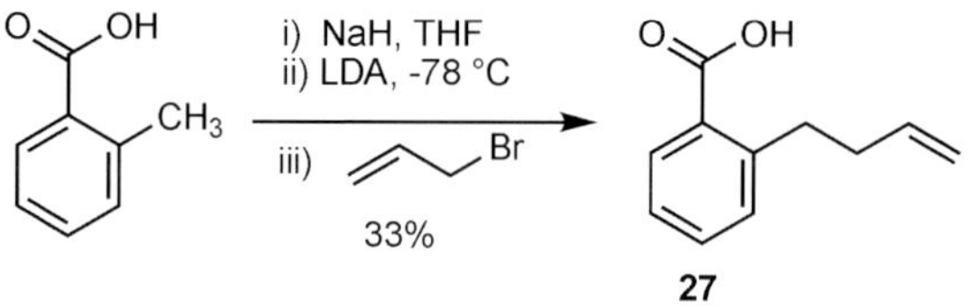

Scheme 7.13 Synthesis of a substrate for a three-point interaction with a hydrogenation catalyst.

7.5 Catalysis Experiments

Hydrogenation was chosen as the transition metal-mediated process since the catalysis can be considered as virtually irreversible, and, more importantly, non-activated C=C double bonds as well as the reagent hydrogen do not interfere with the secondary interaction based on hydrogen bonding to the substrate. Furthermore, no leaving groups that produce salts are involved. Initial experiments revealed that rhodium complexes of NH,NR-stabilized carbene ligands with triphenylphosphine as co-ligands are already active in catalysis. The readily available tricyclohexylphosphine modified complex **15** (Scheme 7.11) was not active under ambient conditions (25 °C, H_2, 1.0 bar) or under drastic conditions (60 °C, H_2, 100.0 bar).

For proof of the principle, a competition experiment with an equimolar ratio of two substrates was envisioned. In the case of a supramolecular interaction, the two substrates should be converted at different rates. The selectivity should drop along with the depletion of the preferred substrate. Therefore, comparable values should be given after about 30% of the total conversion of all substrates. In the hydrogenation of 1-dodecene (**23**) with the isomer mixture **14a,b** a catalytic activity higher than that of Wilkinson's catalyst was observed (Scheme 7.14). When the mixture **14a,b** was used in a competitive hydrogenation experiment employing **23** and a 3-butenoic acid ester (**24**), the substrate **24** with the supramolecular recognition unit is clearly preferred. Up to a conversion of 80% a significant preference for the substrate bearing the supramolecular key was observed. Control experiments with Wilkinson's catalyst unequivocally eliminate the Lewis acidic properties of Rh^I as the source of this selectivity since only a very slight preference for the oxygenated substrate was found in this case (see Table 7.1, entry 2). Besides the hydrogenation reaction, Rh^{I} is also capable of isomerizing multiple bonds [43]. Consequently, the conversion of the individual substrate refers to the sum of the hydrogenated derivative and the products of isomerization.

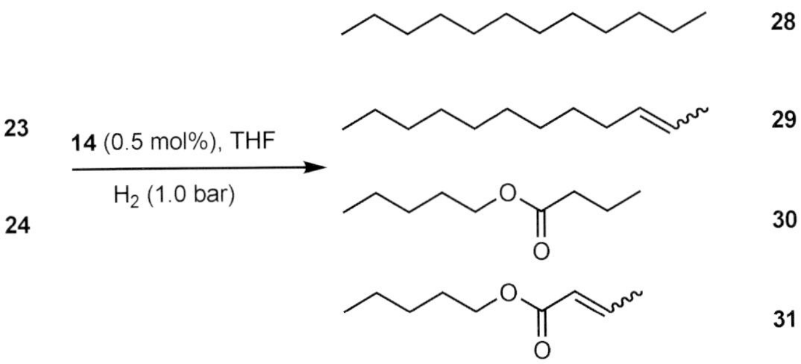

23, 24 — 14 (0.5 mol%), THF; H_2 (1.0 bar) → 28, 29, 30, 31

Scheme 7.14 Competitive substrate-selective hydrogenation and observed products.

Table 7.1 Substrate selectivity for various catalysts.

Entry	Catalyst	Substrate pair	S
1	14 (~80% 14a)	23/24	4.0 ± 0.2
2	Wilkinson's Catalyst [Rh(PPh$_3$)$_3$Cl]	23/24	1.3 ± 0.1
3	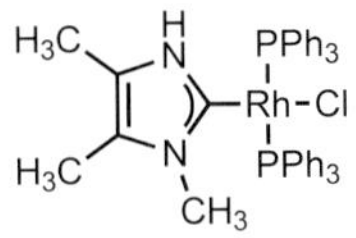 H_3C, H_3C, H, N, N, CH_3, PPh_3, Rh–Cl, PPh_3	23/24	3.7 ± 0.3
4	14 (~80% 14a)	23/25	2.6 ± 0.2
5	14 (~80% 14a)	23/26	1.9 ± 0.3

a) Catalysis conditions: 0.5 mol% catalyst, THF, 25 °C, H_2 (1.0 bar)

The conversion of **24** also involves hydrogenation to pentyl butyrate (**30**) and the isomerized product pentyl crotonate (**31**). However, double-bond shifting to the conjugated product **31** is not as dominant as might be anticipated. Therefore, isomerization can be excluded as a driving force for the selectivity. When a total consumption of about 30% is reached the selectivity is determined by the ratio of conversions for the individual substrates. This selectivity *S* shows a high degree of correlation. In Figure 7.8 the competitive hydrogenation of **23** and **24** is displayed. The catalyst precursor employed was **14** as a mixture of both redox tautomers with a content of about 80% of complex **14a** bearing the NH,NR-stabilized *N*-heterocyclic carbene ligand. At a total conversion of 30% the selectivity $S = 4.0 \pm 0.2$ was determined. The errors originate by standard deviation from multiple runs of catalysis.

Other Rh^{I} complexes equipped with NH,NR-stabilized *N*-heterocyclic carbene ligands were also employed in catalysis. The benzannulation seems to have no significant influence on the conversion. When substrates are applied with a hydrogen bonding acceptor separated from the olefin by a larger distance, then the selectivity

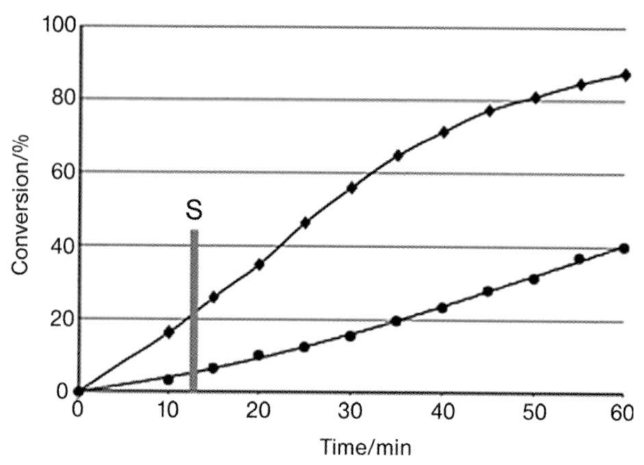

Figure 7.8 Competitive hydrogenation of 1-dodecene (23, •) and a 3-butenoic acid ester (**24**, ♦) using catalyst **14a,b**, with indicated area where the selectivity *S* was determined.

for the preferred substrate drops along with the increasing distance. 4-Pentenoate (**25**) is still the better substrate compared to the alkene **23**, whereas 5-hexenoate (**26**) only shows a slight preference upon conversion with **14a,b**. With the latter substrate, a non-favored 10-membered ring system had to be formed in the two-point interaction, which diminishes the selectivity dramatically.

Table 7.2 Substrate-selective hydrogenation by *in situ* preparation of catalysts.

Entry		Precursor mixture	Substrate pair	*S*
1		$[RhCl(coe)_2]_2$, PPh_3	**23/24**	3.6 ± 0.4
2		$[RhCl(coe)_2]_2$, PPh_3	**23/24**	1.5 ± 0.4
3		$[RhCl(coe)_2]_2$, PPh_3	**23/24**	1.2 ± 0.2[a]
4		$[RhCl(coe)_2]_2$, PPh_3	**23/24**	3.1 ± 0.3
5		$[RhCl(coe)_2]_2$, PPh_3	**23/25**	2.3 ± 0.4

Catalysis conditions: 0.5 mol% precursor mixture, THF, 25 °C, H_2 (1.0 bar).
[a] Hydrogenation proceeds very slowly.

Inspired by the redox tautomeric equilibrium, an *in situ* preparation of the catalyst mixture was tested. Surprisingly, the supramolecular recognizable substrate is also preferred in this catalysis and very similar selectivities were found (Table 7.2). However, in general, the *in situ* preparation leads to lower selectivity. This is attributed to a significant induction period of several minutes, wherein some unselective hydrogenation activity diminishes the selectivity (Table 7.2, entries 1, 4, 5). If the C^2 position at the benzimidazole is blocked by a methyl group an almost unselective conversion is found (Table 7.2, entry 2). Addition of benzimidazole with no *N*-alkyl substituents slows down the hydrogenation process tremendously, and this is accompanied by a loss of preference for a substrate.

7.6
Conclusions and Summary

We have developed different methods for the synthesis of complexes bearing *N*-heterocyclic carbene ligands with free NH moieties. These can be prepared in a straightforward synthesis from complexes with isocyanide ligands, in a one-pot synthesis from *N*-alkyl benzimidazoles (e.g. **13**), or from a transition metal complex fragments by oxidative addition. Catalytically active metal centers such as Rh^I bearing an NH,NR-stabilized NHC ligand exhibit a redox tautomeric behavior. An equilibrium between the complex with the *N*-heterocyclic carbene ligand (**14a**) and the *N*-coordinated imidazole congeners (**14b**) is most likely. The NH group of an NH,NR-stabilized NHC ligand is able to act as a strong hydrogen-bonding donor. However, hydrogen bonding does not lead to self-aggregation in the case of **13**. The supramolecular interaction can be studied in detail by ^1H NMR spectroscopy. The bond energy of the N–H···O hydrogen bond observed for **13** (2.2 kcal mol^{-1}) is low enough to allow a fast dissociation of the substrate from the complex after completion of the catalytic reaction.

The NH group of an NH,NR-stabilized *N*-heterocyclic carbene ligand can act as a supramolecular key in the vicinity of a catalytically active rhodium center, which enabled a substrate-selective hydrogenation. This was demonstrated by the selective hydrogenation of differently oxygenated substrates exhibiting terminal C=C double bonds. Substrates which offer an acceptor for hydrogen bonds at an appropriate distance were converted preferentially. If the length of the spacer between the reactive moiety and the supramolecular key increases the selectivity drops. Most interestingly, the oxidative addition of the C^2–H bond of an *N*-alkyl benzimidazole to Rh^I can be employed to generate catalytically active species with striking similar properties to those of complexes of type **14**. Control experiments using rhodium complexes with C^2-methyl-substituted benzimidazole ligands showed no selectivity. The Lewis-acidic character of the metal center as the source of selectivity can also be excluded since Wilkinson's catalyst showed almost no substrate selectivity.

We present here a novel concept for regioselective hydrogenation catalysis using complexes with NH,NR-stabilized NHC ligands. Currently, we are studying complexes of various catalytically active metal centers with the intention of employing the supramolecular recognition of suitable substrates for selective catalytic transformations such as regioselective C–C coupling reactions.

References

1 (a) Hahn, F.E. and Jahnke, U.C. (2008) *Angewandte Chemie*, **120**, 3166–3216; (2008) *Angewandte Chemie – International Edition*, **47**, 3122–3172; (b) Hahn, F.E. (2006) *Angewandte Chemie*, **118**, 1374–1378; (2006) *Angewandte Chemie – International Edition*, **45**, 1348–1352; (c) Bourissou, D., Guerret, O., Gabbaï, F.P. and Bertrand, G. (2000) *Chemical Reviews*, **100**, 39–91; (d) Herrmann, W.A. and Köcher, C. (1997) *Angewandte Chemie*, **109**, 2256–2282; (1997) *Angewandte Chemie – International Edition*, **36**, 2162–2187.

2 (a) Nolan, S. (ed.) (2006) *N–Heterocyclic Carbenes in Synthesis*, Wiley-VCH, Weinheim; (b) Glorius, F. (ed.) (2006) *N–Heterocyclic Carbenes in Transition Metal Catalysis, Topics in Organometallic Chemistry 21*, Springer-Verlag; (c) Herrmann, W.A. (2002) *Angewandte Chemie*, **114**, 1342–1363; (2002) *Angewandte Chemie-International Edition*, **41**, 1290–1309.

3 Trnka, T.M. and Grubbs, R.H. (2001) *Accounts of Chemical Research*, **34**, 18–29.

4 Hamilton, A.D. (ed.) (1996) *Supramolecular Control of Structure and Reactivity*, Wiley, New York.

5 (a) Coolen, H.K.A.C., van Leeuwen, P.W.N.M. and Nolte, R.J.M. (1992) *Angewandte Chemie*, **104**, 906–909; (1992) *Angewandte Chemie – International Edition*, **31**, 905–907.

6 Reetz, M.T. and Waldvogel, S.R. (1997) *Angewandte Chemie*, **109**, 870–873; (1997) *Angewandte Chemie – International Edition*, **36**, 865–867.

7 (a) Breit, B. and Zahn, S.K. (1999) *Angewandte Chemie*, **111**, 1022–1024; (1999) *Angewandte Chemie – International Edition*, **38**, 969–971; (b) Breit, B., Dauber, M. and Harms, K. (1999) *Chemistry - A European Journal*, **5**, 2819–2827; (c) Breit, B. and Zahn, S.K. (2000) *Polyhedron*, **19**, 513–515; (d) Breit, B. (2000) *Chemistry - A European Journal*, **6**, 1519–1524; (e) Breit, B. and Zahn, S.K. (2001) *Angewandte Chemie*, **113**, 1964–1967; (2001) *Angewandte Chemie – International Edition*, **40**, 1910–1913; (f) Breit, B. and Demel, P. (2001) *Advanced Synthesis and Catalysis*, **343**, 429–432.

8 Ohkuma, T., Kitamura, M. and Noyori, R. (2000) *Asymmetric Hydrogenation*, in *Catalytic Asymmetric Synthesis* (ed. I. Ojima), Wiley-VCH, New York.

9 (a) Tschugajeff, L. and Skanawy-Grigorjewa, M. (1915) *Journal of the Russian Chemical Society*, **47**, 776; (b) Tschugajeff, L., Skanawy-Grigorjewa, M. and Posnjak, A. (1925) *Zeitschrift fur Anorganische und Allgemeine Chemie*, **148**, 37–42.

10 Fischer, E.O. and Maasböl, A. (1964) *Angewandte Chemie*, **76**, 645; (1964) *Angewandte Chemie – International Edition*, **3**, 580–581.

11 (a) Rouschias, G. and Shaw, B. (1970) *Journal of the Chemical Society. Chemical Communications*, 183; (b) Burke, A., Balch, A.L. and Enemark, J.H. (1970) *Journal of the American Chemical Society*, **92**, 2555–2557; (c) Wanniarachchi, Y.A. and Slaughter, L.M. (2007) *Chemical Communications*, 3294–3296; (d) Moncada, A.I., Manne, S., Tanski, J.M. and Slaughter, L.M. (2006) *Organometallics*, **25**, 491–505; (e) Stork, J.R., Olmstead, M.M., Fettinger, J.C. and Balch, A.L. (2006) *Inorganic Chemistry*, **45**, 849–857; (d) Stork, J.R., Olmstead, M.M. and Balch, A.L. (2004) *Inorganic Chemistry*, **43**, 7508–7515.

12 (a) Beck, W., Weigand, W., Nagel, U. and Schaal, M. (1984) *Angewandte Chemie*, **96**, 377–378; (1984) *Angewandte Chemie – International Edition*, **23**, 377–378; (b) Bär, E., Völkl, A., Beck, F., Fehlhammer, W.P. and Robert, A. (1986) *Journal of The Chemical Society-Dalton Transactions*, 863–868; (c) Kunz, R., Le Grel, P. and Fehlhammer, W.P. (1996) *Journal of The Chemical Society-Dalton Transactions*, 3231–3236; (d) Bertani, R., Mozzon, M. and Michelin, R.A. (1988) *Inorganic*

Chemistry, **27**, 2809–2815; (e) Bertani, R., Mozzon, M., Michelin, R.A., Benetollo, F., Bombieri, G., Castilho, T.J. and Pombeiro, A.J.L. (1991) *Inorganica Chimica Acta*, **189**, 175–187; (f)Beck, G. and Fehlhammer, W.P. (1988) *Angewandte Chemie*, **100**, 1391–1394; (1988) *Angewandte Chemie – International Edition*, **27**, 1344–1347; (g) Fehlhammer, W.P., Ahn, S. and Beck, G. (1991) *Journal of Organometallic Chemistry*, **411**, 181–191; (h) Fehlhammer, W.P. and Beck, G. (1989) *Journal of Organometallic Chemistry*, **369**, 105–116; (i) Liu, C.-Y., Chen, D.-Y., Lee, G.-H., Peng, S.-M. and Liu, S.-T. (1996) *Organometallics*, **15**, 1055–1061; (j) Ku, R.-Z., Chen, D.-Y., Lee, G.-H., Peng, S.-M. and Liu, S.-T. (1997) *Angewandte Chemie – International Edition*, **36**, 2631–2632, (1997) *Angewandte Chemie*, **109**, 2744–2746; (k) Motschi, H. and Angelici, R.J. (1982) *Organometallics*, **1**, 343–349; (l) Michelin, R.A., Zanotto, L., Braga, D., Sabatino, P. and Angelici, R.J. (1988) *Inorganic Chemistry*, **27**, 93–99; (m) Ito, Y., Hirao, T., Tsubata, K. and Saegusa, T. (1978) *Tetrahedron Letters*, 1535–1538; (n) Fehlhammer, W.P., Bliß, T., Fuchs, J. and Holzmann, G. (1992) *Zeitschrift fur Naturforschung*, **47b**, 79–89; (o) Ruiz, J., Garcia, G., Mosquera, M.E.G., Perandones, B.F., Gonzalo, M.P. and Vivanco, M. (2005) *Journal of the American Chemical Society*, **127**, 8584–8585; (p) Ruiz, J., Perandones, B.F., García, G. and Mosquera, M.E.G. (2007) *Organometallics*, **26**, 5687–5695; (q) Grundy, K.R. and Roper, W.R. (1975) *Journal of Organometallic Chemistry*, **91**, C61–C64; (r) Fehlhammer, W.P., Bartel, K. and Petri, W. (1975) *Journal of Organometallic Chemistry*, **87**, C34–C36; (s) Fehlhammer, W.P., Völkl, A., Plaia, U. and Beck, G. (1987) *Chemische Berichte*, **120**, 2031–2040; (t) Fehlhammer, W.P., Zinner, G., Beck, G. and Fuchs, J. (1989) *Journal of Organometallic Chemistry*, **379**, 277–288.

13 (a) Bartel, K. and Fehlhammer, W.P. (1974) *Angewandte Chemie*, **86**, 588–589; (1974) *Angewandte Chemie – International Edition*, **13**, 599–600; (b) Kernbach, U. and Fehlhammer, W.P. (1995) *Inorganica Chimica Acta*, **235**, 299–305; (c) Fehlhammer, W.P., Bartel, K., Plaia, U., Völkl, A. and Liu, A.T. (1985) *Chemische Berichte*, **118**, 2235–2254; (d) Plaia, U., Stolzenberg, H. and Fehlhammer, W.P. (1985) *Journal of the American Chemical Society*, **107**, 2171–2172.

14 Ferris, J.P., Antonucci, F.R. and Trimmer, R.W. (1973) *Journal of the American Chemical Society*, **95**, 919–920.

15 Jutzi, P. and Gilge, U. (1983) *Journal of Organometallic Chemistry*, **246**, 159–162.

16 (a) Hahn, F.E. (1993) *Angewandte Chemie*, **105**, 681–696; (1993) *Angewandte Chemie – International Edition*, **32**, 650–665.

17 (a) Hahn, F.E. and Tamm, M. (1993) *Journal of Organometallic Chemistry*, **456**, C11–C14; (b) Hahn, F.E., Hein, P. and Lügger, T. (2003) *Zeitschrift fur Anorganische und Allgemeine Chemie*, **629**, 1316–1321.

18 Tamm, M. and Hahn, F.E. (1999) *Inorganica Chimica Acta*, **288**, 47–52.

19 (a) Kernbach, U., Lügger, T., Hahn, F.E. and Fehlhammer, W.P. (1997) *Journal of Organometallic Chemistry*, **541**, 51–55; (b) Hahn, F.E. and Lügger, T. (1994) *Journal of Organometallic Chemistry*, **481**, 189–193.

20 Tamm, M., Lügger, T. and Hahn, F.E. (1996) *Organometallics*, **15**, 1251–1256.

21 Hahn, F.E. and Tamm, M. (1993) *Journal of the Chemical Society. Chemical Communications*, 842–844.

22 Hahn, F.E. and Imhof, L. (1997) *Organometallics*, **16**, 763–769.

23 Hahn, F.E. and Tamm, M. (1995) *Organometallics*, **14**, 2597–2600.

24 Hahn, F.E. and Tamm, M. (1995) *Journal of the Chemical Society. Chemical Communications*, 569–570.

25 (a) Tamm, M. and Hahn, F.E. (1999) *Coordination Chemistry Reviews*, **182**, 175–209; (b) Michelin, R.A., Pombeiro, A.J.L., Fátima, M. and Guedes da Silva, C. (2001) *Coordination Chemistry Reviews*, **218**, 75–112.

26 Hahn, F.E., Langenhahn, V., Meier, N., Lügger, T. and Fehlhammer, W.P. (2003) *Chemistry - A European Journal*, **9**, 704–712.

27 Staudinger, H. and Meyer, J. (1919) *Helvetica Chimica Acta*, **2**, 635–646.

28 Hahn, F.E., Wittenbecher, L., Boese, R. and Bläser, D. (1999) *Chemistry - A European Journal*, **5**, 1931–1935.

29 (a) Basato, M., Facchin, G., Michelin, R.A., Mozzon, M., Pugliese, S., Sgarbossa, P. and Tassan, A. (2003) *Inorganica Chimica Acta*, **356**, 349–356; (b) Basato, M., Benetollo, F., Facchin, G., Michelin, R.A., Mozzon, M., Pugliese, S., Sgarbossa, P., Sbovata, S.M. and Tassan, A. (2004) *Journal of Organometallic Chemistry*, **689**, 454–462.

30 Hahn, F.E., García Plumed, C. Münder, M. and Lügger, T. (2004) *Chemistry - A European Journal*, **10**, 6285–6293.

31 Hahn, F.E., Langenhahn, V. and Pape, T. (2005) *Chemical Communications*, 5390–5392.

32 (a) Hahn, F.E., Langenhahn, V., Lügger, T., Pape, T. and Le Van, D. (2005) *Angewandte Chemie*, **117**, 3825–3829; (2005) *Angewandte Chemie – International Edition*, **44**, 3759–3763; (b) Kaufhold, O., Stasch, A., Edwards, P.G. and Hahn, F.E. (2007) *Chemical Communications*, 1822–1824.

33 Meier, N., Hahn, F.E., Pape, T., Siering, C. and Waldvogel, S.R. (2007) *European Journal of Inorganic Chemistry*, 1210–1214.

34 Schalley C. (ed.) (2007) *Analytical Methods in Supramolecular Chemistry*, Wiley-VCH, Weinheim.

35 Fitting of the data was performed with SPECFIT, version 2.12, Spectrum Software Associates, Chapel Hill, NC, USA; Gampp, H., Maeder, M., Meyer, C.J. and Zuberbühler, A.D. (1986) *Talanta* **33**, 943 and references cited therein. The given error is based on the applied method. Multiple determinations gave a good reproducibility. Prolonged exposure of **13** to ambient conditions led to some decomposition. Employed conditions: 2×10^{-2} M **13** in CD_2Cl_2 treated with DMPU.

36 (a) Lewis, J.C., Wiedemann, S.H., Bergman, R.G. and Ellman, J.A. (2004) *Organic Letters*, **6**, 35–38; (b) Tan, K.L., Bergman, R.G. and Ellman, J.A. (2002) *Journal of the American Chemical Society*, **124**, 3202–3203; (c) Wiedemann, S.H., Lewis, J.C., Ellman, J.A. and Bergman, R.G. (2006) *Journal of the American Chemical Society*, **128**, 2452–2462.

37 Ruiz, J. and Perandones, B.F. (2007) *Journal of the American Chemical Society*, **129**, 9298–9299.

38 (a) Kunz, D. (2007) *Angewandte Chemie*, **119**, 3473–3476; (2007) *Angewandte Chemie – International Edition*, **46**, 3405–3408.

39 Jeffery, G.H. and Vogel, A.I. (1948) *Journal of the Chemical Society*, 658–673.

40 Fukuda, H. and Kitazume, T. (1997) *Heterocylces*, **46**, 275–285.

41 Juaristi, E. and Jiménez-Vázquez, H.A. (1991) *The Journal of Organic Chemistry*, **56**, 1623–1630.

42 Roux, M.-C., Paugam, R. and Rousseau, G. (2001) *The Journal of Organic Chemistry*, **66**, 4304–4310.

43 (a) Fang, F.G., Xie, S. and Lowery, M. (1994) *The Journal of Organic Chemistry*, **59**, 6142–6143; (b) Hulme, A.N. and Meyers, A.I. (1994) *The Journal of Organic Chemistry*, **66**, 952–953.

8
Functionalized Cycloheptatrienyl-Cyclopentadienyl Sandwich Complexes as Building Blocks in Metallo-supramolecular Chemistry

Matthias Tamm

8.1
Introduction

Since the serendipitous synthesis and discovery of ferrocene [1], its functionalization has become an important task in organometallic chemistry, and ferrocene-containing compounds are nowadays ubiquitous and indispensable to the development of research areas such as homogeneous catalysis and materials science [2, 3]. In stark contrast, the modification of other sandwich complexes is significantly less developed, and, for instance, little use has been made of sandwich complexes containing cycloheptatrienyl (Cht) ligands, although mixed cycloheptatrienyl-cyclopentadienyl (Cht-Cp) complexes of the type $[(\eta^7\text{-}C_7H_7)M(\eta^5\text{-}C_5H_5)]$ (M = group 4, 5 or 6 metal) have been known for more than three decades [4]. The mono- and diphosphines $[(\eta^7\text{-}C_7H_6PR_2)Ti(\eta^5\text{-}C_5H_5)]$ and $[(\eta^7\text{-}C_7H_6PR_2)Ti(\eta^5\text{-}C_5H_4PR_2)]$ (R = Me, Ph) are among the very few examples of functionalized Cht-Cp complexes, and they have been obtained by mono- or dilithiation of $[(\eta^7\text{-}C_7H_7)Ti(\eta^5\text{-}C_5H_5)]$ (troticene) followed by reaction with the respective chlorophosphine $ClPR_2$ [5, 6]. In a similar fashion, a number of functionalized 17-electron vanadium complexes have been obtained from $[(\eta^7\text{-}C_7H_7)V(\eta^5\text{-}C_5H_5)]$ (trovacene) and used to study the intermetallic communication (exchange coupling) between the resulting paramagnetic sandwich moieties [7]. Only recently, the interest in early-transition metal Cht-Cp complexes has become revitalized by independent reports from the groups of *Elschenbroich*, *Braunschweig* and *Tamm* on the preparation of *ansa*-Cht-Cp complexes (Figure 8.1), which could be obtained from the 16-, 17- and 18-electron sandwich compounds $[(\eta^7\text{-}C_7H_7)M(\eta^5\text{-}C_5H_5)]$ (M = Ti, V, Cr) [8–10]. It is the scope of this review to summarize the latest progress in this area, which has led to a number of new functionalized Cht-Cp sandwich complexes, and to demonstrate that there are in fact signs of life beyond ferrocene and other conventional metallocenes, if it comes to applications in materials science and homogeneous catalysis. Since the research contributed to this field from the author's group comprises mostly group 4 sandwich complexes, this article has its focus on the use of 16-electron Cht-Cp complexes containing the metals titanium, zirconium and hafnium.

Activating Unreactive Substrates: The Role of Secondary Interactions.
Edited by Carsten Bolm and F. Ekkehardt Hahn

ISBN: 978-3-527-31823-0

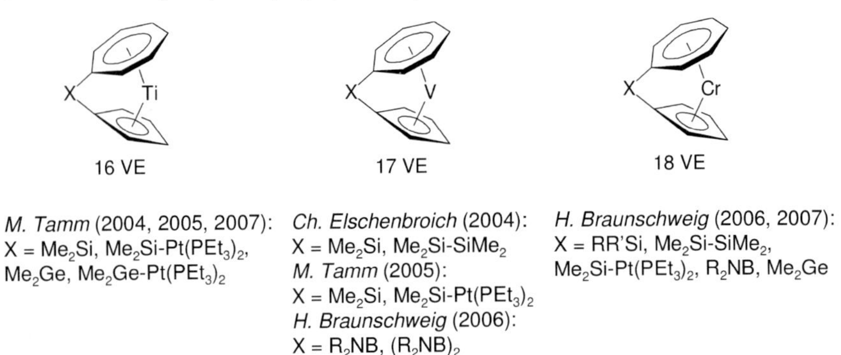

Figure 8.1 *Ansa*-Cycloheptatrienyl-cyclopentadienyl transition metal complexes.

8.2 Syntheses and Electronic Structures of Group 4 Cycloheptatrienyl-Cyclopentadienyl Sandwich Complexes

The group 4 sandwich complexes $[(\eta^7\text{-}C_7H_7)M(\eta^5\text{-}C_5H_5)]$ (M = Ti, **1**; Zr, **2**; Hf, **3**) can be prepared in moderate yields by reduction of $[(\eta^5\text{-}C_5H_5)MCl_3]$ with magnesium in the presence of an excess of cycloheptatriene, C_7H_8 (Scheme 8.1) [11–13]. After sublimation, the complexes are obtained as blue (**1**), purple (**2**) or orange-red crystalline solids (**3**), which have been characterized by means of X-ray diffraction analyses [13–15]. The three complexes, troticene (**1**), trozircene (**2**) and trohafcene (**3**) [16], are isostructural and crystallize in the orthorhombic space group *Pnma*. In each, the metal atom and one carbon in each ring reside on a crystallographic mirror plane. As these two carbon atoms adopt a cis orientation, the conformation of the two rings in **1**–**3** can be regarded as being perfectly eclipsed. The rings are virtually coplanar, and the centroid-metal-centroid angles are close to linearity (178.3° in **1**, 175.4° in **2**, and 176.6° in **3**).

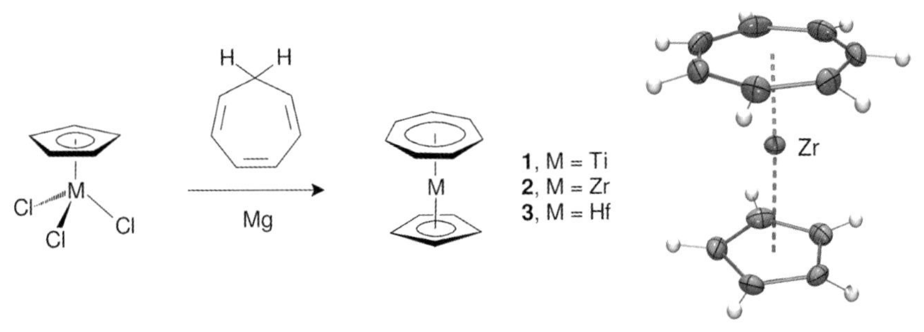

Scheme 8.1 Preparation of cycloheptatrienyl-cyclopentadienyl group 4 metal complexes and molecular structure of trozircene (**2**).

The bond lengths in **1–3** are summarized in Table 8.1. In all cases, the metal–carbon bonds to the seven-membered ring are significantly shorter than those to the five-membered ring, revealing a much stronger interaction between the metal center and the cycloheptatrienyl ring (see below). The differences in the metal–centroid distances are significantly more pronounced, which can be mainly attributed to the larger diameter of the seven-membered ring, naturally allowing the larger ring to more closely approach the metal atom. Unsurprisingly, substitution of titanium for zirconium and hafnium leads to considerably longer metal–carbon bond lengths. Also, the shortening of the metal–carbon distances upon going from trozircene (**2**) to trohafcene (**3**) is in agreement with the trend of the metal radii [17]. Related structural features have also been observed for the closely related complexes $[(\eta^7\text{-}C_7H_7)M(\eta^5\text{-}C_5Me_5)]$ (M = Ti, Zr, Hf) containing the pentamethylcyclopentadienyl (Cp*) ligand [18].

Table 8.1 Selected bond distances [Å] in complexes $[(\eta^7\text{-}C_7H_7)M(\eta^5\text{-}C_5H_5)]$ (M = Ti, Zr, Hf)

	M = Ti (**1**)	M = Zr (**2**)	M = Hf (**2**)
M-C_7	2.202(1)–2.2165(6)	2.323(2)–2.342(2)	2.293(3)–2.307(2)
M-C_7 (av.)	2.212	2.335	2.303
M-Ct_7 [a]	1.487	1.664	1.612
M-C_5	2.3213(8)–2.3375(8)	2.494(2)–2.504(2)	2.457(3)–2.471(2)
M-C_5 (av.)	2.333	2.501	2.466
M-Ct_5 [b]	1.988	2.195	2.151

[a] Ct_7 = centroid of the seven-membered ring.
[b] Ct_5 = centroid of the five-membered ring.

The bonding of cycloheptatrienyl ($\eta^7\text{-}C_7H_7$) rings to early transition metals has been the subject of extensive experimental and theoretical investigation [19–24]. Originally, this ligand was regarded as a coordinated aromatic tropylium ion, $[\eta^7\text{-}C_7H_7]^+$, and was thus classified as a six-electron donor with the assignment of a +1 formal charge in the resulting Cht metal complexes. The qualitative MO diagram in Figure 8.2 is in agreement with this assignment, and the relative energies of the metal d orbitals and of the Cp and Cht e_1 and e_2 orbitals have been chosen such that the d orbitals are higher in energy than the filled ligand e_1 orbitals and lower in energy than the ligand e_2 orbitals, which remain empty if the ligands are both considered as six-electron ligands, i.e. as Cp^- and Cht^+, respectively. From the Frost-Musulin diagrams shown on the left, it can easily be deduced that both the e_1 and e_2 orbitals of the Cht ligand are lower in energy than those of the Cp ligand. Consequently, their relative energies with regard to the metal d orbitals indicate that the ligand-to-metal π-donation can be expected to be much more pronounced for the Cp ligand, whereas metal-to-ligand δ-back-donation represents the most significant contribution to metal-Cht binding. For 16-electron group 4 metal complexes, the resulting $1e_2$ orbitals are filled with four electrons and represent the highest occupied molecular orbitals (HOMOs), whereas the $1a_1$ orbital, which should be essentially metal d_{z^2} in character, represents the lowest unoccupied molecular orbital (LUMO). In contrast, this orbital contains one or two electrons in Cht-Cp complexes of group 5 (M = V, Nb, Ta) or group 6 (M = Cr, Mo, W) transition metals, respectively.

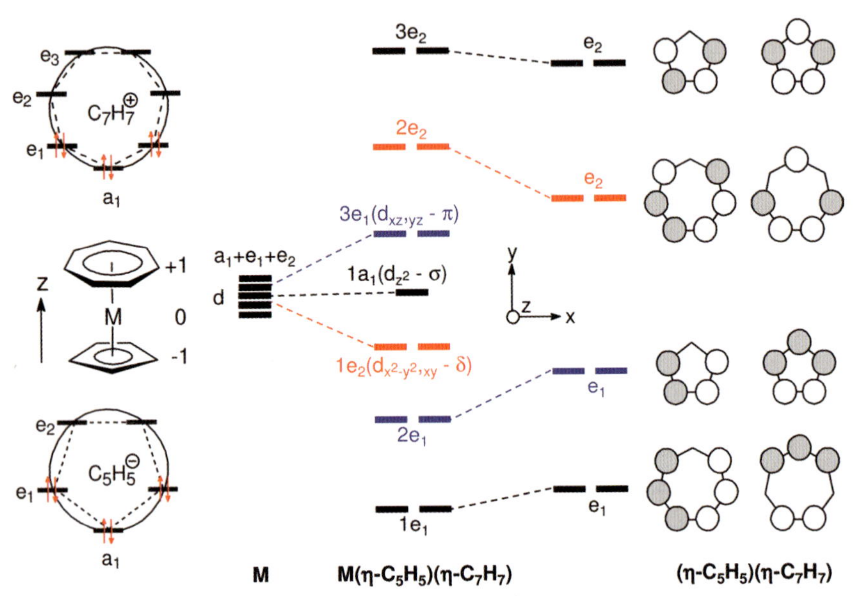

Figure 8.2 Qualitative frontier molecular orbital diagram for $[(\eta^7\text{-}C_7H_7)M(\eta^5\text{-}C_5H_5)]$; the symmetry labels refer to the idealized point group $C_{\infty v}$, in which infinite axes of rotation are assumed for the carbocyclic rings [23].

Increase of the d orbital energies could eventually result in the bonding situation depicted in Figure 8.3, where the Cht e_2 orbitals have become lower in energy than the metal d orbitals, so that the Cht-metal interaction must be formally regarded as ligand-to-metal δ-donation from a $[\eta^7\text{-}C_7H_7]^{3-}$ trianion, which would also satisfy the Hückel $4n + 2$ rule. Thereby, the ionic character of the Cp-metal π-interaction would become even pronounced. The decision whether the MO diagram in Figure 8.2 or that in Figure 8.3 is better suited for such formal bonding considerations can be derived from the relative metal d and Cht fragment contributions to the $1e_2$ frontier molecular orbitals. Table 8.2 summarizes the eigenvalues of the $1a_1$ (LUMO), $1e_2$ (HOMOs), $2e_1$ and $1e_1$ frontier orbitals in complexes **1–3** together with their metal, Cht and Cp fragment contributions [25]. For the HOMOs, the strong contributions from both the metal d and the Cht e_2 orbitals reveal a strongly covalent interaction with significant mixing of metal and ligand orbital character. Since the electrons in these orbitals are more ligand- than metal-localized, the bonding situation is in agreement with the representation in Figure 8.3 and thus consistent with assigning a −3 formal charge to the C_7H_7 ligand. This situation becomes even more pronounced upon going from Ti (Cht/metal fragment contribution of 58/40%) to Zr (64/34%) and

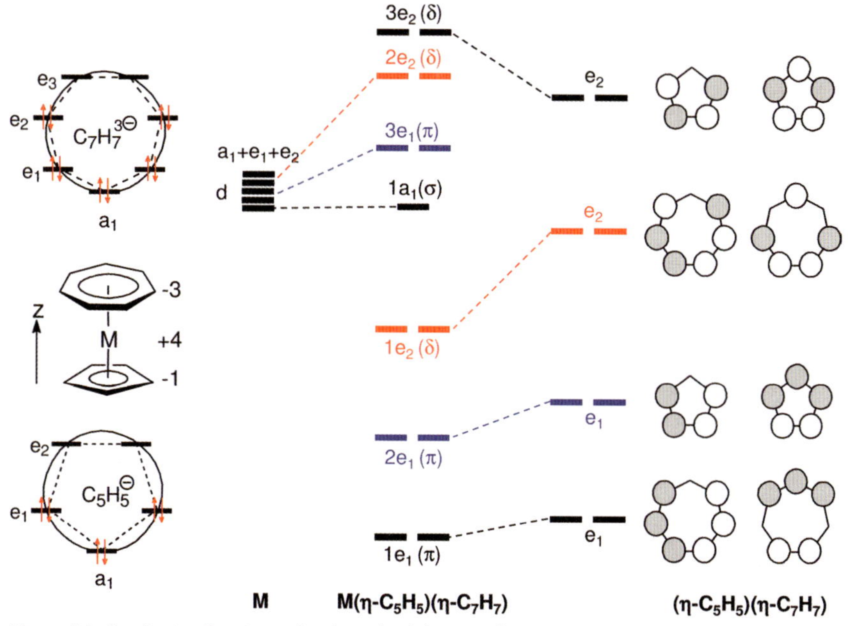

Figure 8.3 Qualitative frontier molecular orbital diagram for $[(\eta^7\text{-}C_7H_7)M(\eta^5\text{-}C_5H_5)]$; the symmetry labels refer to the idealized point group $C_{\infty v}$, in which infinite axes of rotation are assumed for the carbocyclic rings.

Table 8.2 Metal (M), Cht and Cp fragment contributions to the frontier orbitals in complexes $[(\eta^7\text{-}C_7H_7)M(\eta^5\text{-}C_5H_5)]$ (M = Ti, Zr, Hf).

Metal	Orbital	Symmetry[a)]	Eigenvalue (eV)	M (%)	Cht (%)	Cp (%)
Ti			−1.74	87	10	3
Zr	49 (LUMO)	$1a_1$	−1.58	90	7	3
Hf			−1.33	93	7	0
Ti			−4.99	40	58	2
Zr	48, 47 (HOMOs)	$1e_2$	−4.92	34	64	2
Hf			−4.86	32	66	2
Ti			−6.96	14	2	84
Zr	46, 45	$2e_1$	−6.93	9	6	85
Hf			−7.03	9	6	85
Ti			−8.31	10	89	1
Zr	44, 43	$1e_1$	−8.36	13	84	3
Hf			−8.33	12	85	3

a) The symmetry labels refer to the idealized point group $C_{\infty v}$, in which infinite axes of rotation are assumed for the carbocyclic rings.

Hf (66/32%). The next two levels, the $2e_1$ and $1e_1$ orbitals, correspond to the π-interaction with the Cp and Cht rings, respectively, and their predominant ligand orbital character reveals only a little mixing with the metal d orbitals. Accordingly, the interaction between the metal and the Cp ring, which lacks a δ-component, can be regarded as being mainly ionic. For the zirconium complex **2**, the contour plots of the relevant frontier molecular orbitals are given in Figure 8.4, nicely illustrating the δ- and π-symmetry of the metal-ring interactions.

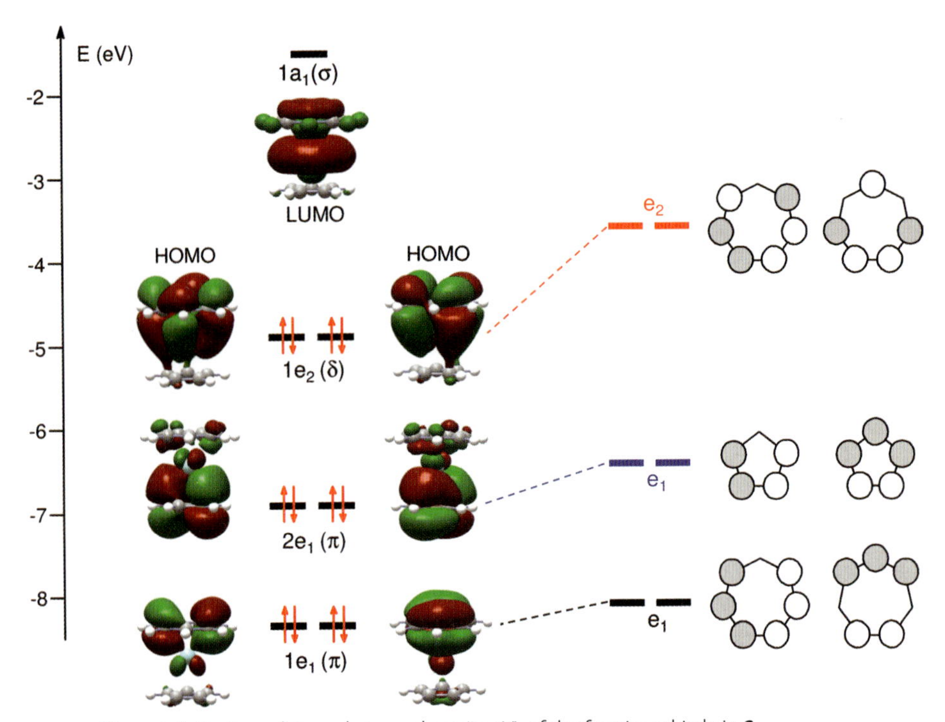

Figure 8.4 Contour plots and eigenvalues (in eV) of the frontier orbitals in **2**.

The $1a_1$ LUMO in complexes **1–3** is principally metal-localized and consists of the respective d_{z^2} orbital with small to marginal Cht and Cp contributions, in a coordinate system in which the metal–ring axis is defined as the *z* axis. The presence of an empty metal d-orbital in each complex suggests that these systems should be susceptible to the addition of a two-electron donor ligand. However, the coordination of additional ligands to the titanium atom has never been observed for unbridged troticene (**1**), whereas the formation of thermally labile phosphine adducts had been previously reported for related zirconium and hafnium indenyl complexes of the type $[(\eta^7\text{-}C_7H_7)M(\eta^5\text{-}C_9H_7)]$. Only for M = Hf could a dinuclear complex containing a bridging 1,2-bis(dimethylphosphino)ethane (dmpe) ligand be structurally characterized [26]. This Lewis-acidic behavior could be confirmed for trozircene (**2**), which forms isolable 1 : 1

complexes upon addition of alkyl and aryl isocyanides, 1,3,4,5-tetramethylimidazolin-2-ylidene (IMe) and trimethyl phosphine (PMe_3) (Scheme 8.2) [15, 27]. The isocyanide adducts **4** and **5** exhibit CN stretching vibrations at 2156 cm^{-1} (**4**) and 2134 cm^{-1} (**5**), which are only slightly shifted compared to the values for the free isocyanides. These observations indicate that metal-to-ligand π-back-bonding is weak and significantly less pronounced than in related zirconocene derivatives, in which the zirconium center is formally considered to be in the +II oxidation state. On the other hand, for d^0-configured Zr^{IV} complexes, higher values, even above 2200 cm^{-1}, have been observed [15].

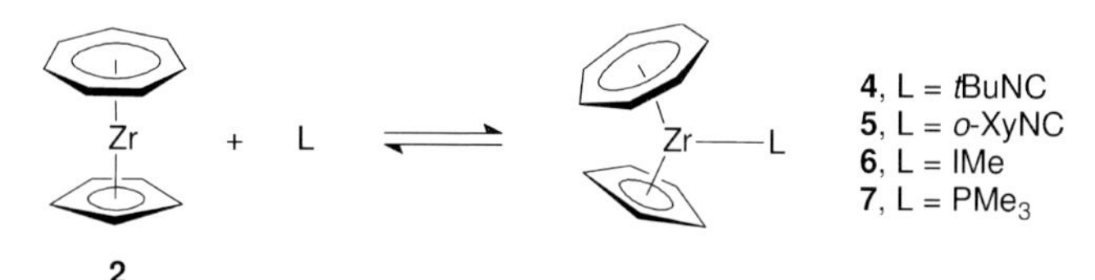

Scheme 8.2 Reaction of **2** with *tert*-butyl isocyanide (*t*BuNC), 2,6-dimethylphenyl isocyanide (*o*-XyNC), 1,3,4,5-tetramethylimidazolin-2-ylidene (IMe) and trimethyl phosphine (PMe_3).

Thermodynamic data for the reaction of **2** with *tert*-butyl isocyanide could be established by means of an NMR titration study [15], revealing that the reaction is only slightly exergonic with an average value of $\Delta G^\circ = -0.85 \pm 0.54$ kJ mol^{-1}. The average standard enthalpy for the formation of complex **4** and the associated entropy change amount to $\Delta H^\circ = -39.6 \pm 6.6$ kJ mol^{-1} and $\Delta S^\circ = -132.2 \pm 25.8$ J mol^{-1} K^{-1}, respectively, which clearly indicates a rather weak zirconium-isocyanide bond in comparison with conventional transition metal isocyanide complexes [28]. The experimental data are in excellent agreement with theoretical studies, and a calculated enthalpy of formation of $\Delta H^\circ = -36.5$ kJ mol^{-1} could be derived for **4**. The formation of the 2,6-dimethylphenyl isocyanide complex **5** is computed to be slightly more exothermic ($\Delta H^\circ = -38.2$ kJ mol^{-1}). *N*-Heterocyclic carbenes such as IMe are considered to be significantly stronger σ-donors than isocyanides, and accordingly the calculated reaction enthalpy of formation of carbene complex **6** ($\Delta H^\circ = -56.3$ kJ mol^{-1}) suggests a stronger interaction, which was also confirmed by NMR spectroscopy [27]. In contrast, the formation of the phosphine complex **7** is almost thermoneutral ($\Delta H^\circ = -2.3$ kJ mol^{-1}), which is also in accord with our inability to isolate this species from solution [27]. The molecular structures of **4** and **6** are shown in Figure 8.5. The Zr–C bond length of 2.376(3) Å in **4** is slightly shorter than that in **6** [2.445(2) Å]. Both values fall in the range previously observed for Zr^{IV} complexes containing isocyanide or *N*-heterocyclic carbene ligands. The angles at the metal defined by the ring centroids are 149.4° (**4**) and 147.5° (**6**), and it should be noted that this is significantly more acute than the angle δ observed in *ansa*-Cht-Cp titanium complexes (see below).

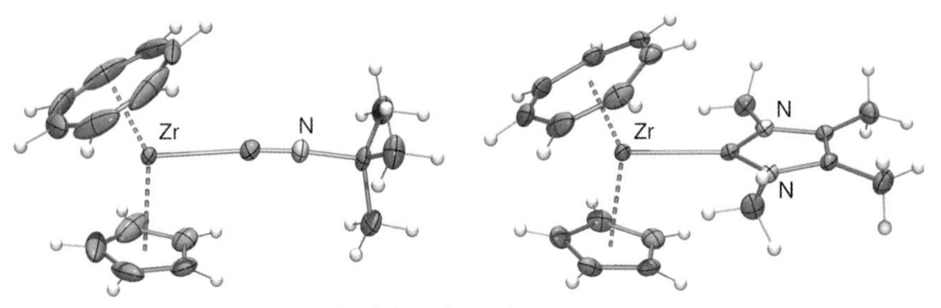

Figure 8.5 Molecular structures of **4** (left) and **6** (right).

8.3
Syntheses and Reactivity of *ansa*-Cycloheptatrienyl-Cyclopentadienyl Complexes

Double lithiation of troticene (**1**) can be achieved with *n*-butyllithium/*N,N,N′,N′*-tetramethylethylenediamine (tmeda) resulting in the formation of the tmeda-stabilized complex [(η^7-C_7H_6Li)Ti(η^5-C_5H_4Li)] [6]. The reaction of the dilithio complex with Me_2SiCl_2 or Me_2GeCl_2 afforded the sila- or germa[1]troticenophanes **8** and **9** as blue crystalline solids in moderate yield [8a, 8c]. The molecular structures of both complexes were established by X-ray diffraction analyses, and Scheme 8.3 shows an ORTEP presentation of the germanium derivative **9**. Despite the considerable strain imposed by the introduction of the Me_2E-bridges (E = Si, Ge), the Cht and Cp rings in **8** and **9** are virtually planar and can still be regarded as being essentially η^7- or η^5-coordinated, respectively. The deviation from an unstrained sandwich structure with an ideal coplanar ring orientation, as observed in troticene [14], can be characterized by the angles α, β, β', θ and δ, which are defined in Table 8.3. Since germanium has a larger covalent radius than silicon, one would expect the germa[1]troticenophane **9** to be less distorted than the sila[1]troticenophane **8**, and this is confirmed by the observation of a smaller tilt angle α (22.9° in **9** *versus* 24.1° in **8**) together with a larger angle δ (161.0° in **9** versus 160.5° in **8**) at the metal center defined by the ring centroids. The same trend has been observed for other structurally characterized Si/Ge couples, the ferrocenophanes **10**/**11** [29, 30], the chromarenophanes **12**/**13** [31, 32] and the trochrocenophanes **15**/**16** [10b,10d] (Figure 8.6, Table 8.3).

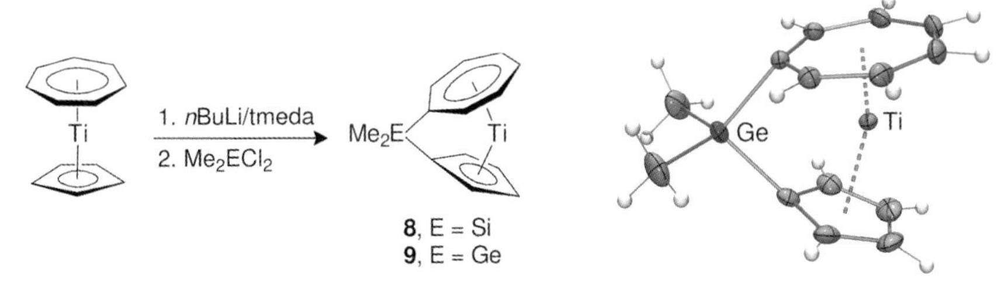

Scheme 8.3 Preparation of silicon- and germanium-bridged *ansa*-cycloheptatrienyl-cyclopentadienyl titanium complexes and molecular structure of **9**.

It can be clearly concluded that both **8** and **9** represent the most strongly distorted sandwich complexes in comparison to the silylene- and germylene-bridged derivatives shown in Figure 8.6. For instance, a pronounced decrease in strain is observed on going from the 16-electron sila[1]troticenophane **8** ($\alpha = 24.1°$, $\delta = 160.5°$) to the 17-electron sila[1]troticenophane **14** ($\alpha = 17.3°$, $\delta = 167.0°$) [9] and to the sila[1]-troticenophane **15** ($\alpha = 15.6°$, $\delta = 168.4°$) [10b], which is undoubtedly a consequence of the larger interannular distance in $[(\eta\text{-}C_7H_7)Ti(\eta\text{-}C_5H_5)]$ (troticene, 3.48 Å) compared to that in $[(\eta\text{-}C_7H_7)V(\eta\text{-}C_5H_5)]$ (trovacene, 3.38 Å) and in $[(\eta\text{-}C_7H_7)Cr(\eta\text{-}C_5H_5)]$ (trochrocene, 3.26 Å) [14a]. Large angles β and β' are observed for **8** (42.3°, 29.2°) and **9** (41.4°, 28.5°), indicating a strong distortion from planarity at the *ipso*-carbon atoms (Table 8.3). These values are only exceeded by the corresponding angles found in the vanadium and chromium congeners **14** (48.3° and 32.6°) and **15** (47.7° and 30.6°). This trend, however, needs not necessarily impose higher strain on these sandwich molecules in comparison to **8**, since a significant out-of-plane displacement of the substituents of the Cht ring can also be partially attributed to a reorientation of the large seven-membered ring for a better overlap with the smaller vanadium and chromium atoms [4, 33].

Table 8.3 Structural comparison of silylene- and germylene-bridged *ansa*-complexes.

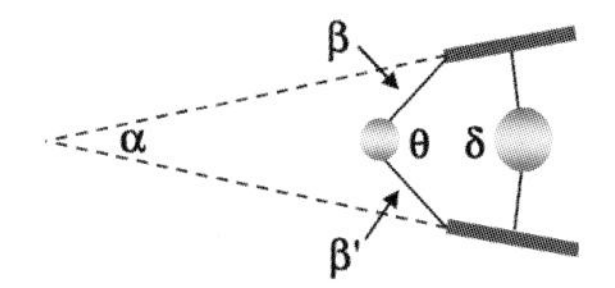

Compound	α/°	β/°, β'/°	θ/°	δ/°
8	24.1	42.3, 29.2	95.6	160.5
9	22.9	41.4, 28.5	92.8	161.0
10	20.8	37.0	95.7	164.7
11	19.1	37.8, 35.9	91.7	n/a
12	16.6	38.2, 37.9	92.9	167.6
13	14.4	38.7	91.8	n/a
14	17.3	48.3, 32.6	98.2	167.0
15	15.6	47.7, 30.6	93.9	168.4
16	15.1	43.8, 29.8	90.5	168.3

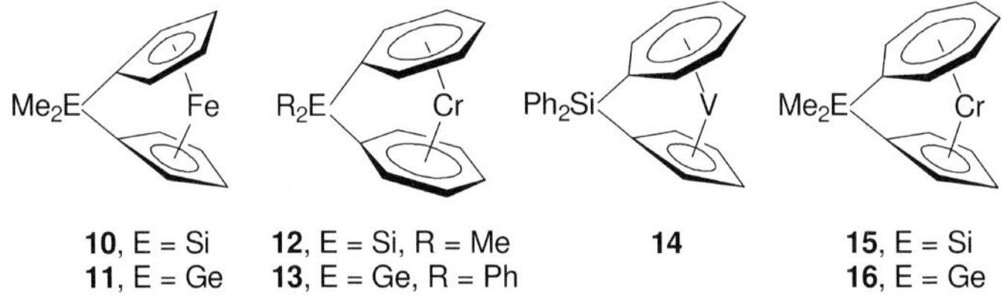

Figure 8.6 Selected silylene- and germylene-bridged *ansa*-complexes.

As mentioned before, coordination of additional ligands has never been observed for troticene (**1**). However, bridging and bending of the two rings in **8** and **9** creates a gap at the titanium atom, which might thereby become accessible to "slender" monodentate ligands such as carbon monoxide and isocyanides. To identify suitable frontier orbitals for such metal–ligand interactions, the structure of the silicon-bridged troticenophane **8** was optimized using DFT methods [8a]. Owing to the distortion of the symmetric sandwich structure, the highest occupied molecular orbitals in **8** have given up their degeneracy; their energy positions are nonetheless very close to those in **1**. A closer inspection (Figure 8.7 and Table 8.2) reveals that the introduction of the *ansa*-bridge leads to a slight increase of the HOMO-LUMO gap. Since the lowest energy band in the optical absorption spectrum of **1** has been assigned to a one-electron HOMO-LUMO transition, which is partly a d–d transition and partly a ligand-to-metal charge transfer (LMCT) [34], UV/Vis measurements could be used to quantify these effects: owing to a lowering of the symmetry and as a consequence of relaxation of the Laporte selection rule, a blue shift of the lowest energy band in the visible spectrum with enhanced intensity was observed (from $\lambda = 696$ nm and $\varepsilon = 31$ L mol^{-1} cm^{-1} in **1** to $\lambda = 663$ nm and $\varepsilon = 105$ L mol^{-1} cm^{-1} in **8**). It is worth noting that the introduction of an Si bridge in troticene has an opposite effect on the splitting and energy of the molecular orbitals to that observed for sila[1]ferrocenophanes and other [1]ferrocenophanes [3g,35].

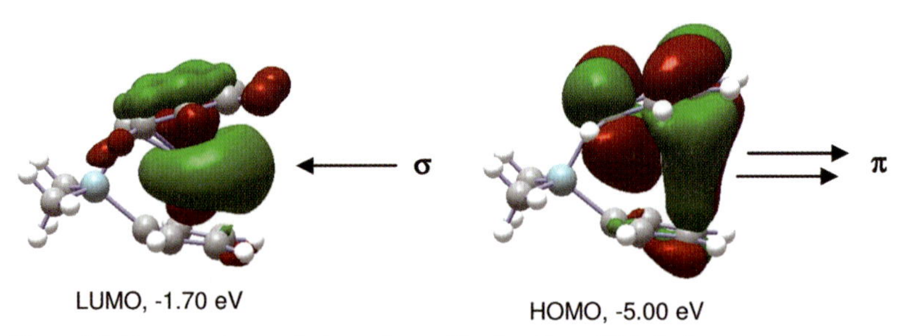

Figure 8.7 Contour plots and eigenvalues of the frontier orbitals in **8**.

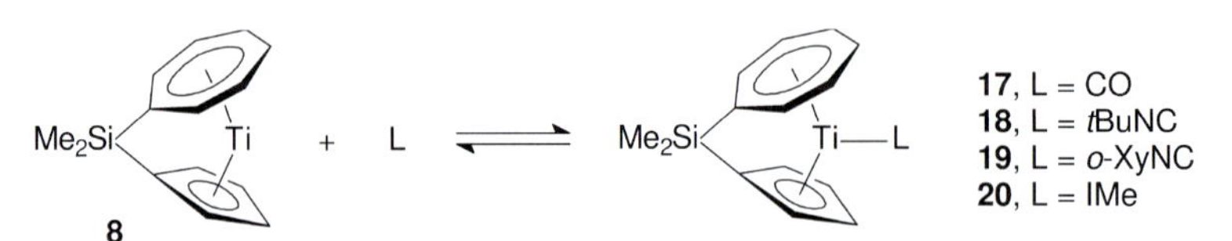

Scheme 8.4 Reaction of **8** with carbon monoxide (CO), *tert*-butyl isocyanide (*t*BuNC), 2,6-dimethylphenyl isocyanide (*o*-XyNC) and 1,3,4,5-tetramethylimidazolin-2-ylidene (IMe).

The electronic structure calculation of **8** reveals that this 16-electron complex contains a LUMO and a HOMO (Figure 8.7), which seem to be suitably oriented for σ- and π-interaction with one additional ligand. However, the formation of a stable carbonyl complex of type **17** (Scheme 8.4) could not be detected at ambient pressure. An NMR spectroscopic study of **8** under elevated CO pressure in the temperature range from $+20\,°C$ to $-70\,°C$ indicates that the metal–CO interaction is very weak and that CO is quickly exchanged on the NMR time scale [8a]. An exchange to more σ-donating, less π-accepting isocyanide ligands produced the adducts **18** and **19** as brown crystalline solids in nearly quantitative yields (Scheme 8.4). As also observed for the trozircene-isocyanide complexes **4** and **5** (Scheme 8.2), the CN stretching vibrations at $2153\,cm^{-1}$ (**18**) and at $2112\,cm^{-1}$ (**19**) are only slightly shifted compared to the values of the free isocyanides, again indicating only weak metal-to-ligand π-back-donation. In principle, it should be possible to quantify the interaction between **8** and the CO and isocyanide ligands and to produce thermodynamic and kinetic data for these equilibrium reactions by employing NMR titration techniques as described for the trozircene-*tert*-butyl isocyanide adduct **4** [15]. Figure 8.8 shows excerpts from the ^{1}H NMR spectra of **8** at $20\,°C$ with variation of the isocyanide concentration. Upon increase of the isocyanide concentration, the resonances for the ring hydrogen atoms are clearly shifted, although in different directions and to a different extent, depending on their position in the rings. Monitoring the change of the chemical shifts as a function of the isocyanide concentration and application of an NMR adaptation [36] of the Benesi–Hildebrand treatment, originally used in optical spectroscopic studies [37], allows us to establish the equilibrium contant K_C for a given temperature. Determination of K_C at different temperatures produces $\Delta H°$ and $\Delta S°$ from a van't Hoff plot of $\ln K_C$ versus $1/T$. Unfortunately, the data for the formation of **18** do not seem to be fully reliable, since the almost thermoneutral nature of this reaction ($\Delta H° = +6.7 \pm 3.4$ kJ mol^{-1}) leaves K_C hardly affected by variation of the temperature. However, it can be concluded that the interaction is actually very weak, which is also confirmed by the calculated enthalpy of formation: $\Delta H° = -14.5\,kJ\,mol^{-1}$.

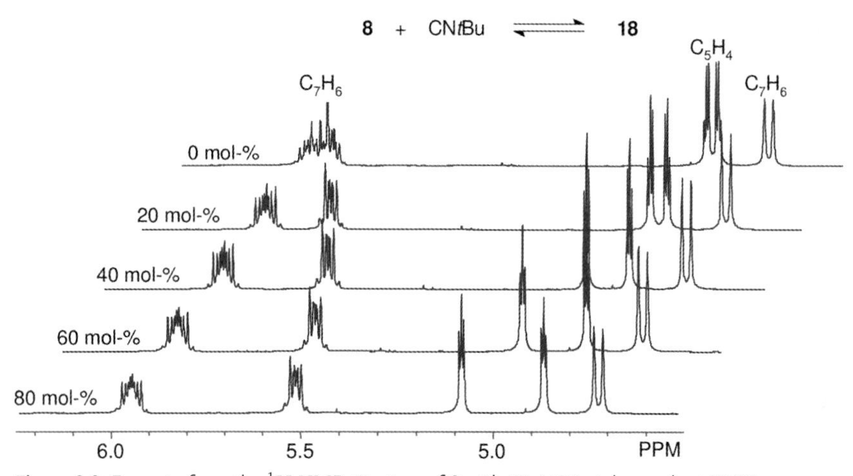

Figure 8.8 Excerpts from the ^{1}H NMR titration of **8** with *t*BuNC in toluene-d_8 at $20\,°C$.

The molecular structures of **8** and its *t*BuNC **18** adduct are shown in Figure 8.9. Coordination of the isocyanide ligand leads to pronounced elongation of the metal–carbon bonds, in particular to the seven-membered ring (from 2.170(3)–2.256(2) Å to 2.248(2)–2.412(3) Å). Hence, the distance between titanium and

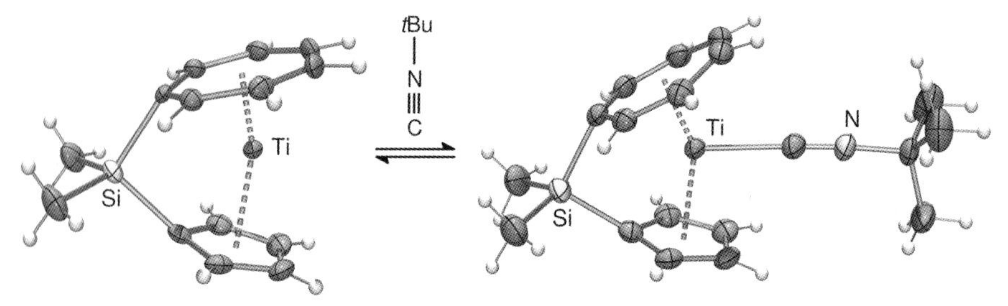

Figure 8.9 Molecular structures of **8** and **18**.

the centroid of the seven-membered ring increases from 1.496 Å to 1.649 Å. As a result, modified coordination is observed in the solid state, indicating a distortion from a symmetric η^7- towards an open η^5-bonding mode [38]. In agreement with the small Ti–isocyanide bond dissociation energy (see above), the Ti–CN*t*Bu distance is long (2.223(2) Å) and falls in the range observed for Ti^{+IV} complexes, where the d^0-electron configuration prevents effective metal-to-ligand back-donation. Despite its weakness, however, π-interaction can be held responsible for the observation that the 2,6-dimethylphenyl isocyanide ligand in **19** adopts a vertical conformation with a coplanar orientation toward the mirror plane including titanium (Figure 8.10), silicon and the centroids of the Cht and Cp rings. This orientation, which was also experimentally and theoretically confirmed for the corresponding trozircene *o*-XyNC complex **5** [15], is electronically favorable, since it allows the alignment of the LUMO of the isocyanide ligand in an antisymmetric fashion with respect to the mirror plane in order to optimally interact with the HOMO of the troticenophane **8** (Figure 8.7). In contrast, a horizontal conformation is observed for complex **20**, which contains an imidazolin-2-ylidene ligand with stronger σ-donor and weaker π-acceptor characteristics (Figure 8.10). As expected, the calculated enthalpy of formation of $\Delta H^\circ = 27.8\,kJ\,mol^{-1}$ indicates a stronger metal–ligand interaction than in **18**.

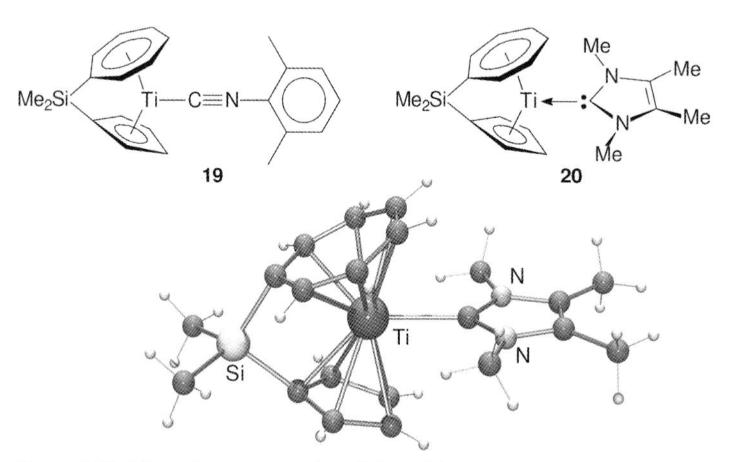

Figure 8.10 Schematic representation of the solid-state conformations of **19** and **20** (top); molecular structure of **20** (below).

In view of the similar reactivities of sila[1]troticenophane (**8**) and trozircene (**2**) toward carbon monoxide or isocyanides, it can be concluded that these 16-electron complexes have only a small propensity to efficiently interact with σ-donor/π-acceptor ligands. In this respect, **8** and **2** do not act like complexes containing titanium or zirconium in a lower oxidation state but rather bear a closer resemblance to Lewis-acidic Ti^{+IV} and Zr^{+IV} complexes, respectively. Based on theoretical calculations, this behavior can be mainly attributed to the strong and appreciably covalent metal–cycloheptatrienyl interaction leading to highly stabilized frontier orbitals and consequently to a diminishing π-electron release capability. Therefore, these experimental results support the conclusion that the cycloheptatrienyl ring functions more as a −3 ligand than as a +1 ligand in these mixed-ring complexes [24], a description which was also found to be valid for cycloheptatrienyl sandwich compounds of actinides [39].

8.4 Ring-opening Reactions of *ansa*-Cycloheptatrienyl-Cyclopentadienyl Complexes

Poly(ferrocenes) represent the most important class of transition metal-containing macromolecules and are playing an important role in the development of novel polymers with intriguing structural, conductive, magnetic, optical, or redox properties [3]. In most cases, these polymers are accessed by ring-opening polymerization (ROP) of strained [1]ferrocenophanes such as **10** (Scheme 8.5), which can be initiated thermally or by anionic or transition metal catalysis. In contrast to the Cp-Si-Cp linked poly(ferrocenylsilane) obtained from **10**, the ROP of the troticenophanes **8** and **9** is expected to give an irregular polymer, which contains all possible Cp-X-Cp, Cht-X-Cht and Cp-X-Cht ($X = Me_2Si$, Me_2Ge) linkages (Figure 8.11). In fact, it could be demonstrated that **8** exothermically polymerizes at about 170 °C, whereas **9** ring-opens at a lower temperature of about 130 °C. This different behavior can be attributed to the weaker germanium–carbon bond in **9** compared to the silicon–carbon bond in **8**. The strain energies of both molecules were estimated to be approximately 45 kJ mol^{-1} by means of differential scanning calorimetry (DSC) studies [8c], and Figure 8.11 shows the DSC trace of **8**.

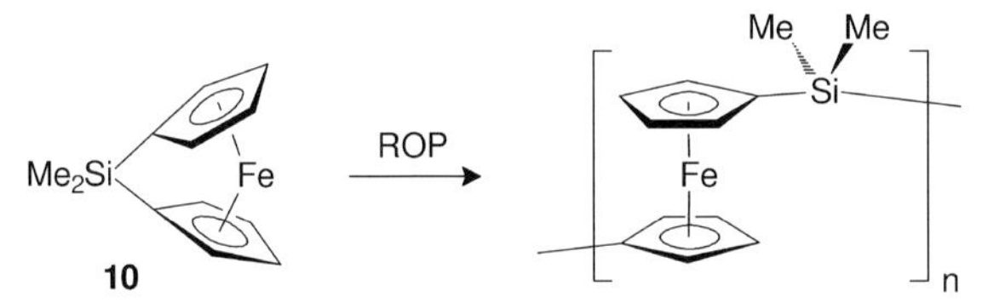

Scheme 8.5 Ring-opening polymerization of sila[1]ferrocenophane **10**.

Interestingly, the strain energies of **8** and **9** are substantially smaller than the values reported for related [1]ferrocenophanes, e.g. 80 kJ mol^{-1} for **10** [40], although the structural parameters of **8** and **9** (Table 8.3) suggest that these molecules are strongly distorted from an unstrained sandwich structure. An explanation could be derived from a theoretical comparison of the energy content of ferrocene (Cp_2Fe) and troticene (**1**) as a function of the bending angle δ (Figure 8.12). As expected, bending of both sandwich molecules results in an exponential increase in energy [41]. For ferrocene, however, a significantly steeper rise in energy is observed upon lowering δ, and it can be clearly deduced that distortion of the [(η-C_5H_5)$_2$Fe] moiety requires significantly more energy than that of [(η^7-C_7H_7)Ti(η^5-C_5H_5)]. As a consequence, the energy released by conversion of a strained into an unstrained molecule is in fact expected to be higher for ferrocenophanes than for troticenophanes.

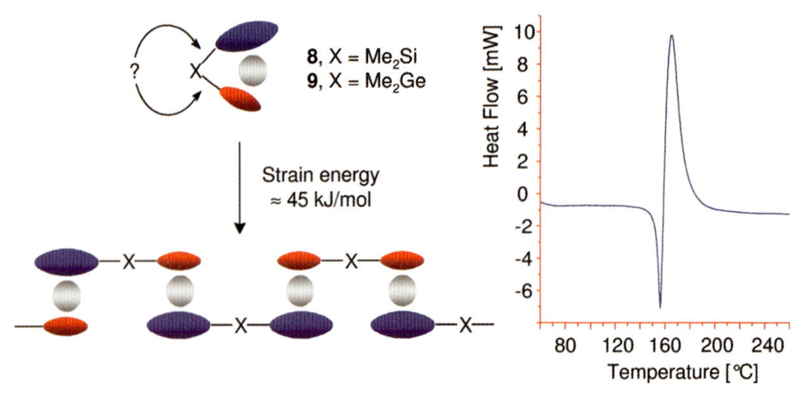

Figure 8.11 Schematic presentation of the ring-opening polymerization of **8** and **9** (blue = Cht, red = Cp, grey = Ti); DSC thermogram of **8** with a melt endotherm at 155 °C and a polymerization exotherm at 170 °C.

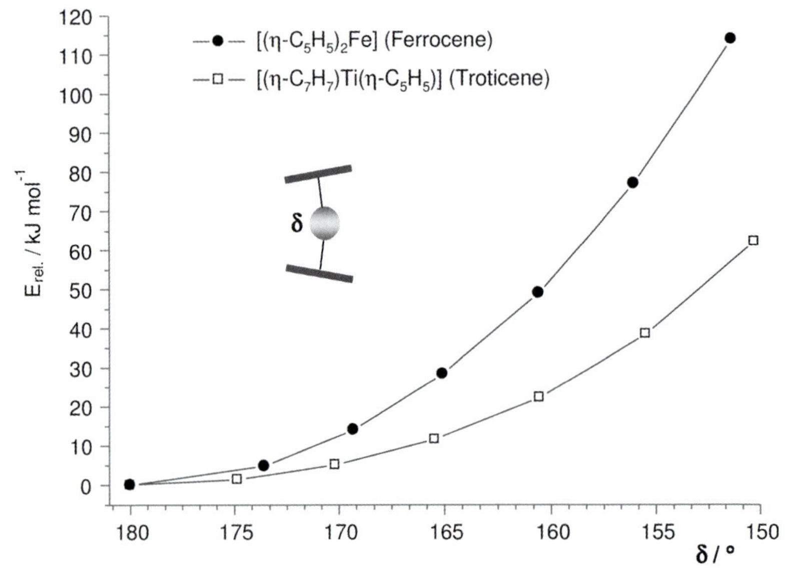

Figure 8.12 Relative potential energy for ferrocene and troticene as a function of the angle δ.

Since the thermal ring-opening polymerization of *ansa*-Cht-Cp complexes leads to the formation of irregular polymers (Figure 8.11), the possibility of regioselective Si–C bond cleavage in **8** was investigated. Treatment of **8** with ethereal HCl exclusively affords the troticene derivative **21** with a chlorodimethylsilyl substituent attached to the Cp ring [8c], indicating that a regioselective protonolysis of the silicon–carbon bond to the seven-membered ring must have occurred. In a similar fashion, the reaction with ethereal HBF_4 produces the fluorosilane **23**. The reaction presumably proceeds via a highly reactive silyl cation intermediate **22**, which is able to abstract F^- from the tetrafluoroborate anion. It should be noted that this mechanism had already previously been established for the same reaction employing sila[1]ferrocenophanes such as the dimethylsilyl-bridged derivative **10** [42]. The regioselective Si–C bond cleavage and exclusive protonation of the Cht ring could have been anticipated from the structural constraints in **8** (Table 8.3), where the unusually large angle β (42.3°) at the Si-C_7H_6 site implies that this Si–C bond or rather the *ipso*-C_7H_6 carbon atom represents the most highly strained part of this molecule (Scheme 8.6).

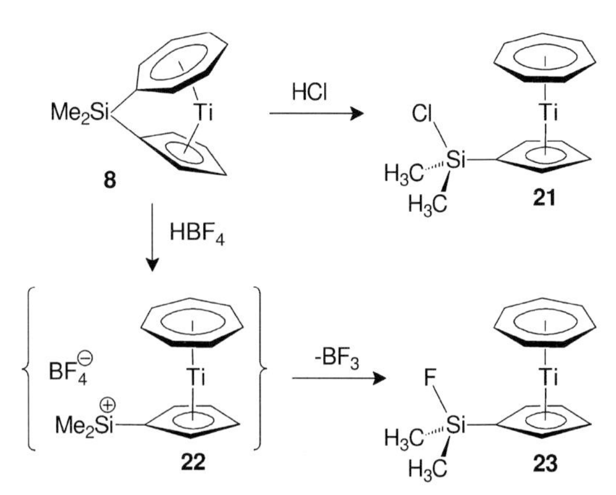

Scheme 8.6 Acid-induced ring opening of **8**.

Regioselective Si–C bond cleavage could also be observed on treatment of **8** with equivalent amounts of the tris(triethylphosphine)platinum(0), which results in the formation of the platinasila[2]troticenophane **(24)** by oxidative addition and insertion of a $[Pt(PEt_3)_2]$ moiety into the Si–C bond to the seven-membered ring (Scheme 8.7) [8b, 8c]. In a similar fashion, the analogous vanadium and chromium complexes **26** and **27** could be isolated from the respective sila[1]trovacenophane **24** [8b] and sila[1]trochrocenophane **15** [10b]. The molecular structures of **25–27** were established by X-ray diffraction analyses, and Scheme 8.7 shows the structure of the Ti derivative **25**. All three compounds are isotypic and crystallize in the orthorhombic space group $P2_12_12_1$. Selected bond lengths and angles are summarized in Table 8.4. As expected from the crystal structures of the parent complexes $[(\eta^7\text{-}C_7H_7)M(\eta^5\text{-}C_5H_5)]$ (M = Ti, V, Cr) [14a], the metal–carbon bond distances

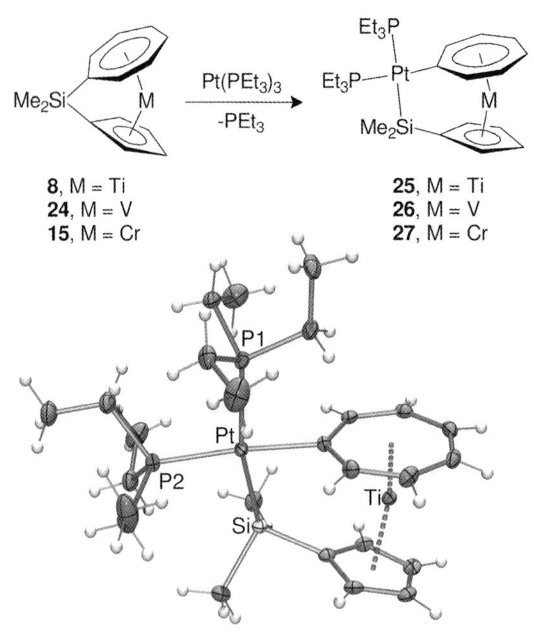

Scheme 8.7 Synthesis of Si–Pt-bridged *ansa*-Cht-Cp complexes.

Table 8.4 Selected bond distances [Å] and angles [°] in complexes of the type $[(PEt_3)_2PtSiMe_2(\eta^7\text{-}C_7H_6)M(\eta^5\text{-}C_5H_4)]$.

	M = Ti (25)	M = V (26)	M = Cr (27)
M–C_7	2.184(3)–2.216(3)	2.168(3)–2.192(3)	2.141(4)–2.173(4)
M–C_7 (av.)	2.203	2.184	2.149
M–Ct_7[a)]	1.468	1.446	1.411
M–C_5	2.280(3)–2.333(3)	2.209(3)–2.274(3)	2.150(4)–2.201(4)
M–C_5 (av.)	2.308	2.242	2.174
M–Ct_5[b)]	1.967	1.888	1.808
Pt–C	2.096(3)	2.092(2)	2.087(4)
Pt–P1	2.3874(6)	2.3849(6)	2.3745(10)
Pt–P2	2.2962(9)	2.3020(7)	2.3057(10)
Pt–Si	2.3881(8)	2.3900(6)	2.4229(13)
α	13.5	10.6	7.5
δ	169.1	171.9	174.9

a) Ct_7 = centroid of the seven-membered ring.
b) Ct_5 = centroid of the five-membered ring.

decrease in the order Ti > V > Cr. Accordingly, **25** shows the strongest deviation from an unstrained sandwich structure with $\alpha = 13.5°$ and $\delta = 169.1°$. Smaller angles α together with larger angles δ are consequently observed for the vanadium ($\alpha = 10.6°$, $\delta = 171.9$) and chromium analogs ($\alpha = 7.5°$, $\delta = 174.9°$). In all complexes, the platinum centers are in a slightly distorted square-planar environment with two different PEt_3 ligands. Because of the strong trans influence of the silyl substituents, the Pt–P1 distances are significantly longer than Pt–P2. These distances and also the Pt–C and Pt–Si bond lengths are in reasonable agreement with the values obtained from structural characterization of related platinasila[2]ferrocenophanes [43]. Finally, it should be noted that Pt insertion could also be observed for the germa[1] troticenophane **9**, resulting in the formation of a metallocycle with a rare Ge–Pt bond [8c].

The Pt–Si-bridged derivative **25** could be used as a single-source catalyst for the metal-catalyzed ROP of the original strained molecule **8**, and the reaction of **25** with an excess of **8** at elevated temperature led to the formation of regioregular poly(troticenylsilanes), in which the Cp-Cht-sandwich moieties are exclusively linked via Cp-Si-Cht bridges [8b]. In analogy to the metal-catalyzed polymerization of ferrocenophanes such as **10** and **11** (Figure 8.6) [43, 44], propagation may proceed by sequences of oxidative-addition and reductive-elimination processes involving the Si–C bonds or via σ-bond metathesis of the Pt–Si and Si–C bonds. Reductive elimination from the intermediate platinacyclic oligomers **28** eventually leads to cyclic oligotroticenes **29** (Scheme 8.8). MALDI-TOF mass spectrometric characterization produced molecular peaks for oligomers **29** between $m/z = 1301$ ($n = 5$) and 5985 ($n = 23$). Unfortunately, the high reactivity of the mixture **29** precluded the isolation and purification of specific oligomers in a similar manner to that recently described for dimeric, pentameric and hexameric oligo(ferrocenylsilanes) [45]. The calculated structure of a corresponding hexa(troticenylsilane) of type **29** is shown in Scheme 8.8.

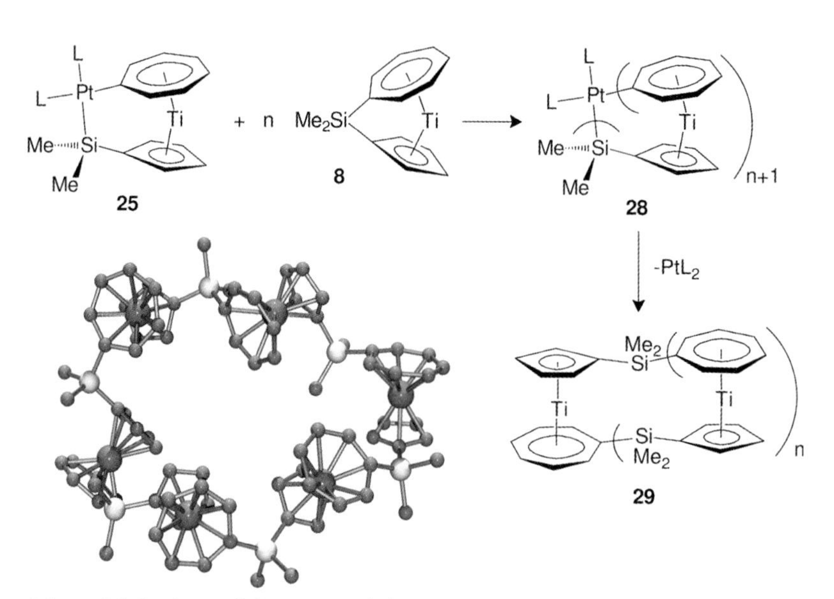

Scheme 8.8 Synthesis of oligo(troticenylsilanes).

8.5
Phosphine-functionalized Cycloheptatrienyl-Cyclopentadienyl Sandwich Complexes

In contrast to ferrocenylphosphines, P-functionalized Cht-Cp sandwich complexes are scarce, and the mono- and diphosphines [(η^7-$C_7H_6PR_2$)Ti(η^5-C_5H_5)] (**30**) and [(η^7-$C_7H_6PR_2$)Ti(η^5-$C_5H_4PR_2$)] (**31**) (R = Me, Ph) are among the very few examples of functionalized Cht-Cp complexes; they have been obtained by mono- or dilithiation of [(η^7-C_7H_7)Ti(η^5-C_5H_5)] (troticene) followed by reaction with the respective chlorophosphine $ClPR_2$ (Scheme 8.9). Several transition metal complexes have been synthesized, in which **30** or **31** act as mono- or bidentate phosphine ligands, respectively [5, 6]. Taking into account the different reactivity of the Cht-Cp complexes **1** – **3** towards σ-donor/π-acceptor ligands, it could be assumed that related phosphorus-functionalized zirconium and hafnium complexes should feature intermolecular metal-phosphine contacts (see above) [26]. Since the lithiation of trozircene (**2**) and trohafcene (**3**) with *n*-BuLi/tmeda in an analogous manner to troticene (**1**) proved unsuccessful, we aimed toward the syntheses of the monophosphine derivatives [(η^7-C_7H_7)M(η^5-$C_5H_4PR_2$)] (M = Ti, **35**; Zr, **36**; Hf, **37**). The titanium complexes **35** could be obtained from the reduction of [(η^5-$C_5H_4PR_2$)$TiCl_3$] [46] with magnesium in the presence of cycloheptatriene, whereas the corresponding zirconium and hafnium complexes required the use of the metallocenes [(η^5-$C_5H_4PR_2$)$_2MCl_2$] **33** (M = Zr) and **34** (M = Hf) [47] with loss of one Cp ligand (Scheme 8.10) [48].

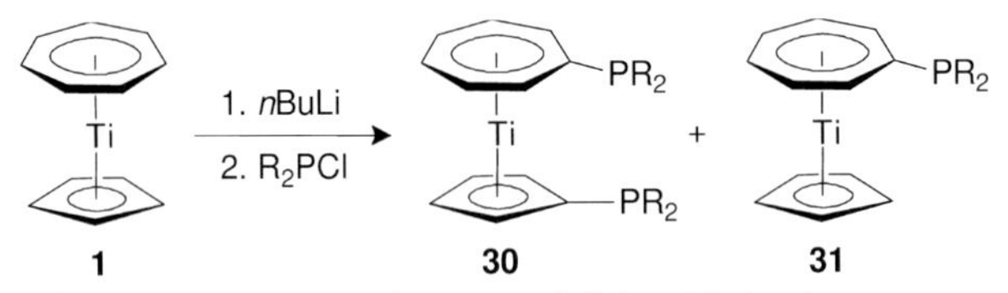

Scheme 8.9 Preparation of mono- and diphosphinyltroticenes.

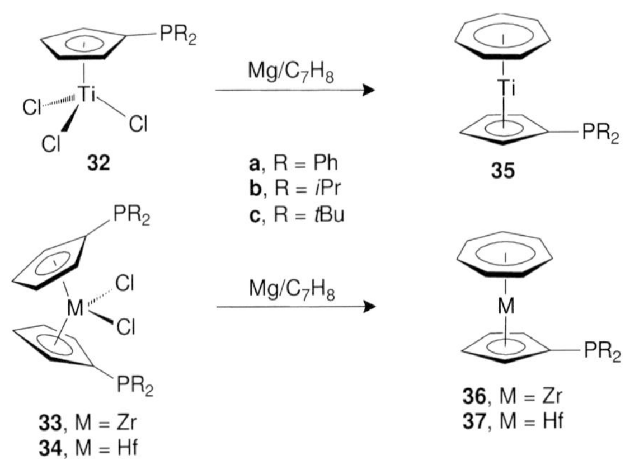

Scheme 8.10 Preparation of Cp-functionalized Cht-Cp sandwich complexes.

The molecular structures of **35a**, **35c**, **36a**, **36b** and **37a** were established by means of X-ray diffraction analyses, and Figure 8.13 shows the molecular structures of **35a** and **36a** as representative examples. As expected, no Ti–P contacts could be observed in the solid state for the troticenes **35**, and the Cht and Cp rings remain virtually coplanar. In contrast, the formation of dimeric structures with Zr–P and Hf–P contacts is observed for the trozircenes **36** and trohafcenes **37**. In all cases, the dimerization leads to centrosymmetric metallacycles with a M-P-Ct_5-M-P-Ct_5 sequence (Ct_5 = centroid of the Cp ring; M = Zr, Hf), and the six-membered rings adopt undistorted chair conformations. The metal–phosphorus distances are very long, e.g. Zr–P = 2.9833(3) Å in **36b**, indicating weak interactions as expected from the study of the PMe_3 adduct **7** (see above) [27]. In agreement with the trend of the metal radii [17], the Zr–P bond of 2.9305(4) Å in **36a** is longer than the Hf–P bond of 2.8034(6) Å in **37a**.

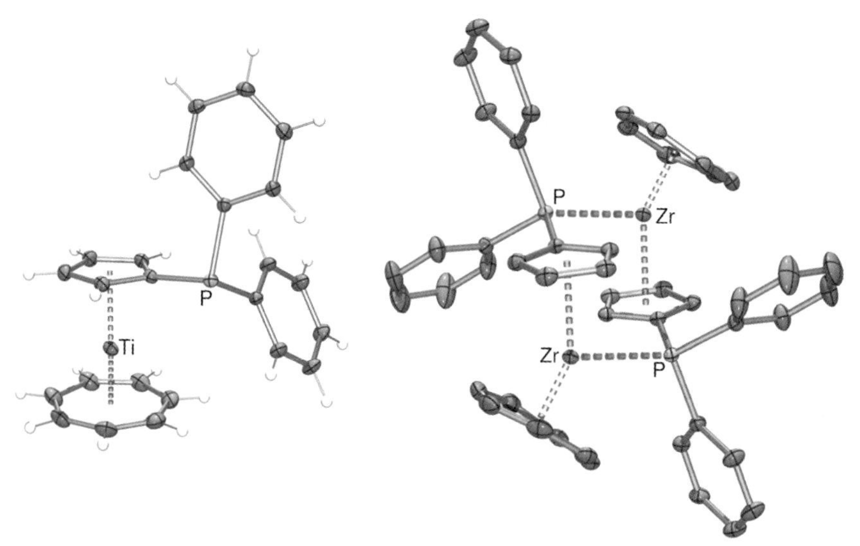

Figure 8.13 Molecular structures of **35a** (left) and **36a** (right).

The presence of weak P–Zr and P–Hf bonds in **36** and **37** suggests that these dimers can easily be cleaved and used for the synthesis of transition metal–phosphine complexes, which could exhibit interesting secondary interactions due to the presence of a Lewis-acidic Cht-Cp metal site. In fact, any structurally characterized bimetallic complex obtained from reactions of **36** or **37** features an unusual intramolecular interaction involving the zirconium or hafnium atoms, respectively. Scheme 8.11 shows two representative examples derived from **36b**. For instance, the reaction with dimeric $[M(\eta^4\text{-}C_8H_{12})Cl]_2$ (C_8H_{12} = 1,5-cyclooctadiene; M = Rh, Ir) furnishes complexes **38** with long Zr–Cl distances of 2.7976(4) Å (M = Rh) and 2.8265(6) Å (M = Ir). A particularly interesting complex can be isolated from the reaction of **36b** with $[Pd(\eta^4\text{-}C_8H_{12})Cl_2]$, which leads to reduction of Pd(II) and formation of the diphosphine-Pd(0) complex **39**. The molecular structure reveals a T-shaped palladium complex with an unprecedented Pd–Zr bond of 2.9709(3) Å (Figure 8.14). The angles at palladium are P1-Pd-P2 = 131.83(2)°, P1-Pd-Zr1 73.652(11)° and P2-Pd-Zr1 = 152.935(11)°, indicating a strong deviation from a linear arrangement normally observed for dicoordinate structures of the type $[Pd(0)(PR_3)_2]$; for instance P-Pd-P = 180° in $[Pd(FcPtBu_2)_2]$, in which the di-*tert*-butylferrocenylphosphine ligand ($FcPtBu_2$) is incapable of developing a secondary Fe–Pd interaction [49].

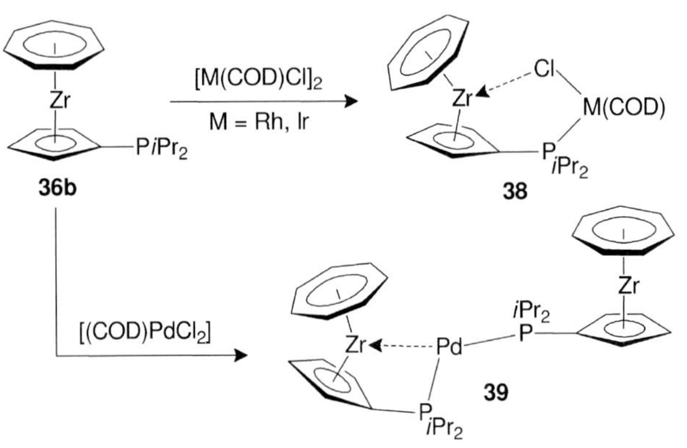

Scheme 8.11 Synthesis of trozircenylphosphine complexes featuring secondary interactions.

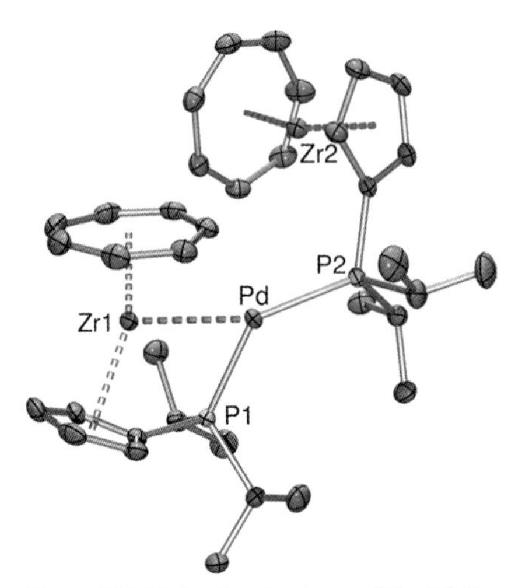

Figure 8.14 Molecular structures of the Pd–Zr complex **39**.

With these complexes, we have finally arrived at the heart of the focus program SPP 1118 "Secondary Interactions as a Steering Principle for the Selective Functionalization of Non-Reactive Substrates", and the secondary interactions available in complexes containing the phosphines **36** and related species will certainly have an impact on the activation of substrates, for instance by oxidative addition to complexes such as **39**. A number of related complexes await their publication [48], and the full potential of the zirconium- and hafnium-phosphine ligands **36** and **37** as ligands for applications in homogeneous catalysis needs yet to be investigated. Thereby, the Cht-Cp metal moiety can be regarded as a pendant Lewis-acidic acceptor ligand in a similar fashion to that described for ambiphilic phosphine ligands featuring pendant borane and alane moieties, which have also been investigated as promising candidates for organotransition metal catalysis, notably via intramolecular activation of M–X bonds [50].

Acknowledgments

This work was supported by the Deutsche Forschungsgemeinschaft through the focus program SPP 1118 "Secondary Interactions as a Steering Principle for the Selective Functionalization of Non-Reactive Substrates". I wish to thank all my coworkers, in particular Dr. Andreas Kunst, Dipl.-Chem. Susanne Büschel and Dr. Thomas Bannenberg, who have made the most significant contributions to the results presented in this overview.

References

1 Special Issue: 50th Anniversary of the Discovery of Ferrocene, Adams, R.D. Ed. (2001) *Journal of Organometallic Chemistry*, **637–639**, 1.

2 Selected reviews: (a) Togni, A. and Hayashi, T. (eds) (1995) *Ferrocenes*, VCH, Weinheim. (b) Bandoli, G. and Dolmella, A. (2000) *Coordination Chemistry Reviews*, **209**, 161. (c) Siemeling, U. and Auch, T.-C. (2005) *Chemical Society Reviews*, **34**, 284. (d) Arrayás, R.G., Adrio, J. and Carretero, J.C. (2006) *Angewandte Chemie – International Edition*, **45**, 7674. (e) Osakada, K., Sakano, T., Horie, M. and Suzaki, Y. (2006) *Coordination Chemistry Reviews*, **250**, 1012. (f) Colacot, T.J. (2003) *Chemical Reviews*, **103**, 3101. (g) Štěpnička, P. (2005) *European Journal of Inorganic Chemistry*, 3787. (h) Atkinson, R.C.J., Gibson, V.C. and Long, N.J. (2004) *Chemical Society Reviews*, **33**, 313.

3 Selected reviews: (a) Chandrasekhar, V. (2005) *Inorganic and Organometallic Polymers*, Springer-Verlag GmbH & Co. KG, Heidelberg. (b) Nguyen, P., Gomez-Elipe, P. and Manners, I. (1999) *Chemical Reviews*, **99**, 1515. (c) Manner, I. (2004) *Synthetic Metal Containing Polymers*, Wiley-VCH, Weinheim. (d) Abd-el-Aziz, A.S. Carraher, C.E., Jr., Pittman, C.U., Jr., Sheats, J.E. and Zeldin, M. (eds) (2004) *Macromolecules Containing Metal and Metal-Like Elements*, vol. **1–3**, Wiley, New York. (e) Archer, R.D. (2001). *Inorganic and Organometallic Polymers*, Wiley, New York. (f) Manners, I. (2001) *Science*, **294**, 1664. (g) Herbert, D.E., Mayer, U.F.J. and Manners, I. (2007) *Angewandte Chemie – International Edition*, **46**, 5060. (h) Bellas, V. and Rehahn, M. (2007) *Angewandte Chemie – International Edition*, **46**, 5082

4 Green, M.L.H. and Ng, D.K.P. (1995) *Chemical Reviews*, **95**, 439.

5 (a) Demerseman, B. and Dixneuf, P.H. (1981) *Journal of Organometallic Chemistry*,

210, C20. (b) Demerseman, B., Dixneuf, P.H., Douglade, J. and Mercier, R. (1982) *Inorganic Chemistry*, **21**, 3942.

6 (a) Rausch, M.D., Ogasa, M., Ayers, M.A., Rogers, R.D. and Rollins, A.N. (1991) *Organometallics*, **10**, 2481. (b) Kool, L.B., Ogasa, M., Rausch, M.D. and Rogers, R.D. (1989) *Organometallics*, **8**, 1785. (c) Ogasa, M., Rausch, M.D. and Rogers, R.D. (1991) *Journal of Organometallic Chemistry*, **403**, 279.

7 (a) Elschenbroich, C., Plackmeyer, J., Nowotny, M., Behrendt, A., Harms, K., Pebler, J. and Burghaus, O. (2005) *Chemistry - A European Journal*, **11**, 7427. (b) Elschenbroich, C., Plackmeyer, J., Nowotny, M., Harms, K., Pebler, J. and Burghaus, O. (2005) *Inorganic Chemistry*, **44**, 955. (c) Elschenbroich, C., Wolf, M., Pebler, J. and Harms, K. (2004) *Organometallics*, **23**, 454. (d) Elschenbroich, C., Plackmeyer, J., Harms, K., Burghaus, O. and Pebler, J. (2003) *Organometallics*, **22**, 3367. (e) Elschenbroich, C., Wolf, M., Schiemann, O., Harms, K., Burghaus, O. and Pebler, J. (2002) *Organometallics*, **21**, 5810. (f) Elschenbroich, C., Lu, F. and Harms, K. (2002) *Organometallics*, **21**, 5152. (g) Elschenbroich, C., Schiemann, O., Burghaus, O., Harms, K. and Pebler, J. (1999) *Organometallics*, **18**, 3273. (h) Elschenbroich, C., Wolf, M., Burghaus, O., Harms, K. and Pebler, J. (1999) *European Journal of Inorganic Chemistry*, 2173.

8 (a) Tamm, M., Kunst, A., Bannenberg, T., Herdtweck, E., Sirsch, P., Elsevier, C.J. and Ernsting, J.M. (2004) *Angewandte Chemie – International Edition*, **43**, 5530. (b) Tamm, M., Kunst, A. and Herdtweck, E. (2005), *Chemical Communications*, 1729. (c) Tamm, M., Kunst, A., Bannenberg, T., Randoll, S., Jones, P.G., (2007) *Organometallics*, **26**, 417.

9 Elschenbroich, C., Pganelli, F., Nowotny, M., Neumüller, B. and Burghaus, O. (2004) *Zeitschrift fur Anorganische und Allgemeine Chemie*, **630**, 1599.

10 (a) Braunschweig, H., Lutz, M. and Radacki, K. (2005) *Angewandte Chemie – International Edition*, **44**, 5647. (b) Bartole-Scott, A., Braunschweig, H., Kupfer, T., Lutz, M., Manners, I., Nguyen, T., Radacki, K. and Seeler, F. (2006) *Chemistry - A European Journal*, **12**, 1266. (c) Braunschweig, H., Lutz, M., Radacki, K., Schaumlöffel, A., Seeler, F. and Unkelbach, C. (2006) *Organometallics*, **25**, 4433. (d) Braunschweig, H., Kupfer, T., Lutz, M. and Radacki, K. (2007) *Journal of the American Chemical Society*, **129**, 8893.

11 van Oven, H.O. and Liefde Meijer, H.J. (1970) *Journal of Organometallic Chemistry*, **23**, 158.

12 van Oven, H.O., Groenenboom, C.J. and Liefde Meijer, H.J. (1974) *Journal of Organometallic Chemistry*, **81**, 379.

13 Glöckner, A., Büschel, S. and Tamm, M. unpublished results.

14 (a) Lyssenko, K.A., Antipin, M.Y. and Ketkov, S.Y. (2001) *Russian Chemical Bulletin - International Edition*, **50**, 130. (b) Zeinstra, J.D. and de Boer, J.L. (1973) *Journal of Organometallic Chemistry*, **54**, 207.

15 Tamm, M., Kunst, A., Bannenberg, T., Herdtweck, E. and Schmid, R. (2005) *Organometallics*, **24**, 3163.

16 For the sandwich compounds $[(\eta^7\text{-}C_7H_7)M(\eta^5\text{-}C_5H_5)]$ (M = Ti, Zr, Hf), we wish to introduce the trivial names *troticene* [8a], *trozircene* and *trohafcene*, in accord with the naming of $[(\eta\text{-}C_7H_7)V(\eta\text{-}C_5H_5)]$ as *trovacene*, which stands for [(η^7-*tro*pylium) *va*nadium(η^5-cyclopenta*d*ienyl)], see for instance: Elschenbroich, Ch., Schiemann, O., Burghaus, O. and Harms, K., (1997) *Journal of the American Chemical Society*, **119**, 7452–7457. Accordingly, the compound $[(\eta^7\text{-}C_7H_7)Cr(\eta^5\text{-}C_5H_5)]$ has recently been named *trochrocene* [10].

17 Shannon, R.D. (1976) *Acta Crystallographica. Section A, Crystal Physics, Diffraction, Theoretical and General Crystallography*, **32**, 751.

18 Rogers, R.D. and Teuben, J.H. (1988) *Journal of Organometallic Chemistry*, **354**, 169.

19 (a) Groenenboom, C.J., Liefde Meijer, H.J. and Jellinek, F. (1974) *Journal of*

Organometallic Chemistry, **69**, 235. (b) Groenenboom, C.J., Sawatzky, G., Liefde Meijer, H.J. and Jellinek, F. (1974) *Journal of Organometallic Chemistry*, **76**, C4.

20 Zeinstra, J.D. and Nieuwpoort, W.C. (1978) *Inorganica Chimica Acta*, **30**, 103.

21 Clack, D.W. and Warren, K.D. (1977) *Theoretica Chimica Acta*, **46**, 313.

22 Anderson, J.E., Maher, E.T. and Kool, L.B. (1991) *Organometallics*, **10**, 1248.

23 (a) Green, J.C., Green, M.L.H., Kaltsoyannis, N., Mountford, P., Scott, P. and Simpson, S.J. (1992) *Organometallics*, **11**, 3353. (b) Green, J.C., Kaltsoyannis, N., Sze, K.H. and MacDonald, M. (1994) *Journal of the American Chemical Society*, **116**, 1994.

24 (a) Menconi, G. and Kaltsoyannis, N. (2005) *Organometallics*, **24**, 1189. (b) Kaltsoyannis, N. (1995) *Journal of The Chemical Society-Dalton Transactions*, 3727.

25 All DFT calculations have been performed with the Gaussian 03 program suite using the B3LYP density functional and the 6-311G(d,p) basis set combination for all main group elements. For the transition metal, double zeta basis sets optimised for use with effective core potentials (ECP) in combination with the corresponding Stuttgart ECP were employed (see original publications for further details).

26 (a) Green, M.L.H. and Walker, N.M. (1989) *Journal of the Chemical Society. Chemical Communications*, 1865. (b) Diamond, G.M., Green, M.L.H., Mountford, P., Walker, N.M. and Howard, J.A.K. (1992) *Journal of The Chemical Society-Dalton Transactions*, 417.

27 Baker, R.J., Bannenberg, T., Kunst, A., Randoll, S. and Tamm, M. (2006) *Inorganica Chimica Acta*, **395**, 4797.

28 Wang, K., Rosini, G.P., Nolan, S.P. and Goldman, A.S. (1995) *Journal of the American Chemical Society*, **117**, 5082.

29 Finckh, W., Tang, B.-Z., Foucher, D.A., Zamble, D.B., Ziembinski, R., Lough, A. and Manners, I. (1993) *Organometallics*, **12**, 823.

30 Foucher, D.A., Edwards, M., Burrow, R.A., Lough, A.J. and Manners, I. (1994) *Organometallics*, **13**, 4959.

31 Hultzsch, K., Nelson, J.M., Lough, A.J. and Manners, I. (1995) *Organometallics*, **14**, 5496.

32 Elschenbroich, C., Schmidt, E., Gondrum, R., Metz, B., Burghaus, O., Massa, W. and Wocadlo, S. (1997) *Organometallics*, **16**, 4589.

33 Elian, M., Chen, M.M.L., Mingos, D.M.P. and Hoffmann, R. (1976) *Inorganic Chemistry*, **15**, 1148.

34 Gourier, D. and Samuel, D. (1988) *Inorganic Chemistry*, **27**, 3018.

35 Rulkens, R., Gates, D.P., Balaishis, D., Pudelski, J.K., McIntosh, D.F., Lough, A.J. and Manners, I. (1997) *Journal of the American Chemical Society*, **119**, 10976

36 Fielding, L. (2000) *Tetrahedron*, **56**, 6151.

37 Benesi, H.A. and Hildebrand, J.H. (1949) *Journal of the American Chemical Society*, **71**, 2703.

38 Tamm, M., Dreßel, B., Fröhlich, R. and Bergander, K. (2000) *Chemical Communications*, 1731.

39 Li, J., Bursten, B.E., (1997) *Journal of the American Chemical Society*, **119**, 9021.

40 Foucher, D.A., Tang, B.Z. and Manners, I. (1992) *Journal of the American Chemical Society*, **114**, 6246.

41 Green, J.C. (1998) *J Chem Soc Rev*, **27**, 263.

42 Bourke, S.C., MacLachlan, M.J., Lough, A.J. and Manners, I., (2005) *Chemistry - A European Journal*, **11**, 1989.

43 (a) Sheridan, J.B., Lough, A.J. and Manners, I. (1996) *Organometallics*, **15**, 2195. (b) Reddy, N.P., Choi, N., Shimada, S. and Tanaka, M. (1996) *Chemistry Letters*, 649. (c) Sheridan, J.B., Temple, K., Lough, A.J. and Manners, I. (1997) *Journal of The Chemical Society-Dalton Transactions*, 711.

44 (a) Temple, K., Lough, A.J., Sheridan, J.B. and Manners, I., (1998) *Journal of The Chemical Society-Dalton Transactions*, 2799. (b) Temple, K., Jäkle, F., Sheridan, J.B. and Manners, I., (2001) *Journal of the American*

Chemical Society, **123**, 1355. (c) Chan, W.Y., Berenbaum, A., Clendenning, S.B., Lough, A.J. and Manners, I., (2003) *Organometallics*, **22**, 3796.

45 Chan, W.Y., Lough, A.J. and Manners, I., (2007) *Angewandte Chemie – International Edition*, **46**, 9069.

46 Flores, J.C., Hernandez, R., Royo, P., Butt, A., Spaniol, T.P. and Okuda, J. (2000) *Journal of Organometallic Chemistry*, **593–594**, 202.

47 Cornelissen, C., Erker, G., Kehr, G. and Fröhlich, R. (2005) *Organometallics*, **24**, 214.

48 Büschel, S., Jungton, A.-K., Hrib, C. and Tamm, M. unpublished results.

49 Mann, G., Incarvito, C., Rheingold, A.L. and Hartwig, J.F. (1999) *Journal of the American Chemical Society*, **121**, 3224.

50 (a) For selected recent examples, see: (a) Sircoglou, M., Bontemps, S., Mercy, M., Saffon, N., Takahashi, M., Bouhadir, G., Maron, L. and Bourissou, D. (2007) *Angewandte Chemie – International Edition*, **46**, 8583. (b) Bontemps, S., Bouhadir, G., Miqueu, K. and Bourissou, D. (2006) *Journal of the American Chemical Society*, **128**, 12056.

9
Monosaccharide Ligands in Organotitanium and Organozirconium Chemistry

Peter Kitaev, Daniela Zeysing, and Jürgen Heck

9.1
Introduction

Despite their advantageous properties, monosaccharide-derived ligands have rarely been applied in the synthesis of organometallic compounds and thus little is known about their coordination capabilities. Most of the established monosaccharide–metal complexes are based on late transition metals such as Rh, Pd and Pt [1]. Among early transition metals, monosaccharide complexes containing Ti and Zr are of importance [2]. Although the information pertaining to their metal–ligand binding sites is scarce, both late and early transition metal complexes of modified monosaccharides have been successfully applied as chiral reagents and catalysts in stereoselective synthesis [3].

9.2
Synthesis of Organotitanium Carbohydrate Compounds

Within this research topic we are interested in the ability of pyranosides to coordinate to varying titanium precursors and the use of the resulting complexes in stereoselective synthesis. In particular we are investigating the 2- and 3- positions of pyranosides as coordination sites for organometallic complexes of the early transition metal Ti. The torsion angle between the oxygen functions of C2 and C3 of the pyranosidato ligand may determine whether a mononuclear chelate complex or a dinuclear compound with a bridging coordination between two Ti atoms is formed.

Our interest in the coordination chemistry of Ti when reacted with carbohydrate ligands has led to previously reported reactions [4, 5]. The reaction of methyl-4,6-*O*-benzylidene-β-D-glucopyranoside (β-MeBnGluH_2) with the organotitanium complex Cp^*TiCl_3 in the presence of triethylamine gave the dinuclear titanium complex (T-4-R; T-4-R)bis[chlorido(η^5-pentamethylcyclopentadienyl)][μ-(methyl-4,6-*O*-benzylidene-β-D-glucopyranosidato-1κO^2, 2κO^3)][μ-(methyl-4,6-*O*-benzylidene-β-D-glucopyranosi-

Activating Unreactive Substrates: The Role of Secondary Interactions.
Edited by Carsten Bolm and F. Ekkehardt Hahn

ISBN: 978-3-527-31823-0

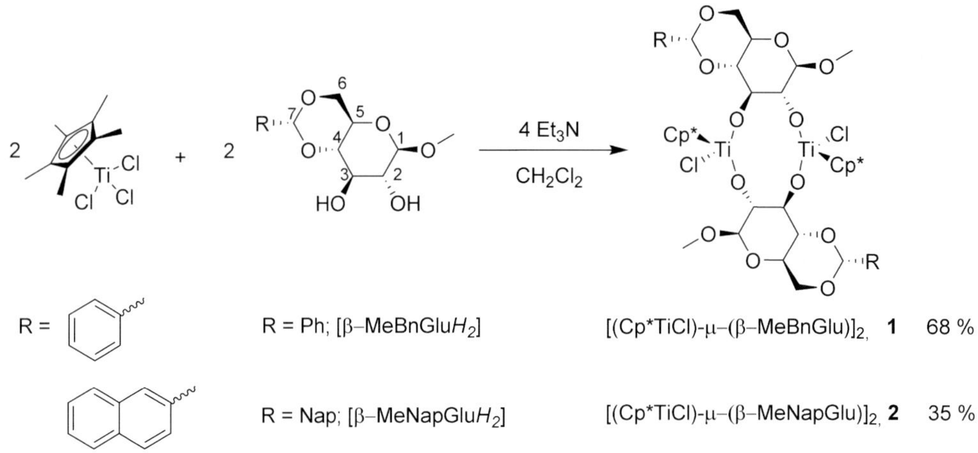

R = Ph; [β–MeBnGluH_2] [(Cp*TiCl)-μ–(β–MeBnGlu)]$_2$, **1** 68 %

R = Nap; [β–MeNapGluH_2] [(Cp*TiCl)-μ–(β–MeNapGlu)]$_2$, **2** 35 %

Scheme 9.1 Reaction of protected glucopyranosides with Cp^*TiCl_3 to form **1** and **2**.

dato-1κO^3, 2κO^2)]dititanium(IV) ([(Cp*TiCl)-μ-(β-MeBnGlu)]$_2$ (**1**), with two glucopyranosidato ligands (Scheme 9.1). Under similar conditions, a reaction was carried out with Cp^*TiCl_3 using methyl (4,6-*O*-β-naphthyl-2′-methylidene)-β-D-glucopyranoside (β-MeNapGluH_2) as the sugar ligand. The result was compound **2**, (T-4-R; T-4-R) bis[chlorido(η^5-pentamethylcyclopentadienyl)][μ-(methyl-4,6-*O*-naphthyl-2'-methylidene-β-D-glucopyranosidato-1κO^2, 2κO^3)][μ-(methyl-4,6-*O*-naphthyl-methylidene-β-D-glucopyranosidato-1κO^3, 2κO^2)]dititanium(IV) ([(Cp*TiCl)-μ-(β-MeNapGlu)]$_2$) as shown in Scheme 9.1. Molecular structures of both complexes could be obtained and are depicted in Figure 9.1.

The analysis of these compounds indicates that the dihedral angle between the trans-orientated oxygen functions of C2 and C3 of the glucopyranosidato ligand is too large for a chelating coordination of a small titanium(IV) ion, thus forming a dinuclear complex. The uncoordinated glucopyranoside β-MeBnGluH_2 has a dihedral angle O2-C2-C3-O3 of 64.2° [6]. Compound **1** has dihedral angles of 66.97° and 68.81°. These angles have different values because the solid-state structure deviates slightly from the C_2 symmetry found in solution by means of NMR spectroscopy. Compound **2**, on the other hand, exhibits perfect C_2 symmetry and thus both dihedral angles have the same value (66.42°). The slight increase in the dihedral angle of both compounds, when compared to the uncoordinated glucopyranoside ligand, prevents the formation of the originally desired chelate complexes.

We sought to extend the efforts of obtaining chelate complexes by using a carbohydrate ligand with a different configuration at the coordinating hydroxyl functions of C2 and C3. In comparison to the *trans* orientation found in the glucopyranoside, a *cis* orientation of the binding oxygen atoms O2 and O3 may result in a smaller dihedral angle and thus in a mononuclear organotitanium complex with a chelating vicinal diolato function. A properly protected allopyranoside fulfills this stereochemical prerequisite. Initial experiments using this sugar were carried out with $CpTiCl_3$ and methyl-4,6-*O*-benzylidene-α-D-allopyranoside (α–MeBnAllH_2) in the presence of triethylamine (Scheme 9.2). Using a ratio of

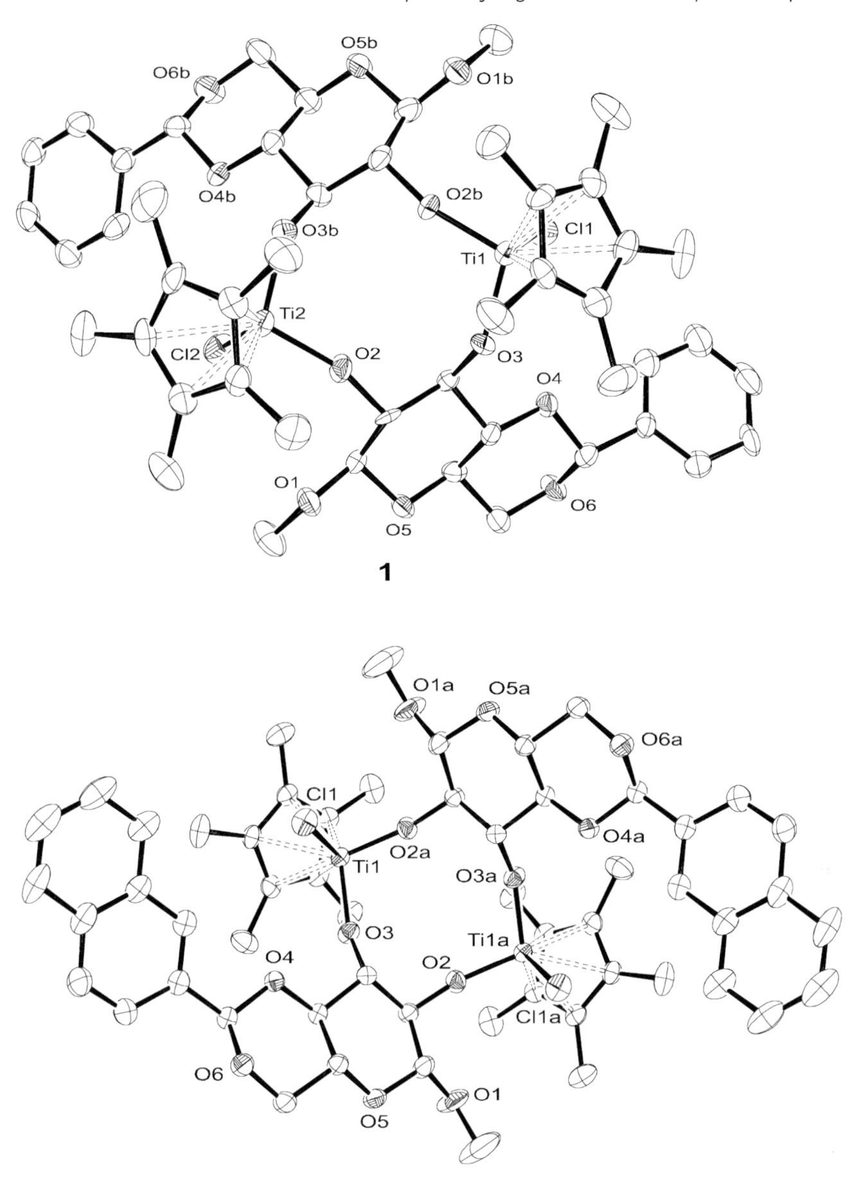

Figure 9.1 ORTEP plot of the molecular structures of **1** and **2** (hydrogen atoms are omitted for clarity; displacement ellipsoids are drawn at the 50% probability level).

two equivalents of the titanium precursor to one equivalent of the sugar ligand, Tetrachlorido-1κ^2Cl, 2κ^2Cl-bis(η^5-cyclopentadienyl)[μ-(methyl-4,6-*O*-benzylidene-α-D-allopyranosidato-1κO^2, 2κO^3]dititanium(IV)] **3** was formed and crystals suitable for X-ray diffraction measurements were obtained [5].

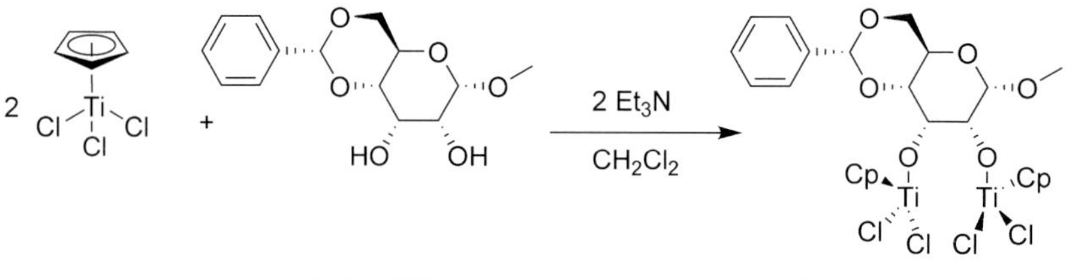

2 Cl–Ti(Cp)Cl–Cl + α–MeBnAllH_2 → (2 Et_3N, CH_2Cl_2) → [$(CpTiCl_2)_2$–μ–(α–MeBnAll)], **3**

Scheme 9.2 Reaction of $CpTiCl_3$ with methyl-4,6-*O*-benzylidene-α-D-allopyranoside to form **3**.

The crystal structure of the protonated ligand α-MeBnAllH_2 has been previously published and the dihedral angle O2-C2-C3-O3 measures 55.5° [7]. This value is considerably smaller than that measured for the dihedral angle in β-MeBnGluH_2. In addition a reduction of the dihedral angle O2-C2-C3-O3 of the free ligand upon coordination has been shown for the glucopyranosidato ligand in the zirconate complex [6]. A coordination of one titanium atom to one chelating allopyranosidato ligand was therefore plausible, but, as the results indicate, such as structure was not obtained (Figure 9.2). The coordination of the $CpTiCl_2$ fragments led to a slight widening of the dihedral angle (58.44°). Again, the pyranosidato ligand is a bidentate bridging ligand despite the smaller dihedral angles of both the free ligand and the resulting compound **3**.

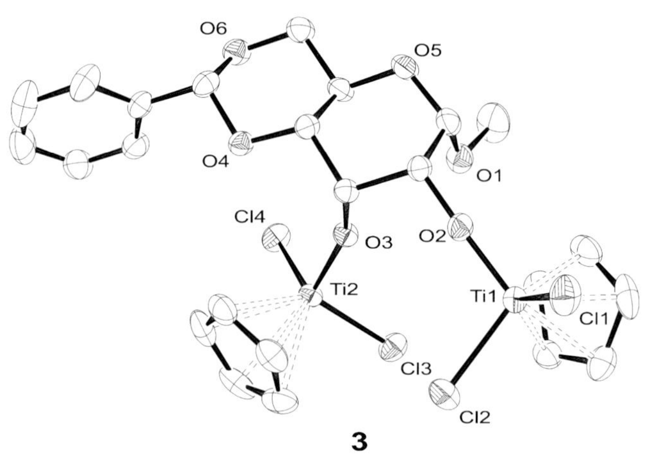

3

Figure 9.2 ORTEP plot of the molecular structure of **3** (hydrogen atoms are omitted for clarity; displacement ellipsoids are drawn at the 50% probability level).

After determining that the chosen gluco- and allopyranosidato ligands are not suitable for chelate coordination to the titanium(IV) ion, another approach was undertaken to synthesize a potential catalyst precursor. Starting from the 4,6-protected glucopyranoside, an additional protecting group was added to the hydroxyl function at C2. $CpTiCl_3$ could then be reacted with one equivalent of the monosaccharide in the presence of triethylamine. With the hydroxyl function at C3 being the only remaining reactive site, we were able to obtain dichlorido-(η^5-cyclopentadienyl)-

(methyl-2-*O*-benzoyl-4,6-*O*-benzylidene-α-D-glucopyranosidato-1κO^3)titanium (**4**) and the crystal structure thereof (Figure 9.3). The coordination of the ligands of **4** resembles a typical precursor compound for catalytic reactions.

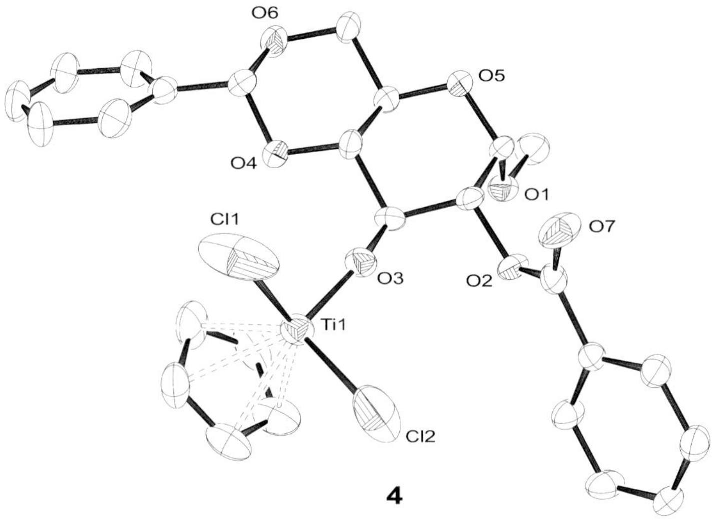

Figure 9.3 ORTEP plot of the molecular structure of **4** (hydrogen atoms are omitted for clarity; displacement ellipsoids are drawn at the 50% probability level).

In order to apply any of these compounds in catalytic reactions, however, the chlorido ligands must be exchanged for ligands that are suitable for the targeted reactions. In particular, we are looking to utilize these carbohydrate–titanium compounds for the hydroamination of primary and secondary aminoalkenes as well as for the polymerization of monomers such as ethylene and styrene. For both reactions, the existing ligands need to be exchanged against alkyl ligands. This reaction was first carried out using compound **1**. For this reaction, **1** was dissolved in THF and a slight excess of methyl lithium was added at a temperature between −50 °C and −70 °C (Scheme 9.3).

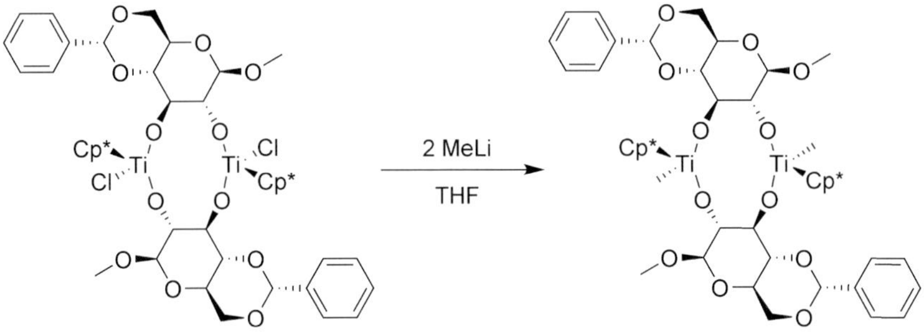

Scheme 9.3 Reaction of **1** with methyl lithium to form the precatalyst **5**.

The product bis[(η^5-pentamethylcyclopentadienyl)methyl-1κ*C*, 2κ*C*][μ-(methyl-4,6-*O*-benzylidene-β-D-glucopyranosidato-1κO^2, 2κO^3)]][μ-(methyl-4,6-*O*-benzylidene-β-D-glucopyranosidato-1κO^3, 2κO^2)]dititanium(IV)([(Cp*TiMe)-μ-(β-MeBnGlu)]$_2$, **5** was obtained in quantitative yield and could be used as a precursor compound for catalytic reactions.

9.3 Organotitanium Carbohydrate Compounds for Use in Catalytic Reactions: Polymerization of Ethylene

For initial experiments pertaining to the polymerization of ethylene, compound **1** was used as a pre-catalyst. These experiments were carried out in collaboration with the group of Kaminsky at the University of Hamburg. The best experimental results were obtained when the catalyst precursor **1** (2.5×10^{-4} mmol) in toluene (400 mL) was combined with a large excess of MAO (400 mg) in order to activate the compound. The monomer ethylene (1 mol L^{-1}) was added at a constant rate during the reaction time of 15 min and at a temperature of 45 °C. The resulting polymer had a molecular weight of 6.6×10^4 g mol^{-1} with a molecular weight distribution of the polymer, the polydispersity, determined to be 2.14. The activity of the catalyst had a value of 18 560 kg_{poly} $(h \cdot c_{cat} \cdot c_{mono})^{-1}$, which is one order of magnitude larger than that of well-established catalysts like derivatives of CpTi(OAr)Cl_2 [8].

Further experiments were carried out using the same reaction conditions, but varying the amounts of ethylene concentration. The results are summarized in Table 9.1 below.

Table 9.1 Molecular weight distribution of the the polymers A and B with varying monomer concentration.

		Polymer A		Polymer B	
Sample	$c_{ethylene}$ [mol L^{-1}]	M_w [g mol^{-1}]	M_n [g mol^{-1}]	M_w [g mol^{-1}]	M_n [g mol^{-1}]
———	1	n.d.	n.d.	66 000	33 100
———	0.75	833 000	798 000	47 300	23 400
———	0.5	809 000	764 000	38 500	16 900
———	0.25	627 000	495 000	16 100	8 200
———	0.15	385 000	213 400	8 100	4 660

According to this data and the results of the gel permeation chromatography (Figure 9.4) to determine the distribution of the molecular weights, there is a tendency toward a bimodal distribution.

This distribution could imply that there are two catalytically active sites in the reaction mixture. These could allow the polymerization to proceed in two different physical states and with varying mobility in the polymer chains. As a result, the polymers are formed with differing molecular weights [9]. This may be a consequence of the harsh reaction conditions due to the use of MAO as an activator and methylating agent [10]. We were unable to determine which catalytic species was

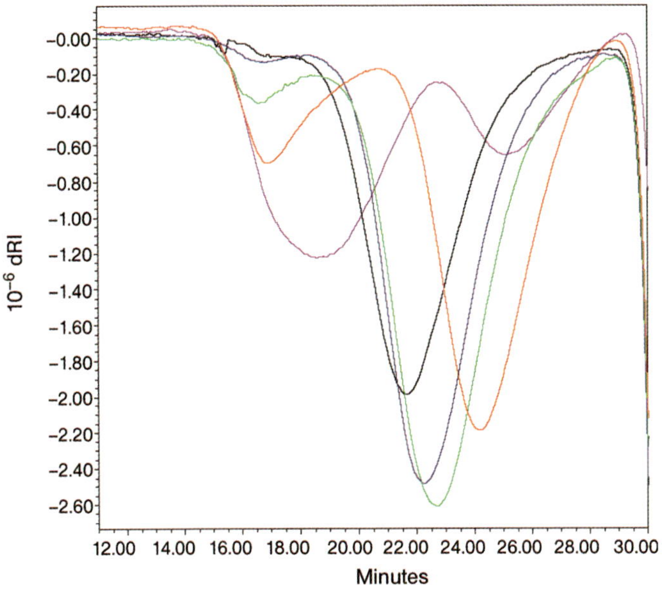

Figure 9.4 Gel permeation chromatography depicting a bimodal distribution of polymers.

actually formed and was thus responsible for the formation of the polyethylene. The use of compound **5** as a pre-catalyst may make the addition of MAO unnecessary and thus create a milder atmosphere for this polymerization reaction.

In an effort to obtain a polymer with stereoselectivity, we also attempted to polymerize propene. Reactions were undertaken with varying monomer concentrations at different temperatures with a catalyst amount of 0.005 mmol, but no activity was detected. The added propene was not used up during the reaction.

9.4 Intramolecular Hydroamination of Aminoalkenes

In recent years, research groups have been investigating the use of titanium compounds as catalysts for the intramolecular hydroamination of aminoalkenes, a reaction leading to the formation of C–N bonds in an enantioselective fashion (Scheme 9.4). This has been a significant challenge for the synthesis of pharmaceuticals and fine chemicals [11, 12].

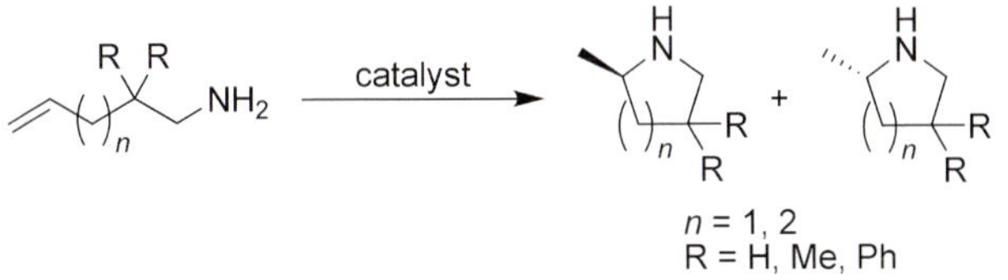

Scheme 9.4 Intramolecular hydroamination of aminoalkenes.

Previously, this field has been dominated by lanthanide and late transition metal complexes [13], but, because of their lower toxicity and higher abundance, the use of group 4 metal compounds is of great interest for these reactions. Thus far, however, only a few zirconium compounds have led to products with the desired stereoselectivity [14]. With this in mind, we investigated the use of our dinuclear sugar compound **5** as a catalyst precursor for these reactions. Initial hydroamination experiments were performed with *N*-benzyl-4-pentenylamine as starting material. In order to activate compound **5** for the hydroamination of a secondary amine, it must be treated with dimethylanilinium tetrakis(pentafluorophenyl)borate ($[PhNMe_2H][B(C_6F_5)_4]$). A cationic species is then the active catalyst for this reaction [12]. Under differing reaction conditions and varying amounts of catalyst, the cyclic amine was synthesized in an NMR-scale reaction using Young tubes. The results are summarized in Table 9.2.

Table 9.2 Results of the intramolecular hydroamination of *N*-benzyl-4-pentylamine using **5** as a pre-catalyst:

5

$[PhNMe_2H][B(C_6F_5)_4]$

Catalyst [mol %]	Solvent	*T* [°C]	*t* [h]	Yield [%]
7	Benzene-d_6	95	52	90
10	Toluene-d_8	100	87	>90
10	Toluene-d_8	rt, 150 W, microwave	4	89

These first investigations show that in the case of *N*-benzyl-4-pentenylamine the catalyst works best in benzene at 95 °C. In order to separate the enantiomers and determine whether an enantiomeric excess was achieved, (R)-(+)-α-methoxy-α-(trifluoromethyl)phenylacetic acid [(+)-Mosher's acid] was added. The residue was dissolved in $CDCl_3$, and at −20 °C a ^1H NMR spectrum revealed that the two possible products were obtained as a racemic mixture. Although monosaccharide titanium complex **5**, which was used as a catalyst precursor, contains stereoinformation, this information was not carried over to the product. It is possible that the catalyst decomposes upon reaction with the activator, and thus a combination of smaller compounds or fragments allows for the initial success of the hydroamination reaction.

Doye and Hultzsch have used titanocene-based catalysts for investigations into similar intramolecular hydroamination reactions. These were achiral, and products with enantioselectivity were not obtained. In order to further understand this reaction, we synthesized a chiral titanocene compound for use as a pre-catalyst for intramolecular hydroaminations. This idea was based on the results of the lanthanocene-derived compounds used by Marks for hydroamination reactions [15]. Starting from (+)-neomenthol, we synthesized a menthyl-substituted titanocene compound for use in test reactions. With the exception of the last step in which the chlorido complex is methylated using methyl lithium (Scheme 9.5), all of the reactions were taken from literature procedures [16]. Compound **6**, bis{η^5-{(1*R*,2*S*,5*R*)-5-methyl-2-[prop-2-yl]-cyclohex-1-yl}cyclopentadienyl}-dimethyltitanium (MCp_2TiMe_2) was obtained in good yield.

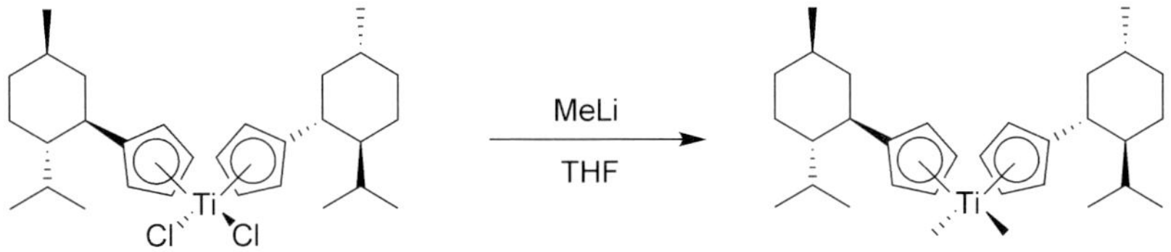

MeLi

THF

6, 56 % yield

Scheme 9.5 Synthesis of **6** using methyl lithium.

These reactions were carried out with the idea that we could eventually synthesize a chiral cyclopentadienyl ligand using carbohydrate derivatives in place of the menthyl substituent. In such a compound the chiral information may be closer to the active metal atom of the catalyst and thus allow for the transfer of stereoselectivity onto the hydroamination product.

Using 10 mol% of **5** with $[PhNMe_2H][B(C_6F_5)_4]$ to activate the pre-catalyst, *N*-benzyl-2,2-diphenyl-4-pentenylamine underwent an intramolecular hydroamination. This amine was selected for initial examination as it was expected that cyclization of this substrate would be facilitated by a geminal Thorpe–Ingold effect [17]. The reaction took place in toluene-d_8 at 100 °C and was completed in 3.5 h. The reaction work-up was similar to that described for the previous reaction and (+)-Mosher's acid was used to convert the enantiomers into diastereomers and thus make them detectable for ^1H NMR spectroscopy. The best resolution was obtained by recording the spectra at room temperature. Although the peaks could not be completely separated it was detectable that a racemic mixture of the cyclic product was formed. It is plausible that the cyclopentadienyl ligand is liberated in varying amounts during the reaction [18]. This may also be true when compound **5** is used as precatalyst. If the reactions do not take place with the intended catalyst, but instead with fragments thereof, enantioselective products cannot be formed.

9.5 Organozirconium Carbohydrate Compounds

The reaction of $CpZrCl_3$ with the modified methyl glucopyranoside also reveals a dinuclear metal complex, μ-chloro-bis{chloro(η^5-cyclopentadienyl)(methyl-4,6-*O*-benzylidene-β-D-glucopyranosidato-1κO^2,1:2κO^3-zirconate} (7^-), composed of two bridging sugar ligands and two chlorido (cyclopentadienyl)metal units [6] (Scheme 9.6 and Figure 9.5).

In contrast to the described neutral dititanium compound, the dizirconium complex bears a negative charge due to an additional coordination of a bridging chlorido ligand. The zirconate anion $\mathbf{7}^-$ represents a C_2 symmetric chiral cavity containing triethylammonium as a counter ion. In this 'host-guest' arrangement the triethylammonium cation is fixed inside the cavity through a hydrogen bond to one of the terminal chlorido ligands in the cavity.

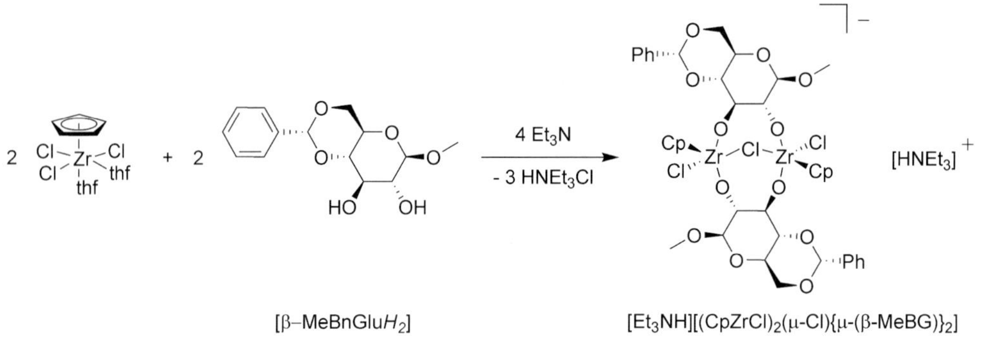

4 Et_3N

- 3 $HNEt_3Cl$

[β-MeBnGluH_2]

$[Et_3NH][(CpZrCl)_2(\mu\text{-Cl})\{\mu\text{-}(\beta\text{-MeBG})\}_2]$

$[Et_3NH]$**7**

Scheme 9.6 Synthesis of the glucopyranosidato zirconium complex $[Et_3NH]$**7**.

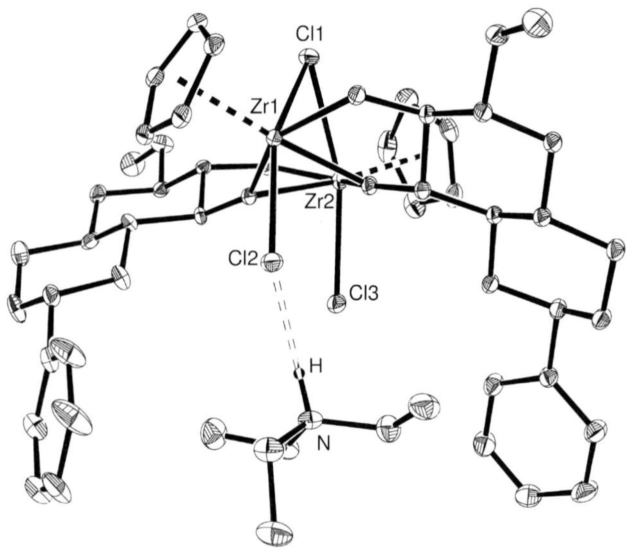

Figure 9.5 ORTEP plot of the molecular structure of $[Et_3NH]$**7** (hydrogen atoms are omitted for clarity; displacement ellipsoids are drawn at the 50% probability level).

9.6
Amine Exchange

It has been found in ^1H NMR experiments that the encapsulated triethylammonium cation is rapidly exchanged with external $[Et_3NH]^+$ [6]. Recent results demonstrate that external amines also undergo a rapid exchange with the amine of the hosted ammonium cation. In ^1H NMR spectra a noticeable shift of the resonance signals of the amines could be observed when the amines were added to a solution of the zirconium complex $[Et_3NH]$7. This shift is caused by the transformation of the amines into ammonium cations and by the inclusion into the cavity. The exchange reaction is rapid enough to generate an average signal for the amine inside and outside the cavity. This exchange reaction also occurs when various primary, secondary and tertiary amines replace the triethylamine, forming ammonium salts within the zirconate anion **7**$^-$ (Scheme 9.7).

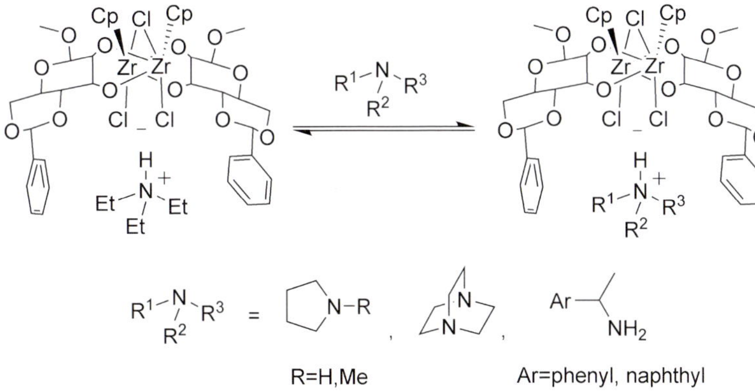

R=H,Me

Ar=phenyl, naphthyl

Scheme 9.7 Amine exchange in **7**$^-$.

9.7
Chiral Recognition

Using racemic mixtures of amines, the resonance signals of the different enantiomers are resolved by the diastereotopic interactions of the enantiomeric amines with the chiral cavity. For example, in Figure 9.6b the ^{1}H NMR signal (1.31 ppm) of the methyl group of D-1-phenyl-1-aminoethane, which is included in the cavity, is depicted. In Figure 9.6c, the chiral cavity was used to separate the signals of a racemic mixture. Comparable diastereotopic interactions have been reported from amino acids and amines [19]. In fact, for all used chiral amines (naphthylethylamine, 2-aminopentane, tetrahydronaphthylamine) the splitting of the corresponding signals of the enantiomers was sufficient to determine the enantiomeric excess. Thus,

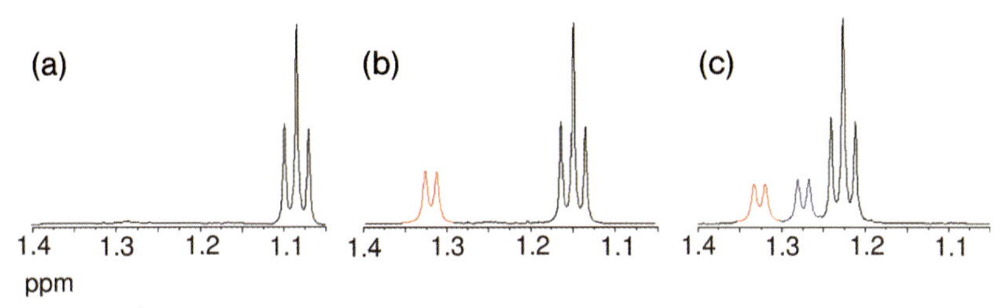

Figure 9.6 ^{1}H NMR signals of the methyl groups of amines, coordinated in **7**$^-$ as ammonium cations. (a) triethylamine; (b) triethylamine and D-1-phenyl-1-aminoethane; (c) triethylamine and D,L-1-phenyl-1-aminoethane.

the zirconate complex could be used as a type of chiral shift reagent for optically active amines.

Regarding the diastereotopic interaction between a chiral substrate (amine) and the chiral cavity, which is reminiscent of the diastereotopic activity of an enzyme pocket, it is an obvious possibility to use the chiral cavity of the zirconate complex for chiral resolution of amines. The rationale behind this is the difference in binding energies coupled with the inclusion of an R- or an S-isomer into the chiral cavity. This should lead to preferential protonation inside the cavity for either the R- or S-isomer. If one enantiomer is preferentially protonated and incorporated into the pocket, the other enantiomer should remain dissolved and available for typical amine reactions such as the formation of amides.

With this in mind, two equivalents of a racemic mixture of D,L-1-phenyl-1-aminoethane were mixed with one equivalent of the zirconate complex $\mathbf{7}^-$ at low temperature (−78 °C). After some minutes the mixture was allowed to react with one equivalent of pivaloyl chloride (Scheme 9.8). As a result, the pivaloylamide is formed with the external amine whereas the cavity-incorporated amine is deactivated for amide formation by protonation. From a ^{1}H NMR sample of the product mixture, which involves the zirconate complex $\mathbf{7}^-$ with the chiral ammonium cation and the pivaloylamide, an enantiomeric excess of about 30% can be calculated for the S-isomer preferentially enclosed in the cavity.

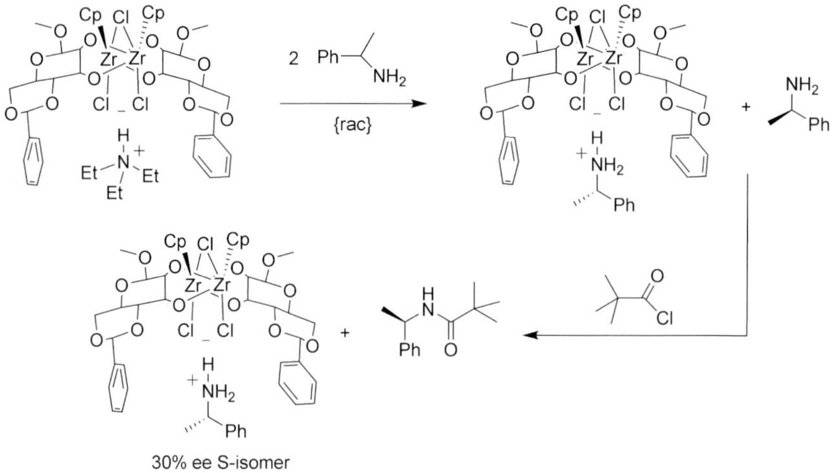

Cp Cl Cp Zr Zr Cl Cl Et N Et Et H + 2 Ph NH2 {rac} NH2 Ph + NH2 H + Ph O Cl H N Ph O

30% ee S-isomer

Scheme 9.8 Racemic resolution with $\mathbf{7}^-$.

The diastereotopic interaction between the chiral cavity and the chiral amine can even be used for the synthesis of the zirconate fragment $\mathbf{7}^-$. In close analogy to the original synthesis of [Et_3NH]**7**, where NEt_3 is used as a trap for the hydrogen chloride formed [6], the reaction was performed with a racemic mixture of 1-phenyl-1-aminoethane as the amine component (Scheme 9.9). From the ^{1}H NMR spectrum of the zirconate product an excess of the incorporated S-isomer of about 65% can be calculated.

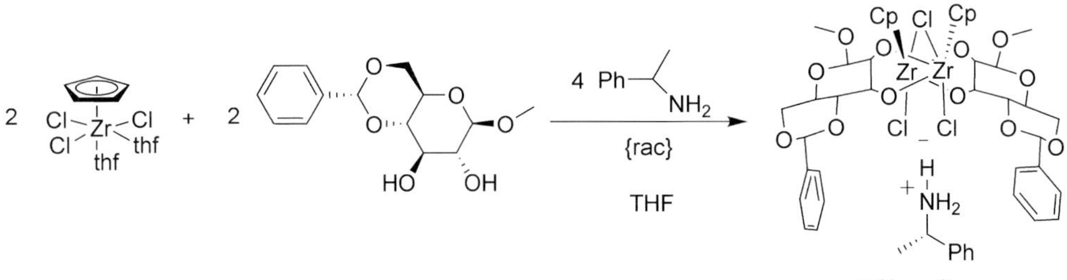

Scheme 9.9 Chiral resolution during the synthesis of the zirconate.

9.7.1 Diels–Alder Reaction

Another attempt to use the chirality of the zirconate complex for enantioselective synthesis was the exchange of the amine of the ammonium cation for an imine in order to activate the Schiff base for hetero Diels–Alder reactions by the formation of the iminium cation [20]. However, because of the considerably weaker basicity of imines compared to amines, the attempts at hetero Diels–Alder reactions with imines failed.

Enamines are another option for nitrogen-containing dienophiles suitable for Diels–Alder reactions. Aside from the activation through the coordination of Lewis acids on the imine nitrogen atom, enamines can be activated by protonation at the nitrogen atom of the enamine within the chiral cavity [21]. However, adding enamines to a solution of the zirconate complex $[Et_3NH]$**7** yields the corresponding iminium cation quantitatively for most enamines used as a result of the protonation at the β-carbon atoms of the enamine.

9.7.2 Nucleophilic Addition

The formation of the iminium cation inside the chiral pocket led to the idea of an enantioselective nucleophilic addition at the α-carbon atom of the iminium cation, for example with a cyanide anion. This reaction results in the formation of α-aminonitriles, which reveal α-aminoacids after hydrolysis. For this reaction trimethylsilylcyanide is commonly used as a cyanide nucleophile, also known as an 'iminium salt trap' (Scheme 9.10) [22].

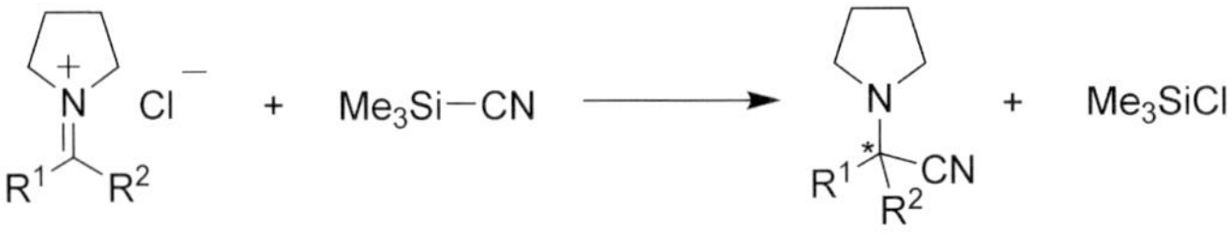

Scheme 9.10 Nucleophilic addition of the cyano group to iminium cation ('iminium salt trap').

Starting with an enamine containing different R^1 and R^2 substituents at the α-carbon atom (Scheme 9.10), a stereogenic center will be formed upon nucleophilic addition. The *in situ* synthesis of iminium salts by addition of $[HNEt_3]Cl$ to a solution of the enamine and Me_3SiCN results in a racemic mixture of the α-aminonitriles **9** (Scheme 9.11).

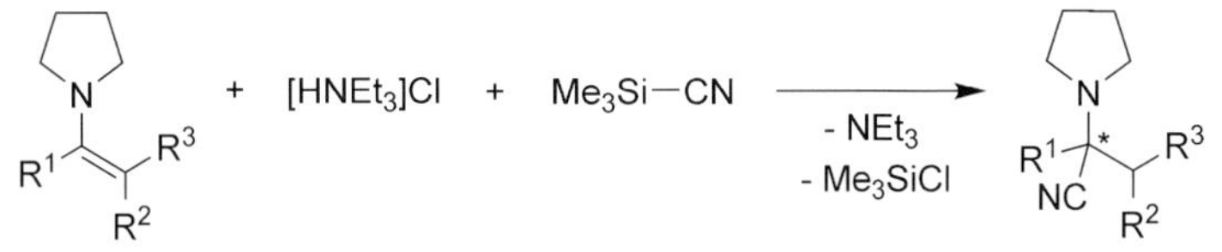

8 + $[HNEt_3]Cl$ + Me_3Si-CN → (- NEt_3, - Me_3SiCl) 9

8	9
a) R^1,R^2 = -$(CH_2)_3$-, R^3=H	a) R^1,R^2 = -$(CH_2)_3$-, R^3=H
b) R^1,R^2 = -$(CH_2)_4$-, R^3=H	b) R^1,R^2 = -$(CH_2)_4$-, R^3=H
c) R^1=Ph, R^2=R^3=H	c) R^1=Ph, R^2=R^3=H
d) R^1=H, R^2=R^3=Me	d) R^1=H, R^2=R^3=Me
e) R^1=H, R^2=Me, R^3=Ph	e) R^1=H, R^2=Me, R^3=Ph

Scheme 9.11 Synthesis of α-aminonitriles **9** from the corresponding enamines **8**.

When using the C_2 symmetric pocket of the zirconate as a chiral matrix, it should be possible to obtain chiral induction. In fact, by consecutive addition of the enamine **8** and trimethylsilyl cyanide to a solution of the zirconate complex $[Et_3NH]$**7** the targeted α-aminonitrile **9** could be obtained. Since the zirconate complex decomposes during the reaction, the nucleophilic addition could only be conducted in an equimolar amount.

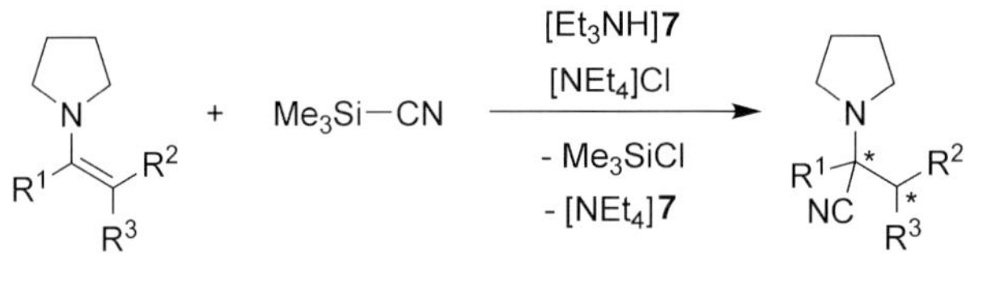

8 + Me_3Si-CN → ($[Et_3NH]$**7**, $[NEt_4]Cl$; - Me_3SiCl, - $[NEt_4]$**7**) 9

Scheme 9.12 Use of the zirconate **7**⁻ for stereoselective cyano addition.

The stabilization of the zirconate was successful using tetraethylammonium chloride, and the revealed complex $[NEt_4]$**7** (Scheme 9.12) can be used once again for this reaction sequence, starting with $[NEt_4]$**7**, adding $[HNEt_3]Cl$, the corresponding enamine and nucleophile (Scheme 9.13).

Unfortunately, no crystals suitable for an X-ray structure analysis have yet been obtained. Therefore, a DFT calculation of the conjectural structure has been performed, confirming the possibility of incorporating the tetraethylammonium ion in the chiral cavity (Figure 9.7).

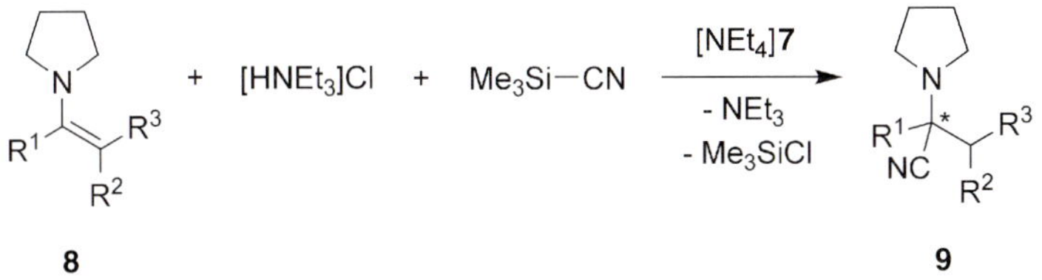

8 + $[HNEt_3]Cl$ + Me_3Si-CN → ($[NEt_4]$**7**; $- NEt_3$; $- Me_3SiCl$) 9

Scheme 9.13 Revealed after cyano addition complex $[NEt_4]$**7** as a chiral catalyst.

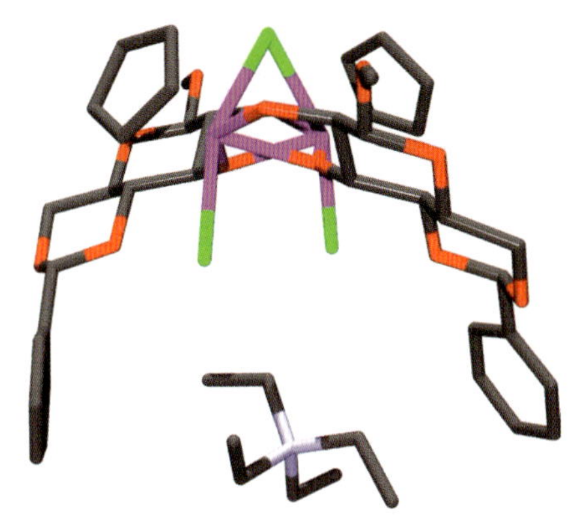

Figure 9.7 Calculated structure of the compound $[NEt_4]$**7** (DFT B3LYP [23] method with LANL2DZ basis set).

In order to investigate the stereochemical induction of the chiral pocket of $\mathbf{7}^-$ the reaction shown in Scheme 9.12 was performed with the enamine **8e**, which after transformation into the aminonitrile **9e** must have two stereogenic centers, thus forming two diastereomers. Two different diastereomers can be assigned by means of ^{1}H NMR spectroscopy. In a reaction without the chiral pocket a diastereomeric ratio of 5:1 can be observed (Figure 9.9b). For one diastereomer it was possible to determine the molecular structure by X-ray diffraction analysis (Figure 9.8).

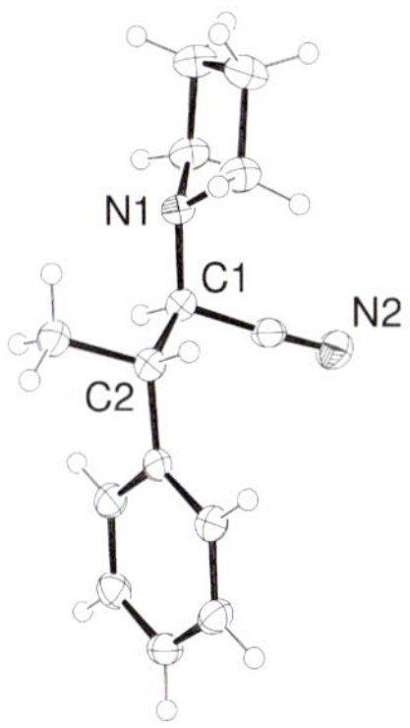

N1 C1 N2 C2

Figure 9.8 ORTEP plot of the molecular structure of **9e** (displacement ellipsoids are drawn at the 50% probability level).

The unit cell contains both of the enantiomers with the combination of the configurations (RS) and (SR), respectively, on C1 and C2. A comparison with ^{1}H NMR spectra indicates that the low-field signals of the α-protons can be assigned to this diastereomer. In contrast to this is the result when using the chiral pocket for the reaction described in Scheme 9.12. In this reaction the diastereomeric ratio has completely changed to 1:1.7, showing the diastereomer with the low-field-shifted α-proton signals (Figure 9.9c) as the minor product whereas the other diastereomer is the major product. Hence, there must be a stereochemical induction.

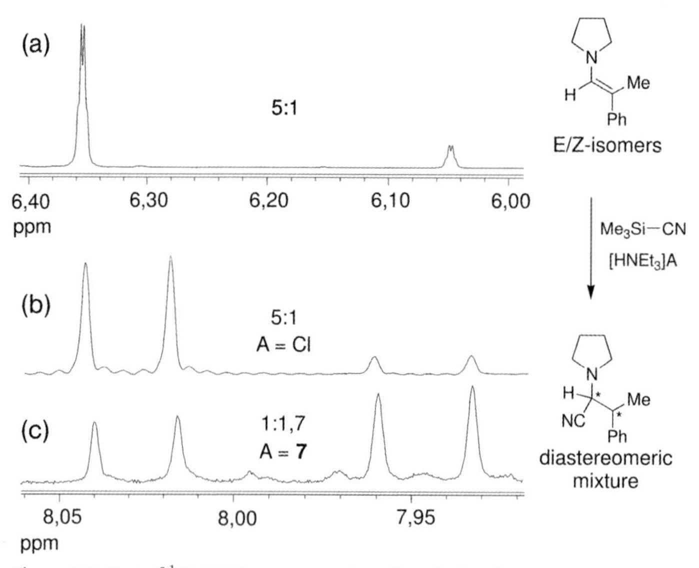

Figure 9.9 Part of ^{1}H NMR spectra assigned to depicted α-hydrogen atoms. (a) E/Z forms of initial enamine: diastereomers of **9e** obtained from the reaction with triethylammonium chloride (b) and zirconate complex $[Et_3NH]$**7**(c).

9.8 Conclusions

Through this research we have been able to gain further insight into the coordination of monosaccharides to the group 4 metals Ti and Zr. Properly protected allo- and glucopyranosidato ligands were reacted with $CpTiCl_3$ and Cp^*TiCl_3 precursor molecules to obtain organotitanium carbohydrate complexes. With the help of X-ray structure analyses, the coordination properties of the synthesized compounds could be studied. Furthermore, initial catalytic polymerization and hydroamination reactions were carried out using the dititanium carbohydrate compounds **1** and **5** as pre-catalysts. These experiments were carried out successfully. The stereoinformation of the catalyst precursor compounds, however, could not be conveyed to the products.

For the glucopyranosidato zirconium complex **7**$^-$ the exchange of the amine of the ammonium cation was investigated. The possibility of using the chiral zirconate cavity for separation of optically active amines and performance of stereoselective nucleophilic addition has been shown.

References

1 Klüfers, P. and Kunte, T. (2003) *Chemistry - A European Journal*, **78**, 2013–2018; Klüfers, P. and Kunte, T. (2001) *Angewandte Chemie*, **113**, 4356–4358; (2001) *Angewandte Chemie (International Edition in English)*, **40**, 4210–4212; Diéguez, M., Pàmies, O., Ruiz, A., Castillón, S. and Claver, C. (2001) *Chemistry - A European Journal*, **7**, 3086–3094; Pàmies, O., van Strijdonck, G.P.F., Diéguez, M., Deerenberg, S., Net, G., Ruiz, A., Claver, C., Kamer, P.C.J. and van Leeuwen, P.W.N.M. (2001) *The Journal of Organic Chemistry*, **66**, 8867–8871; Junicke, H., Bruhn, C., Kluge, R., Serianni, A.S. and Steinborn, D. (1999) *Journal of the American Chemical Society*, **121**, 6236–6241; Steinborn, D., Junicke, H. and Bruhn, C. (1997) *Angewandte Chemie*, **109**, 2803–2805; Steinborn, D., Junicke, H. and Bruhn, C. (1997) *Angewandte Chemie (International Edition in English)*, **36**, 2686–2688.

2 You, J.-S., Shao, M.-Y. and Gau, H.-M. (2000) *Organometallics*, **19**, 3368–3373; Riediker, M. and Duthaler, R.O. (1989) *Angewandte Chemie*, **101**, 488–490; Riediker, M. and Duthaler, R.O. (1989) *Angewandte Chemie (International Edition in English)*, **28**, 494–495; Duthaler, R.O., Herold, P., Lottenbach, W., Oertle, K. and Riediker, M. (1989) *Angewandte Chemie*, **101**, 490–491; Duthaler, R.O., Herold, P., Lottenbach, W., Oertle, K. and Riediker, M. (1989) *Angewandte Chemie (International Edition in English)*, **28**, 495–497.

3 Huang, H., Zhang, Z., Luo, H., Bai, C., Hu, X. and Chen, H. (2004) *The Journal of Organic Chemistry*, **69**, 2355–2361; Cossy, J., BouzBouz, S., Pradaux, F., Willis, C. and Bellosta, V. (2002) *Synlett*, 1595–1606; Diéguez, M., Ruiz, A. and Caver, C. (2001) *Chemical Communications*, 2702–2703; Hafner, A., Duthaler, R.O., Marti, R., Rihs, G., Rothe-Streit, P. and Schwarzenbach, F. (1992) *Journal of the American Chemical Society*, **114**, 2321–2336.

4 Küntzer, D., Jessen, L. and Heck, J. (2005) *Chemical Communications*, 5653–5655.

5 Küntzer, D., Tschersich, S. and Heck, J. (2007) *Zeitschrift fur Anorganische und Allgemeine Chemie*, **663**, 43–45.

6 Jessen, L., Haupt, E.T.K. and Heck, J. (2001) *Chemistry - A European Journal*, **7**, 3791–3797.

7 Muddasani, P.R., Bernet, B. and Vasella, A. (1994) *Helvetica Chimica Acta*, **77**, 334–350.

8 Nomura, K., Naga, N., Miki, M., Yanagi, K. and Imai, A. (1998) *Organometallics*, **17**, 2152–2154.

9 Watanabe, T., Wada, T., Machi, S. and Takehisa, M. (1972) *Journal of Polymer Science, Part B: Polymer Letters*, **10**, 741–745.

10 Kaminsky, W. (2004) *Journal of Polymer Science, Part A: Polymer Chemistry*, **42**, 3911–3921.

11 Müller, C., Loos, C., Schulenberg, N. and Doye, S. (2006) *European Journal of Organic Chemistry*, 2499–2503.

12 Gribkov, D.V. and Hultzsch, K.C. (2004) *Angewandte Chemie*, **116**, 5659–5663.

13 Hultzsch, K.C. (2005) *Advanced Synthesis and Catalysis*, **347**, 367–391; Hultzsch, K.C. (2005) *Organic and Biomolecular Chemistry*, **3**, 1819–1824; Roesky, P.W. and Müller, T.E. (2003) *Angewandte Chemie – International Edition*, **42**, 2708–2710; Müller, T.E. and Beller, M. (1998) *Chemical Reviews*, **98**, 675–703.

14 Wood, M.C., Leitch, D.C., Yeung, C.S., Kozak, J.A. and Schafer, L.L. (2007) *Angewandte Chemie*, **119**, 358–362; Watson, D.A., Chiu, M. and Bergman, R.G. (2006) *Organometallics*, **25**, 4731–4733; Knight, P.D., Munslow, I., O'Shaughnessy, P.N. and Scott, P. (2004) *Chemical Communications*, 894–895.

15 Gagné, M.R., Brard, L., Conticello, V.P., Giardello, M.A., Stern, C.L. and Marks, T.J. (1992) *Organometallics*, **11**, 2003–2005.

16 Gansäuer, A., Narayan, S., Schiffer-Ndene, N., Bluhm, H., Oltra, J.E., Cuerva, J.M., Rosales, A. and Nieger, M. (2006) *Journal of Organometallic Chemistry*, **691**, 2327–2331; Cesarotti, E., Kagan, H.B., Goddard, R. and Krüger, C. (1978) *Journal of Organometallic Chemistry*, **162**, 297–309.

17 Kaneti, J., Kirby, A.J., Koedjikov, A.H. and Pojarlieff, I.G. (2004) *Organic and Biomolecular Chemistry*, **2**, 1098–1103.

18 Johnson, J.S. and Bergman, R.G. (2001) *Journal of the American Chemical Society*, **123**, 2923–2924.

19 Luo, Z., Li, B., Fang, X., Hu, K., Wu, X. and Fu, E. (2007) *Tetrahedron Letters*, **48**, 1753–1756.

20 (a) Hedberg, C., Pinho, P., Roth, P. and Andersson, P.G. (2000) *The Journal of Organic Chemistry*, **65**, 2810–2812; (b) Domingo, L.R., Oliva, M. and Andres, J. (2001) *The Journal of Organic Chemistry*, **66**, 6151–6157.

21 Jung, M.E. (1988) *Journal of the American Chemical Society*, **110**, 3965–3969.

22 Santamaria, J., Kaddachi, M.T. and Ferroud, C. (1992) *Tetrahedron Letters*, **33**, 781–784.

23 (a) Becke, A.D. (1993) *Journal of Chemical Physics*, **98**, 5648–5652; (b) Lee, C., Yang, W. and Parr, R.G. (1998) *Physical Review B-Condensed Matter*, **37**, 785–789.

10
Reactions of C–F Bonds with Titanocene and Zirconocene: From Secondary Interaction via Bond Cleavage to Catalysis

Uwe Rosenthal, Vladimir V. Burlakov, Perdita Arndt, Anke Spannenberg, Ulrike Jäger-Fiedler, Marcus Klahn, and Marko Hapke

Dedicated to Professor Dr. mult. Dietmar Seyferth on the occasion of his 80th birthday.

10.1
Introduction and Background

Coordinatively and electronically unsaturated organometallic complex fragments form the basis for stoichiometric and catalytic reactions of a large array of substrates. These core complexes are able to coordinate to the substrates, activate them, and direct the subsequent reaction to the desired products. This is nicely illustrated in the multifaceted reactions of the complex fragments titanocene (Cp_2Ti) and zirconocene (Cp_2Zr) as unstable 14-electron compounds with a d^2 configuration (M(II)) [1]. The interactions between occupied and unoccupied orbitals can explain why these metallocenes (Cp_2M) react with a variety of unsaturated compounds with formation of metallacycles, which have a high potential to undergo diverse conversions with further substrates. A question of great importance in this context is the choice of the appropriate precursor compound. In other words, the ligands should have the feature of sufficiently stabilizing the metallocene fragment but be able to be released under mild conditions to generate the unstable and very reactive metallocene core.

Many systems are very well suited to generate titanocene Cp_2Ti or zirconocene Cp_2Zr. The applicability and the success of these systems clearly depend on their preparative accessibility, on the selectivity of the conversions, and on the inertness of the stabilizing ligands in the intended reactions. As reagents we described the stable and well-defined metallocene bis(trimethylsilyl)acetylene complexes $Cp_2M(L)(\eta^2\text{-}Me_3SiC_2SiMe_3)$ ($Cp = \eta^5\text{-}C_5H_5$, M = Ti, without L (**1**); M = Zr, L = THF (**2a**); M = Zr, L = pyridine (**2b**)), the pentamethylcyclopentadienyl complexes $Cp^*{}_2M(\eta^2\text{-}Me_3SiC_2\text{-}SiMe_3)$ ($Cp^* = \eta^5\text{-}C_5Me_5$; M = Ti (**3**), Zr (**4**)) and the ethene-bis-tetrahydroindenyl complexes *rac*-(ebthi)$M(\eta^2\text{-}Me_3SiC_2SiMe_3)$ (M = Ti (**5**), Zr (**6**)) (Scheme 10.1) [1].

All these complexes were synthesized under anaerobic conditions by the reaction of the corresponding dichlorides Cp'_2MCl_2 (Cp′ = substituted Cp; M = Ti, Zr) with magnesium in THF at room temperature in the presence of bis(trimethylsilyl) acetylene. We have already reviewed the vast chemistry of these complexes in several

Activating Unreactive Substrates: The Role of Secondary Interactions.
Edited by Carsten Bolm and F. Ekkehardt Hahn

ISBN: 978-3-527-31823-0

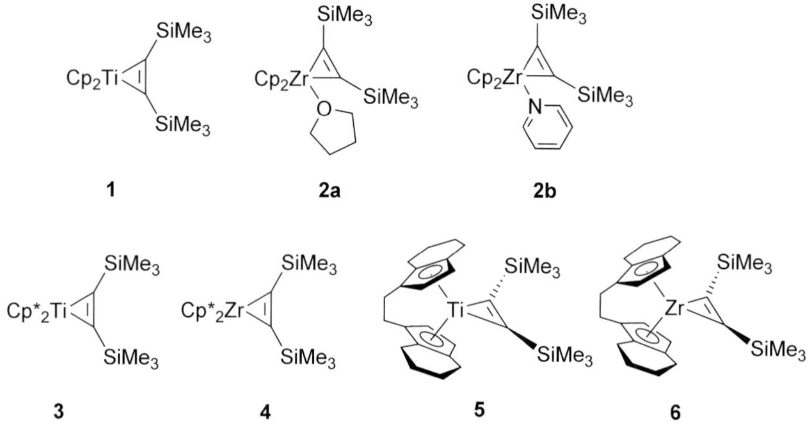

Scheme 10.1 Selected stable and well-defined titanocene and zirconocene reagents incorporating bis(trimethylsilyl)acetylene.

reports [1]. By using different Cp′ ligands (Cp, Cp*, *rac*-ebthi, etc.), additional ligands (THF, pyridine) and metals (Ti, Zr) a fine tuning of the reactivity was realized. A remarkable outcome from these modifications was the activation of unreactive bonds, which had already been described in some of the reviews. For example, cleavage of C—H, C—C (Cp-ring opening, cleavage of butadiynes), Si—C, P—C, N—H, N—C, N—N, Si—O, N—O, C—O, C—S, and C—B bonds gave complexes of potential applicability in stoichiometric and catalytic reactions [1].

This reactivity can in most cases be attributed to a single ‘primary’ interaction which selectively leads to the products. However, in some cases there is the possibility for a ‘secondary’ interaction to occur with the coordinated substrates and which could thereupon lead to further reactions. Following the concept of using these secondary interactions with the metal for the activation of unreactive bonds we came across the question whether the investigated systems could realize stoichiometric or catalytic cleavage of the strongest carbon–heteroatom connection: the C—F bond. The inertness of fluorocarbons as a consequence of the great strength of the C—F bond arises from the small size and the high electronegativity of the fluorine atom. Nevertheless, the possibility of the activation of several types of carbon–fluorine bonds by titanocene and zirconocene complexes has been summarized in a number of reviews [2].

10.2 Secondary Interactions with C—F Bonds

10.2.1 Reactions of Metallacyclopropenes with $B(C_6F_5)_3$

The reaction of **1** with $B(C_6F_5)_3$ in toluene produced the zwitterionic titanium(III) complex $CpTi[\eta^5\text{-}C_5H_4B(C_6F_5)_3]$ (**7**) by electrophilic substitution of a hydrogen atom at one of the Cp rings by the $B(C_6F_5)_3$ molecule (Scheme 10.2) [3a].

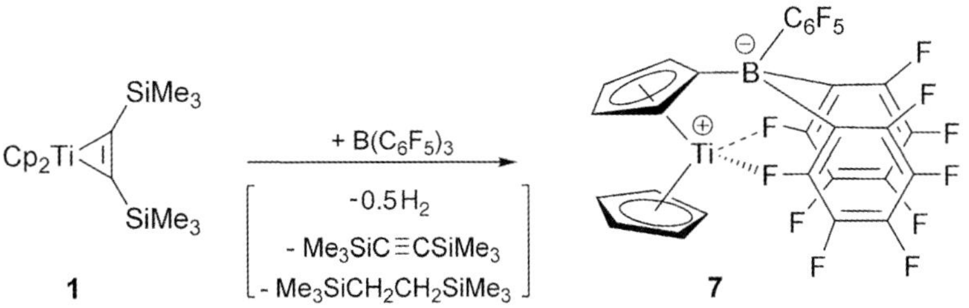

Scheme 10.2 Treatment of **1** with $B(C_6F_5)_3$ gives, after an electrophilic substitution, the zwitterionic titanium(III) complex **7**.

A characteristic feature of this zwitterionic complex is the presence of coordinative bonds between the *ortho*-F atoms of two C_6F_5-substituents and the positively charged titanium center. The air oxidation of this complex gave the dimeric titanoxane $\{Cp[\eta^5\text{-}C_5H_4B(C_6F_5)_3]Ti\}_2O$ (**8**) with two zwitterionic centers in the molecule [3b]. Only one of the *ortho*-F atoms of a $B(C_6F_5)_3$ group coordinates to the titanium atom in each of these units. With acetone the complex $CpTi[\eta^5\text{-}C_5H_4B(C_6F_5)_3]$ (**7**) gave the zwitterionic adduct $Cp[\eta^5\text{-}C_5H_4B(C_6F_5)_3]Ti(O{=}CMe_2)$ (**9**) wherein a molecule of acetone is coordinated to the Ti(III) center via the oxygen atom. Also, in this adduct only one *ortho*-F atom of a $B(C_6F_5)_3$ group coordinates to the metal (Scheme 10.3) [3b].

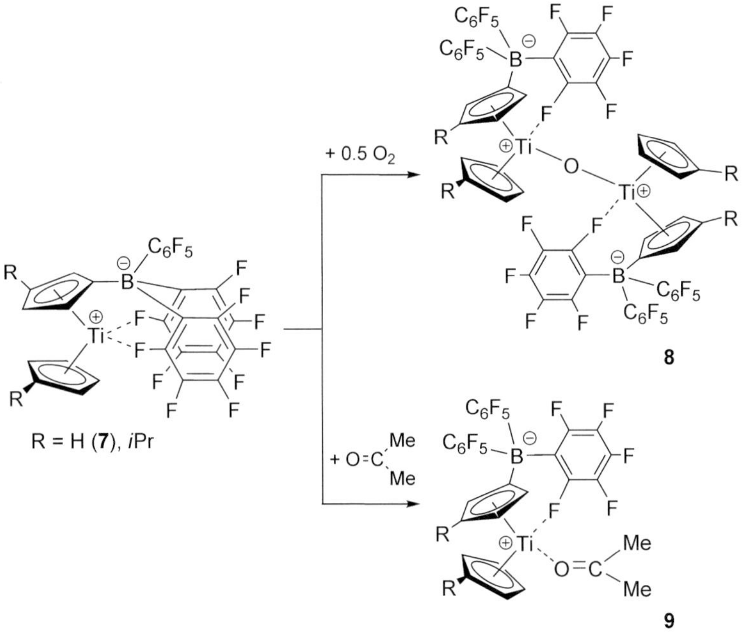

Scheme 10.3 Reactions of **7** with oxygen and acetone, respectively.

Different reactions of the permethylmetallocene alkyne complexes $Cp^*{}_2M(\eta^2\text{-}PhC_2SiMe_3)$ with M = Ti and Zr were found with $B(C_6F_5)_3$, giving a functionalization of the pentamethylcyclopentadienyl ligand [3c]. The compound [Cp*Ti{$C_5Me_4CH_2B(C_6F_5)_3$}] (**10**) is formed by dissociation of the alkyne and an electrophilic substitution of a hydrogen atom at one methyl group of the Cp* ligand together with the liberation of hydrogen (Scheme 10.4).

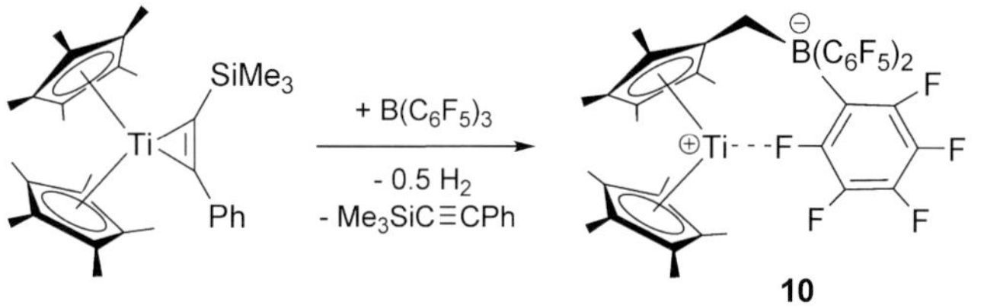

Scheme 10.4 Starting from $Cp^*{}_2Ti(\eta^2\text{-}PhC_2SiMe_3)$ and $B(C_6F_5)_3$, a zwitterionic titanium(III) complex (**10**) is formed.

In complex **10** the *ortho*-F atom of only one of the perfluorphenyl groups coordinates to the titanium (Ti–F distance 2.406(3) Å). This Ti–F distance is longer than either of the two Ti–F distances (2.223(3) Å and 2.248(2) Å) found in [CpTi{$\eta^5\text{-}C_5H_4B(C_6F_5)_3$}] (**7**). The alkenyl compound [Cp*Zr{C(Ph) = CH($SiMe_3$)}{$C_5Me_4CH_2B(C_6F_5)_3$}] (**11**) is also formed by an electrophilic substitution of a hydrogen atom at one methyl group of the Cp* ligand. In this case the product was formed without dissociation of the alkyne and without liberation of hydrogen (Scheme 10.5) [3c].

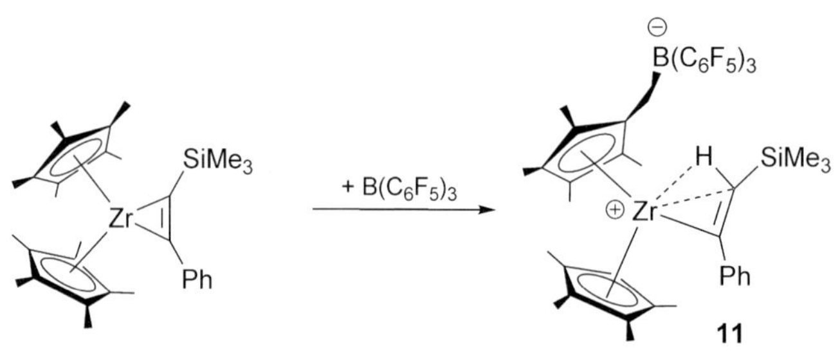

Scheme 10.5 In the reaction of $Cp^*{}_2Zr(\eta^2\text{-}PhC_2SiMe_3)$ with $B(C_6F_5)_3$ the alkenyl complex **11** is obtained.

The reactions of complexes *rac*-(ebthi)M($\eta^2\text{-}Me_3SiC_2SiMe_3$) with $B(C_6F_5)_3$ proceed completely differently for M = Ti (**5**) and M = Zr (**6**) [3d]. With the zirconium complex **6**, again an electrophilic substitution at one of the Cp rings occurs, forming the alkenyl compound **12** (Scheme 10.6).

The molecular structure shows the (ebthi)Zr-moiety, which is substituted in 3-position by the boranate. A single *ortho*-F atom coordinates the zirconium, which is accompanied by the elongation of the corresponding C–F bond of the C_6F_5-ring. This coordination is also accompanied by the ^{19}F-NMR signal at −200.2 ppm being shifted to extremely high field. The alkenyl group is σ-bonded to Zr and has an additional agostic interaction. In solution at 100 °C, the alkyne is eliminated and one of the B–C bonds is cleaved, with a subsequent transfer of a pentafluorophenyl group

Scheme 10.6 Reaction of *rac*-(ebthi) zirconium alkyne complex **6** with $B(C_6F_5)_3$.

from boron to zirconium, whereupon complex **13** with one bridging μ-H-atom between boron and zirconium is formed. The compound obtained can be described either as Zr(IV)-σ-C_6F_5-hydrido complex with an electroneutral borane ligand or as a zwitterionic betaine-boranate. After this thermal interconversion the high field-shifted ^{19}F-NMR signal disappeared, indicating that no Zr-F interaction in the product exists any more.

The Ti-complex *rac*-(ebthi)Ti(η^2-$Me_3SiC_2SiMe_3$) (**5**) reacts, quite surprisingly, totally differently with $B(C_6F_5)_3$ compared to the Zr-compound (**6**) [3d]. Neither the electrophilic substitution of H-atoms nor an elimination of the alkyne was observed, but following a C—Si bond cleavage and subsequent B—C bond formation the first zwitterionic titanocene(III)-η^2-trimethylsilylalkynyl-boranate complex [*rac*-(ebthi) Ti]$^+$[η^2-$Me_3SiC_2B(C_6F_5)_3$]$^-$ (**14**) was obtained (Scheme 10.7).

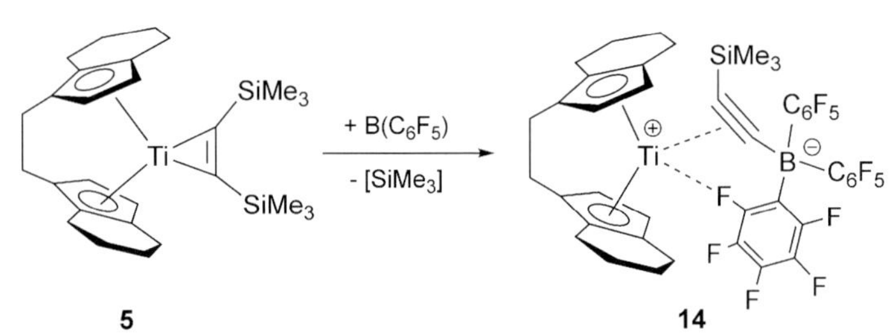

Scheme 10.7 *rac*-(ebthi)Ti(η^2-$Me_3SiC_2SiMe_3$) (**5**) forms, together with $B(C_6F_5)_3$, the zwitterionic titanocene(III)-η^2-trimethylsilylalkynyl-boranate complex **14**.

The trimethylsilylalkynyl-boranate moiety [$Me_3SiC{\equiv}CB(C_6F_5)_3$]$^-$ shows an unsymmetrical coordination mode and is only weakly attached. One of the *ortho*-F atoms is coordinated at titanium, and therefore the situation can be described as kind of a bidentate secondary coordination.

In addition to these representative examples, other complexes were isolated and studied, in which the molecular structures showed more or less distinctive interactions of one or more *ortho*-F atoms to the metal, which we describe here in terms of a 'secondary' interaction [3e–i].

10.2.2
Reactions of Five-membered Metallacycles with $B(C_6F_5)_3$

Certain ion pairs $[Cp'_2ZrR]^+[RB(C_6F_5)_3]^-$, which can be generated by the formal abstraction of methyl anions from Group 4 dimethyl complexes Cp'_2MMe_2 (M = Ti, Zr) with the help of the strong Lewis acid $B(C_6F_5)_3$, are active catalysts for the polymerization of α-olefins. Several metallacycles, such as metallacyclopropanes $Cp'_2M(\eta^2\text{-}1,2\text{-}C_2R_4)$, -propenes $Cp'_2M(\eta^4\text{-}1,2\text{-}C_2R_2)$, -pentanes $Cp'_2M(\eta^2\text{-}1,4\text{-}C_4R_8)$, -pentenes $Cp'_2M(\eta^2\text{-}1,4\text{-}C_4R_6)$, and pentadienes $Cp'_2M(\eta^2\text{-}1,4\text{-}C_4R_4)$ (R = H, alkyl, aryl etc.) also undergo similar activation reactions for the catalytic olefin polymerization. What all the mentioned metallacycles have in common are the two M–C σ-bonds like those found in the dimethyl complexes described above.

Not in the scope of this project but for reasons of comparison the reactions of rather unusual five-membered metallacycles, such as zirconacyclocumulenes $Cp'_2M(\eta^4\text{-}1,2,3,4\text{-}RC_4R)$ [1i] and 1-metallacyclopent-3-ynes $Cp'_2M(\eta^4\text{-}1,2,3,4\text{-}H_2C_4H_2)$, [1i] with $B(C_6F_5)_3$ were investigated. Interestingly, the five-membered zirconacyclocumulenes $Cp^*_2Zr(\eta^4\text{-}1,2,3,4\text{-}RC_4R)$ (R = Me or Ph) showed varying reactions with $B(C_6F_5)_3$ (Scheme 10.8) [4a].

For R = Me the $B(C_6F_5)_3$ attacks the β-C-atom of the starting cumulene, forming complex **15** with a hex-2-ene-4-yne-2-yl-3-[tris(pentafluorophenyl)borate] ligand at the permethylzirconocene center. With R = Ph the $B(C_6F_5)_3$ attacks the α-C-atom of the starting cumulene and yields complex **16** with a 1,4-diphenylbuta-1,2,3-triene-1-yl-4-[tris(pentafluorophenyl)borate] ligand coordinated at the permethylzirconocene center.

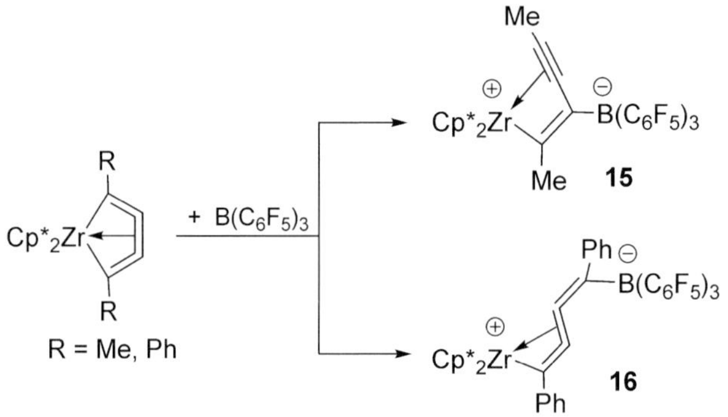

Scheme 10.8 Five-membered zirconacyclocumulenes showed variable reactions with $B(C_6F_5)_3$ depending on the substituent R of the cumulenic ring.

The 1-metalla-cyclopent-3-ynes $Cp_2Ti(\eta^4\text{-}1,2,3,4\text{-}H_2C_4H_2)$ and *rac*-(ebthi)$Zr(\eta^4\text{-}1,2,3,4\text{-}H_2C_4H_2)$ react with $B(C_6F_5)_3$ under ring opening of the metallacycle and formation of the zwitterionic complexes $[Cp_2Ti]^+[-CH_2C{\equiv}CCH_2-B(C_6F_5)_3]^-$ (**17**) and [*rac*-(ebthi)Zr]$^+$ $[-CH_2C{\equiv}CCH_2-B(C_6F_5)_3]^-$ (**18**) (Scheme 10.9) [4b,c]. The molecular structures of these and similar complexes [4c] show a but-2-yne-1,4-diyl ligand bridging the metallocenium center and the formed boranate.

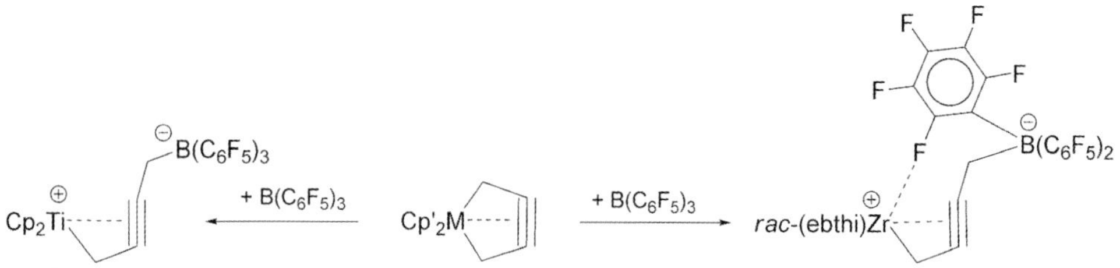

Scheme 10.9 1-Metalla-cyclopent-3-ynes react with $B(C_6F_5)_3$ with ring-opening of the metallacycle and formation of zwitterionic complexes **17** and **18**.

For the saturated zirconacyclopentane *rac*-(ebthi)Zr(η^2-1,4-C_4H_8) in the reaction with $B(C_6F_5)_3$ a similar ring opening of the metallacycle and formation of the zwitterionic complex $[rac\text{-(ebthi)Zr}]^+[-CH_2-CH_2-CH_2-CH_2-B(C_6F_5)_3]^-$ was assumed, which is active in ethene polymerization [4d]. These examples emphasize the importance of the proper sequence of the addition of reagents to achieve catalytic activity. If the alkyne complex *rac*-(ebthi)Zr(η^2-$Me_3SiC_2SiMe_3$) (**6**) is pretreated with ethene to yield the zirconacyclopentane *rac*-(ebthi)Zr(C_4H_8), which is subsequently activated by $B(C_6F_5)_3$, an active catalyst for polymerization results. If on the other hand the borane is added to the alkyne compound firstly, complex (**12**) or its thermolysis product (**13**) is formed (see Scheme 10.6). Neither of these reacts with ethene and, accordingly, they are inactive in polymerization reactions [3d].

10.3 Formation of M–F Bonds

10.3.1 Stoichiometric Cleavage of C–F Bonds

Beckhaus and coworkers published excellent examples of reactions of titanocene bis (trimethylsilyl)acetylene complexes with different *N*-heterocyclic compounds [5a]. In these reactions the complexes $Cp_2Ti(\eta^2\text{-}Me_3SiC_2SiMe_3)$ (**1**) and $Cp^*{}_2Ti(\eta^2\text{-}Me_3SiC_2SiMe_3)$ (**3**) react differently with fluorinated pyridines [5b,c]. The reactions of **1** with several pyridines of different degrees of fluorination basically showed identical results. In each case the higher 'concentration' of potentially accessible hydrogen atoms when going from the perfluorinated pyridine to the 2-fluoropyridine had no effect on the outcome, and in all cases F-bridged binuclear titanium complexes were obtained (Scheme 10.10). It is quite remarkable that no product of a C–H activation was found in this study at all.

With $Cp^*{}_2Ti(\eta^2\text{-}Me_3SiC_2SiMe_3)$ (**3**) the pentafluoropyridine forms, after C–F activation, a mononuclear titanium(IV) mono-fluoro complex $Cp^*{}_2Ti(2\text{-}C_5NF_4)F$ (**22**) as an intermediate as proved by ^1H- and ^{19}F-NMR measurements. Complex **22** reacts

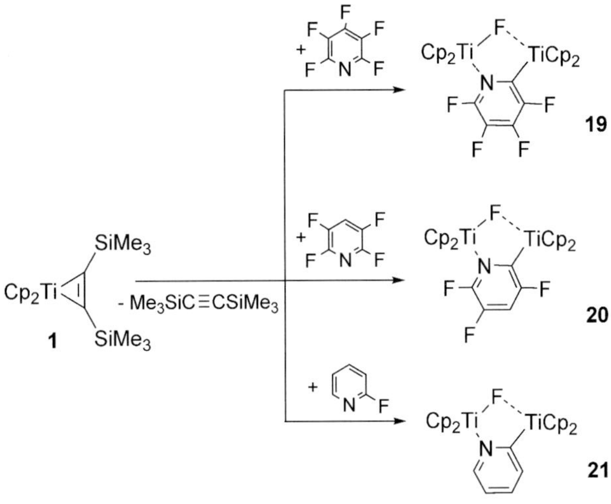

Scheme 10.10 In reactions of pyridines possessing different degrees of fluorination with the titanocene complex **1** the C–F bond activation is favored.

with an excess of the substrate to give the difluoride $Cp^*{}_2TiF_2$. A similar reaction was found for the cyanuric fluoride (Scheme 10.11).

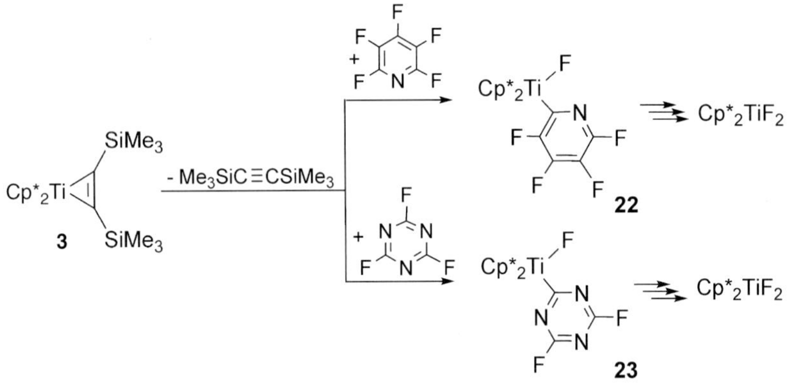

Scheme 10.11 $Cp^*{}_2Ti(\eta^2\text{-}Me_3SiC_2SiMe_3)$ (**3**) gives with fluorinated heterocycles an elimination of the alkyne and a C–F bond activation.

The analogous zirconocene complexes gave totally different reactions [6]. For example, $Cp_2Zr(THF)(\eta^2\text{–}Me_3SiC_2SiMe_3)$ (**2a**) and the complex *rac*-(ebthi)Zr $(\eta^2\text{–}Me_3SiC_2SiMe_3)$ (**6**) react with 2,3,5,6-tetrafluoropyridine under C–H bond activation to furnish the fluorosubstituted pyridyl complexes, metalated in 4-position, with agostic alkenyl groups $Cp'_2Zr(4\text{-}C_5NF_4)[(\text{–}C(SiMe_3)=CH(SiMe_3)]$ ($Cp'_2 = Cp_2$, ebthi (**24**)) (Scheme 10.12). Complex **6** yields, with 2,3,4,6-tetrafluoropyridine after

C–H bond activation, two isomers of the corresponding fluorosubstituted pyridyl complexes, metalated in 3-position, *rac*-(ebthi)Zr(3-C_5NF_4)[(–C($SiMe_3$) = CH($SiMe_3$)] (**25a**, **25b**) with agostic alkenyl groups in a 1:2 ratio (Scheme 10.12).

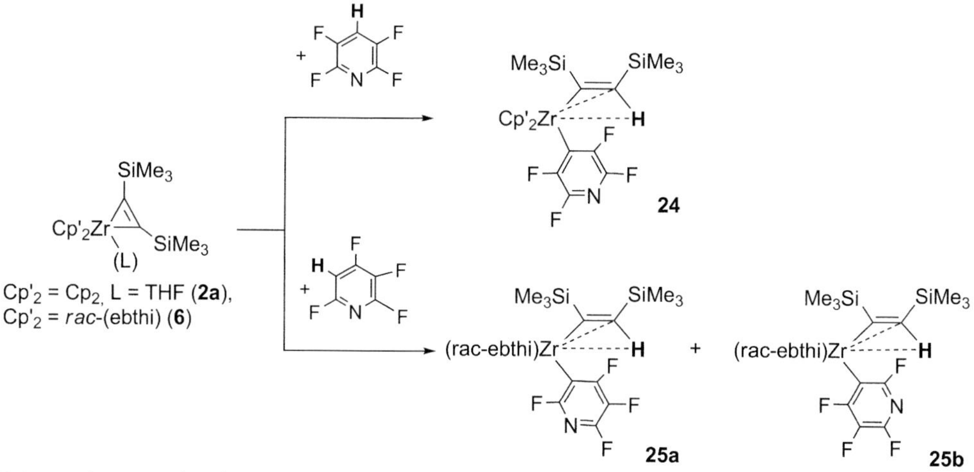

Scheme 10.12 C–H bond activation is preferred in reactions of the zirconocene complexes **2a** and **6** with different substituted tetrafluoropyridines.

In the reaction of pentafluoropyridine and **2b** an elimination of bis(trimethylsilyl) acetylene and a C–F bond activation in 4-position with the formation of Cp_2Zr(4-C_5NF_4)F (**26**) were found (Scheme 10.13) [6].

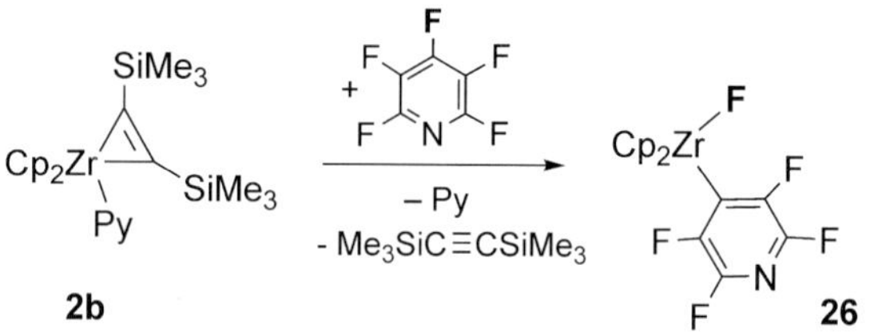

Scheme 10.13 In the reaction of the zirconocene complex **2b** and pentafluoropyridine a C–F bond activation is found after elimination of the alkyne.

For zirconium such a species was isolated, whereas in the above-mentioned reaction of $Cp^*{}_2Ti(\eta^2\text{-}Me_3SiC_2SiMe_3)$ (**3**) with pentafluoropyridine the similar mononuclear titanium(IV) mono-fluoro complex $Cp^*{}_2Ti$(2-C_5NF_4)F was only assumed as an intermediate which proceeds to the difluoride $Cp^*{}_2TiF_2$ (see Scheme 10.11).

In summary, for zirconium C–H activation is preferred rather than C–F activation. The attack at C–F bonds occurs for zirconium in the 4-position and not in the 2-position as described for titanium. Both positions are activated, but in the case of zirconium the 4-position is favored.

10.3.2
Stoichiometric Formation by M–C Bond Cleavage and Exchange Reactions

Starting from zirconocene bis(trimethylsilyl)acetylene complexes $Cp'_2Zr(L)(\eta^2\text{-}Me_3SiC_2\text{-}SiMe_3)$ via the 2-vinylpyridine complexes $Cp'_2Zr(2\text{-vipy})$ [7a] a novel synthetic method for zirconocene difluoride and alkyl-monofluoride complexes was elaborated [7b]. We have been able to synthesize a number of examples with selected different ligand systems. Different methods for the preparation of the zirconocene difluorides Cp'_2ZrF_2 (**27**) and $Cp'_2Zr(F)(CH_2CH_2\text{-}2\text{-}Py)$ (**28**) were reported and compared (Scheme 10.14) [7b].

+ $NEt_3 \cdot 3HF$ or $HBF_4 \cdot$ether

- et-py

Cp'_2ZrF_2 **27**

Cp'_2Zr

$NEt_3 \cdot 3HF$ | - et-py

+ $NEt_3 \cdot 3HF$

Cp'_2Zr **28**

Scheme 10.14 Preparation of zirconocene difluoride and alkyl-monofluoride complexes.

10.4
Stoichiometric Formation of Zr–H Bonds

10.4.1
From Zr–F/Al–H to Zr–H/Al–F Bonds

In the reaction of *rac*-(ebthi)ZrF_2 with two equiv. of $i Bu_2AlH$ the fluoride substituents were replaced by hydrogen and the dimeric complex containing two terminal and two bridging H atoms [*rac*-(ebthi)ZrH(μ-H)]$_2$ (**29**) was formed (Scheme 10.15) [8].

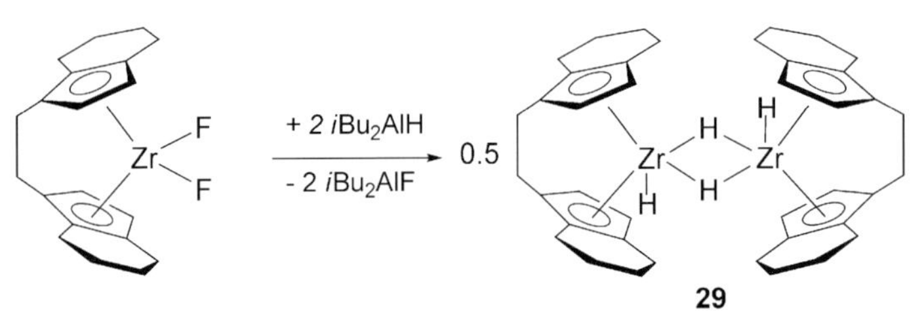

Scheme 10.15 *rac*-(ebthi)ZrF_2 reacts with two equiv. of $i Bu_2AlH$ yielding the dimeric [*rac*-(ebthi)ZrH(μ-H)]$_2$ (**29**).

In contrast to this behavior the chloride *rac*-(ebthi)$ZrCl_2$ did not react at all with $i$$Bu_2AlH$ under analogous conditions and only unchanged starting material was isolated.

In reactions of the zirconocene 2-vinylpyridine complex Cp_2Zr(2-vipy) with $i$$Bu_2AlH$ and $i$$Bu_2AlF$ the isostructural complexes [$Cp_2Zr(\mu\text{-}\eta^1:\eta^2\text{-}2\text{-}CH_2CH\text{-}C_5H_4N)(\mu\text{-}H)$][$i$$Bu_2Al$] (**30**) and [$Cp_2Zr(\mu\text{-}\eta^1:\eta^2\text{-}2\text{-}CH_2CH\text{-}C_5H_4N)(\mu\text{-}F)$][$i$$Bu_2Al$] (**31**) were formed (Scheme 10.16) [8]. These complexes can serve as a very nice model for the transfer of fluoride from zirconium to aluminum (Zr–F/Al–H to Zr–H/Al–F) found in the reaction of *rac*-(ebthi)ZrF_2 with $i$$Bu_2AlH$ to form the dimeric complex [*rac*-(ebthi)ZrH(μ-H)]$_2$ (**29**). These compounds are important for the understanding of the activation of zirconocene fluoro complexes and the role of $i$$Bu_2AlH$ in activation processes for the catalytic polymerization of olefins.

Scheme 10.16 In the reactions of a zirconocene 2-vinylpyridine complex with $i$$Bu_2AlH$ or $i$$Bu_2AlF$ the isostructural complexes [$Cp_2Zr(\mu\text{-}\eta^1:\eta^2\text{-}2\text{-}CH_2CH\text{-}C_5H_4N)(\mu\text{-}H)$][$i$$Bu_2Al$] (**30**) and [$Cp_2Zr(\mu\text{-}\eta^1:\eta^2\text{-}2\text{-}CH_2CH\text{-}C_5H_4N)(\mu\text{-}F)$][$i$$Bu_2Al$] (**31**) are formed.

In addition, the interaction of metallacyclopropenes and five-membered metallacyclocumulenes with $i$$Bu_2AlH$ was investigated, leading to bi- or trimetallic complexes. Some of these were identified to be active catalysts in the polymerization of ethene and also in ring-opening polymerization [9].

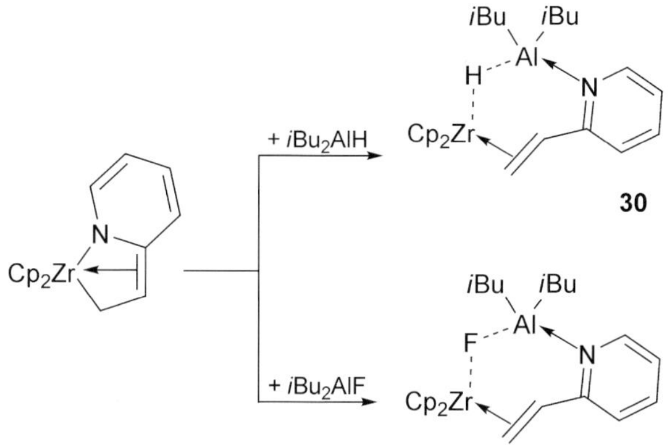

10.5 Catalytic Formation of Zr–H Bonds

10.5.1 From Zr–F using Al–H to Zr–H and Al–F Bonds

In the light of all these results obtained for zirconocene difluorides a novel catalytic cycle (Scheme 10.17) was realized in which cleavage of the Zr–F bond by interaction with Al–H sources yields Al–F and Zr–H containing compounds of which the latter reacts with C–F bonds to form C–H and again original Zr–F bonds. The driving force for this cycle is the formation of strong Al–F bonds. These investigations are especially relevant in terms of deactivation steps in the olefin polymerization [10, 11] and catalytic dehydrofluorination of fluorocarbons [12].

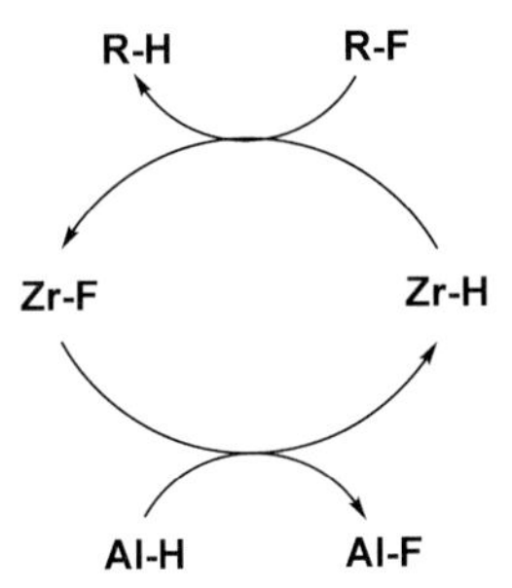

Scheme 10.17 Catalytic cycle of the activation of Cp'_2ZrF_2 with Al–H sources.

10.5.2
Catalytic Ethene Polymerization

Two of the primarily discussed deactivating side reactions, leading to poisoning of the catalyst and thereby decreasing its productivity, feature either the C_6F_5-group transfer to the cationic metal center or the fluorine transfer to the metal. It was thought that zirconocene fluoro complexes are not active catalysts and that their formation is an important deactivation process [10]. Nevertheless, we found that in contrast to the dichloride *rac*-(ebthi)$ZrCl_2$, the analogous difluoride *rac*-(ebthi)ZrF_2 is very active as a precatalyst for ethene polymerization, with *i*Bu_3Al (or *i*Bu_2AlH) as the sole activating cocatalyst. This finding was corroborated by the reaction of *rac*-(ebthi)ZrF_2 with two equiv. of *i*Bu_2AlH (Scheme 10.15), yielding the dimeric complex [*rac*-(ebthi)ZrH (μ-H)]$_2$ (**29**). This complex indeed showed activity when applied in polymerizations. As mentioned above, in the reaction of *rac*-(ebthi)$ZrCl_2$ with *i*Bu_2AlH under analogous conditions the chloride ligands were not replaced by hydrogen. For comparison of the ability of zirconium and aluminum to coordinate with F-atoms we studied the reactions of zirconium 2-vinylpyridine complexes with *i*Bu_2AlH and *i*Bu_2AlF (Scheme 10.16). The easier the abstraction of a fluoride ligand from the metal center of the starting precatalyst, the more reactive is the resulting system. The cleavage of the Zr–F bond by interaction with Al–H sources, including the transfer of the fluoride substituent from zirconium to aluminum and formation of Zr–H and Al–F bonds, was found to be the reason for this 'fluoride effect'.

On the other hand, it was shown that zirconium hydride complexes were frequently used in C–F bond activation reactions. In the light of the above-mentioned results for zirconocene difluorides in the catalytic polymerization of olefins, we came across the question whether the complex [*rac*-(ebthi)ZrH(μ-H)]$_2$ (formed by reactions of Zr–F and Al–H containing compounds) can cleave the C–F bonds of $B(C_6F_5)_3$ as well [10].

Interestingly we found in the reaction of [*rac*-(ebthi)ZrH(μ-H)]$_2$ (**29**) with B(C_6F_5)$_3$ a nearly quantitative formation of *rac*-(ebthi)ZrF_2 (Scheme 10.18, top right). In order to collect more information on the reaction pathway, the reaction was investigated by NMR spectroscopy. During the course of these investigations we have found the appearance of several intermediates that were relatively difficult to characterize due to their inherent stability (Scheme 10.18, bottom).

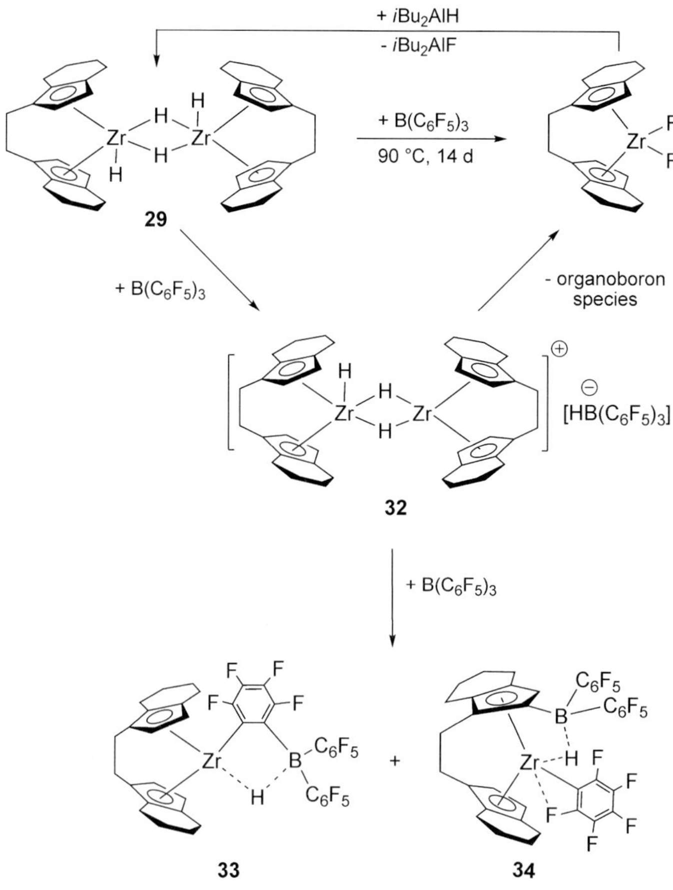

Scheme 10.18 Elementary steps of C–F bond activation in the reaction of the zirconocene hydride complex **29** with B(C_6F_5)$_3$.

The first isolated intermediate was the dinuclear zwitterionic contact ion pair [*rac*-(ebthi)Zr(H)(μ-H)$_2$Zr-*rac*-(ebthi)]$^+$ [HB(C_6F_5)$_3$]$^-$ (**32**). Treatment of **32** under thermolytic conditions resulted in the formation of *rac*-(ebthi)ZrF_2 by nucleophilic aromatic substitution. Complex **32** reacts slowly with additional B(C_6F_5)$_3$ and forms a mononuclear complex with *ortho*-metalated borane [*rac*-(ebthi)Zr(μ-H)][*o*-C_6F_4B(C_6F_5)$_2$] (**33**). A secondary product (**34**) emerges after migration of a C_6F_5 group and an electrophilic substitution at the ebthi-ligand. Both of them are considered to be relevant for catalytic olefin polymerization. The latter is totally inactive, representing a clear deactivation process, whereas the *rac*-(ebthi)ZrF_2 (obtained from [*rac*-(ebthi)ZrH (μ-H)]$_2$ (**29**) and $B(C_6F_5)_3$ by C–F bond cleavage) is reactivated by *i*Bu_2AlH.

In the olefin polymerization, zirconocene halides such as Cp'_2ZrCl_2, MAO (methylalumoxane as an activator and scavenger) and fluorosubstituted arylboranes like $B(C_6F_5)_3$ are mixed together to obtain the catalytically active systems. Additional hydrogen is added to control the molecular weight of the polymer. The formed methyl complexes $Cp^*_2ZrMe_2$ yielded with hydrogen, followed by elimination of methane, the hydrides $Cp^*_2ZrH_2$, which then react with $B(C_6F_5)_3$ to form the active ion pair $[Cp^*_2ZrH]^+[HB(C_6F_5)_3]^-$ [11]. After this step the formation of fluoro complexes by C–F bond cleavage was only discussed in terms of a deactivation process. Our results clearly show that these formed Zr–F bonds are effectively reactivated by the interaction with aluminum hydrides to give zirconium hydride complexes which are very active catalysts in olefin polymerizations [10].

10.5.3
Catalytic Hydrodefluorination of Activated C–F Bonds

Reaction mixtures consisting of zirconocene difluorides Cp'_2ZrF_2 as precatalysts and diisobutylaluminumhydride *i*Bu_2AlH as activator are active catalysts in the room-temperature hydrodefluorination (HDF) of fluorinated pyridines. Evaluation of these systems established *rac*-(ebthi)ZrF_2 and Cp_2ZrF_2 together with *i*Bu_2AlH as active catalysts in the room-temperature HDF of pentafluoropyridine (Scheme 10.19) [12].

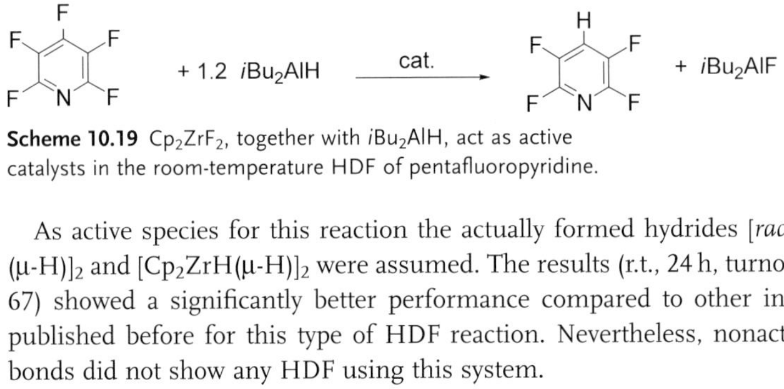

C_5F_5N + 1.2 *i*Bu_2AlH —cat.→ 4-H-C_5F_4N + *i*Bu_2AlF

Scheme 10.19 Cp_2ZrF_2, together with *i*Bu_2AlH, act as active catalysts in the room-temperature HDF of pentafluoropyridine.

As active species for this reaction the actually formed hydrides [*rac*-(ebthi)ZrH (μ-H)]$_2$ and $[Cp_2ZrH(\mu\text{-}H)]_2$ were assumed. The results (r.t., 24 h, turnover number 67) showed a significantly better performance compared to other investigations published before for this type of HDF reaction. Nevertheless, nonactivated C–F bonds did not show any HDF using this system.

10.5.4
Hydrodefluorination of Nonactivated C–F Bonds by Diisobutylaluminumhydride via the Aluminum Cation [*i*$Bu_2Al]^+$

During the course of our investigations we have found a novel system for the hydrodefluorination of nonactivated C–F bonds at room temperature [13]. The reaction of *i*Bu_2AlH with $[Ph_3C]^+[B(C_6F_5)_4]^-$, $[Ph_3C]^+[Al(C_6F_5)_4]^-$ and $[Ph_3C]^+[Al\{OC(CF_3)_3\}_4]^-$ as precatalysts leads, with formation of triphenylmethane, to the aluminum cation [*i*$Bu_2Al]^+$ and the noncoordinating anions $[M(C_6F_5)_4]^-$ (M = B, Al) and $[Al\{OC(CF_3)_3\}_4]^-$. The formed aluminum cation is very reactive toward C–F bonds and easily forms iBu_2AlF releasing a carbocation that abstracts the hydride of excess iBu_2AlH and yields the corresponding hydrocarbon. Thereby the active species $[iBu_2Al]^+$ is regenerated and can participate in the catalytic cycle (Scheme 10.20).

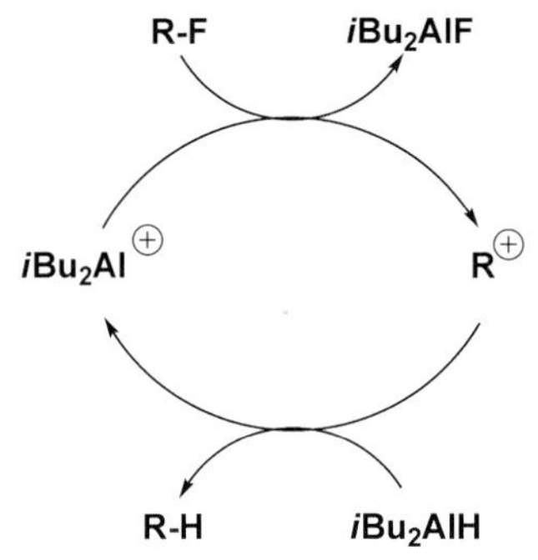

Scheme 10.20 Hydrodefluorination of nonactivated C–F bonds by iBu_2AlH via the aluminum cation $[iBu_2Al]^+$.

For 1-fluorohexane as a substrate with a nonactivated C–F bond different activities were found (turnover numbers: $[Ph_3C]^+[B(C_6F_5)_4]^-$ 20; $[Ph_3C]^+[Al(C_6F_5)_4]^-$: 12; $[Ph_3C]^+[Al\{OC(CF_3)_3\}_4]^-$: 30). There seems to be a trend in which the decreasing coordination power of the anions is responsible for this behavior. The results we obtained (r.t., 24 h) did show smaller turnover numbers compared to other investigations published before for this HDF reaction [14, 15]. A deeper discussion and comparison of the system $iBu_2AlH/[iBu_2Al]^+$ with similar systems like $Et_3SiH/[Et_3Si]^+$ [14] or a disilyl cation [15] is not justified because of the sometimes different reaction conditions and the limited data available. Nevertheless, the question arises whether the hydrides $[Cp'_2ZrH(\mu\text{-}H)]_2$ (formed in the above-mentioned mixtures from Cp'_2ZrF_2 and iBu_2AlH [12]) are really the active species. In principle they could form with additional iBu_2AlH yielding ate-complexes of the type $[iBu_2Al]^+[Cp'_2ZrH_3]^-$ in which $[iBu_2Al]^+$ as the active cation is formed also.

10.6 Conclusion

Our results are in agreement with the statement that 'In many instances organometallic fluorides have been the unexpected products of fluorine abstraction in the decomposition of complexes of fluorinated anions [16]'. Exactly this was investigated in the case of reactions of fluorinated boranes $B(C_6F_5)_3$ and boranates $[RB(C_6F_5)_3]^-$ with different zirconocene sources yielding difluorides Cp'_2ZrF_2. In addition, we have been able to show that these complexes are not responsible for deactivation processes.

With regard to the reactivity of $B(C_6F_5)_3$, there are many examples of *electrophilic substitution*, but only a very small number of examples of *nucleophilic aromatic substitution* reactions exist. The electron-releasing inductive effect of boron is sometimes overcompensated by the influence of the five F-substituents, and makes this special Lewis acid sensitive to nucleophilic attack at the electron-deficient aromatic unit [17].

The unexpected (but here explained) activity of metallocene difluoride complexes Cp'_2ZrF_2 as precatalysts in olefin polymerization and hydrodefluorination consists with what was outlined for other systems, namely that 'early transition metal fluoride complexes do *not* preclude catalytic chemistry [18]'.

Acknowledgments

This work was supported by the Deutsche Forschungsgemeinschaft, Priority Program 1118, and the federal state of Mecklenburg-Western Pomerania. Funding and facilities provided by the Leibniz-Institut für Katalyse e.V. at the University of Rostock are gratefully acknowledged. The work reported in this contribution would not have been possible without the excellent efforts by various former PhD students, the postdoctoral scientists, and technical staff, in particular Petra Bartels and Regina Jesse. We are grateful to our coworker Dr. habil. Wolfgang Baumann and many other colleagues whose names appear in the list of references. We thank Professor Rüdiger Beckhaus for many useful suggestions and discussions.

References

1 Reviews and book contributions with references cited therein: (a) Ohff, A., Pulst, S., Lefeber, C., Peulecke, N., Arndt, P., Burlakov, V.V. and Rosenthal, U. (1996) *Synlett*, 111–118; (b) Rosenthal, U., Pellny, P.-M., Kirchbauer, F.G. and Burlakov, V.V. (2000) *Accounts of Chemical Research*, **33**, 119–129; (c) Rosenthal, U. and Burlakov, V.V. (2002) in *Titanium and Zirconium in Organic Synthesis*, (ed. I. Marek), Wiley-VCH, Chapter 10, pp. 355–389; (d) Rosenthal, U., Burlakov, V.V., Arndt, P., Baumann, W. and Spannenberg, A. (2003) *Organometallics*, **22**, 884–900; (e) Rosenthal, U., Arndt, P., Baumann, W., Burlakov, V.V. and Spannenberg, A. (2003) *Journal of Organometallic Chemistry*, **670**, 84–96; (f) Rosenthal, U. (2003) *Angewandte Chemie*, **115**, 1838–1842; (2003) *Angewandte Chemie – International Edition*, **42**, 1794–1798; (g) Rosenthal, U. (2004) *Angewandte Chemie*, **116**, 3972–3977; (2004) *Angewandte Chemie – International Edition*, **43**, 3882–3887; (h) Rosenthal, U. (2003) in *Acetylene Chemistry II. Chemistry, Biology and Material Science*, (eds F. Diederich, P.J. Stang and R.R. Tykwinski), Wiley-VCH, Chapter 4, pp. 139–171; (i) Rosenthal, U., Burlakov, V.V., Arndt, P., Baumann, W., Spannenberg, A. and Shur, V.B. (2004) *European Journal of Inorganic Chemistry*, **22**, 4739–4749; (j) Rosenthal, U., Burlakov, V.V., Arndt, P., Baumann, W. and Spannenberg, A. (2005) *Organometallics*, **24**, 456–471; (k) Rosenthal, U., Burlakov, V.V., Bach, M.A. and Beweries, T. (2007) *Chemical Society Reviews*, **36**, 719–728.

2 *Examples:* (a) Jones, W.D. (2003) *Dalton Transactions*, 3991–3995; (b) Kiplinger, J.L., Richmond, T.G. and Osterberg, C.E. (1994) *Chemical Reviews*, **94**, 373–431; (c) Richmond, T.G. (2000) *Angewandte Chemie*, **112**, 3378–3380;Richmond, T.G. (2000) *Angewandte Chemie – International Edition*, **39**, 3241–3244; (d) Richmond, T.G. (1999) in *Topics in Organometallic Chemistry*, (ed. S. Murai), Vol. 3, Springer,

New York, pp. 243–269; (e) Mazurek, U. and Schwarz, H. (2003) *Chemical Communications*, 1321–1326; (f) Deck, P.A., Konate, M.A., Kelly, B.V. and Slebodnik, C. (2004) *Organometallics*, **23**, 1089–1097; (g) Kiplinger, J.L. and Richmond, T.G. (1996) *Chemical Communications*, 1115–1116; (h) Kim, B.-H., Woo, H.-G., Kim, W.-G., Yun, S.-S. and Hwang, T.-S. (2000) *Bulletin of the Korean Chemical Society*, **21**, 211–214; (i) Kraft, B.M. and Jones, W.D. (2002) *Journal of Organometallic Chemistry*, **658**, 132–140; and references cited therein.

3 (a) Burlakov, V.V., Troyanov, S.I., Letov, A.V., Strunkina, L.I., Minacheva, M.Kh., Furin, G.G., Rosenthal, U. and Shur, V.B. (2000) *Journal of Organometallic Chemistry*, **598**, 243–247; (b) Burlakov, V.V., Arndt, P., Baumann, W., Spannenberg, A., Rosenthal, U., Letov, A.V., Lyssenko, K.A., Korlyukov, A.A., Strunkina, L.I., Minacheva, M.Kh. and Shur, V.B. (2001) *Organometallics*, **20**, 4072–4079; (c) Burlakov, V.V., Shur, V.B., Pellny, P.-M., Arndt, P., Baumann, W., Spannenberg, A. and Rosenthal, U. (2000) *Chemical Communications*, 241–242; (d) Arndt, P., Baumann, W., Spannenberg, A., Rosenthal, U., Burlakov, V.V. and Shur, V.B. (2003) *Angewandte Chemie*, **115**, 1455–1458; (2003) *Angewandte Chemie – International Edition*, **42**, 1414–1418; (e) Burlakov, V.V., Letov, A.V., Arndt, P., Baumann, W., Spannenberg, A., Fischer, C., Strunkina, L.I., Minachova, M.Kh., Vygodskii, Ya.S., Rosenthal, U. and Shur, V.B. (2003) *Journal of Molecular Catalysis*, **200**, 63–67; (f) Strunkina, L.I., Minachova, M.Kh., Lisenko, K.A., Chaustova, O.I., Burlakov, V.V., Rosenthal, U. and Shur, V.B. (2003) *Izvestiya Akademii Nauk SSSR-Seriya Khimicheskaya*, 737–738; (2003) *Russian Chemical Bulletin-International Edition*, **52**, 773–774; (g) Spannenberg, A., Burlakov, V.V., Arndt, P., Baumann, W., Shur, V.B. and Rosenthal, U. (2002) *Zeitschrift fur Kristallographie, NCS*, **217**, 546–548; (h) Strunkina, L.I., Minacheva, M.Kh., Lyssenko, K.A., Petrovskii, P.V., Burlakov, V.V., Rosenthal, U. and Shur, V.B. (2006) *Russian Chemical Bulletin International Edition*, **55**, 174–176; (i) Strunkina, L.I., Minacheva, M.Kh., Lyssenko, K.A., Petrovski, P.V., Mysova, N.E., Strunkin, B.N., Burlakov, V.V., Rosenthal, U. and Shur, V.B. (2007) *Journal of Organometallic Chemistry*, **692**, 4321–4326.

4 (a) Burlakov, V.V., Arndt, P., Baumann, W., Spannenberg, A. and Rosenthal, U. (2004) *Organometallics*, **23**, 5188–5192; (b) Bach, M.A., Beweries, T., Burlakov, V.V., Arndt, P., Baumann, W., Spannenberg, A., Rosenthal, U. and Bonrath, W. (2005) *Organometallics*, **24**, 5916–5918; (c) Beweries, T., Bach, M.A., Burlakov, V.V., Arndt, P., Baumann, W., Spannenberg, A. and Rosenthal, U. (2007) *Organometallics*, **26**, 241–244; (d) Mansel, S., Thomas, D., Lefeber, C., Heller, D., Kempe, R., Baumann, W. and Rosenthal, U. (1997) *Organometallics*, **16**, 2886–2890.

5 (a) Kraft, S., Beckhaus, R., Haase, D. and Saak, W. (2004) *Angewandte Chemie*, **116**, 1609–1614; (2004) *Angewandte Chemie – International Edition*, **43**, 1583–1587; (b) Piglosiewicz, I.M., Kraft, S., Beckhaus, R., Haase, D. and Saak, W. (2005) *European Journal of Inorganic Chemistry*, **5**, 938–945; (c) Kraft, S., Hanuschek, E., Beckhaus, R., Haase, D. and Saak, W. (2005) *Chemistry - A European Journal*, **11**, 969–978.

6 (a) Jäger-Fiedler, U., Arndt, P., Baumann, W., Burlakov, V.V., Spannenberg, A. and Rosenthal, U. (2005) *European Journal of Inorganic Chemistry*, 2842–2849; (b) Spannenberg, A., Jäger-Fiedler, U., Arndt, P. and Rosenthal, U. (2005) *Zeitschrift fur Kristallographie, NCS*, **220**, 253–254.

7 (a) Thomas, D., Baumann, W., Spannenberg, A., Kempe, R. and Rosenthal, U. (1998) *Organometallics*, **17**, 2096–2102; (b) Spannenberg, A., Arndt, P., Baumann, W., Burlakov, V.V., Rosenthal, U., Becke, S. and Weiß, T. (2004)

Organometallics, **23**, 3819–3825, and references cited therein.

8 Spannenberg, A., Arndt, P., Baumann, W., Burlakov, V.V., Rosenthal, U., Becke, S. and Weiß, T. (2004) *Organometallics*, **23**, 4792–4795.

9 (a) Arndt, P., Spannenberg, A., Baumann, W., Becke, S. and Rosenthal, U. (2001) *European Journal of Inorganic Chemistry*, 2885–2890; (b) Burlakov, V.V., Arndt, P., Baumann, W., Spannenberg, A. and Rosenthal, U. (2004) *Organometallics*, **23**, 4160–4165; (c) Burlakov, V.V., Arndt, P., Baumann, W., Spannenberg, A. and Rosenthal, U. (2006) *Organometallics*, **25**, 519–522; (d) Burlakov, V.V., Arndt, P., Baumann, W., Spannenberg, A. and Rosenthal, U. (2006) *Organometallics*, **25**, 1317–1320; (e) Burlakov, V.V., Bach, M.A., Klahn, M., Arndt, P., Baumann, W., Spannenberg, A. and Rosenthal, U. (2006) *Macromolecular Symposia*, **236**, 48–53.

10 Arndt, P., Jäger-Fiedler, U., Klahn, M., Baumann, W., Spannenberg, A., Burlakov, V.V. and Rosenthal, U. (2006) *Angewandte Chemie*, **118**, 4301–4304; (2006) *Angewandte Chemie – International Edition*, **45**, 4195–4198, and references cited therein.

11 Marks, T.J., Yang, X. and Stern, C.L. (1992) *Angewandte Chemie*, **104**, 1406–1408; (1992) *Angewandte Chemie (International Edition in English)*, **31**, 1375–1377.

12 Jäger-Fiedler, U., Klahn, M., Arndt, P., Baumann, W., Spannenberg, A., Burlakov, V.V. and Rosenthal, U. (2007) *Journal of Molecular Catalysis A-Chemical*, **261**, 184–189.

13 Klahn, M., Spannenberg, A., Fischer, C., Rosenthal, U. and Krossing, I. (2007) *Tetrahedron Letters*, **48**, 8900–8903.

14 Scott, V.J., Celenligil-Cetin, R. and Ozerov, O.V. (2005) *Journal of the American Chemical Society*, **127**, 2852–2853.

15 Panisch, R., Bolte, M. and Müller, T. (2006) *Journal of the American Chemical Society*, **128**, 9676–9682.

16 Murphy, E.F., Murugavel, R. and Roesky, H.W. (1997) *Chemical Reviews*, **97**, 3425–3468.

17 *Examples:* (a) Döring, S., Erker, G., Fröhlich, R., Meyer, O. and Bergander, K. (1998) *Organometallics*, **17**, 2183–2187; (b) Vagades, D., Kehr, G., König, D., Wedeking, K., Fröhlich, R., Mück-Lichtenfeld, C. and Grimme, S. (2002) *European Journal of Inorganic Chemistry*, 2015–2021; (c) Chernega, A.N., Graham, A.J., Green, M.L.H., Haggitt, J., Lloyd, J., Mehnert, C.P., Metzler, N. and Souter, J. (1997) *Journal of the Chemical Society-Dalton Transactions*, 2293–2304.

18 Kiplinger, J.L. and Richmond, T.G. (1996) *Journal of the American Chemical Society*, **118**, 1805–1806.

11
Bisazines in the Coordination Sphere of Early Transition Metals*

Ruediger Beckhaus

11.1
Introduction

The design of highly ordered supramolecular structures has gained more and more interest within the last few decades. The concept of self-assembly chemistry takes a key position in this field and a multitude of supramolecular compounds have been synthesized by combining simple building blocks to obtain two- and three-dimensional structures [1–3]. Because of their electronic and steric versatility, aromatic *N*-heterocycles play a prominent role as classical ligands in coordination compounds [4, 5], as bridging ligands in binuclear derivatives [6–8] and as building blocks for supramolecular compounds [9–17]. Beyond their ability to connect metal centers by forming ligand-to-metal bonds they provide the opportunity for π-backbonding and thereby may affect delocalization and transport of electrons [18]. Compared with the highly developed late transition metal supramolecular chemistry, few attempts have been made to use the well-defined coordination modes and reducing properties of the early transition metals [19, 20].

The formation of molecular squares and rectangles requires 90° angles at the vertices, as are typical for square-planar or octahedral late transition metal species [1]. Hence, only a few examples are known using distorted tetrahedral geometries at the corners [21–23].

In the course of our studies on the reactions of low-valent titanium nitrogen complexes (**1**) [24], which are characterized by strong magnetic coupling of both titanium centers leading to diamagnetic properties of **1** [25], we are interested in the behavior of complexes of type **2** exhibiting bisazines as bridging ligands between low-valent titanium centers.

* A List of abbreviations is provided at the end of this chapter.

Activating Unreactive Substrates: The Role of Secondary Interactions.
Edited by Carsten Bolm and F. Ekkehardt Hahn

ISBN: 978-3-527-31823-0

Here we wish to report on the syntheses of these novel self-assembled polynuclear titanium complexes and their properties, employing different types of low-valent titanium 'corners' (Scheme 11.1) and typical bisazine ligands (Scheme 11.2). The different titanocene (d^2) fragments are generated *in situ* by liberation of bistrimethylacetylene from the corresponding acetylene complexes [26], whereas the titanium (d^3) fragments become available from the nitrogen complex **1**. The titanium (d^4) species can be prepared in the form of the bisfulvene complexes [27] and can be used in a direct manner [28].

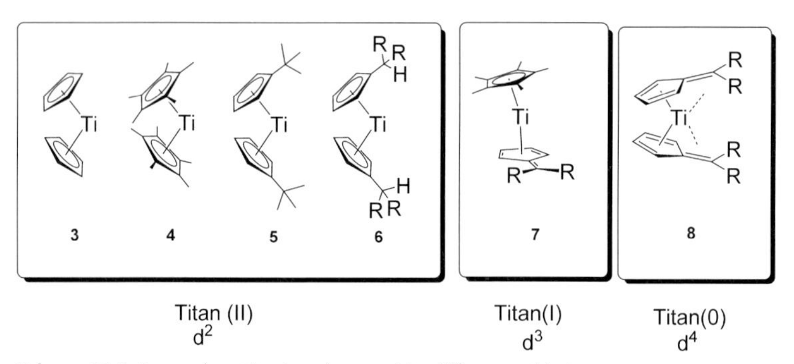

Scheme 11.1 Low-valent titanium 'corners' in different oxidation states.

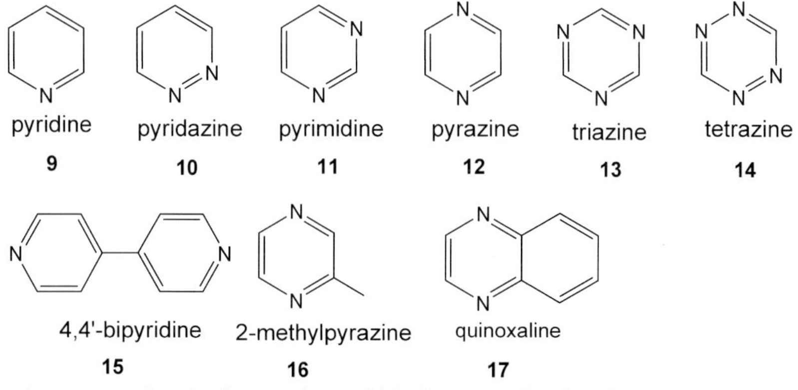

Scheme 11.2 Selected *N*-heterocycles useful for formation of multinuclear titanium complexes.

11.2
Results and Discussion

It was found in our investigations that reactions of low-valent early transition metal fragments with potentially bridging bisazines led to the formation of well-defined molecular architectures (see (A) in Scheme 11.3) [29], to accompanying radical-induced C–C coupling reactions due to the strong reducing properties (B) [21], and by primary C–H bond activation reactions to multifold dehydrogenative C–C couplings forming large-surface aromatic systems (C) [30, 31].

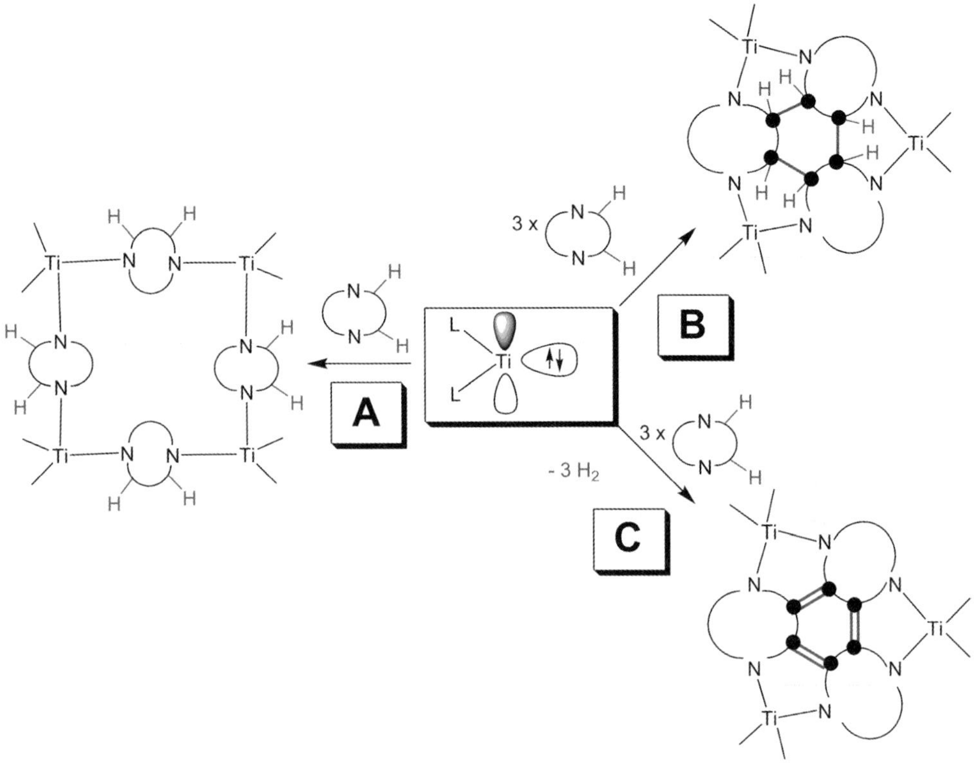

Scheme 11.3 Reaction pathways of low-valent titanium complexes with *N*-heterocyles (A: formation of molecular architectures, B: radical induced C–C coupling reactions, and C: dehydrogenative coupling).

11.2.1
Formation of Molecular Architectures

We recently reported on the reaction of the excellent titanocene precursor $[Cp_2Ti\{\eta^2\text{-}C_2(SiMe_3)_2\}]$ for Cp_2Ti (**3**) [26] with pyrazine (**12**) that leads to the formation of the first structurally characterized molecular square with titanocene(II) corner units $[Cp_2Ti(\mu\text{-}C_4H_4N_2)]_4$ (**20**) [21]. Using different starting materials (**5**) for the metal compound as well as for the ligand (**14, 15**), further neutral molecular squares with titanocene corner units can be synthesized. Scheme 11.4 shows the formation of the molecular squares **18, 19, 20** and **21**, which have been characterized by single-crystal X-ray analysis, elemental analysis and IR. All compounds are intensely colored and highly sensitive to air and moisture.

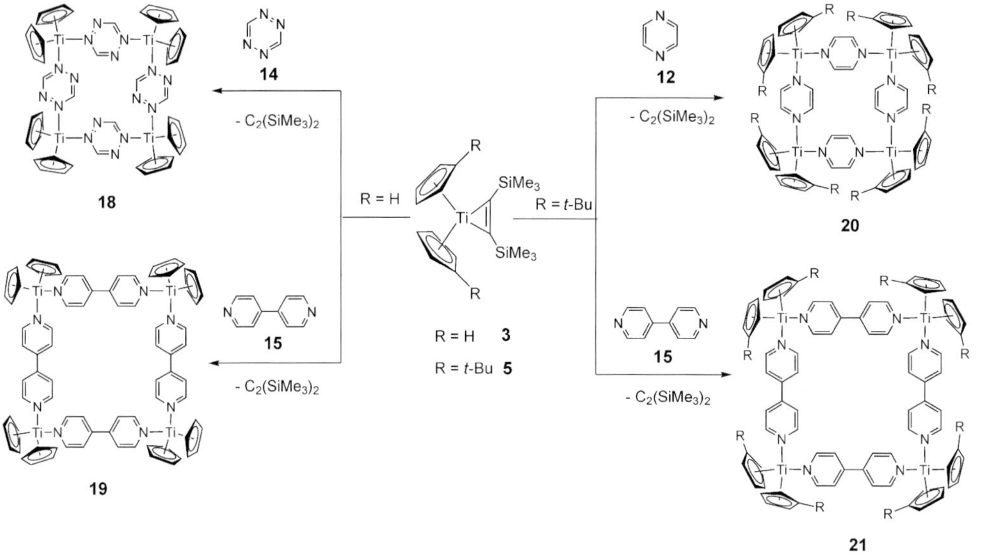

Scheme 11.4 Reactions of the titanocene complexes **3** and **5** with pyrazine (**12**), bipyridine (**15**) and tetrazine (**14**).

The reaction of $[Cp_2Ti\{\eta^2\text{-}C_2(SiMe_3)_2\}]$ with 4,4-bipyridine (**15**) in toluene leads after a few minutes to a color change from yellow to dark blue, and after 48 h at 60 °C dark blue crystals of **19** can be isolated in yields of about 50%. The tetrazine-bridged complex **18** can be isolated from a dilute reaction solution of **3** and tetrazine (**14**) in toluene after 48 h as a crystalline solid. The dark blue crystals of both complexes show an intense metallic lustre. They are only sparingly soluble in aliphatic and aromatic solvents and ethers and do not melt below 250 °C. In the mass spectra (EI 70 eV) no molecular peaks could be observed. Suitable single crystals for X-ray analysis were grown from the reaction solutions.

The molecular structure (**19a**) of **19**, crystallized from tetralin, is shown in Figure 11.1.

19a crystallizes in the space group $P4_2/n$ with four solvent molecules per tetramer. The metal atoms are coordinated tetrahedrally by two Cp ligands and two heterocycles. As the titanium atoms are located in one plane the complex forms a nearly perfect square with the bent metallocene moieties as corner units.

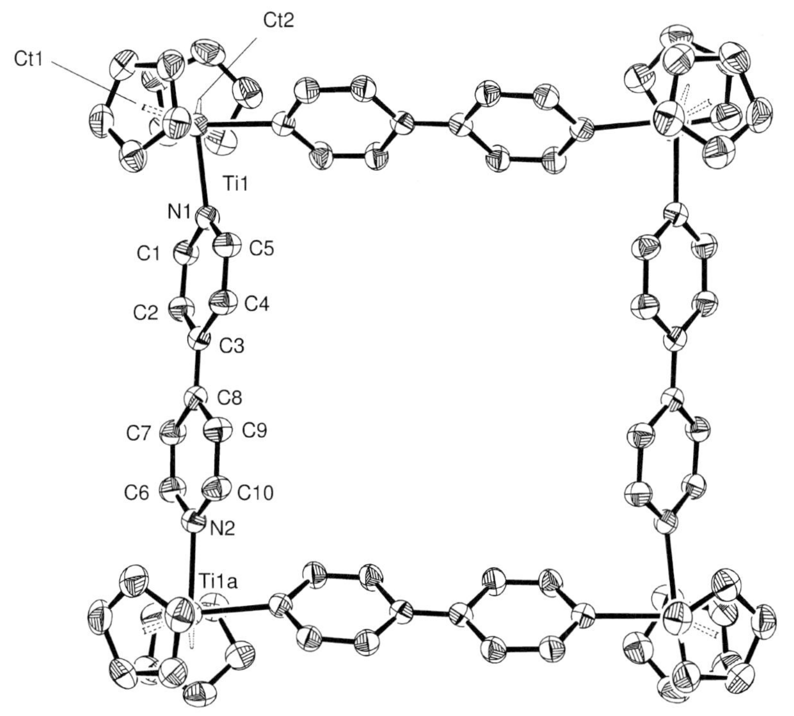

Figure 11.1 Structure of **19a** (50% probability, without H-atoms). Selected bond lengths in Å and angles in °: Ti1–N1 2.2132(17), Ti1–N2a 2.1976(17), Ti1–Ct1 2.086, Ti1–Ct2 2.092, N1–C1 1.358(3), N1–C1 1.374(3), N2–C6 1.363(3), N2–C10 1.370(3), C1–C2 1.366(3), C2–C3 1.418(3), C3–C4 1.418(3), C3–C8 1.424(3), C4–C5 1.366(3), C6–C7 1.368(3), C7–C8 1.424(3), C8–C9 1.423(3), C9–C10 1.364(3), N1–Ti–N2a 84.83(6), Ct1–Ti–Ct2 132.39, Ct1 = ring centroid of C11–C15, Ct2 = ring centroid of C16–C20, symmetry transformation for the generation of equivalent atoms: a = −y + 1/2, x, −z + $^{1}/_{2}$.

Single crystals of the tetrazine-bridged complex **18** can be grown from dilute reaction mixtures in toluene. Figure 11.2 shows the molecular structure of **18**.

18 crystallizes in the space group $P42_1c$ and the crystal contains no solvent molecules. In contrast to the analogous tetrameric pyrazine-bridged complex $[Cp_2Ti(\mu\text{-}C_4H_4N_2)]_4$ (**20**) [21] the tetrazine complex **18** does not really form a molecular square since the four titanium atoms do not lie in one plane but rather form a tetrahedron. As is mostly observed with tetrazine the heterocycle coordinates as a bismonodentate ligand similar to pyrazine and not as a bisbidentate ligand [7].

In order to obtain analogous complexes with higher solubilities $[(t\text{-BuCp})_2Ti\{\eta^2\text{-}C_2(SiMe_3)_2\}]$ was used as a source for a titanocene fragment (**4**) with bulky substituted Cp ligands. The reactions of **4** with pyrazine and bipyridine proceed more slowly but show the same color changes to violet and blue that are observed when using $[Cp_2Ti\{\eta^2\text{-}C_2(SiMe_3)_2\}]$. If the reactions with pyrazine and 4,4′-bipyridine are carried out in *n*-hexane, crystals of **20** and **21** can be isolated from the reaction mixture in yields of 65% and 79%, respectively. Compared to the analogous complexes with

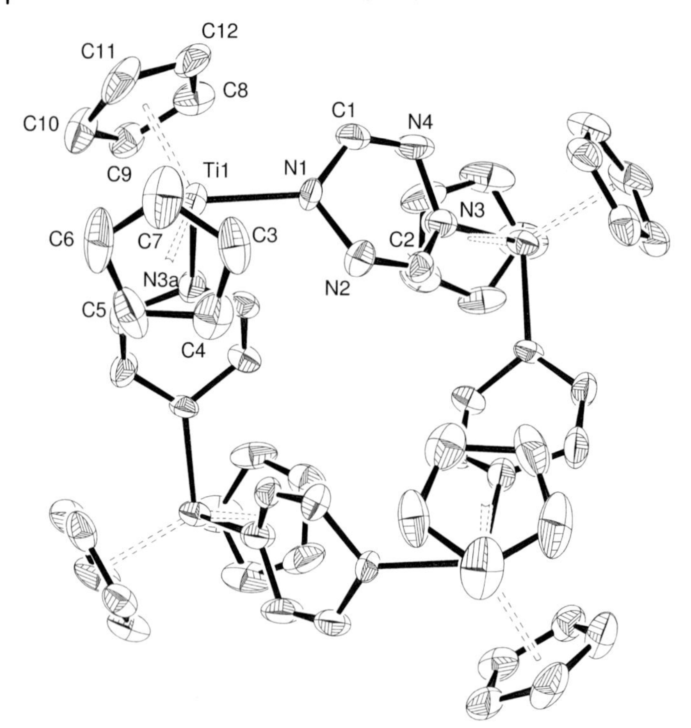

Figure 11.2 Structure of **18** in the crystal (50% probability, without H–atoms). Selected bond lengths in Å and angles in °: Ti1–N1 2.028(5), Ti1–N3a 2.132(5), Ti1–Ct1 2.086, Ti1–Ct2 2.075, N1–C1 1.377(7), N1–N2 1.420(5), N2–C2 1.305(7), N3–C2 1.337(7), N3–N4 1.412(7), N4–C1 1.298, N1–Ti1–N3a 88.84(19), Ct1–Ti–Ct2 130.24. Ct1 = ring centroid of C3–C7, Ct2 = ring centroid of C8–C12, symmetry transformation for the generation of equivalent atoms: a = −y + 1, x + 1, −z + 2.

unsubsituted Cp ligands they show a considerably increased solubility in aromatic solvents and THF. Furthermore they have lower melting points (**20** 197–200 °C; **21** 203–206 °C).

Single crystals of **20** could be obtained from *n*-hexane; single crystals of **21** were grown by slow diffusion of *n*-hexane into a THF solution. Figure 11.3 shows the molecular structure of **20**.

20 crystallizes in the space group $P2_1/n$ and the crystal contains two *n*-hexane molecules per molecular square. **21** crystallizes in the space group P-1 and contains 11 molecules THF per tetranuclear unit. Both complexes show a more or less square configuration. The sterically demanding *t*-butyl groups take nearly the same position in both complexes. The Ti−N distances in **19**, **20** and **21** lie in the upper limit for Ti−N bonds and correspond to values expected for titanium-coordinated *N*-heterocycles [21]. Bond lengths and angles of the titanocene units correspond to known values for tetrahedral coordination geometry.

The successful syntheses of molecular squares with the different bridging ligands lead to the attempt to synthesize a mixed-bridged complex that contains bridging

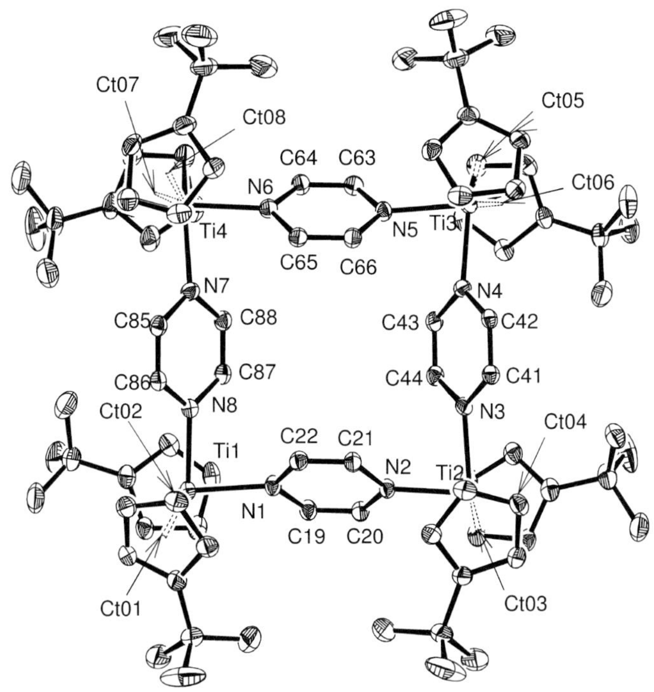

Figure 11.3 Structure of **20** in the crystal (50% probability, without H-atoms). Selected bond lengths in Å and angles in °: Ti1–N8 2.132(3), Ti1–N1 2.186(3), Ti1–Ct1 2.115, Ti1–Ct2 2.118, N1–C19 1.378(4), N1–C22 1.381(4), N2–C21 1.388(4), N2–C20 1.391(4), C19–C20 1.352(4), C21–C22 1.359(4), N8–Ti1–N1 84.30(10), Ct1–Ti1–Ct2 133,49, N2–Ti2–N3 85.08(10).

ligands of different lengths and exhibits the structure of a molecular rectangle. Generally, only a few molecular rectangles are so far known because most attempts to synthesize them in one-step reactions resulted in the preferred formation of the two homobridged molecular squares [3, 14]. Therefore, a reaction with two subsequent steps was used to coordinate the two different ligands to the titanocene moiety. Scheme 11.5 shows possible synthetic routes starting from a titanocene-chlorine complex in the oxidation state +III (**11**).

In the first reaction step the first bridging ligand is coordinated between two [Cp_2TiCl] units whose last coordination site is blocked by the chlorine atom. This reaction can be carried out successfully with pyrazine as well as bipyridine, and complexes **24** and **23** can be isolated as green crystals in yields of 55% and 43%, respectively, and characterized by X-ray analysis, IR and elemental analysis [28]. In the mass spectra of **24** and **23** only peaks of the free ligands and [Cp_2TiCl] are observed showing the low stability of the dimeric compounds. For similar monomeric compounds [Cp_2TiClL] with L = pyridine or $PPhMe_2$ a complete dissociation into the ligand and **22** has been observed at higher temperature (130 °C in vacuum

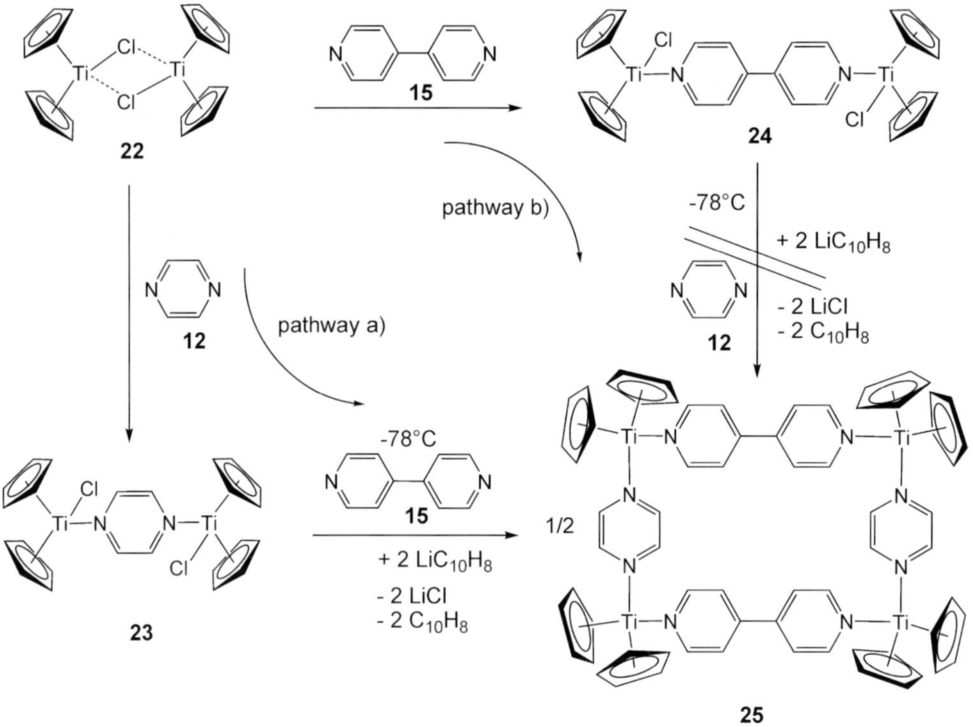

Scheme 11.5 Synthetic routes to the molecular rectangle **25**.

for $[Cp_2TiClPPhMe_2]$) [32]. In the second step an abstraction of the chloride ligand by reduction of the titanocene(III)complexes **24** and **23** and a coordination of the second bridging ligand has to take place. Lithium naphthalenide is used as a soluble reducing agent and the sparingly soluble rectangle **25** precipitates from the reaction mixture. To inhibit dissociation of complexes **24** and **23** and avoid an exchange of the ligands during the reduction the reaction was carried out at −78 °C. If pathway b) is used and **24** is reduced in presence of pyrazine the reaction does not lead to the rectangular complex **25**; instead the formation of the molecular square **20** (R : H) is observed accompanied by a further product (probably **19**). However, if **23** is reduced in presence of 4,4′-bipyridine (pathway a) **25** can be isolated as needle-shaped blue-violet crystals with an intense metallic luster. The complex was characterized by X-ray analysis, elemental analysis and IR spectroscopy. The synthesis of **25** can be further simplified so that starting from $[Cp_2TiCl_2]$ neither **22** nor **24** have to be isolated and the molecular rectangle is easily accessible from simple, commercially available starting materials. If titanocene dichloride is reduced in presence of pyrazine with one equivalent of lithium naphthalenide and then a 4,4′-bipyridine solution and the second equivalent of lithium naphthalenide are added after cooling to −78 °C, **25** can be isolated in 45% yield. Crystallization from THF yields crystals of **25** that are suitable for X-ray diffraction. The molecular structure of **25** is shown in Figure 11.4.

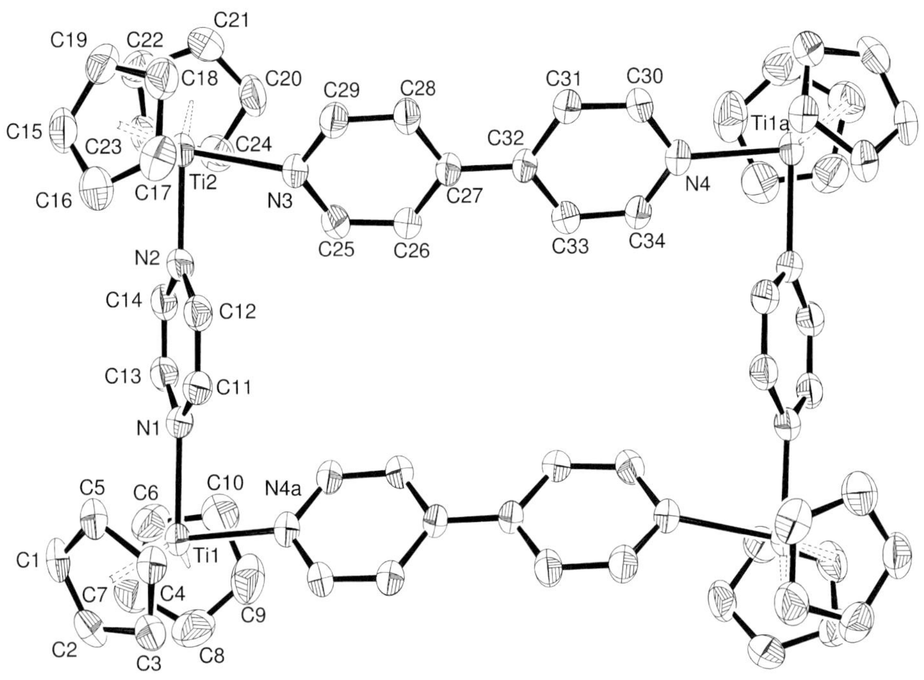

Figure 11.4 Structure of **25** in the crystal (50% probability, without H-atoms). Selected bond lengths in Å and angles in °: Ti1–N1 2.166(4), Ti1–N4a 2.215(4), Ti1–Ct1 2.111, Ti1–Ct2 2.068, Ti2–Ct3 2.095, Ti2–Ct4 2.094 Ti2–N2 2.128(4), Ti2–N3 2.220(4), N1–C13 1.375(6), N1–C11 1.385(6), N2–C14 1.379(6), N2–C12 1.396(6), N3–C29 1.353(6), N3–C25 1.359(6), N4–C30 1.352(7), N4–C34 1.356(6), C11–C12 1.359(7), C13–C14 1.351(7), C25–C26 1.371(7), C26–C27 1.426(7), C27–C28 1.425(7), C27–C32 1.432(7), C28–C29 1.359(7), C30–C31 1.347(7), C31–C32 1.427(7), C32–C33 1.411(7), C33–C34 1.363(7), N1–Ti1–N1a 83.68(15), N2–Ti2–N3 84.10(15), Ct1–Ti–Ct2 131.48, Ct3–Ti2–Ct4 131.00. Ct1 = ring centroid of C1–C5, Ct2 = ring centroid of C6–C10, Ct3 = ring centroid of C15–C19, Ct4 = ring centroid of C20–C24, symmetry transformation for the generation of equivalent atoms: a = x − y + 1, −y + 2, −z + 2/3.

25 crystallizes from THF in the trigonal space group $P3_121$ containing two solvent molecules per tetrameric molecule in the crystal. Each titanocene unit is coordinated by a pyrazine molecule and a 4,4′-bipyridine molecule, and with the planar configuration of the four titanium atoms a rectangular geometry results for the complex. A similar but more soluble rectangular complex **26** can be obtained by the same procedure using $[(t\text{-BuCp})_2TiCl_2]$ instead of $[Cp_2TiCl_2]$ as starting material. Again the blue-violet crystals show an intense metallic luster. The coordination geometry of **26** shows no significant differences from the already discussed structures of the other tetranuclear complexes.

In contrast to most of the known molecular rectangles with basically different sides of the rectangle [3] the titanium-based compounds **25** and **26** contain two bridging ligands of similar type. Molecular rectangles with pyrazine and 4,4′-bipyridine bridges have become available in the case of octahedrally coordinated rhenium corners [33], exhibiting comparable L-M-L angles (83.5°, **25**: 83.9°, **26**: 83.8°)

and sizes of the cavities (7.21 × 11.44 Å, **25**: 7.20 × 11.52 Å, **26**: 7.22 × 11.38 Å). However, in **25** and **26** the rectangular geometry is realized by tetrahedrally coordinated corner atoms. Except for **18** all complexes contain solvent molecules that can be removed by drying the crystalline solid in a vacuum. In **19a** the tetralin molecules are located in canals that are formed by the molecular squares.

The importance of the solvent molecules for the solid-state structure and the relatively great conformational freedom of the tetranuclear compounds is shown by the structures of **19** obtained by crystallization from tetralin (**19a**) and toluene (**19b**). Figure 11.5 shows the configuration of the four titanium atoms in **19a** and **19b** in the side view onto the tetramers.

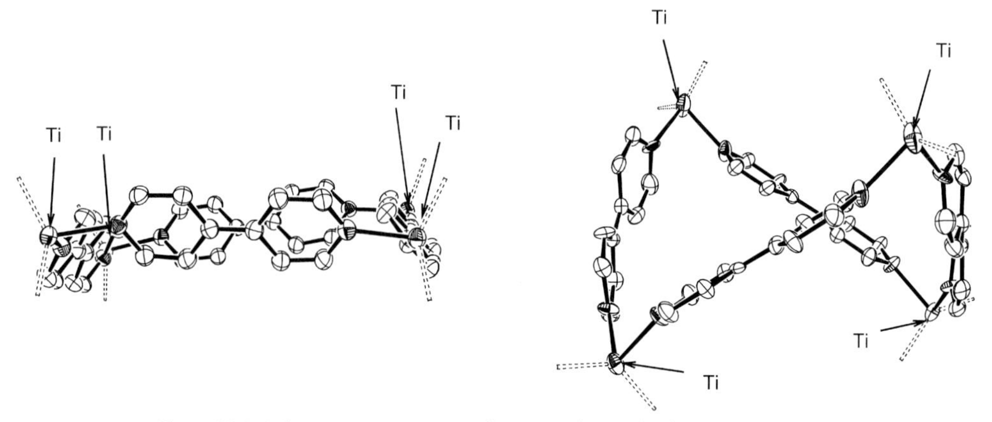

Figure 11.5 Side view on a structure of **19a** (tetralin) and **19b** (toluene), Cp rings are omitted for clarity.

Whereas the configuration of the bicyclic bridged complexes **19a** and **19b** is influenced by the solvents used for crystallization, the monocyclic bridged complexes **18** and **20** exhibit different configurations as well.

Although all N-Ti-N angles lie in the relatively small range between 83.2° and 86.6° the complexes exhibit the quite different conformations shown in Figures 11.8 and 11.9. This is indicated by the different Ti–Ti–Ti angles and their sum. If all corner atoms lie in the same plane a sum of 360° results for the quadrangle, whereas every distortion of the planar configuration leads to a decrease of the sum. Table 11.1 gives the average values for the N–Ti–N angles, the Ti–Ti–Ti angles and their sum and the Ti–Ti distances for all complexes.

The values of the sum of the Ti–Ti–Ti angles show that **19a**, **20**, **21**, **25** and **26** take a nearly planar configuration whereas **19b** and **18** are distorted from this planarity.

As the titanocene fragment has proved to reduce some aromatic *N*-heterocycles by C–C coupling reactions [21] and C–H and C–F bond cleavage [34] as well as to form complexes with stable heterocyclic radical anions by electron transfer [35, 36] it seems reasonable to presume an electron transfer to the ligand also for these reactions, and some proof of this is given by the molecular structures in the solid state.

Table 11.1 Selected geometric data of complexes **19a**, **19b**, **18**, **20**, **21**,**25** and **26**.

	Ti–Ti [Å]	N–Ti–N [°]	Ti–Ti–Ti [°]	Σ Ti–Ti–Ti [°]
19a	11.586	84.83(6)	89.75	359.0
19b	11.500	83.4(4)	68.05	272.2
18	6.691	84.3(3)	71.24	285.0
20	7.203	84.23(10)	89.82	359.3
21	11.521	83.45(13)	88.15	352.6
25	7.203 11.516	83.89(15)	89.37	357.5
26	7.219 11.380	83.79(18)	89.95	359.8

In numerous complexes of low-valent metals and heterocyclic ligands the change of bond lengths and angles that are sensitive to the transfer of electrons has been used to support the electronic structure. With tetrazine derivatives as bridging ligands, for example, a lengthening of the N–N bond of 0.068 Å and 0.073 Å has been observed by Kaim et al. indicating the formation of a radical anion in two copper complexes [37, 38]. In an Fe(0)-complex of 2,2′-bipyridine the electron transfer to the heterocycle leads to a decrease of the bond length between the two pyridyl rings of 0.083 Å. In a dimeric Yb(II)-complex of 2,2′-bipyrimidine with a dianionic ligand a shortening of 0.142 Å is observed [6, 39, 40]. Scheme 11.6 shows the two reduction steps for tetrazine and bipyridine [7, 8].

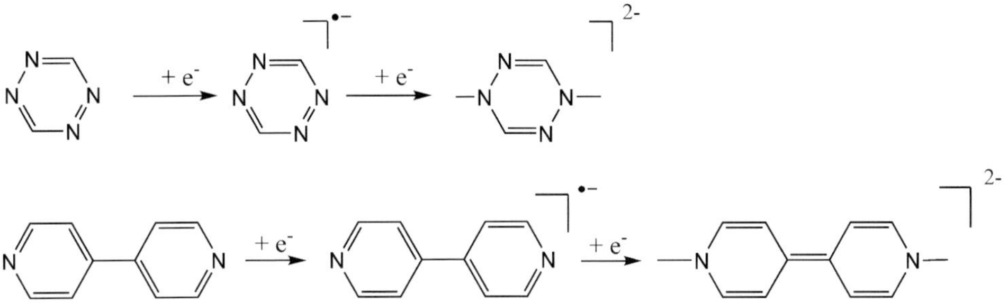

Scheme 11.6 Two-electron reductions of tetrazine and bipyridine.

A two-electron reduction of tetrazine leads to a single bond between the two nitrogen atoms, and the reduction of 4,4-bipyridine to a double bond between the two aromatic rings and an angle of 0° between the two pyridyl rings. These bonds and angles, respectively, show the greatest change upon reduction. To compare our data to the values of the free ligand and a fully reduced species the crystal structures of free 4,4′-bipyridine (**15**) and the two-electron reduced bis(trimethylsilyl)dihydro-4,4′-bipyridine (**27**) [28] were determined. The asymmetric units of both structures contain two independent molecules and for 4,4′-bipyridine the important values are given for both since the angle between the pyridyl rings differ significantly in the structures of the two molecules. In **27** no significant differences exist so that here the average is given. Table 11.2 shows the angles and bond lengths between the two pyridyl rings in 4,4′-bipyridine for the free ligand, for the two-electron reduced species **27** and for the titanium-coordinated ligands.

Table 11.2 Bond lengths and angles for free, reduced and coordinated 4,4′-bipyridine.

	C–C′ [Å]	Angle of twist [°]
15	1.4842(19)	34.39(6)
	1.4895(18)	18.14 (8)
27[a)]	1.381(3)	0
19a	1.424(3)	4.23
21	1.425(30)	7.60
25	1.432(7)	4.89
26	1.438(7)	9.80

a) 1)

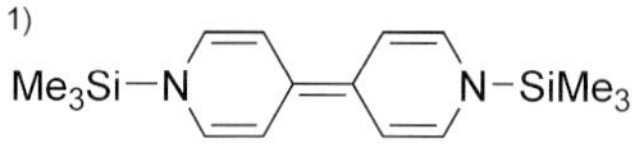

Me_3Si-N $N-SiMe_3$

27

These values show a significant shift toward the two-electron reduced species for the coordinated ligands. Nevertheless the decrease is less than that observed for the two-electron reduced bipyrimidine in the ytterbium complex of Andersen that contains bipyrimidine as a dianionic ligand [6]. So the changes in bond lengths and angles may indicate for the 4,4′-bipyridine complexes an electronic structure with a radical ligand and titanium centers in the oxidation state +III. Scheme 11.7 shows the effect of shortening and lowering the twist angles in 4,4′-bipyridine complexes depending on the oxidation states of titanium centers [28]. A change of bond lengths toward the reduced species is also observed in the lengthening of the N–N bond in tetrazine from 1.321 Å in free tetrazine [41] to 1.416(6) Å in **18**. Another sign of the reduction of the ligand is given by the change in the conformation of the ligand in **18** that no longer exhibits the planarity of the free ligand (Figure 11.2).

The electronic structure of the radical complexes formed by electron transfer from titanium to *N*-heterocyclic ligands has been thoroughly investigated with the 2,2′-bipyridine complexes, which show a ground singlet and a thermally accessible triplet state [35, 36, 42]. Like the monomeric 2,2′-bipyridine complexes, for the 4,4′-bipyridine complex **6** antiferromagnetic behavior is observed for the temperature dependence of the magnetic susceptibility. Owing to the special geometry of the frontier orbitals in metallocene units [43, 44], the 2,2′-bipyridine cannot act as a classical π-acceptor ligand but forms complexes with two remote electrons in interaction that are not diamagnetic, like the π-acceptor complexes of titanocene with carbonyl or phosphane ligands. If an overlap of metal and ligand orbitals is possible, a ground singlet state results for the two remote electrons [42]. This situation occurs if the ligand orbitals lie in the L–Ti–L plane. Because of the low solubility of ′**19**, **18** and **25**, NMR spectra are not available. The more soluble

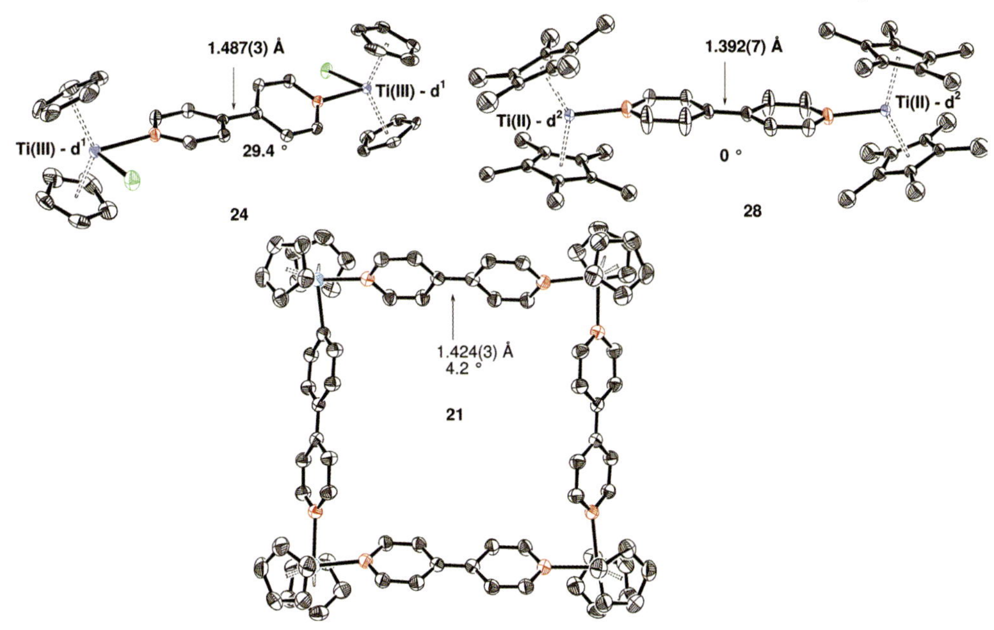

Scheme 11.7 Comparison of internal C–C distances and twist angles of 4,4′-bipyridine ligands in **21**, **24** and **28**.

complexes **20** (R: t-Bu), **21** (R: t-Bu) and **26** exhibit sharp ^{1}H NMR signals of the *t*-Bu groups (δ: **21** 1.16; **20** 1.12; **26** 1.19, 1.44). Further signals show more or less strong broadenings and chemical shifts over a wide range as are characteristic of non-diamagnetic derivatives.

11.2.2
Molecular Architectures Accompanied by Radical-induced C–C Coupling Reactions

Low-valent titanium complexes are characterized by strong reducing properties. Here we wish to report on the reactions of six-membered N-heterocycles, which leads, by the route of selective C–C coupling, to the formation of polynuclear titanium compounds. Reactions of the titanocene complexes $Cp_2Ti(\eta^2\text{-}C_2(SiMe_3)_2$ (**3**) and $Cp^*_2Ti(\eta^2\text{-}C_2(SiMe_3)_2$ (**4**) with triazine (**13**) at 25 (**3**) or 60 °C (**4**), respectively, led, after 48 h, to the dinuclear chelate complexes **29** or **30**, respectively, which can be isolated in 45% or 32% yield (Scheme 11.8) [21]. Complexes **29** and **30** are sparingly soluble in aliphatric and aromatic hydrocarbons as well as in THF and decompose above 300 °C, without melting. They display the extended molecular peak at m/z 518 (90) (**29**) and 799(40) (**30**) in the mass spectrum (EI, 70 eV). Owing to the low solubility, single crystals of **29** and **30** were grown directly from the reaction solutions. Figure 11.6 shows the molecular structure of **29** determined by X-ray crystallography.

The reactions of the alkyne complexes **3** and **4** with pyrazine (**12**) displayed varied behavior. As expected, in both cases the alkyne ligand is displaced: the tetranuclear

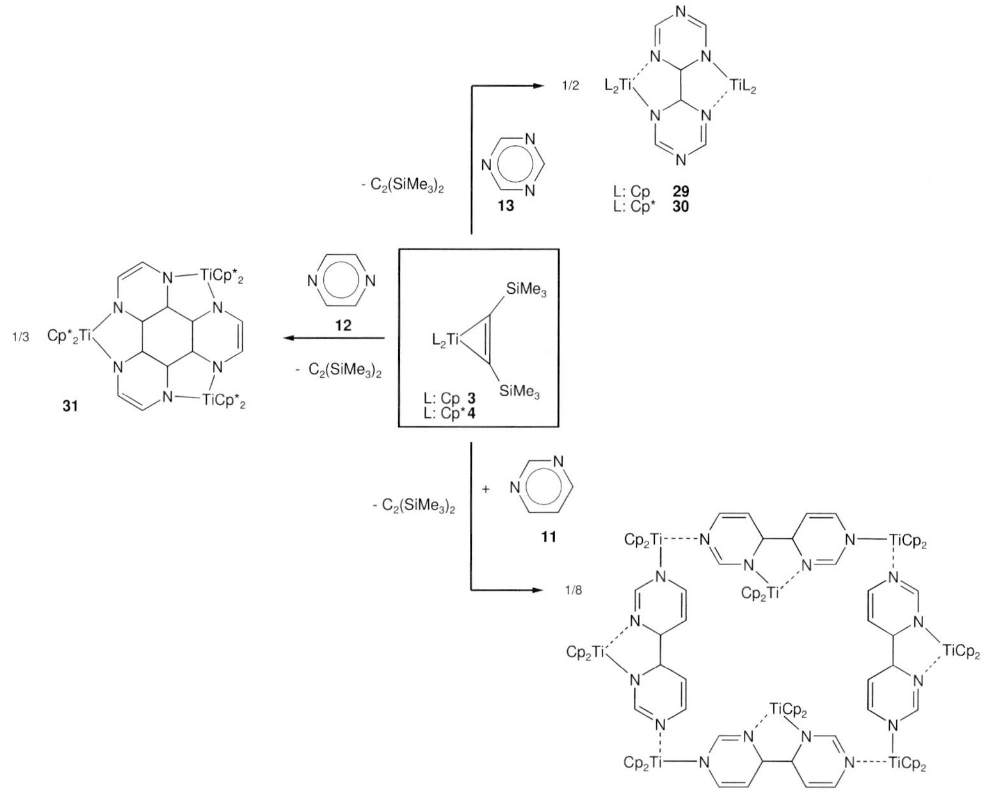

Scheme 11.8 Molecular architectures accompanied by radical-induced C–C coupling reactions.

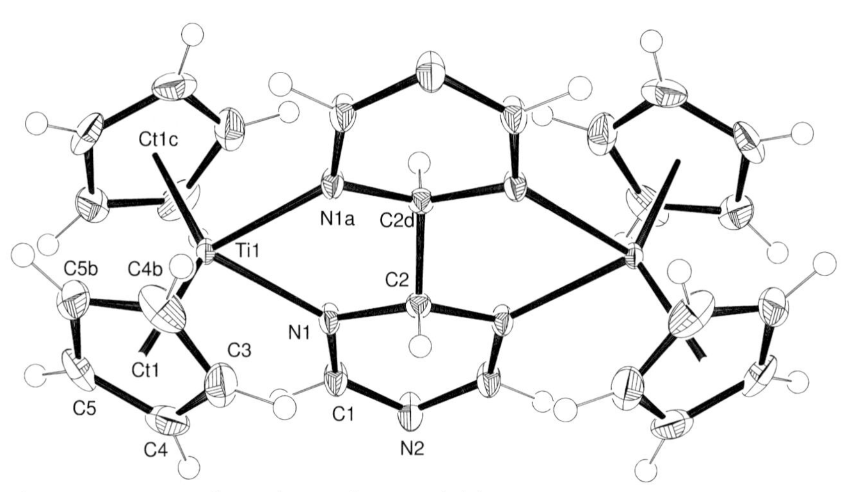

Figure 11.6 Structure of **29** in the crystal (50% probability) Selected bond lengths [Å] and angles [°]: Ti1–N1 2.2108(16), Ti1–Ct1 2.072, N1–C1 1.299(3), N1–C2 1.468(2), N2–C1 1.352(2), C2–C2c 1.501(6), N1–Ti–N1a 76.13(9) Ct1–Ti1–Ct 133.09.

pyrazine-bridged titanium complex **20** is formed starting from **3**, whereas **4** reacts under spontaneous threefold C–C coupling to give the trinuclear chelate complex **31** (Scheme 11.8). The reaction of **3** with **12** occurs in THF at room temperature, as is evident from the color change from yellow-brown to deep purple; after 48 h crystalline **20** can be obtained in 43% yield from dilute reaction solutions. Complex **31** is formed at 60 °C in THF or toluene in 60% yield. The molecular structure of **31** is shown in Figure 11.7. Like **20**, **31** is also C_2-symmetrical: Ti2 lies on the twofold axis. The three new C–C bonds generated lead to the formation of an ideal cyclohexane ring in the chair conformation.

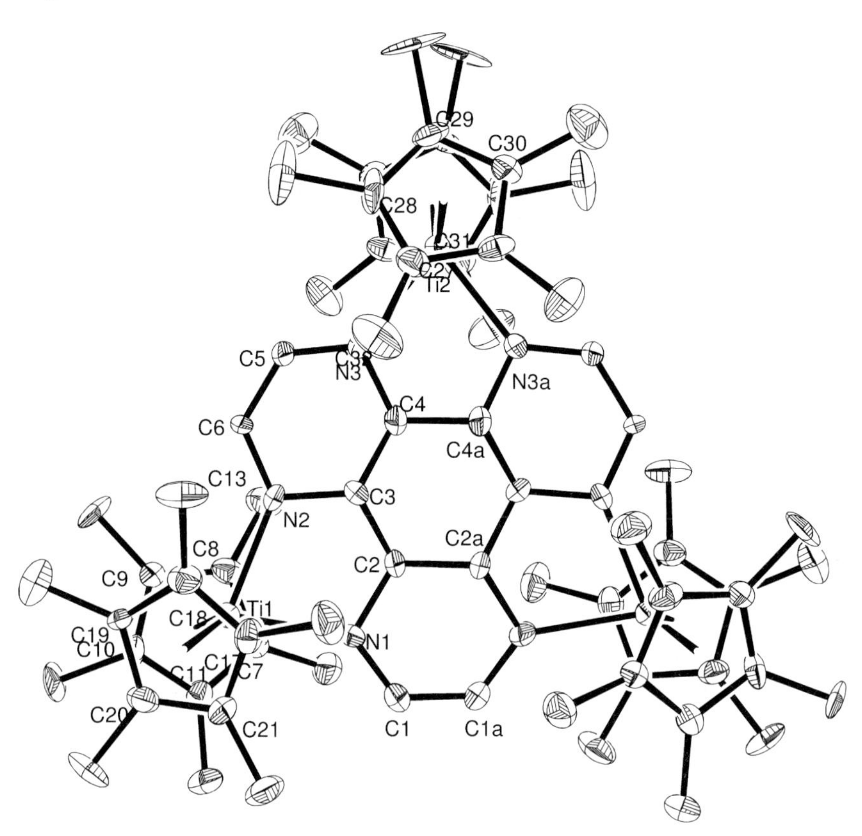

Figure 11.7 Structure of **31** in the crystal (50% probability, without H-atoms). Selected bond lengths [Å] and angles [°]: Ti1–N1 2.212 (3), Ti1–N2 2.193 (2), Ti2–N3 2.184 (2), Ti1–Ct1 2.149, Ti1–Ct2 2.146, Ti2–Ct3 2.158, N1–C1 1.343(4), N1–C2 1.458 (4), N2–C3 1.446 (4), N2–C6 1.349 (4), N3–C4 1.460 (4), N3–C5 1.342 (4), C1–C1a 1.413 (6), C2 –C2a 1.508 (7), C2–C3 1.496 (4), C3–C4 1.510 (4), C4–C4a 1.484 (7), C5–C6 1.413 (4), N1–Ti–N2 76.97 (9), N3–Ti–N3a 77.89 (12), Ct1–Ti1–Ct2 137.53, Ct3–Ti2–Ct3a 140.18.

The reaction of **3** with pyrimidine (**11**) after 2 h at 64 °C gives the octanuclear titanium complex **32** in the form of yellow, needle-like crystals in 80% yield (Scheme 11.8). Crystals suitable for X-ray structure analysis were obtained by crystallization from toluene. The resulting molecular structure is shown in

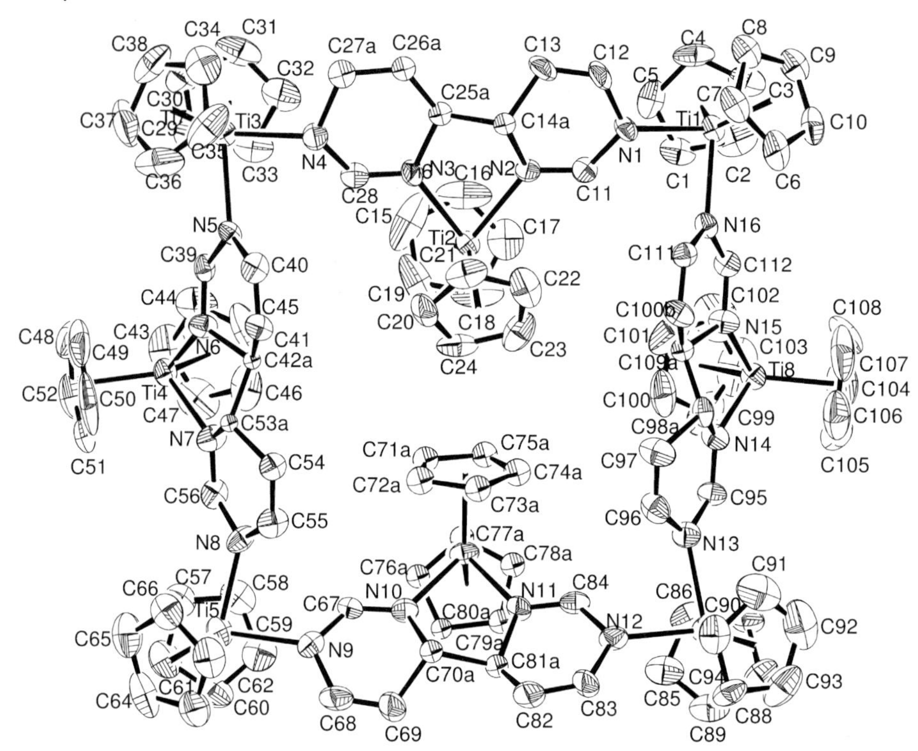

Figure 11.8 Structure of **32** in the crystal (50% probability, without H-atoms). Selected bond lengths [Å] and angles at the terminal Ti5 and Ti7 centers as well as at the chelate positions Ti6 and Ti8: Ti5–N9 2.193(6), Ti5–N8 2.205(6), Ti6–N10 2.155(6), Ti6–N11 2.157(5), Ti7–N12, 2.182(6), Ti7–N13 2.190(6), Ti8–N15 2.153(6), Ti(8)–N(14) 2.158(6), N9–C67 1.334(9), N9–C68 1.430(9), N10–C67 1.272(9), N10–C70B 1.526(16), N10–C70A 1.569(15), N11–C84 1.301(9), N11–C81A 1.468(9), N11–C81B 1.565(10), N12– C(84) 1.367(8), N12–C83 1.378(9), N13–C95 1.361(9), N13–C96 1.415(9), N14–C95 1.313(9), N14–C98B 1.54(3), N14–C98A 1.552(10), N15–C112 1.317(9), N15–C109B 1.48(2), N15–C109A 1.496(11), N16–C112 1.338(9), N16–C111 1.422(8), C68–C69 1.302(11), C69–C70A 1.529(18), C70A–C81A 1.602(19), C81A–C82 1.420(9), C82–C83 1.306(10), C70B–C81B 1.462(19), C96–C97 1.287(11), C97–C98A 1.502(12), C98A–C109A 1.500(14), C98B–C109B 1.63(3), C109A–C110 1.481(11), C(110)–C(111) 1.314(10), C113–C119 1.362(16), N9–Ti5–N8 82.0(2), Ct–Ti5–Ct 131.72, N10–Ti6–N11 75.4(2), Ct–Ti6–Ct 133.48, N12–Ti7–N13 83.4(2), Ct–Ti7–Ct 131.88, N15–Ti8–N14 75.8(2), Ct–Ti8–Ct 135.16.

Figure 11.8. Similarly to **29**, **30**, **20**, and **31** no NMR spectra could be recorded for **32** because of its low solubility. In the crystal, five equivalents of toluene were found per molecule of **32**. The octanuclear compound crystallized in the polar space group *Pc*. The C(sp^3) centres formed by the C–C coupling show typical disorders.

The formation of **29**, **30**, and **32** is characterized in each case through the linkage of one C–C bond of the *N*-heterocycle (**11**, **13**) used. Whereas in the reaction with **13**, only **29** or **30**, respectively, is formed even when **3** and **4** are used in excess, that is only the chelate positions are occupied. The reaction of **3** with **11** leads to the saturation of the terminal N atoms and thus to the octanuclear molecular aggregate **32**. In **32** the eight linked titanium centers form a puckered ring (see Figure 11.8), and the chelate centers are located above the plane of the terminal Cp_2Ti units. Two of the chelate positions (Ti2, Ti6) point toward the center of the ring. The formation of **29**, **30**, **31**, and **32** is accompanied by a loss of the aromaticity of the *N*-heterocycle employed. Along the new C–C bonds, the H atoms always adopt a trans position. Particularly notable is the formation of **31** from **4** in comparison to that of **20** from **3**.

The Ti–N distances in **29**, **30**, **31**, and **32** lie at the upper limit for pure Ti–N σ bonds without p_π–d_π interactions [45], and correspond to the values expected for titanium-coordinated *N*-heterocycles.[46, 47, 42] The Ti–N distances within **29**, **30**, and **31** do not differ significantly, so that mesomeric forms can be assumed. Bond lengths and angles of the titanocene units correspond to those for tetrahedral coordination geometry.

The formation of the C–C coupled polynuclear complexes **29**, **30**, and **32** can be explained by the initial reduction of the heterocycles to radical anions, which has been described for several *N*-heterocycles in reactions with low-valent metal complexes [46, 48–51] or by electrochemical reactions [52, 53]. The C–C coupling in the reaction to give **32** occurs regioselectively at thc C4 atom of the heterocycle, and the same is true for the electrochemical coupling [52]. The formation of **31** is particularly surprising, since pyrazine (**12**) is considered to be a typical bridging ligand and no defined C–C coupling at metal centres has been described to date for this heterocycle [8, 54].

11.2.3 Molecular Architectures Based on C–C Coupling Reactions Initiated by C–H Bond Activation Reactions

C–C formation reactions are of fundamental interest in various applications of organometallic substrates. A great deal of work has been devoted to the combination of transition metal-initiated C–H bond activations and subsequent C–C bond formation [55, 56]. The primary C–H bond activation step is well established [57–60]. From the practical point of view, a dehydrogenative coupling reaction from two C–H bonds make synthetic procedures shorter and more efficient [61]. While reductive coupling of aryl ligands is well documented for group 10 biaryls [62], the analogous process on biaryl zirconocene derivatives can only be induced under photochemical conditions [63], and a radical decomposition seems to be preferred when diorganyltitanocenes are irradiated [64]. The most commonly observed carbon–carbon reductive elimination process involving complexes of group 4 metals is the coupling of 1-alkenyl groups [65–70] or iminoacyl ligands [71].

We have found, that by reacting pyrazine, triazine or pyrimidine with the titanocene acetylene complexes $Cp_2Ti\{\eta^2\text{-}C_2(SiMe_3)_2\}$ (**3**) or their permethylated analogs (**4**), as excellent titanocene sources [72, 26], multinuclear titanium complexes are formed [29], often accompanied by simultaneously occurring C–C couplings of the primarily formed radical anions [21] (Scheme 11.9).

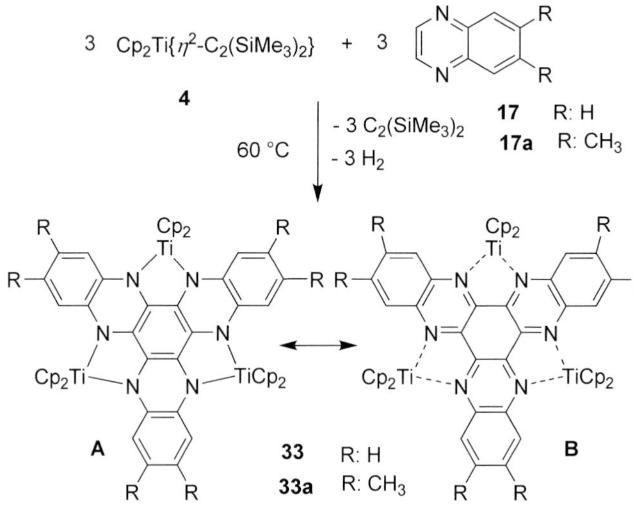

Scheme 11.9 Formation of the trinuclear HATN titanium complexes **33** and **33a**.

In this chapter we report the spontaneous coupling of *N*-heterocycles, initiated by C–H bond activation reactions. The reaction of quinoxalines **17** and their dimethyl analogs **17a** with **4** results in the formation of **33** and **33a**, respectively. The compounds can be isolated in yields of 17% (**33**) and 62% (**33a**) as crystalline products in one-pot syntheses at 60 °C (24 h). These hexaazatrinaphthylene (HATN) titanium complexes are thermally stable (mp > 350 °C **33**, 353 °C **33a**) but very sensitive to air and moisture [30, 31].

The molecular peaks can be observed (**33** m/z 919 (3%) [M^+], **33a**m/z 1002 (4%) [M^+]). **33** is nearly insoluble in common solvents, whereas for **33a** paramagnetic behavior is proved by the Evans method [73]. Products formed by reactions of the free HATN ligands [74] with **4** are in every respect identical to **33** and **33a**, respectively. However, because of the generally poor solubility of HATN ligands, the complexation often ends up in poor yields and reveals significant disadvantages compared to the presented route.

Suitable crystals for X-ray diffraction are obtained directly from the reaction solutions (**33**, Figure 11.9). **33** is *Cm*-symmetrical with the Ti1 centre on the mirror plane.

The HATN ligand of **33** is nearly planar with a slight deviation of the outer fused benzene rings. Bond distances and angles in **33** suggest a reduced *N*-heterocyclic system with characteristic patterns of low-valent *N,N′*-chelated titanium complexes. [21, 36, 42]. Whereas uncomplexed hexaazatrinaphthylene (**39**) shows for the central six-membered ring three long (average 1.479 Å) and three short C–C bonds (average 1.425 Å) [74], for **33** shorter and more balanced distances (1.411(9)–1.438(9) Å) are found. The central C–C bonds of the chelate positions in **33** (1.411(9)–1.426(7) Å) appear shorter compared to the free ligand **39** (1.472(6)–1.491(6) Å). [74] The C–N distances in **33** are elongated (1.396(6)–1.352(6) Å) compared to **39** (1.318(5)–1.382

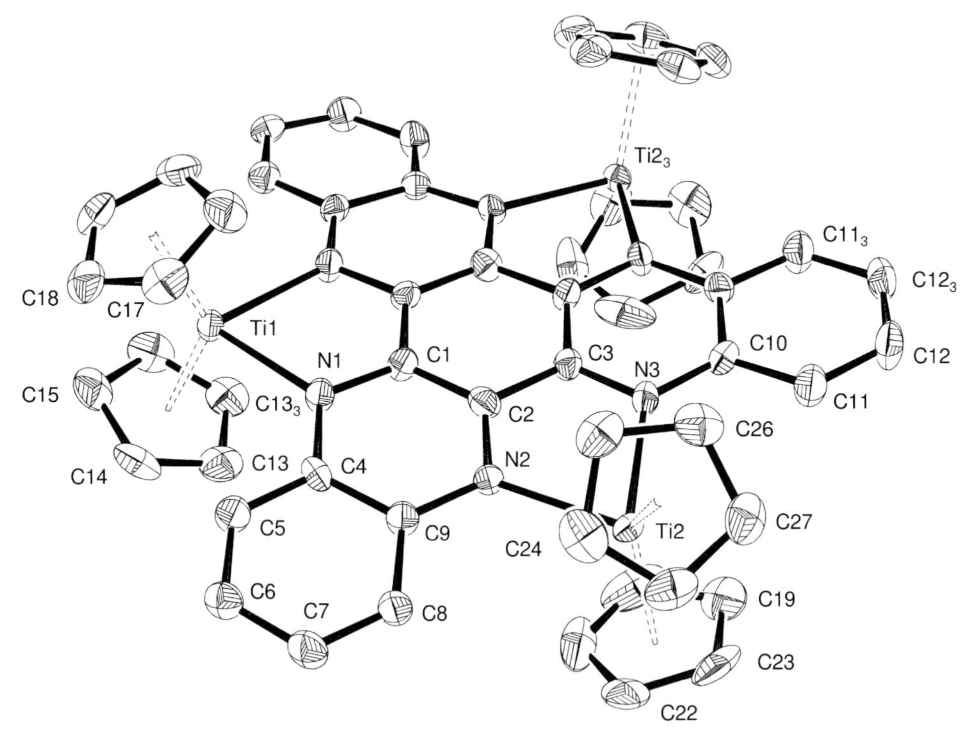

Figure 11.9 Molecular structure of **33** (hydrogen atoms omitted). Ellipsoids are drawn at 50% probability. Selected bond lengths (Å) and angles (deg): Ti1–N1 2.187(4), Ti2–N2 2.170(4), Ti2–N3 2.195(3), C1–C2 1.420(7), C2–C3 1.426(7), C1–N1 1.353(6), C2–N2 1.370(6), C3–N3 1.352(6), C4–N1 1.396(6), C9–N2 1.389(6), C10–N3, 1.384(6), C1–C1′ 1.411(9), C3–C3′ 1.438(9), C10–C10′ 1.435(9), C12–C12′1.359(11), C4–C5 1.379(7), C5–C6 1.386(6), C6–C7 1.390(7), C7–C8 1.369(7), C8–C9 1.406(6), C4–C9 1.434(7), Ti1–Ct1 2.080, Ti1–Ct2 2.096, Ct1–Ti–Ct2 135.57, N1–Ti–N1′ 75.7(2), N2–Ti–N3 75.97(14).

(5) Å [74], 1.323(3)–1.363(3) Å [75]) indicating contributions from the mesomeric form **A** (Scheme 11.3). The Ti–N distances (2.170(4)–2.195(3) Å) lie in the upper limit for Ti–N σ bonds without p_π-d_π interactions and correspond to the values expected for titanium-coordinated *N*-heterocycles in agreement with the mesomeric form **B** [21].

Reactions of pyridine (**9**) and **3** lead to stable binuclear pyridyl titanium hydrides through C–H bond activation and ortho titanation. Subsequent C–C bond formations are not observed [76]. Dehydrogenative coupling proceeds if benzannelated *N*-heterocycles with at least one ortho C–H bond are reacted with **3**. The simplest representative of this type of heterocycle is quinoline (**34**). With the formation of biquinoline **37** we have another example of dehydrogenative coupling, which allows us to present the potential mechanism of the reactions to **33** and **33a** in a concise manner. The assembly of **37** can be explained by a twofold primary C–H bond activation leading to **35** and **36** followed by C–C coupling through reductive elimination (Scheme 11.10).

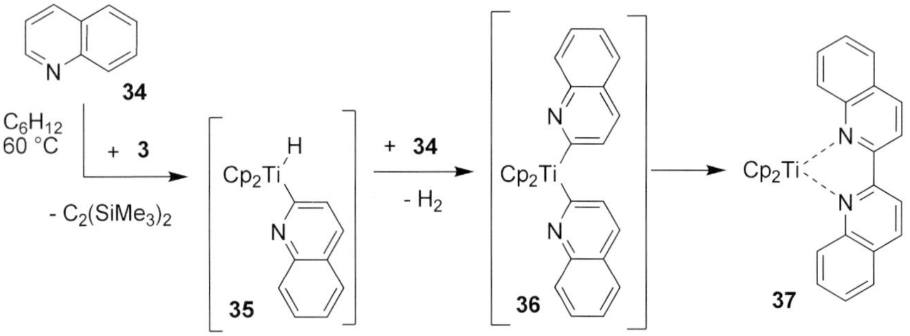

Scheme 11.10 C—C coupling reactions induced by C—H bond activation.

Corresponding intermediates in the reactions of **4** and **17** enable further C—H activation and subsequent C—C bond formation steps to give **33**. However, attempts to isolate intermediates like **35** or **36** even at lower temperatures have not yet been successful. The 2,2′-biquinoline complex **37** can be isolated as crystalline product (61%), exhibiting comparable structural characteristics as 2,2-bipyridine titanium complexes, proved by X-ray diffraction (Figure 11.10) [42]. The shortening of the C9—C10 bond (1.432(2) Å), e.g. compared to free 2,2′-bipyridine (1.50 Å [77], 1.490(3) Å [39]), indicates the reduced nature of the chelating ligand.

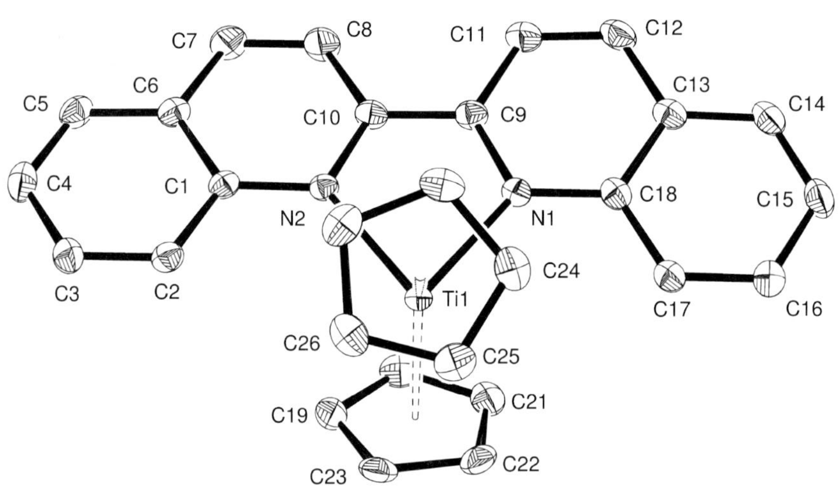

Figure 11.10 Molecule of **37** in the crystal (hydrogen atoms omitted). Ellipsoids are drawn at 50% probability. Selected bond length (Å) and angles (deg): Ti1–N1 2.1920(14), Ti1–N2 2.1960(12), C1–N1 1.390(2), C1–C2 1.417(2), C2–C3 1.380 (2), C3–C4 1.395(3), C4–C5 1.375(2), C5–C6 1.402(2), C1–C6 1.432(2), C6–C7 1.438(2), C7–C8 1.348(3), C8–C9 1.427(2), C9–N1 1.3748(19), C9–C10 1.432(2), N2–C10 1.372(2), C10–C11 1.435(2), C11–C12 1.347(3), C12–C13 1.429(3), C13–C14 1.403(2), C14–C15 1.371(3), C15–C16 1.402(2), C16–C17 1.381(2), C17–C18 1.413(2), C13–C18 1.433(2), C18–N2 1.384(2), Ti–Ct1 2.094, Ti1–Ct2 2.093, Ct1–Ti1–Ct2 133.76, N1–Ti1–N2 76.30(5).

The HATN complex **33** is also formed by dehydrogenative coupling of 2,3-(2′,2″)-diquinoxalylquinoxaline (**38**) with **4**, which is in agreement with the proposed mechanism. Reacting **33** with I_2 (3 equiv.) in *n*-hexane as solvent gives **39** (Scheme 11.11).

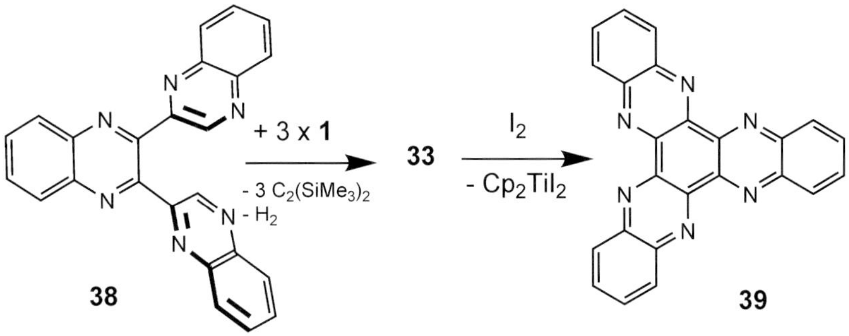

Scheme 11.11 Formation of **33** by dehydrogenative C−C coupling employing **38**.

With the selective formation of carbon–carbon bonds in combination with C−H bond activation processes, particularly using commercial starting materials, an efficient route for the coupling of *N*-heterocycles has been established.

11.3
Conclusions and Future Directions

For the investigations described herein, we employed a novel synthetic method, in which metal-linked complexes of polyvalent *N*-donor ligands are formed from readily accessible building blocks. The good solubility of the starting materials allows a smooth separation of the (in most cases) sparingly soluble products. Further work on these systems should be able to show whether the record linkage of eight [Cp_2Ti] units can be beaten and to what extent the observed self-assembly principles can be developed by using low-valent early transition metals and redox-active acceptor ligands [3, 78]. The synthesis of **31** described herein could be the beginning of a new area of chemistry comparable to the chemistry of the HAT ligands which are difficult to access [79–82].

In recent years research in the area of monometallic compounds has been extended to polymetallic supramolecular systems, which may have a considerable potential in the design of new materials for use in photochemical molecular devices. Furthermore, several polynuclear compounds have been created which possess functionality such as nonlinear optics, molecular magnetism and anion trapping, with possible applications as molecular receptors or DNA photoprobes.

Acknowledgments

This work was supported by the DFG (SPP 1118). I have to express my thanks to my coworkers Dr. Axel Scherer, Dr. Susanne Kraft and Dr. Ingmar Piglosiewicz for excellent preparative work and fruitful discussions. I have also to thank Marion Friedemann, Edith Hanuschek, Wolfgang Saak and Detlev Haase for technical and analytical assistance.

Abbreviations

Cp	C_5H_5 anion
Cp*	C_5Me_5 anion
ct	in single X-ray structures ct = centroid of the carbon atoms of the respective Cp ring
HAT	1,4,5,8,9,12-Hexaazatriphenylene
HATN	1,6,7,12,13,18-Hexaazatrinaphthylene
Me	CH_3

References

1 Leininger, S., Olenyuk, B. and Stang, P.J. (2000) *Chemical Reviews*, **100**, 853–908.

2 Swiegers, G.F. and Malefetse, T.J. (2000) *Chemical Reviews*, **100**, 3483–3537.

3 Holliday, B.J. and Mirkin, C.A. (2001) *Angewandte Chemie (International Edition in English)*, **113**, 2076–2097; (2001) *Angewandte Chemie – International Edition*, **40**, 2022–2043.

4 Schmatloch, S. and Schubert, U.S. (2003) *Chemie in Unserer Zeit*, **37**, 180–187.

5 McWhinnie, W.R. and Miller, J.D. (1969) *Advances in Inorganic Chemistry and Radiochemistry*, **12**, 135–215.

6 Berg, D.J., Boncella, J.M. and Andersen, R.A. (2002) *Organometallics*, **21**, 4622–4631.

7 Kaim, W. (2002) *Coordination Chemistry Reviews*, **230**, 127–139.

8 Kaim, W. (1983) *Angewandte Chemie (International Edition in English)*, **95**, 201–221; (1983) *Angewandte Chemie – International Edition*, **22** 171–191.

9 Kitagawa, S., Kitaura, R. and Noro, S.-I. (2004) *Angewandte Chemie (International Edition in English)*, **116**, 2388–2430; (2004) *Angewandte Chemie – International Edition*, **43**, 2334–2375.

10 Coles, M.P., Swenson, D.C., Jordan, R.F. and Young, V.G. Jr. (1997) *Organometallics*, **16**, 5183–5194.

11 Robson, R. (2000) *Journal of The Chemical Society-Dalton Transactions*, 3735–3744.

12 Fujita, M., Fujita, N., Ogura, K. and Yamaguchi, K. (1999) *Nature*, **400**, 52–55.

13 Baxter, P.N.W., Lehn, J.-M., Baum, G. and Fenske, D. (1999) *Chemistry - A European Journal*, **5**, 102–112.

14 Olenyuk, B., Fechtenkötter, A. and Stang, P. J. (1998) *Journal of The Chemical Society-Dalton Transactions*, 1707–1728.

15 Batten, S.R. and Robson, R. (1998) *Angewandte Chemie (International Edition in English)*, **110**, 1558–1595; (1998) *Angewandte Chemie – International Edition*, **37**, 1460–1494.

16 Stang P.J. and Olenyuk, B. (1997) *Accounts of Chemical Research*, **30**, 502–518.

17 Fujita, M. (1998) *Chemical Society Reviews*, **27**, 417–425.

18 Launay, J.-P. (2001) *Chemical Society Reviews*, **30**, 386–397.

19 Schafer, L.L., Nitschke, J.R., Mao, S.S.H., Liu, F.-Q., Harder, G., Haufe, M. and Tilley, T.D. (2002) *Chemistry - A European Journal*, **8**, 74–83.

20 Schafer, L.L. and Tilley, T.D. (2001) *Journal of the American Chemical Society*, **123**, 2683–2684.

21 Kraft, S., Beckhaus, R., Haase, D. and Saak, W. (2004) *Angewandte Chemie (International Edition in English)*, **116**, 1609–1614; (2004) *Angewandte Chemie – International Edition*, **43**, 1583–1587.

22 Stang, P.J. and Whiteford, J.A. (1996) *Research on Chemical Intermediates*, **22**, 659–665.

23 Schinnerling, P. and Thewalt, U. (1992) *Journal of Organometallic Chemistry*, **431**, 41–45.

24 Scherer, A., Kollak, K., Lützen, A., Friedeman, M., Haase, D., Saak, W. and Beckhaus, R. (2005) *European Journal of Inorganic Chemistry*, 1003–1010; (2005) *European Journal of Inorganic Chemistry*, 1991.

25 Studt, F., Lehnert, N., Wiesler, B.E., Scherer, A., Beckhaus, R. and Tuczek, F. (2006) *European Journal of Inorganic Chemistry*, 291–297.

26 Rosenthal, U., Burlakov, V.V., Arndt, P., Baumann, W. and Spannenberg, A. (2003) *Organometallics*, **22**, 884–900.

27 Diekmann, M., Bockstiegel, G., Lützen, A., Friedemann, M., Haase, D., Saak, W. and Beckhaus, R. (2006) *Organometallics*, **25**, 339–348.

28 Kraft, S. (2004) Dissertation Universität Oldenburg, 1–190.

29 Kraft, S., Hanuschek, E., Beckhaus, R., Haase, D. and Saak, W. (2005) *Chemistry - A European Journal*, **11**, 969–978.

30 Piglosiewicz, I.M., Beckhaus, R., Wittstock, G., Saak, W. and Haase, D. (2007) *Inorganic Chemistry*, **46**, 7610–7620.

31 Piglosiewicz, I.M., Beckhaus, R., Saak, W. and Haase, D. (2005) *Journal of the American Chemical Society*, **127**, 14190–14191.

32 Green, M.L.H. and Lucas, C.R. (1972) *Journal of The Chemical Society-Dalton Transactions*, 1000–1003.

33 Rajendran, T., Manimaran, B., Lee, F.-Y., Lee, G.-H., Peng, S.-M., Wang, C.M. and Lu, K.-L. (2000) *Inorganic Chemistry*, **39**, 2016–2017.

34 Piglosiewicz, I.M., Kraft, S., Beckhaus, R., Haase, W. and Saak, W. (2005) *European Journal of Inorganic Chemistry*, 938–945.

35 McPherson, A.M., Fieselmann, B.F., Lichtenberger, D.L., McPherson, G.L. and Stucky, G.D. (1979) *Journal of the American Chemical Society*, **101**, 3425–3430.

36 Witte, P.T., Klein, R., Kooijman, H., Spek, A.L., Polasek, M., Varga, V. and Mach, K. (1996) *Journal of Organometallic Chemistry*, **519**, 195–204.

37 Schwach, M., Hausen, H.-D. and Kaim, W. (1999) *Inorganic Chemistry*, **38**, 2242–2243.

38 Glockle, M., Hubler, K., Kummerer, H.J., Denninger, G. and Kaim, W. (2001) *Inorganic Chemistry*, **40**, 2263–2269.

39 Chisholm, M.H., Huffman, J.C., Rothwell, I.P., Bradley, P.G., Kress, N. and Woodruff, W.H. (1981) *Journal of the American Chemical Society*, **103** (16), 4945–4947.

40 Radonovich, L.J., Eyring, M.W., Groshens, T.J. and Klabunde, K.J. (1982) *Journal of the American Chemical Society*, **104**, (10), 2816–2819.

41 Eicher, T. and Hauptmann, S. (1995) *The Chemistry of Heterocycles*, Georg Thieme Verlag, Stuttgart, p. 1.

42 Gyepes, R., Witte, P.T., Horacek, M., Cisarova, I. and Mach, K., (1998) *Journal of Organometallic Chemistry*, **551**, 207–213.

43 Lauher, J.W. and Hoffmann, R. (1976) *Journal of the American Chemical Society*, **98**, 1729–1742.

44 Green, J.C. (1998) *Chemical Society Reviews*, **27** (4), 263–272.

45 Feldman, J. and Calabrese, J.C. (1991) *Journal of the Chemical Society. Chemical Communications*, 1042–1044.

46 Ohff, A., Kempe, R., Baumann, W. and Rosenthal, U. (1996) *Journal of Organometallic Chemistry*, **520**, 241–244.

47 Thewalt, U. and Berhalter, K. (1986) *Journal of Organometallic Chemistry*, **302**, 193–200.

48 Corbin, D.R., Stucky, G.D., Willis, W.S. and Sherry, E.G. (1982) *Journal of the American Chemical Society*, **104**, 4298–4299.

49 Corbin, D.R., Willis, W.S., Duesler, E.N. and Stucky, G.D. (1980) *Journal of the American Chemical Society*, **102**, 5969–5971.

50 Evans, W.J. and Drummond, D.K. (1989) *Journal of the American Chemical Society*, **111**, 3329–3335.

51 Durfee, L.D., Fanwick, P.E. Rothwell, I.P., Folting, K. and Huffman, J.C. (1987) *Journal of the American Chemical Society*, **109**, 4720–4722.

52 Tapolsky, G., Robert, F. and Launay, J.P. (1988) *New Journal of Chemistry*, **12**, 761–764.

53 Chien, H.-S. and Labes, M.M. (1986) *J Electrochem Soc Electrochem Science Technol*, **133** (12), 2509–2510.

54 Kaim, W. (1985) *Accounts of Chemical Research*, **18**, 160–166.

55 Konze, W.V., Scott, B.L. and Kubas, G.J. (2002) *Journal of the American Chemical Society*, **124** (42), 12550–12556.

56 Dyker, G. (1999) *Angewandte Chemie (International Edition in English)*, **111** (12), 1808–1822; (1999) *Angewandte Chemie – International Edition*, **38**, 1698–1712.

57 Crabtree, R.H. (2004) *Journal of Organometallic Chemistry*, **689**, 4083–4091.

58 Goldman, A.S. and Goldberg, K.I. (2004) *ACS Symposium Series*, **885**, 1–43.

59 Labinger, J.A. and Bercaw, J.E. (2002) *Nature*, **417**, 507–514.

60 Crabtree, R.H. (2001) *Journal of The Chemical Society-Dalton Transactions*, 2437–2450.

61 Li, Z. and Li, C.-J. (2005) *Journal of the American Chemical Society*, **127**, 3672–3673.

62 Merwin, R.K., Schnabel, R.C. Koola, J.D. and Roddick, D.M. (1992) *Organometallics*, **11**, 2972–2978.

63 Erker, G. (1977) *Journal of Organometallic Chemistry*, **134**, 189–202.

64 Alt, H.G. (1984) *Angewandte Chemie (International Edition in English)*, **96**, 752–769 (1984) *Angewandte Chemie – International Edition*, **23**, 766–782.

65 Beckhaus, R., Oster, J., Sang, J., Strauß, I. and Wagner, M. (1997) *Synlett*, 241–249.

66 Beckhaus, R. and Thiele, K.-H. (1986) *Journal of Organometallic Chemistry*, **317**, 23–31.

67 Beckhaus, R. and Thiele, K.-H. (1984) *Journal of Organometallic Chemistry*, **268**, C7–C8.

68 Stepnicka, P., Gyepes, R., Cisarova, I., Horacek, M., Kubista, J. and Mach, K. (1999) *Organometallics*, **18**, 4869–4880.

69 Rosenthal, U. (2004) *Angewandte Chemie (International Edition in English)*, **116**, 3972–3977; (2004) *Angewandte Chemie – International Edition*, **43**, 3882–3887.

70 Erker, G. (1984) *Accounts of Chemical Research*, **17**, 103–109.

71 Campora, J., Buchwald, S.L., Gutierrez-Puebla, E. and Monge, A. (1995) *Organometallics*, **14**, 2039–2046.

72 Rosenthal, U., Burlakov, V.V., Arndt, P., Baumann, W., Spannenberg, A. and Shur, V.B. (2004) *European Journal of Inorganic Chemistry*, 4739–4749.

73 Evans, D.F. (1959) *Journal of the Chemical Society*, 2003–2005.

74 Alfonso, M. and Stoeckli-Evans, H. (2001) *Acta Crystallographica, Section E: Structure Reports Online*, **E57**, o242–o244.

75 Du, M., Bu, Xian H. and Biradha, K. (2001) *Acta Crystallographica, Section C: Crystal Structure Communications*, **C57** (2), 199–200.

76 Soo, H.S., Diaconescu, P.L. and Cummins, C.C., (2004) *Organometallics*, **23** (3), 498–503.

77 Merritt, L.L. Jr. and Schroeder, E.D. (1956) *Acta Crystallographica*, **9**, 801–804.

78 Stang, P.J., (1998) *Chemistry - A European Journal*, **4** (1), 19–27.

79 Rogers, D.Z. (1986) *The Journal of Organic Chemistry*, **51**, 3904–3905.

80 de Azevedo, C.G. and Vollhardt, K.P. (2002) *Synlett*, 1019–1042.

81 Rademacher, J.T., Kanakarajan, K. and Czarnik, A.W. (1994) *Synthesis*, 378–380.

82 Abrahams, B.F., Jackson, P.A. and Robson, R. (1998) *Angewandte Chemie (International Edition in English)*, **110**, 2801–2804; (1998) *Angewandte Chemie – International Edition*, **37**, 2656–2659.

12
Bifunctional Molecular Systems with Pendant Bis(pentafluorophenyl)boryl Groups: From Intramolecular CH-activation to Heterolytic Dihydrogen Splitting

Michael Hill, Christoph Herrmann, Patrick Spies, Gerald Kehr, Klaus Bergander, Roland Fröhlich, and Gerhard Erker

12.1
Introduction

Tris(pentafluorophenyl)borane [$B(C_6F_5)_3$, **1a**)] was first prepared by Stone, Massey and Park in 1963 [1]. It is a strong Lewis acid, placed close to BF_3 on a scale according to the method described by Childs *et al.* [2]. After not receiving much attention for many years, $B(C_6F_5)_3$ was 'rediscovered' for use as a superb alkyl abstracting agent from methylzirconocene complexes to generate very active homogeneous Ziegler–Natta olefin polymerization catalysts [3, 4]. We contributed to this application of **1a** by reacting $B(C_6F_5)_3$ with a variety of (butadiene)zirconocenes to prepare dipolar single-component 'betaine' zirconocene Ziegler–Natta catalysts (**5**) [5] that could be stored and were ready to initiate the catalytic carbon-carbon coupling sequence when exposed to a reactive α-olefin (Scheme 12.1) [6–8].

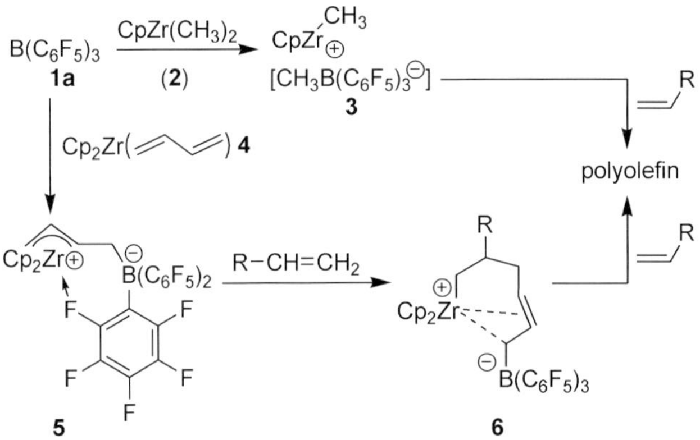

Scheme 12.1

$B(C_6F_5)_3$ has also been used as a general Lewis acid catalyst in organic synthesis [9]. It showed some potential in C−H bond activation and for stabilizing uncommon

Activating Unreactive Substrates: The Role of Secondary Interactions.
Edited by Carsten Bolm and F. Ekkehardt Hahn

ISBN: 978-3-527-31823-0

tautomeric forms. This is nicely exemplified by the formation of the $B(C_6F_5)_3$-stabilized keto isomer of α-naphthol [10] or the rapid formation and stabilization of 2*H*-pyrrole [11], one of the two nonaromatic pyrrole isomers (see Scheme 12.2).

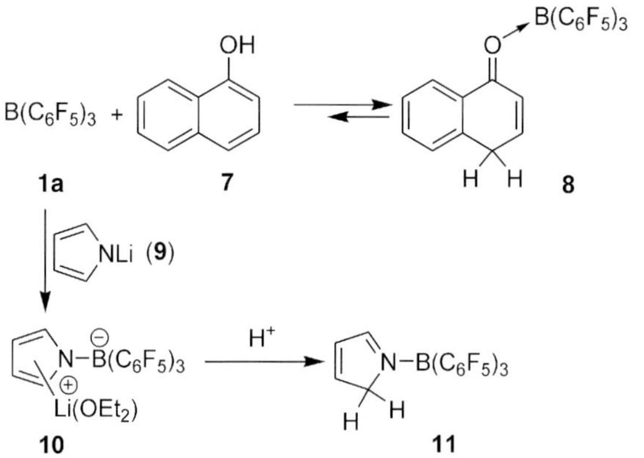

Scheme 12.2

In this project we have determined whether some of these favorable features of $B(C_6F_5)_3$ could be utilized in related bifunctional systems that contain a pendant $-B(C_6F_5)_2$ group. This functionality was readily introduced by a hydroboration route employing the reagent $HB(C_6F_5)_2$ ('Piers' borane', **1b**) [12]. It turned out that the systems prepared featured some very interesting chemistry in themselves, but did not show great catalytic activity, for example in CH activation. However, eventually some very interesting novel dihydrogen activation chemistry was found, when the core transition metal systems were exchanged for phosphine derivatives. Some of these developments are reviewed in the following account.

12.2 Bifunctional Zirconium/Boron Systems

For this first part of the study we have employed an allyl-substituted ansa-metallocene as the starting material. Complex **14** was synthesized conventionally [13] by sequential construction of the functionalized dimethylsilanediyl-linked ligand system followed by transmetalation to zirconium to yield **14** (see Scheme 12.3). As expected, complex **14** undergoes clean regioselective hydroboration reactions [14]. As a typical example its treatment with 9-BBN (**1c**) [15] yields the trimethylene-linked bifunctional Zr/B complex (**15**), which was characterized by X-ray diffraction (see Figure 12.1). It features a typical ansa-zirconocene framework. The trimethylene chain is attached at a Cp ring. It is found in an all anti-periplanar alkane chain conformation. It is oriented away from the bent metallocene core almost perpendicular to the central σ-ligand plane. The bulky 9-BBN unit at the end is thus placed in some distance from the sterically encumbered ansa-metallocene unit [14].

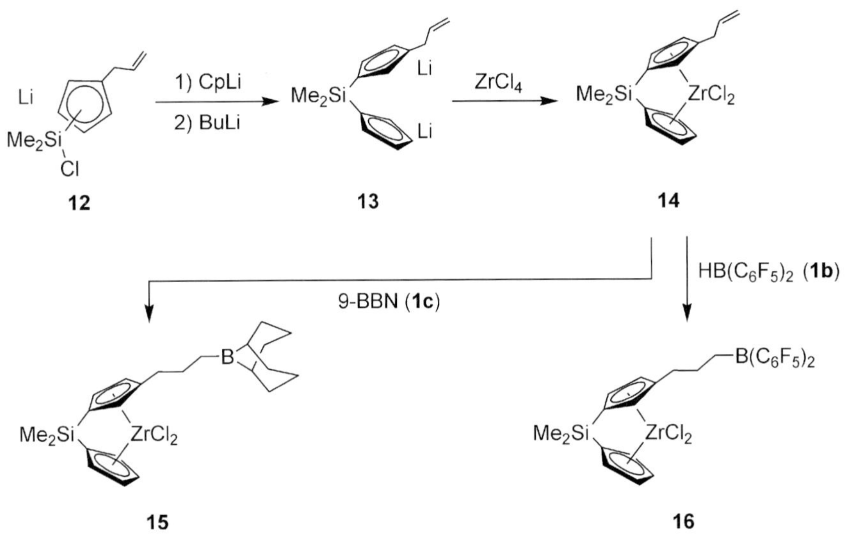

Li Me$_2$Si Cl **12** → 1) CpLi 2) BuLi → Me$_2$Si Li Li **13** → $ZrCl_4$ → Me$_2$Si $ZrCl_2$ **14**

14 → 9-BBN (**1c**) → Me$_2$Si $ZrCl_2$ B **15**

14 → $HB(C_6F_5)_2$ (**1b**) → Me$_2$Si $ZrCl_2$ $B(C_6F_5)_2$ **16**

Scheme 12.3

Complex **14** adds the reactive hydroboration reagent $HB(C_6F_5)_2$ (**1b**) cleanly to give the corresponding Zr/B product **16** in close to quantitative yield. It shows three ^{19}F NMR signals of the ortho (δ −130.2), para (δ −148.6) and meta (δ −160.8) fluorine atoms of the pair of $-C_6F_5$ groups at boron with chemical shifts that are typical of a tricoordinate borane species. Consequently, complex **16** features an ^{11}B NMR resonance at δ 75.6 [$\nu_{1/2}$ 190 Hz; cf $B(C_6F_5)_3$ (**1a**): δ 60.0] [14].

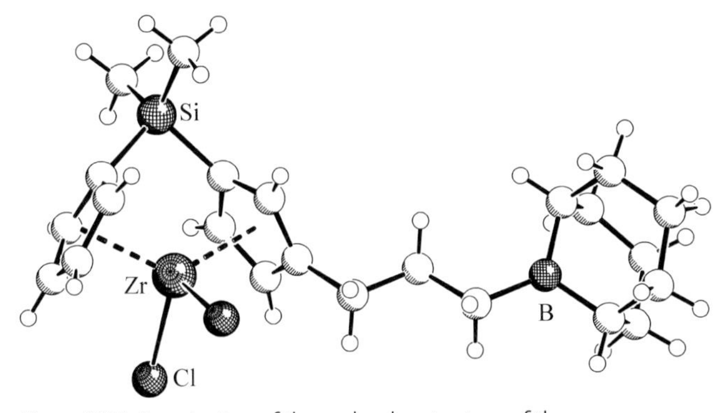

Figure 12.1 A projection of the molecular structure of the bifunctional ansa-zirconocene/trimethylene-9-BBN system **15**.

Complex **16** shows some typical features attributed to a reasonably strong R−B $(C_6F_5)_2$ Lewis acid. For example it reacts slowly with pyrrole to form a 1 : 1 adduct (Scheme 12.4). The hetarene pyrrole itself is not a typical *N*-donor, but its nonaromatic 2*H*-pyrrole isomer is [11]. Consequently, coordination of the C_4H_5N heterocycle requires the Lewis acid-induced tautomerization to the thermodynamically less favored 2*H*-pyrrole isomer so that this will be subsequently coordinated to the

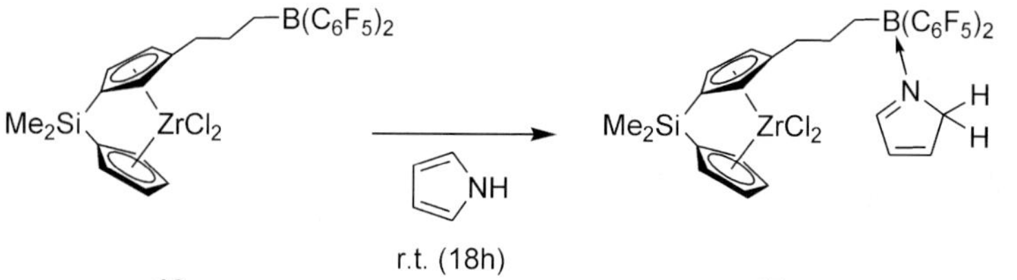

$B(C_6F_5)_2$ Me_2Si $ZrCl_2$ → (pyrrole NH, r.t. (18h)) → $B(C_6F_5)_2$ N H H Me_2Si $ZrCl_2$

16 **17**

Scheme 12.4

electron-deficient boron atom. The resulting bifunctional 'isopyrrole'–borane adduct (**17**) features ^{13}C NMR signals of the five-membered *N*-heterocycle at δ 169.3 (−N=CH−), 127.1, 154.2 (−CH=CH−) and δ 65.0 (−N−CH_2−). The corresponding CH_2 hydrogen atoms are diastereotopic; they give rise to an ^1H NMR AB pattern (δ 3.72, 3.68, $^2J = 27$ Hz) due to the attachment of the 2*H*-pyrrole tautomer to the chiral ansa-zirconocene backbone. The ^{19}F NMR signals of the pair of diastereotopic $-C_6F_5$ substituents at boron are, consequently, in the typical borate range (*o*: δ −132.6/−132.9, *p*: δ −158.1/−158.2, *m*: isochronous at δ −163.3]. The attachment of the 2*H*-pyrrole isomer to the boron Lewis acid center of the pendant $-CH_2-CH_2-CH_2-B(C_6F_5)_2$ group was confirmed by an X-ray crystal structure analysis of **17** (see Figure 12.2) [14].

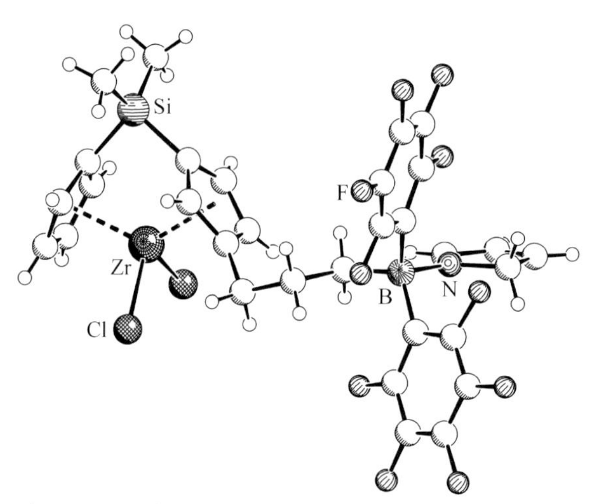

Figure 12.2 Molecular structure of **17**.

Typical *N*-heterocyclic donor ligands, such as oxazole, imidazole or benzimidazole derivatives, form 1 : 1 adducts with the Lewis acidic Zr/B system **16**. A typical example is complex **18**, the reaction product between **16** and 1-methylbenzimidazole. The X-ray crystal structure analysis (see Figure 12.3) has in this case revealed a trimethylene chain conformation that places the *N*-heterocycle in front of the bent metallocene wedge.

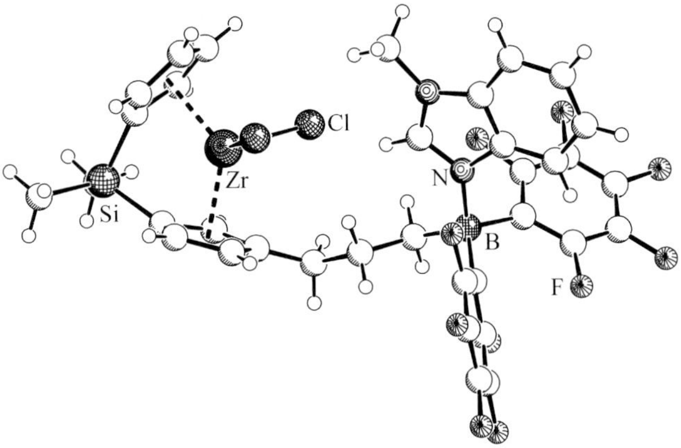

Figure 12.3 A view of the molecular geometry of the 1-methylbenzimidazole adduct **18**.

In this case we have briefly investigated whether a reaction between the zirconocene unit and the coordinated 1-methylbenzimidazole moiety could potentially be achieved under suitable reaction conditions. For that purpose we have treated complex **18** with one equivalent of base (i.e. LDA) to achieve deprotonation at the acidic benzimidazole 2-position. It appears that indeed the reaction of **18** with LDA successfully generated the mono-anionic reactive intermediate **19** (see Scheme 12.5). However, this does not subsequently stabilize itself by carbanion attack on the

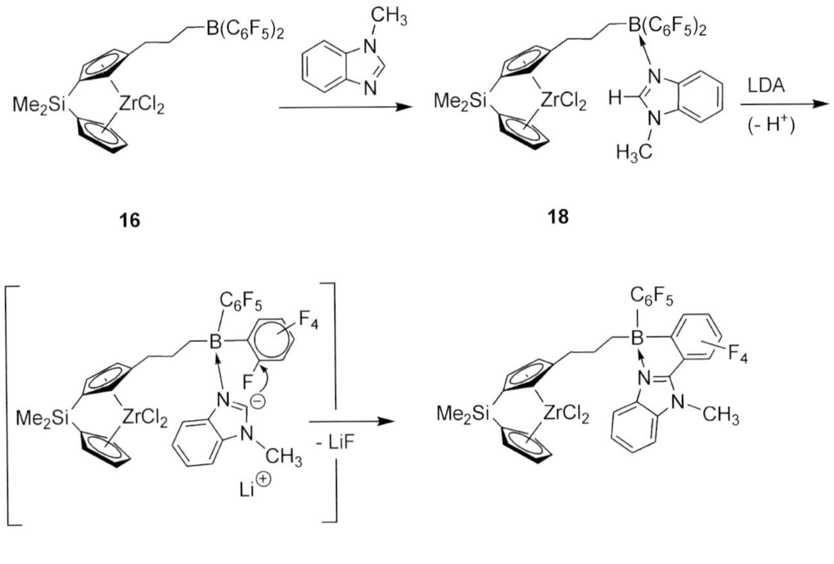

Scheme 12.5

adjacent zirconocene dichloride moiety with formation of a new Zr–C bond, but rather chooses to have the benzimidazole anion attack one of the adjacent $-C_6F_5$ substituents at boron. This favored intramolecular nucleophilic aromatic substitution reaction leads to the formation of the product **20**, which was isolated as a ca. 1 : 1 mixture of two diastereoisomers (due to the presence of a planar-chiral ansa-zirconocene unit and the formation of a newly generated stereogenic center at the four-coordinated boron atom) [16, 17].

This result prompted us to place active σ-ligands at the zirconium center. Treatment of **14** with phenyl lithium gave the corresponding allyl-substituted diphenyl ansa-zirconocene (**21**), which was subsequently subjected to the hydroboration reaction with $HB(C_6F_5)_2$ to yield the corresponding bifunctional metallocene complex **22** (see Scheme 12.6).

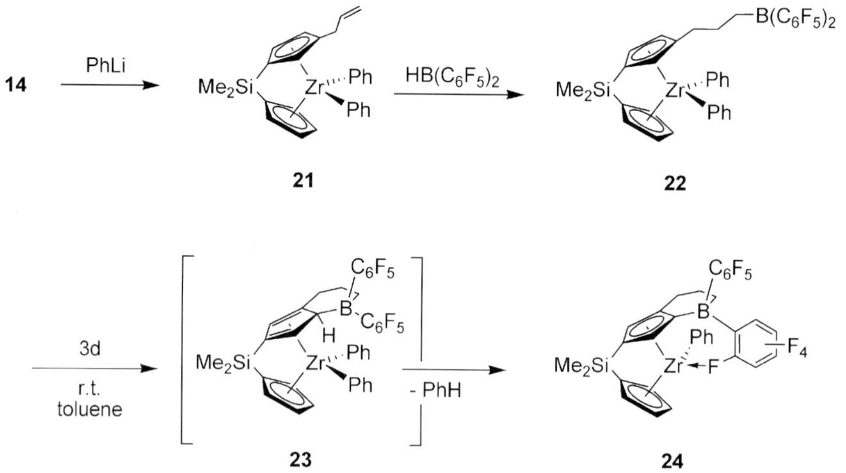

Scheme 12.6

Complex **22** could only be kept unchanged for a limited time at room temperature. Within a period of ca. 3 d it was completely converted to the new system **24** with liberation of one molar equivalent of benzene [16, 18].

Complex **24** was characterized by X-ray diffraction. It revealed that a six-membered boron containing heterocycle had been formed anellated to one of the Cp-ring systems (see Figure 12.3). Consequently, only one σ-phenyl ligand has remained at the zirconium atom. The second available site in the σ-ligand wedge is occupied by a coordination of an ortho-F–C(arene) substituent of the proximal C_6F_5 substituent at boron. The Zr–F distance amounts to 2.250(1) Å, which represents one of the shortest values of a Zr–F–C interaction. The corresponding Zr–F–C(arene) angle at the fluorine atom amounts to 142.8(1)° in complex **24** (Figure 12.4).

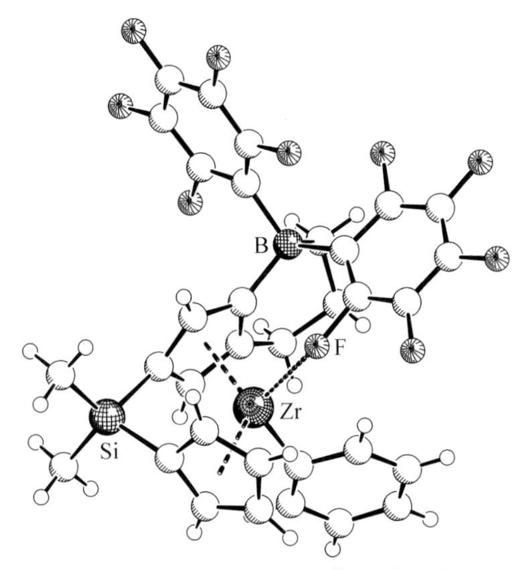

Figure 12.4 Molecular structure of complex **24**.

In solution, complex **24** shows dynamic NMR spectra that indicate equilibration between two half-chair conformers of the boron heterocycle. The major isomer seems to be equivalent to the structure found in the crystal. It features a Zr–(μ-F)–C (arene) ^{19}F NMR signal at δ −175.4. In the 'frozen' low-temperature NMR, complex **24** shows 10 separated ^{19}F NMR resonances [16].

The formation of complex **24** indicates attack of the strongly electrophilic pendant borane in the starting material **22** at the adjacent electron-rich Cp ring system in the sense of an intramolecular electrophilic aromatic substitution reaction. We assume that the acidic endo-H of the alleged intermediate (**23**) is removed by the proximal σ-phenyl ligand at zirconium to directly yield the observed product (**24**) (see Scheme 12.6). We shall see that variants of this characteristic S_{EAr}-type reaction [19] will be observed at the remotely related late transition metal Cp–$CH_2CH_2CH_2$–B $(C_6F_5)_2$ complexes as well (see below). This intramolecular S_{EAr}-type reaction seems to be quite favorable for this general ligand type in various corresponding bifunctional metal complex systems.

Complex **22** forms a 1 : 1 adduct (**25**) with 1-methylbenzimidazole. When this was heated to 80 °C in benzene solution in the presence of one additional molar equivalent of 1-methylbenzimidazole, we observed the formation of the C–H activation product (**26**) (90% yield, isolated as a 3 : 2 mixture of two diastereoisomers). The mechanistic course of this reaction has not yet been investigated in detail, but it may be that we have here observed an example of an intramolecular C–H activation of a boron-bonded (and activated) heterocycle potentially attacked by an *in-situ* generated benzyne ligand at zirconium (see Scheme 12.7) [16].

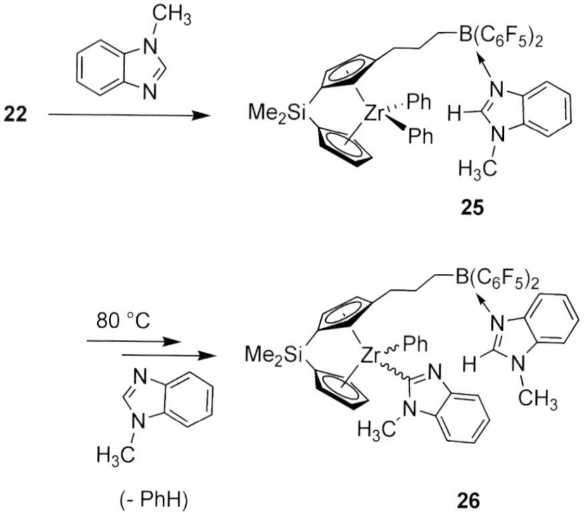

Scheme 12.7

12.3
Bifunctional Group 9 Metal/Boron Systems

In the course of this study we have reacted the lithium allylcyclopentadienide reagent (**27**) with a variety of cobalt(I), rhodium(I), and iridium(I) reagents. The reaction of **27** with $[(CO)_4CoI]$, generated *in situ* by reaction of dicobalt octacarbonyl with I_2 in THF, represents a typical example (see Scheme 12.8). The resulting product (**28a**) was subsequently subjected to the hydroboration reaction with 'Piers' borane' to give the bifunctional Co/B complex (**29a**). Complex **29a** features the typical set of *o* ($\delta -130.5$), *p* ($\delta -147.1$), and *m* ($\delta -160.9$) ^{19}F NMR signals of

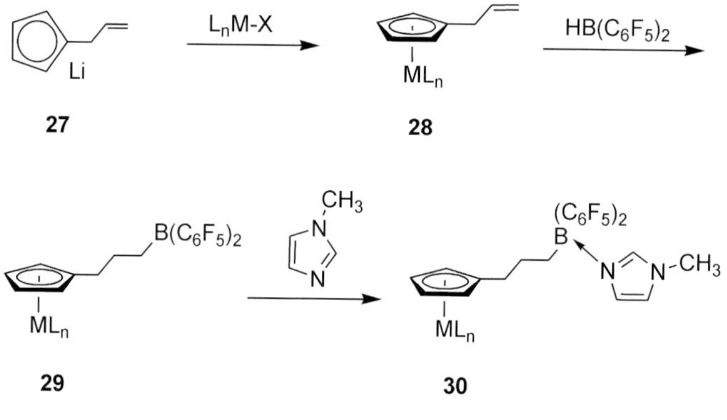

$L_nM = Co(CO)_2$ (**a**); (nbd)Rh (**b**), (cod)Rh (**c**), (cod)Ir (**d**)

Scheme 12.8

a Lewis-acidic tricoordinate R–B(C_6F_5)$_2$ moiety (^{11}B NMR resonance at δ +73). It shows three ^{13}C NMR signals of the trimethylene linker at δ 31.8, 30.9, and 26.7 and a set of three ^{13}C NMR signals of the attached C_5H_4-ring system [δ 106.4 (ipso), δ 84.9 (C2/5), δ 83.3 (C3/4), corresponding ^{1}H NMR signals at δ 4.49, 4.33 (in C_6D_6)]. The IR spectrum of complex **29a** features typical ν(CO) bands at 2036 and 1967 cm^{-1}.

Complex **29a** readily adds heterocyclic donor ligands at the Lewis-acidic borane unit. 1-Methylimidazole addition gave the corresponding adduct **30a**, which was characterized by X-ray diffraction (see Figure 12.5).

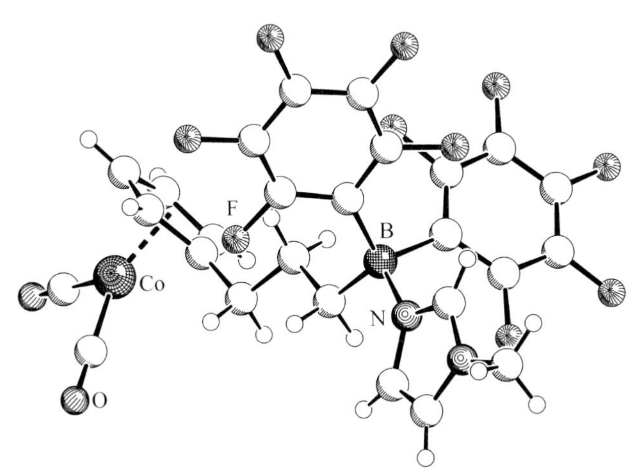

Figure 12.5 A view of the molecular structure of complex **30a**.

Similarly, the reaction of the reagent **27** with (norbornadiene)RhCl dimer [(nbd) RhCl]$_2$ or (1,5-cyclooctadiene)RhCl dimer [(cod)RhCl]$_2$ gave the respective complexes **28b** and **c** which were subsequently treated with the HB(C_6F_5)$_2$ hydroboration reagent (**1b**) to yield the corresponding Rh/B systems **29b** and **29c**, respectively. Both these complexes readily added the 1-methylimidazole donor to yield the respective adducts (**30b**, **30c**).

Both the complexes **30b** and **30c** were characterized by X-ray crystal structure analyses (see Figures 12.6 and 12.7). Complex **30b** features the endo-coordinated norbornadiene ligand at the central rhodium atom (see Figure 12.6) and an extended 'zig-zag' trimethylene chain at its Cp ring with the very bulky –B(C_6F_5)$_2$(1-methylimidazole) unit at its end. The corresponding (cod)Rh complex **30c** features a similar structure (see Figure 12.7). However, the two systems are slightly different with regard to rotation of the CH_2–CH_2 vector of the C3-spacer from the plane of its adjacent Cp ring [**30b**/**30c**: θ C5–C1–C6–C7: 45.8(4)°/−86.4(4)°; θ C2–C1–C6–C7: −130.3(3)°/84.9(4)°].

The related iridium examples turned out to show a slightly more complex chemical behavior. Treatment of [(cod)IrCl]$_2$ with the (allyl-Cp)Li reagent (**27**) gave

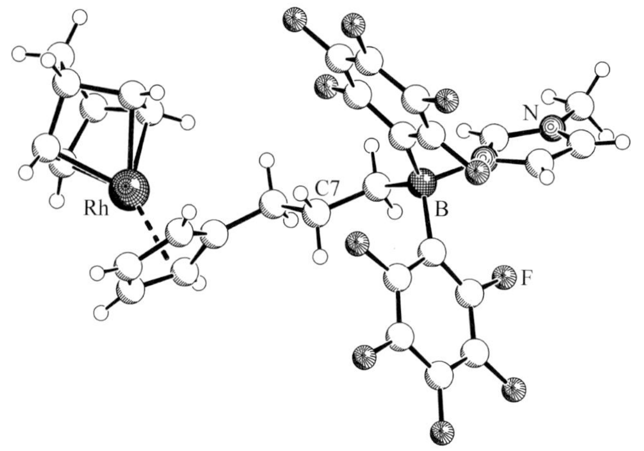

Figure 12.6 Molecular structure of complex **30b**.

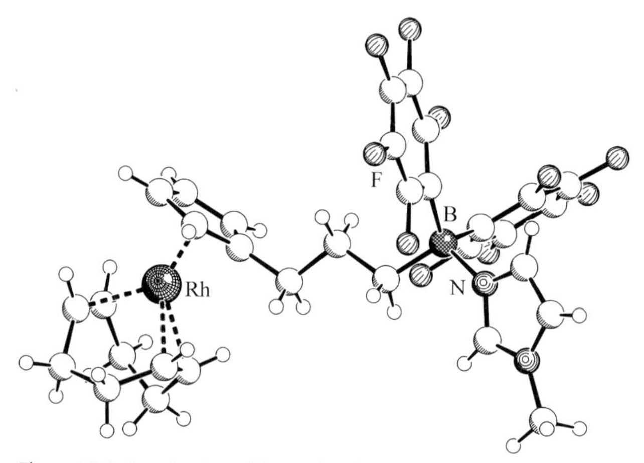

Figure 12.7 A projection of the molecular geometry of the Rh/B-1-methylimidazole adduct **30c**.

the corresponding (allyl-Cp)Ir(cod) complex (**28d**), as expected. This compound shows some interesting thermal behavior. Thermolysis of the neat oil (**28d**) in a Schlenk tube for 1.5 h at 100 °C resulted in the quantitative formation of the isomerized (1-propenyl-Cp)Ir(cod) derivative (**31**). The thermodynamically preferred isomer **31** was characterized by an X-ray crystal structure analysis.

We have reacted complex **31** with $HB(C_6F_5)_2$ (**1b**) and observed a clean hydroboration reaction. However, the assumed product (**32**) of a regioselective hydroboration reaction could not be observed or isolated because of a rapid subsequent S_{EAr} reaction followed by proton transfer to iridium to form the final product (**33**) containing an anellated five-membered borataheterocycle (see Scheme 12.9).

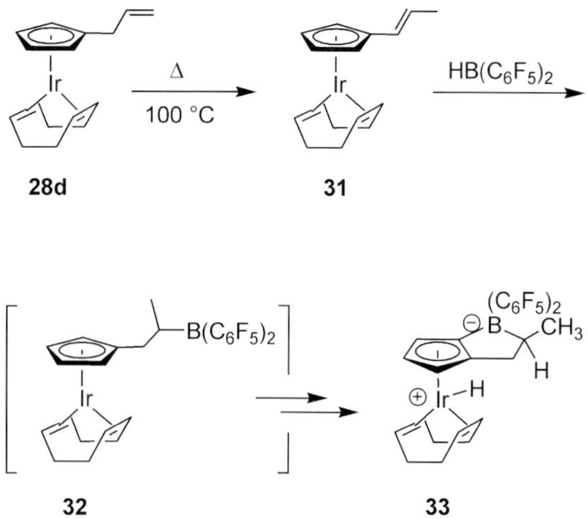

Δ
100 °C
$HB(C_6F_5)_2$
28d
31
$B(C_6F_5)_2$
$(C_6F_5)_2$
B
CH_3
H
Ir–H
32
33

Scheme 12.9

Complex **33** has a rather low solubility in most noncoordinating solvents, but we were able to characterize it by X-ray diffraction (see Figure 12.8). The X-ray crystal structure analysis of complex **33** shows the presence of the anellated five-membered heterocycle in a slightly twisted conformation. The methyl substituent attains a position anti to the iridium center. The (cod)Ir unit carries a hydrogen bonded to the metal with the Ir–H vector oriented toward one $-CH_2-CH_2-$ subunit of the cod ligand and in the direction of the anellated borataheterocycle.

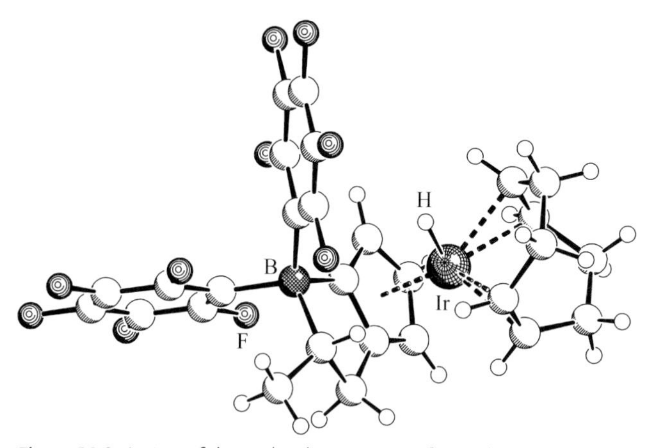

Figure 12.8 A view of the molecular structure of complex **33**.

The hydroboration reaction of the (cod)Ir(allylCp) complex (**28d**) with $HB(C_6F_5)_2$ takes a similar course. After a ca. 1 h reaction time at room temperature in pentane the cyclo-anellated product **36** was isolated in 95% yield. The product was characterized by X-ray diffraction (see Figure 12.9). In the ^1H NMR spectrum it features a very

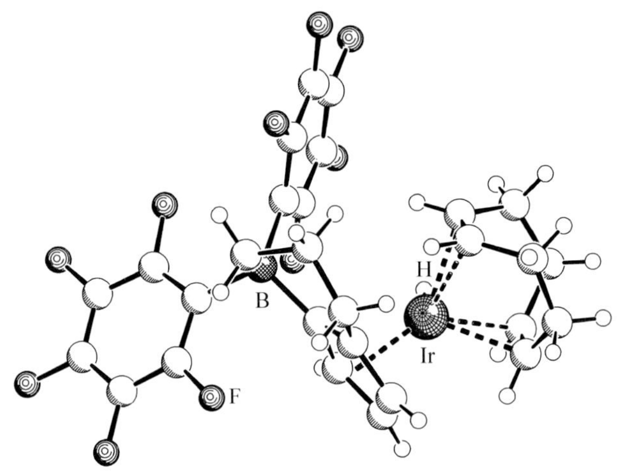

Figure 12.9 Molecular structure of complex **36**.

characteristic Ir–H proton resonance at δ −11.91 and a double set of ^{19}F NMR signals due to $-C_6F_5$ groups at boron oriented cis and trans to iridium, respectively.

We assume that the product **36** is formed starting from the regular hydroboration product (**34**). In this system the electrophilic borane is apparently able to very rapidly attack the adjacent Cp ring [20] to generate **35**. The iridium atom in the reactive intermediate **35** seems to act as a metal base [21] that is able to rapidly abstract the proton from the doubly substituted C_5H_4 moiety to eventually lead to the formation of the observed product (see Scheme 12.10). We have obtained experimental evidence

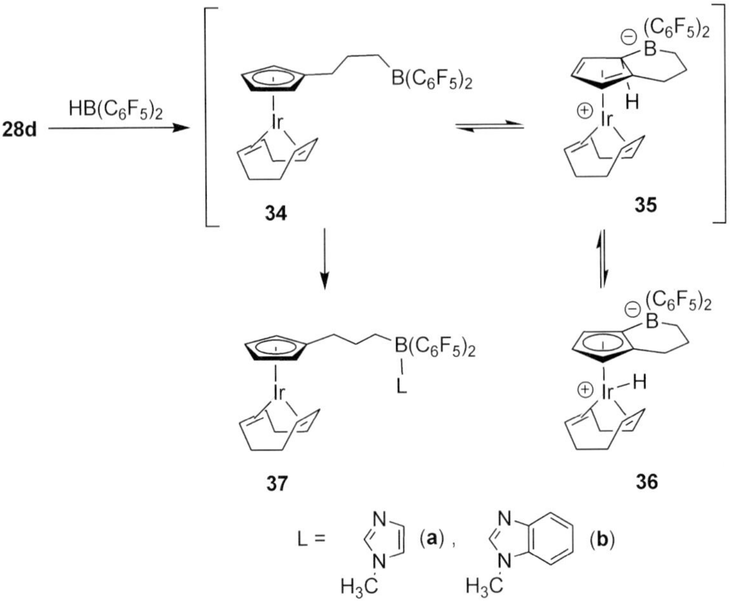

28d — $HB(C_6F_5)_2$ → **34** ⇌ **35**; **34** → **37**; **35** ⇌ **36**

$B(C_6F_5)_2$, Ir, $(C_6F_5)_2$B, H, L

L = (**a**), (**b**); H_3C

Scheme 12.10

that the **34**⇄**35**⇄**36** equilibration takes place rapidly at room temperature. The equilibrium is shifted from the favored S_{EAr}-anellated product **36** to the open-chain intermediate **34** by adding suitable trapping agents. This treatment of **36** with 1-methylbenzimidazole for 10 min in toluene solution at ambient temperature gave a 68% isolated yield of the adduct (**37b**), which was characterized by an X-ray crystal structure analysis (see Figure 12.10). Treatment of **36** with 1-methylimidazole under similar reaction conditions also resulted in a rapid proton shift from Ir to carbon, followed by ring opening to regenerate **34**, which then was trapped by the added donor ligand to form the corresponding (cod)Ir[Cp$-CH_2CH_2CH_2B(C_6F_5)_2-$1-methylimidazole] adduct (**37a**).

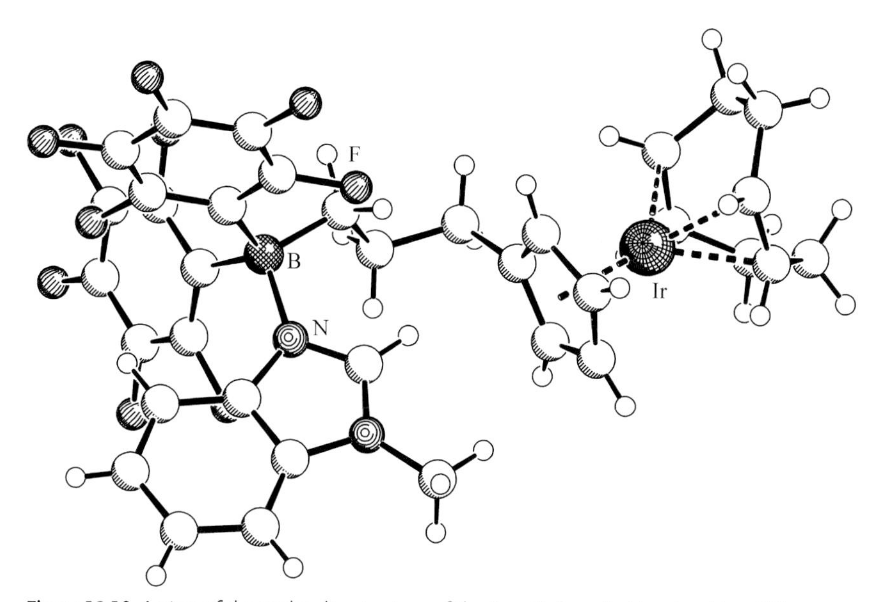

Figure 12.10 A view of the molecular structure of the 1-methylbenzimidazole adduct **37b**.

Complex **36** also reacts (out of the equilibrium with **34**) with pyrrole (Scheme 12.11). Similar to the observed analogous reaction of the corresponding Zr/B system (see above) [14], this reaction in the Ir/B case is rather slow because of the necessary tautomerization reaction, which is rate limiting. After 3 d at room temperature in toluene a 71% yield of the 2H-pyrrole adduct (**38**) was isolated. The ^{13}C NMR spectrum of complex **38** shows signals of the trimethylene linker at δ 31.4, 28.4 and 24.5. The CH_2 group inside the 2*H*-pyrrole ligand features a 2H intensity ^{1}H NMR resonance at δ 3.49. Complex **38** shows a set of ^{19}F NMR signals [δ −133.4 (*o*), δ −158.2 (*p*), δ −163.6 (m)] typical of a tetravalent borate structure with an ^{11}B NMR resonance at δ −5.3.

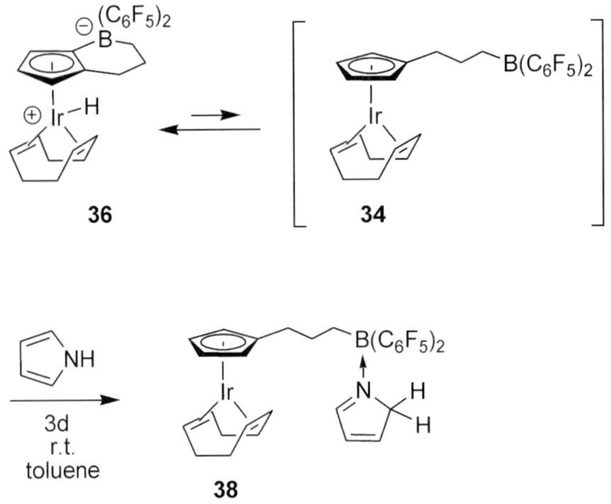

$(C_6F_5)_2$ B

Ir–H

36

$B(C_6F_5)_2$

Ir

34

NH

3d
r.t.
toluene

$B(C_6F_5)_2$

Ir

N H H

38

Scheme 12.11

Complex **38** was also characterized by an X-ray crystal structure analysis (see Figure 12.11). It features a fully extended trimethylene chain at the IrCp unit with a $-B(C_6F_5)_2(2H\text{-pyrrole})$ moiety at its end. The B–N bond length amounts to 1.620(5) Å.

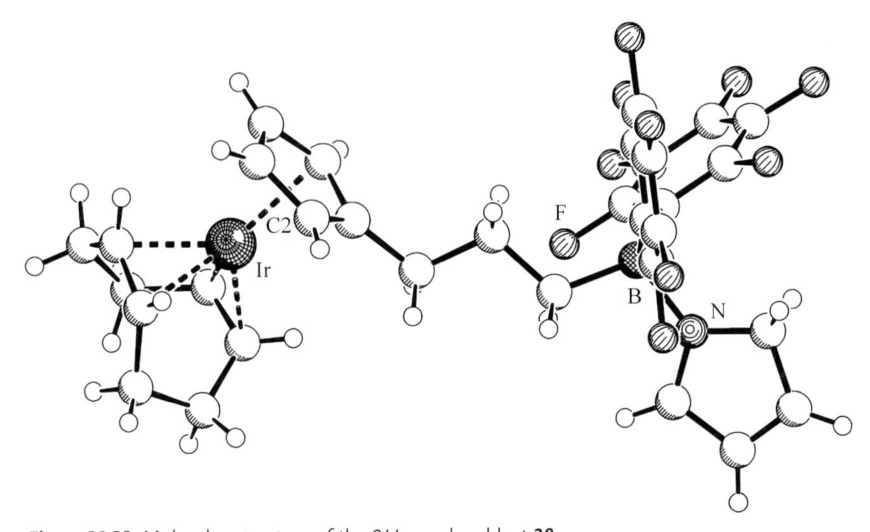

Figure 12.11 Molecular structure of the 2*H*-pyrrole adduct **38**.

We have tried to react complex **36** (via **34**) with dihydrogen in order to potentially form **39** (Scheme 12.12), but could not positively demonstrate heterolytic splitting of the H_2 molecule at this bifunctional Ir/B complex. However, cleavage and activation of dihydrogen was eventually achieved with our bifunctional P/B system described below in chapter 12.4.

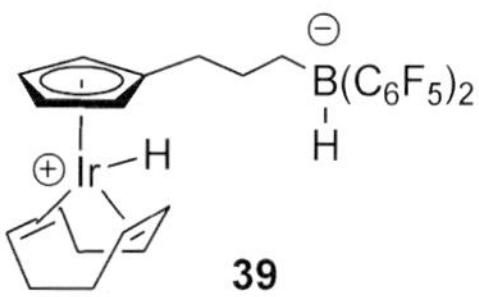

Scheme 12.12

12.4
Bifunctional Phosphorus/Boron Systems

We previously showed that $B(C_6F_5)_3$ (**1a**) undergoes a series of three consecutive reactions with the phosphorus ylide $Ph_3P{=}CHPh$ in an equilibrium situation. At 0 °C, simple adduct formation occurs to yield $Ph_3P^+{-}CH(Ph){-}B^-(C_6F_5)_3$. This reaction is reversible, and at 55 °C (24 h) in toluene we observed the formation of the product of an electrophilic aromatic substitution reaction of $B(C_6F_5)_3$ at the ylidic phenyl group to form $Ph_3P^+{-}CH_2{-}C_6H_4{-}B^-(C_6F_5)_3$. Even this reaction turned out to be reversible and this reaction pathway is eventually overcome under thermodynamic control by a subsequent S_{NAr} reaction, i.e. attack of the ylide nucleophile at one of the $-C_6F_5$ rings at boron to form the final product (**41**) (see Scheme 12.13). The product (**41**) was characterized by an X-ray crystal structure analysis (see Figure 12.12) [22].

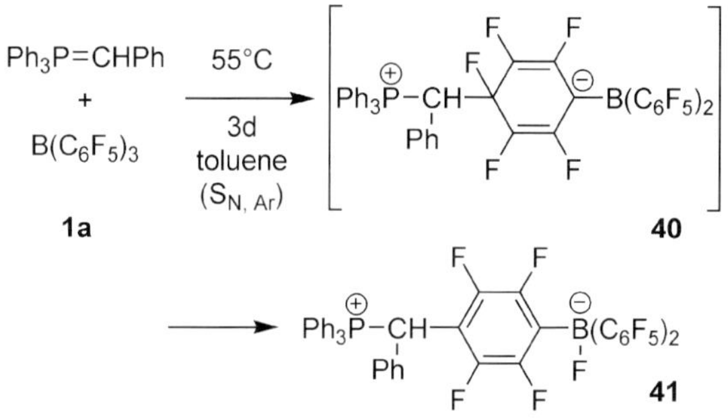

Scheme 12.13

Stephan has shown that secondary phosphines such as $(mesityl)_2PH$ [mes_2PH] (**42**) undergo a similar nucleophilic aromatic substitution reaction at a $-C_6F_5$ ring of **1a** to yield the product **43** (see Scheme 12.14). Subsequent treatment with $Me_2Si(H)Cl$ converts this product to the corresponding hydrido borate system (**44**). Interestingly, the zwitterionic 'inner salt' (**44**) cleaves off dihydrogen when heated to 100 °C to yield the $-C_6F_4-$ linked P/B system **45**. Compound **45**, remarkably, adds hydrogen to re-form the $-C_6F_4-$ linked phosphonium/borate product (**44**) at room temperature [23].

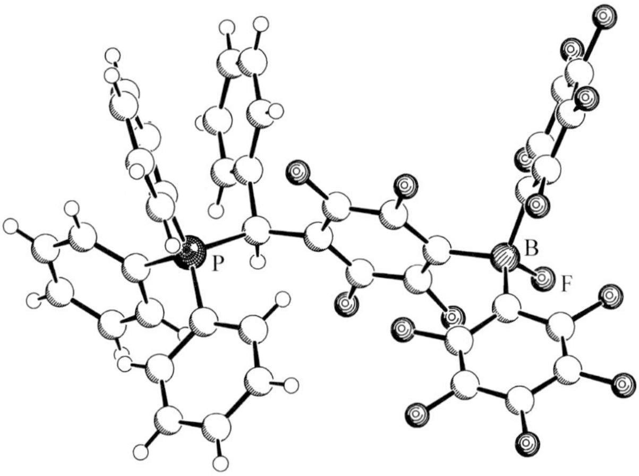

Figure 12.12 A view of the molecular structure of compound **41**.

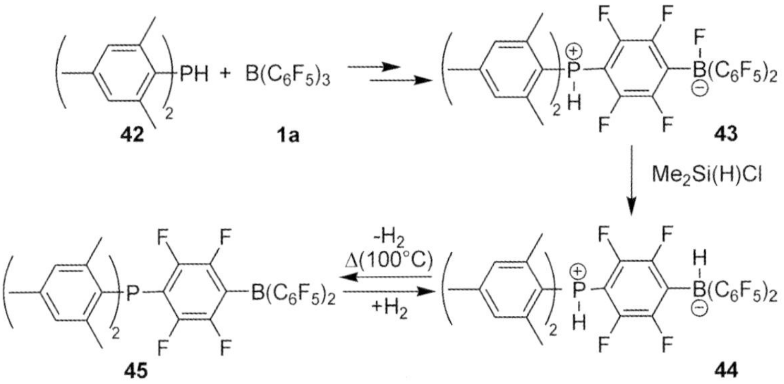

Scheme 12.14

Scheme 12.15

Subsequently, Stephan *et al.* have shown that noncoordinating mixtures of very bulky phosphines R_3P (e.g. mes_3P or $(tert\text{-butyl})_3P$) with $B(C_6F_5)_3$ ('frustrated' phosphine/borane pairs) readily split dihydrogen with formation of the respective phosphonium/borate salts (**46**) (see Scheme 12.15) [24].

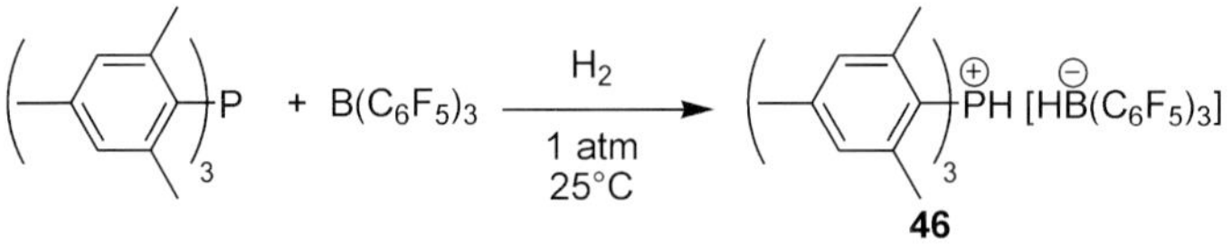

Our recent experiments with the remotely related bifunctional group 9 metal/boron systems (see above) prompted us to prepare a $-CH_2-CH_2-$ linked bulky P/B system and investigate its potential for H_2 activation chemistry. For this purpose we have prepared vinyl-bis(mesityl)phosphine $mes_2P{-}CH{=}CH_2$ (**48**) by treatment of mes_2PCl with vinylmagnesiumbromide. The product **48** then was subjected to treatment with $HB(C_6F_5)_2$ [25]. In this case, this resulted in a clean regioselective hydroboration reaction at ambient temperature to yield **49** (Scheme 12.16) [26]. Compound **49** features a ^{31}P NMR resonance at δ + 20.6 and a ^{11}B NMR signal at δ 8.5. It shows ^{13}C NMR signals of the $P{-}CH_2{-}CH_2{-}B$ unit at δ 29.4 and 18.2. The corresponding ^{1}H NMR features were found at δ 2.87 and δ 2.29 with $^{2}J_{PH} = 8Hz$ and $^{3}J_{PH} = 42Hz$ coupling constants. Together with the typical ^{19}F NMR features [δ −128.8 (*o*), δ −157.0 (*p*), δ −163.6 (m)] these spectroscopic data are in accord with a weakly P−B bonded four-membered internal adduct structure. Unfortunately, we have not yet obtained single crystals of **49** that were suited for a characterization of the system by X-ray diffraction, but the system was adequately described by a DFT calculation. This characterizes compound **49** by featuring a P−B bond length of 2.21 Å and shows the presence of a π-stacking interaction between a mesityl group (π-donor) with a $-C_6F_5$ substituent (π-acceptor) [27] across the P−B vector (see Figure 12.13). In addition, we were able to characterize the P-oxidized derivative (**50**) by an X-ray crystal structure analysis (see Figure 12.14).

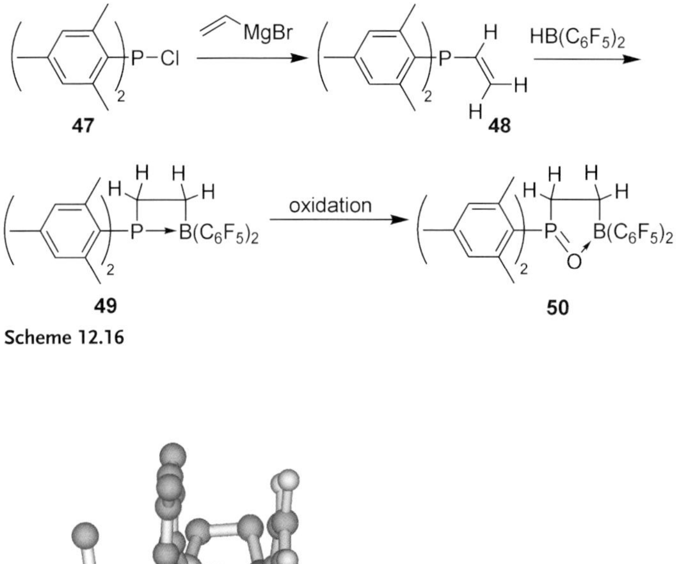

Scheme 12.16

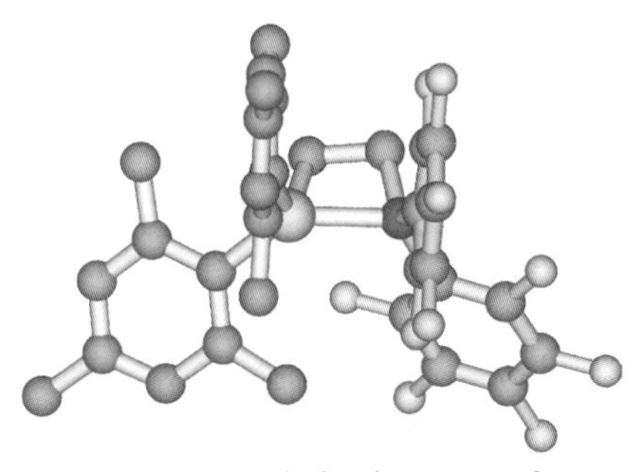

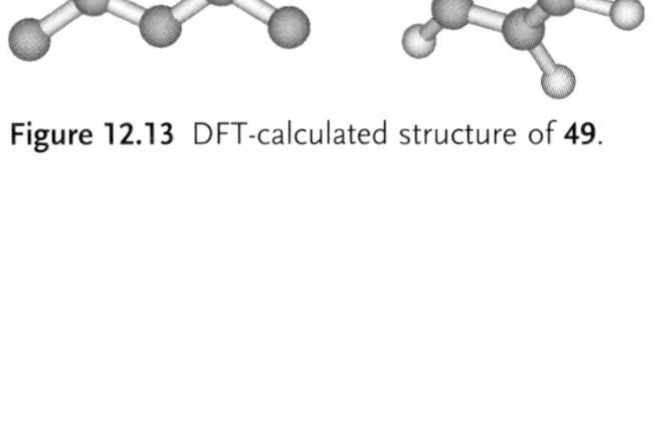

Figure 12.13 DFT-calculated structure of **49**.

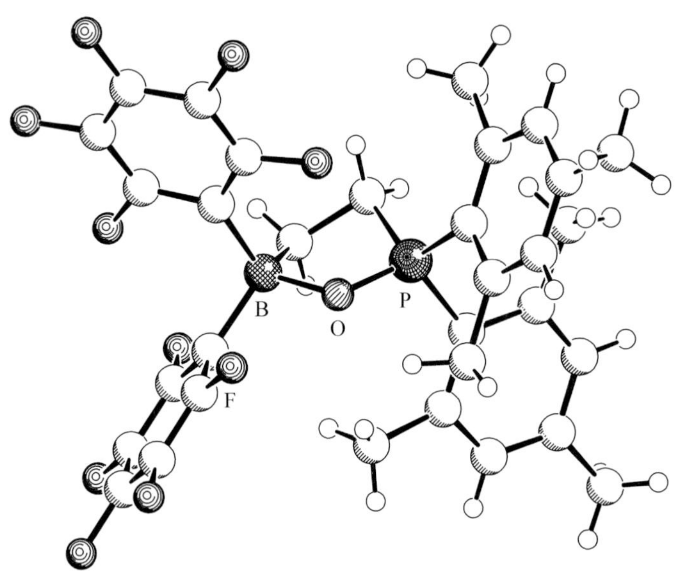

Figure 12.14 A view of the molecular geometry of the phosphinoxide derivative **50**.

We have reacted compound **49** with dihydrogen. At ambient temperature and pressure it adds one equivalent of H_2 to form the $-CH_2-CH_2-$ bridged zwitterionic phosphonium/borate product (**51**) (see Scheme 12.17) [26]. Compound **51** features a ^{31}P NMR PH doublet at δ −6.5 with a typical coupling constant of $^1J_{PH} = 486$Hz and a broad ^{11}B NMR doublet at δ –20.1 with $^1J_{BH} = 88$Hz. The $P-CH_2-CH_2-B$ $^{13}C/^1H$ NMR signals occur at δ 26.9, 17.8/δ 2.80, 1.25. Compound **51** exhibits a set of ^{19}F NMR signals [δ −133.4 (*o*), −165.8 (*p*), −168.0 (*m*)] that is very typical for $-C_6F_5$ groups at a four-coordinated boron center. The corresponding reaction of **49** with D_2 gave the corresponding $mes_2P^+(D)-CH_2-CH_2-B^-(D)(C_6F_5)_2$ product (**51**-d_2, with a ca. 30% P(H) contamination). From the ^{2}H NMR spectrum the B(*D*) resonance was located at δ 2.94, the P(*D*) doublet at δ 7.87 (d, $^1J_{PD} = 75$Hz). The zwitterionic compound **51** was also characterized by an X-ray crystal structure analysis (see Figure 12.15) [26].

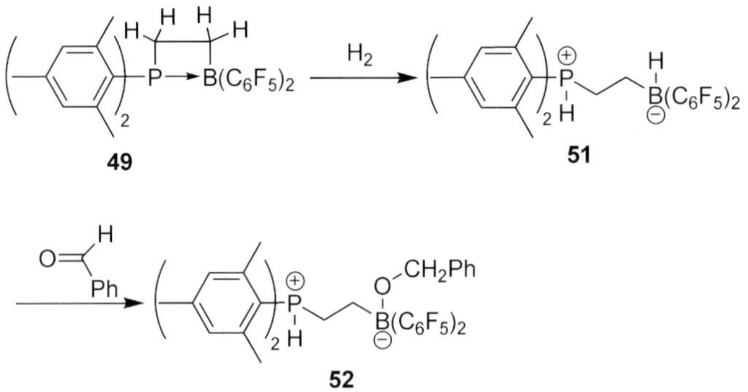

H_2

49 → **51** → **52** (PhCHO)

Scheme 12.17

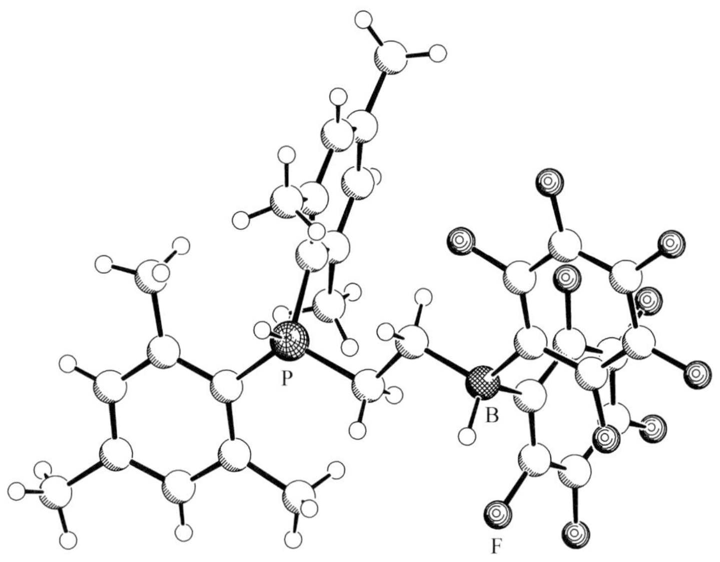

Figure 12.15 Molecular structure of the zwitterionic compound **51**.

The H_2 addition to **49** corresponds to a stoichiometric heterolytic dihydrogen splitting reaction [28]. We have shown that the resulting borohydride functionality of **51** can be employed as a stoichiometric reducing agent. Treatment of **51** with benzaldehyde thus led to the formation of **52**, which was characterized by an X-ray crystal structure analysis (see Scheme 12.17 and Figure 12.16).

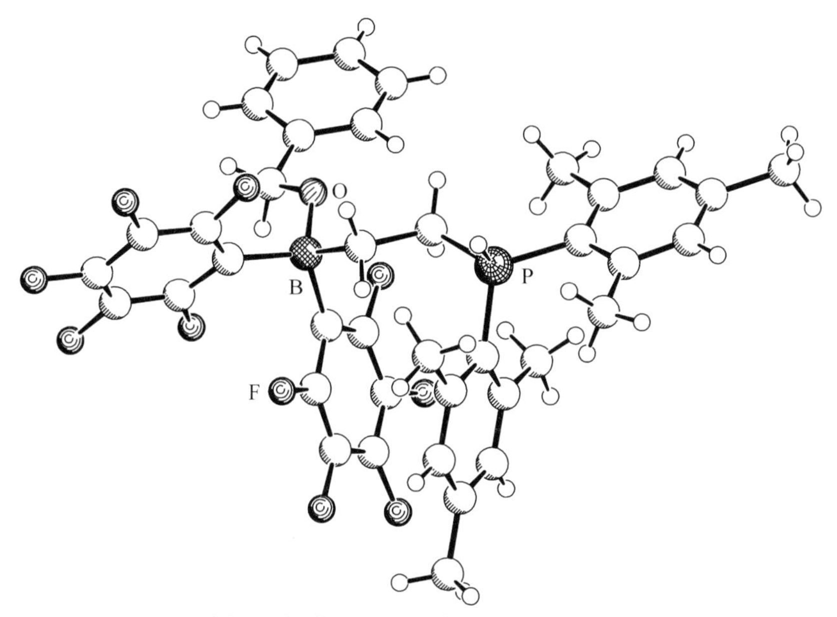

Figure 12.16 A view of the molecular structure of compound **52**.

12.5 Conclusions

The bifunctional transition metal/borane title complexes show some interesting chemical reactivity. They all readily form adducts with a variety of heterocyclic donor ligands. Initially we had hoped to use them as substrates for internal activation, for example of C−H bonds. However, our study has shown that such reactivity is only rarely observed. More common seems to be 'self-attack' of the Lewis-acidic pendant boryl group at its adjacent Cp ligand. This intramolecular electrophilic aromatic substitution reaction very often appears to take place, sometimes even in a reversible fashion. We hope that in the case of the respective zirconium systems this reaction type may eventually lead to developments toward novel single-component Ziegler–Natta catalyst systems. In some examples of the Group 9 metal/borane systems this heterocycle formation is reversible and the resulting anellated heterocycles represent stabilized resting stages for the reactive bifunctional metal/borane systems, whose characteristic reactivities are just beginning to emerge from our studies.

The synthesis of the ethylene-linked phosphine/borane system may be regarded as a logical extension of the bifunctional late metal/borane work. The respective study in this series shows that the activation of strong σ-bonds must no longer be regarded a domain of transition metal chemistry. We regard our new $mes_2P{-}CH_2{-}CH_2{-}B(C_6F_5)_2$ system as important since its synthesis and chemistry may open unprecedented pathways toward the development of active and hopefully selective hydrogenation catalyst systems that do not require activation of either the dihydrogen molecule or the substrate by a common transition metal center.

Acknowledgments

Financial support from the Deutsche Forschungsgemeinschaft (SPP 1118) and the Fonds der Chemischen Industrie is gratefully acknowledged.

References

1 Massey, A.G., Park, A.J. and Stone, F.G.A. (1963) *Proceedings of the Chemical Society*, 212; Massey, A.G. and Park, A.J. (1964) *Journal of Organometallic Chemistry*, **2**, 245–250; Massey, A.G. and Park, A.J. (1986) *Organometallic Syntheses*, vol. **3**, (eds R.B. King and J.J. Eisch), Elsevier, New York, pp. 461–462.

2 Childs, R.F. and Mulholland, D.L. (1982) *Canadian Journal of Chemistry*, **60**, 801–808; Childs, R.F. and Mulholland, D.L. (1982) *Canadian Journal of Chemistry*, **60**, 809–812; Laszlo, P. and Teston, M. (1990) *Journal of the American Chemical Society*, **112**, 8750–8754. See also Ref. 22.

3 Yang, X. Stern, C.L. and Marks, T.J. (1991) *Journal of the American Chemical Society*, **113**, 3623–3625; Yang, X., Stern, C.L. and Marks, T.J. (1994) *Journal of the American Chemical Society*, **116**, 10015–10031.

4 Marks, T.J. (1992) *Accounts of Chemical Research*, **25**, 57–65; Review: Chen, E.Y.-X.

and Marks, T.J. (2000) *Chemical Reviews*, **100**, 1391–1434.

5 Temme, B., Erker, G., Karl, J., Luftmann, H., Fröhlich, R. and Kotila, S. (1995) *Angewandte Chemie*, **107**, 1867–1869; (1995) *Angewandte Chemie – International Edition*, **34**, 1755–1757.

6 Karl, J., Erker, G. and Fröhlich, R. (1997) *Journal of the American Chemical Society*, **119**, 11165–11173; Karl, J. and Erker, G. (1997) *Chemische Berichte*, **130**, 1261–1267; Karl, J., Erker, G. and Fröhlich, R. (1997) *Journal of Organometallic Chemistry*, **535**, 59–62; Karl, J., Dahlmann, M., Erker, G. and Bergander, K. (1998) *Journal of the American Chemical Society*, **120**, 5643–5652; Dahlmann, M. Erker, G. Nissinen, M. and Fröhlich, R. (1999) *Journal of the American Chemical Society*, **121**, 2820–2828; Dahlmann, M., Erker, G., Fröhlich, R. and Meyer, O. (2000) *Organometallics*, **19**, 2956–2967; Dahlmann, M., Fröhlich, R. and Erker, G. (2000) *European Journal of Inorganic Chemistry*, 1789–1793; Dahlmann, M. Erker, G. and Bergander, K. (2000) *Journal of the American Chemical Society*, **122**, 7986–7998.

7 Erker, G. (2001) *Accounts of Chemical Research*, **34**, 309–317; Erker, G. (2003) *Chemical Communications*, 1469–1476.

8 See also: Erker, G., Kehr, G. and Fröhlich, R. (2004) *Advances in Organometallic Chemistry*, **51**, 109–162.

9 Erker, G. (2005) *Dalton*, 1883–1890.

10 Vagedes, D., Fröhlich, R. and Erker, G. (1999) *Angewandte Chemie*, **111**, 3561–3565; (1999) *Angewandte Chemie – International Edition*, **38**, 3362–3365.

11 Kehr, G., Roesmann, R., Fröhlich, R., Holst, C. and Erker, G. (2001) *European Journal of Inorganic Chemistry*, 535–538; see also: Kehr, G., Fröhlich, R., Wibbeling, B. and Erker, G. (2000) *Chemistry - A European Journal*, **6**, 258–266; Guidotti, S., Camurati, I., Focante, F., Angellini, L., Moscardi, G., Resconi, L., Leardini, R., Nanni, D., Mercandelli, P., Sironi, A., Beringhelli, T. and Maggioni, D. (2003) *The Journal of Organic Chemistry*, **68**, 5445–5465; Bonazza, A., Camurati, I., Guidotta, S., Mascellari, N. and Resconi, L. (2004) *Macromolecular Chemistry and Physics*, **205**, 319–333.

12 Parks, D.J., Spence, R.E.v.H. and Piers, W.E. (1995) *Angewandte Chemie – International Edition*, **107**, 895–897; (1995) *Angewandte Chemie*, **34**, 809–811; Spence, R.E.v.H., Parks, D.J., Piers, W.E., MacDonald, M.A., Zaworotko, M.J. and Rettig, S.J. (1995) *Angewandte Chemie*, **107**, 1337–1340; (1995) *Angewandte Chemie – International Edition*, **34**, 1230–1234; Piers, W.E. and Chivers, T. (1997) *Chemical Society Reviews*, **26**, 345–354.

13 Cano Sierra, J., Hüerländer, D., Hill, M., Kehr, G., Erker, G. and Fröhlich, R. (2003) *Chemistry - A European Journal*, **9**, 3618–3622; Hill, M., Kehr, G., Fröhlich, R. and Erker, G. (2003) *European Journal of Inorganic Chemistry*, 3583–3589.

14 Hill, M., Kehr, G., Fröhlich, R. and Erker, G. (2003) *European Journal of Inorganic Chemistry*, 3583–3589.

15 Erker, G. and Aul, R. (1991) *Chemische Berichte*, **124**, 1301–1310; See also: Spence, R.E.v.H. and Piers, W.E. (1995) *Organometallics*, **14**, 4617–4624.

16 Hill, M., Erker, G., Kehr, G., Fröhlich, R. and Kataeva, O. (2004) *Journal of the American Chemical Society*, **126**, 11046–11057.

17 See for a comparision: Vagedes, D., Kehr, G., König, D., Wedeking, K., Fröhlich, R., Erker, G., Mück-Lichtenfeld, C. and Grimme, S. (2002) *European Journal of Inorganic Chemistry*, 2015–2021; Vagedes, D., Erker, G., Kehr, G., Bergander, K., Kataeva, O., Fröhlich, R., Grimme, S. and Mück-Lichtenfeld, C. (2003) *Journal of The Chemical Society-Dalton Transactions*, 1337–1344.

18 Hill, M., Kehr, G., Erker, G., Kataeva, O. and Fröhlich, R. (2004) *Chemical Communications*, 1020–1021.

19 Ruwwe, J., Erker, G. and Fröhlich, R. (1996) *Angewandte Chemie*, **108**, 108–110;

(1996) *Angewandte Chemie – International Edition*, **35**, 80–82.

20 For related reactions see e.g. Doerrer, L.H., Graham, A.J., Haussinger, D. and Green, M.L.H. (2000) *Journal of The Chemical Society-Dalton Transactions*, 813–820.

21 Werner, H. (1982) *Pure and Applied Chemistry*, **54**, 177–188; Werner, H. (1983) *Angewandte Chemie*, **95**, 932–954; (1983) *Angewandte Chemie – International Edition*, **22**, 927–949; Werner, H., Lippert, F., Peters, K. and von Schnering, H.G. (1992) *Chemische Berichte*, **125**, 347–352 and references cited therein.

22 Döring, S., Erker, G., Fröhlich, R., Meyer, O. and Bergander, K. (1998) *Organometallics*, **17**, 2183–2187.

23 Welch, G.C., Juan, R.R.S., Masuda, J.D. and Stephan, D.W. (2006) *Science*, **314**, 1124–1126.

24 Welch, G.C. and Stephan, D. (2007) *Journal of the American Chemical Society*, **129**, 1880–1881.

25 For related hydroboration products of alkenyl phosphines see e.g.: Spies, P., Fröhlich, R., Kehr, G., Erker, G. and Grimme, S. (2008) *Chemistry - A European Journal*, **14**, 333–343; 779.

26 Spies, P., Erker, G., Kehr, G., Bergander, K., Fröhlich, R., Grimme, S. and Stephan, D.W. (2007) *Chemical Communications*, 5072–5074.

27 Williams, J.H. (1993) *Accounts of Chemical Research*, **26**, 593–598; Cockroft, S.L., Hunter, C.A., Lawson, K.R., Perkins, J. and Urch, C.J. (2005) *Journal of the American Chemical Society*, **127**, 8594–8595; El-Azizi, Y., Schmitzer, A. and Collins, S.K. (2006) *Angewandte Chemie*, **118**, 982–987; (2006) *Angewandte Chemie – International Edition*, **45**, 968–973; and references cited in these articles.

28 See for example: Lin, X.-Y., Vankatesan, K., Schmalle, H.W. and Berke, H. (2004) *Organometallics*, **23**, 3153–3163; Magee, M.P. and Norton, J.R. (2001) *Journal of the American Chemical Society*, **123**, 1778–1779; Noyori, R. and Hashiguchi, S. (1997) *Accounts of Chemical Research*, **30**, 97–102; Noyori, R., Kitamura, M. and Ohkuma, T. (2004) *Proceedings of the National Academy of Sciences of the United States of America*, **101**, 5356–5362; Casey, C.P. Bikzhanova, G.A. and Guzei, I.A. (2006) *Journal of the American Chemical Society*, **128**, 2286–2293.

13
Ruthenium-containing Polyoxotungstates: Structure and Redox Activity

Ulrich Kortz

13.1
Introduction

Polyoxometalates (POMs) are discrete molecular metal–oxygen clusters composed of edge- and corner-shared MO_6 octahedra. The so called addenda atoms M are transition metals of groups 5 and 6 in high oxidation states (e.g. V^{5+}, W^{6+}) [1]. Transition metal-substituted POMs are derivatives which in addition to the addenda atoms contain one or more other d-block metal ions (e.g. Fe^{3+}, Co^{2+}, Ru^{3+}, Pd^{2+}, Pt^{4+}). Besides structural and compositional aspects, these species are highly interesting because of their potential applications in many different areas including magnetochemistry [2], catalysis [3], medicinal chemistry [4], material science [5] and nanotechnology [6].

Ruthenium-containing POMs have attracted much attention owing to the unique redox properties of the element Ru [7]. Hence Ru-POMs have been mostly investigated with respect to their catalytic properties [8]. To date it is well established that Ru-POMs exhibit high reactivity and selectivity in the catalytic oxidation by O_2 and H_2O_2 of a variety of organic substrates [9]. In order to seriously consider industrial applications of Ru-POMs, their synthesis procedures should be simple and straightforward. This is not so easy because it must be realized that commercially available '$RuCl_3$' is not pure. Therefore it is not a surprise that based on this Ru precursor only two structurally characterized Ru-POMs have been reported to date [10]. It can be concluded that the search for novel and well-characterized Ru-POMs will add significant value to this research area.

Organoruthenium-POMs are of interest as they represent molecular models for heterogeneous catalysts based on organometallic coordination complexes adsorbed on metal surfaces. Süss-Fink and Proust have reported the formation of neutral organometallic polyoxomolybdates ($[\{Ru(\eta^6\text{-}p\text{-}MeC_6H_4Pr^i)\}_4Mo_4O_{16}]$, $[(\eta^6\text{-}p\text{-}Pr^iC_6HMe)_2Ru_2Mo_2O_6(OMe)_4]$, $[(\eta^6\text{-}p\text{-}MeC_6H_4Pr^i)_4Ru_4Mo_4O_{16}]$) and polyoxotungstates ($[\{Ru(\eta^6\text{-}p\text{-}MeC_6H_4Pr^i)\}_4W_2O_{10}\}]$, $[\{Ru(\eta^6\text{-}p\text{-}MeC_6H_2Pr^i)\}_4W_4O_{16}\}]$) containing $\{Ru^{II}(arene)\}^{2+}$ units (arene = benzene, toluene, *p*-cymene, mesitylene) [11].

Activating Unreactive Substrates: The Role of Secondary Interactions.
Edited by Carsten Bolm and F. Ekkehardt Hahn

ISBN: 978-3-527-31823-0

Recently, our group reported several Ru^{II}(dmso) and Ru^{II}(benzene) derivatives of mono-, di- and trilacunary Keggin precursors [12], whereas the Nomiya group focused on Wells–Dawson derivatives [13]. On the other hand, the group of Proust prepared a Ru^{II}(*p*-cymene)-containing polyanion by using $\{Sb_2W_{22}O_{74}(OH)_2\}^{12-}$ as precursor [14].

13.2
The Organoruthenium(II)-containing 49-Tungsto-8-Phosphate $[\{K(H_2O)\}_3\{Ru(p\text{-cymene})(H_2O)\}_4P_8W_{49}O_{186}(H_2O)_2]^{27-}$

Despite the fact that the crown-shaped $[H_7P_8W_{48}O_{184}]^{33-}$ (**P_8W_{48}**, see Figure 13.1) has been known for over twenty years [15], only a handful of metal-containing derivatives have been reported to date. These are Kortz's $[Cu_{20}Cl(OH)_{24}(H_2O)_{12}(P_8W_{48}O_{184})]^{25-}$ [16] and Pope's $\{Ln_4(H_2O)_{28}[K \subset P_8W_{48}O_{184}(H_4W_4O_{12})_2Ln_2(H_2O)_{10}]^{13-}\}_x$ (Ln = La, Ce, Pr, Nd) [17].

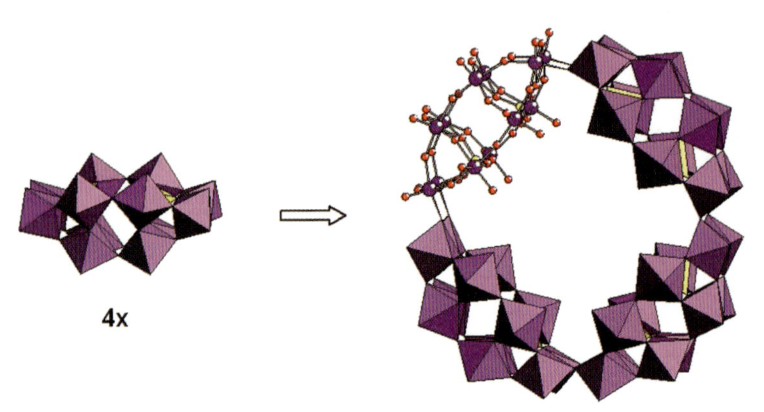

Figure 13.1 The crown-shaped $[H_7P_8W_{48}O_{184}]^{33-}$ (**P_8W_{48}**) can be viewed as being constructed from four $[H_2P_2W_{12}O_{48}]^{12-}$ fragments which are connected via their respective caps. The color code is as follows: WO_6 (violet), PO_4 (yellow), O (red), P (yellow) and W (violet).

Recently, Mialane's Cu_{20}-azide derivative $[P_8W_{48}O_{184}Cu_{20}(N_3)_6(OH)_{18}]^{24-}$ [18] and the Pope/Müller mixed-valence V_{12} derivative $[K_8 \subset \{P_8W_{48}O_{184}\}\{V^{V}{}_4V^{IV}{}_2O_{12}(H_2O)_2\}_2]^{24-}$ [19] have also been reported. Very recently, the 16-iron(III) derivative $[P_8W_{48}O_{184}Fe_{16}(OH)_{28}(H_2O)_4]^{20-}$ has been prepared by Kortz and Muller [20].

Our group has prepared the organoruthenium(II)-containing polyanion $[\{K(H_2O)\}_3\{Ru(p\text{-cymene})(H_2O)\}_4P_8W_{49}O_{186}(H_2O)_2]^{27-}$ (**1**, see Figure 13.2), which has four $\{Ru(p\text{-cymene})(H_2O)\}^{2+}$ fragments grafted onto the cyclic **P_8W_{48}** precursor [21]. Furthermore, **1** contains one extra WO_6 group resulting in an unprecedented P_8W_{49} assembly. Polyanion **1** was prepared by a one-pot reaction of $[Ru(p\text{-cymene})Cl_2]_2$ with P_8W_{48} (10 : 1 ratio) in aqueous 1 M $LiOAc/CH_3COOH$ buffer solution at pH 6.0. We isolated **1** in the form of its mixed potassium-lithium salt $K_{16}Li_{11}[\{K(H_2O)\}_3\{Ru(p\text{-}cymene)(H_2O)\}_4P_8W_{49}O_{186}(H_2O)_2]\cdot 87H_2O$ (**1a**) which was structurally characterized by single-crystal X-ray diffraction. The extra equivalent of tungstate which appears in **1** is most likely formed *in situ* by decomposition of a very minor fraction of the $\mathbf{P_8W_{48}}$ precursor. Compound **1a** has been fully characterized in the solid state by single-crystal XRD, TGA-DSC, IR and elemental analysis.

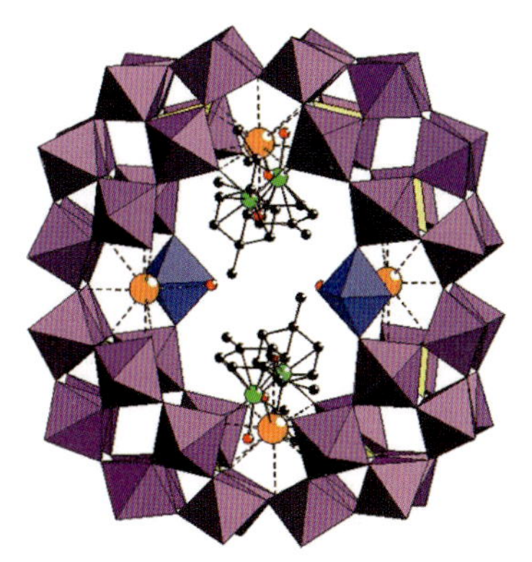

Figure 13.2 Combined polyhedral/ball-and-stick representation of **1**. The color code is as follows: WO_6 (violet, blue), PO_4 (yellow), Ru (green), O (red), C (black), K (orange). No hydrogens are shown for clarity. Note that the two blue WO_6 octahedra and the two adjacent potassium ions have 50% occupancy each.

The molecular structure of **1** reveals that the novel polyanion has four $\{Ru(p\text{-}cymene)(H_2O)\}^{2+}$ groups covalently attached to the cavity of the cyclic $\mathbf{P_8W_{48}}$, resulting in an assembly with C_i symmetry (see Figure 13.2). Each organoruthenium group is bound to $\mathbf{P_8W_{48}}$ via two Ru-O(W) bonds involving a belt-oxygen of each of two adjacent, hexalacunary 'P_2W_{12}' building blocks. The inner coordination sphere of each Ru center is completed by a *p*-cymene ligand and a water molecule (Ru–OH_2 2.143(16) and 2.23(2) Å; Ru–O(W) = 2.043(12), 2.056(17), 2.039(15), and 2.075 (12) Å). In our recently reported $[\{Ru(C_6H_6)(H_2O)\}\{Ru(C_6H_6)\}(\gamma\text{-}XW_{10}O_{36})]^{4-}$ (X = Si, Ge) one of the two Ru^{II}(benzene) groups exhibits the same binding motif [12e]. In this respect, our $[Ru(dmso)_3(H_2O)XW_{11}O_{39}]^{6-}$ (X = Si, Ge) should also be mentioned, as the $Ru^{II}(dmso)_3$ unit is similarly bound to the monolacunary Keggin fragment [12b].

The four organoruthenium units in **1** are grafted near the rim of the central cavity of $\mathbf{P_8W_{48}}$, with the hydrophobic *p*-cymene groups protruding away from the hydrophilic polyanion (see Figure 13.3). This allows for a potassium ion to be accomodated in each of the two opposite hinge areas of 1. Interestingly, two of the waters coordinated to Ru point toward the inside of the cavity of 1 and another two point toward the outside of the cyclic polyanion.

The ruthenium centers in **1** bind to what appear to be the oxygens of $\mathbf{P_8W_{48}}$ which are most sterically accessible and reactive. Also in our copper(II)-containing derivative $[Cu_{20}Cl(OH)_{24}(H_2O)_{12}(P_8W_{48}O_{184})]^{25-}$ the central $\{Cu_{20}(OH)_{24}\}^{16+}$

cluster interacts with the same type of polyanion oxygens, but with all 16 equivalent sites on the inner surface of $\mathbf{P_8W_{48}}$ at the same time.

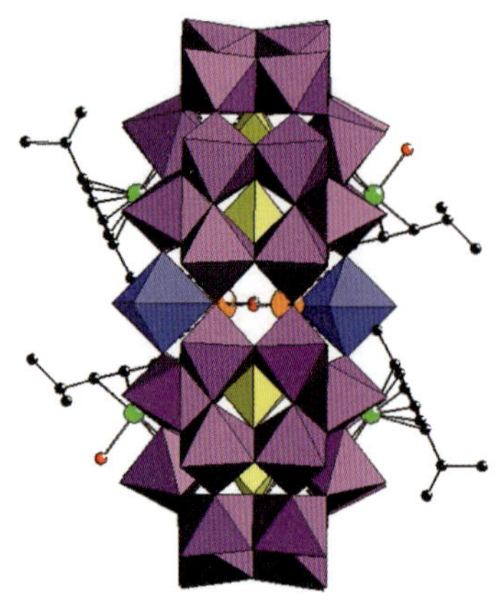

Figure 13.3 Side view of **1** showing that the four organoruthenium units are grafted near the rim of the central cavity, with the hydrophobic *p*-cymene groups protruding away from the hydrophilic polyanion.

At first sight, it seems that steric constraints allow only four rather than the expected eight bulky organoruthenium groups to be accommodated in the cavity of $\mathbf{P_8W_{48}}$. Interestingly, those four groups are all in the same plane, bound on opposite sides of the inner cavity, which indicates that binding cooperativity is at work during the formation of **1** (see below). The remaining four binding sites in **1** are occupied by two potassium ions and, very unexpectedly, also by two additional tungsten atoms (see Figures 13.2 and 13.3).

However, these inversion symmetry-related potassium and tungsten sites exhibit only 50% occupancy, indicating crystallographic disorder. This means that in a given polyanion **1** only two of the four positions are actually occupied. There are no geometrical or steric constraints, therefore allowing for all four possible occupancy scenarios.

We have also performed ^{31}P and ^{13}C solution NMR studies on **1a** redissolved in D_2O, but the results are somewhat difficult to interpret. For example, in ^{31}P NMR we see several signals with different intensities in the expected ppm range (-6.7 to -7.8, see ESI). This result is probably due to the disorder of W25 and K6, which can result in different positional isomers with slightly different NMR spectra. Furthermore, the presence of four potassium ions grafted inside **1** complicates the situation further. In addition, the line broadening somewhat resembles that of our recently reported $[\{Ru(C_6H_6)(H_2O)\}\{Ru(C_6H_6)\}(\gamma\text{-}XW_{10}O_{36})]^{4-}$ (X = Si, Ge), for which we also noticed complex NMR behavior, including the possibility of *in situ*-formed paramagnetic Ru^{III} [12e].

The presence of one extra equivalent of tungsten in **1** is important, as it leads to the first example of a P_8W_{49} framework. Furthermore, the coordination environment of this additional tungsten site is of interest as it exhibits four equatorial, terminal ligands (see Figures 13.2 and 1.3). To our knowledge, this motif has never been observed before in polyoxometalate chemistry. We believe that we are looking at a *cis*-$WO_2(H_2O)_2$ group, rather than a tungsten center with four terminal hydroxo ligands. This is supported by the bond lengths around the unique tungsten atom: two of them are short, presumably oxo bonds (W25–O25A, 1.67(2) Å; W25–O25D, 1.84 (14) Å) and two are long bonds, presumably to water (W25–O25C, 2.27(3) Å; W25–O25B, 2.22(3) Å).

Additional tungstate has also been observed in the lanthanide complexes $\{Ln_4(H_2O)_{28}[K \subset P_8W_{48}O_{184}(H_4W_4O_{12})_2Ln_2(H_2O)_{10}]^{13-}\}_x$ (Ln = La, Ce, Pr, Nd) [17], and probably it too resulted from *in-situ* decomposition of $\mathbf{P_8W_{48}}$. However, in these structures the additional tungstate was attached to individual 'P_2W_{12}' units (rather than bridging them, as here) and the cyclic anions were not distorted.

It can be noticed that the overall shape of the tungsten-oxo framework in **1** is distorted, compared for example with the $\mathbf{P_8W_{48}}$ precursor which has D_{4h} symmetry. The distance across the anion between oxygen atoms bridging adjacent 'P_2W_{12}' fragments in free $\mathbf{P_8W_{48}}$ is 16.69(3) Å. The corresponding distances in **1** are 15.56 (2) Å for the 'P_2W_{12}' units bridged by the added $WO_2(H_2O)_2$ groups and 17.549(17) Å for the units bridged by Ru^{II}(cymene). Thus the coordinated Ru^{II}(cymene) groups appear to draw the 'P_2W_{12}' units together and distort the structure from the square or circle of $\mathbf{P_8W_{48}}$ to a rectangular or oval shape. This distortion provides a rationalization for the observed binding of only four ruthenium atoms in **1**, since the positions containing the added $WO_2(H_2O)_2$ groups are apparently too large for Ru^{II} to bridge. This bonding situation somewhat reflects the allosteric effect of Hervé's well-known 'As_4W_{40}' tungstoarsenate(III) [22]. There are also analogies to the cooperative oxygen binding in hemoglobin. With respect to the mechanism of formation of **1** we speculate that binding of the first Ru^{II}(cymene) group is the slowest step, with concomitant distortion of the $\mathbf{P_8W_{48}}$ framework, allowing for three additional Ru^{II}(cymene) groups to coordinate more readily to the appropriate positions of the distorted structure. The three potassium ions tightly incorporated into the structure of **1** indicate that counterions (here K^+ and Li^+) may also play an important role in the mechanism of the formation of **1** in solution.

Over the last 12 months or so we have prepared several other transition metal- and rare earth-containing derivatives of $\mathbf{P_8W_{48}}$. We successfully incorporated V^V, Mn^{II}, Co^{II}, Ni^{II}, and also U^{VI} in the cyclic precursor [23]. These results are fully consistent with our observations on **1**. For example, we have consistently observed that the most reactive sites of $\mathbf{P_8W_{48}}$ are indeed the 16 terminal oxo groups located at the vacant site (belt region) of the four hexalacunary 'P_2W_{12}' building blocks. Furthermore, we have observed the uptake of additional tungsten atoms in some structures, apparently from *in situ* decomposition of $\mathbf{P_8W_{48}}$.

The structural flexibility of the cyclic $\mathbf{P_8W_{48}}$ precursor is highly interesting and rather unexpected, as this polyanion had in general been considered as a rigid and (maybe for this reason) unreactive entity. However, the results reported here combined with our unpublished work (see above) show that this is not exactly true. On the contrary, the crown-shaped $\mathbf{P_8W_{48}}$ nanoobject uniquely combines high solution stability over a wide pH range with intricate reactivity on its inner surface with an unexpected structural flexibility able to adapt 'intelligently' to the shape, size, charge etc. of the electrophile (e.g. d-block metal ion, lanthanide, organometallic unit). It appears that Müller's recently suggested analogies between super-large polyoxomolybdates and biological systems (e.g. cells) can perhaps be extended to polyoxotungstate chemistry.

13.3
The Mono-Ruthenium(III)-substituted Keggin-Type 11-Tungstosilicate $[\alpha\text{-SiW}_{11}O_{39}Ru^{III}(H_2O)]^{5-}$ and its Dimerization

We prepared the mono-ruthenium(III)-substituted α-Keggin-type tungstosilicate $[\alpha\text{-SiW}_{11}O_{39}Ru^{III}(H_2O)]^{5-}$ (**2**) and we also investigated its solution dimerization to the μ-oxo-bridged dimer $[\{\alpha\text{-SiW}_{11}O_{39}Ru^{m}\}_2O]^{n-}$ (m = III, n = 12, **3a**; m = III/IV, n = 11, **3b**; m = IV, n = 10, **3c**) [24]. The dimeric POM assembly was confirmed by single-crystal XRD on **3c**. The corresponding crystal contains a discrete dimeric Keggin cluster anion, counter cations, and lattice water molecules in the asymmetric unit. The molecular structure of **3c** and the solid-state packing are shown in Figure 13.4. The observed electron densities of the Ru and W atoms are quite different, so that the experimental data allows us to clearly distinguish between the Ru and W sites.

Polyanion **3c** is composed of two mono-ruthenium(IV)-substituted tungstosilicate Keggin units bridged via an Ru–O–Ru bond. The ruthenium atoms are fully incorporated into the Keggin unit, where the Ru–O(Si) bonds (2.126 (12) and 2.098 (11) Å) are significantly shorter than the 11 W–O(Si) bonds (2.325 (12)–2.408 (11) Å). The bond valence sum (BVS) values [25] for the W–O bonds are in the range 5.96–6.40 and for the Si–O bonds 3.95–4.05, confirming a formal valence of +6 for all W and +4 for Si. The BVS values for the oxygen atoms of 3c including the bridging one (2.10) are in the range 1.65–2.10, indicating that none of them is protonated. Furthermore, no acidic protons were detected by acid–base titration of both 3c and 3b. This is the first crystal structure analysis of a dimeric μ-oxo-bridged mono-transition metal-substituted POM.

The single crystal used for XRD was obtained from a solution containing both **3c** and **3b**. The valence of ruthenium in the single crystal is assigned to be +4 because of the following observations [10a]: (i) the BVS values for the Ru–O bonds are 4.00 and 4.06, respectively, suggesting a formal valence of +4 for both Ru atoms, (ii) the Ru–O–Ru bond lengths (1.815 (12) and 1.816 (11) Å) are typical for Ru^{IV}–O–Ru^{IV} (1.787 (11)–1.847 (13) Å) [26], and (iii) the simulated powder XRD pattern using the single-crystal data was similar to the experimental powder XRD pattern of separately prepared $Rb_{10}[\{\alpha\text{-SiW}_{11}O_{39}Ru^{IV}\}_2O]$ and not to $Rb_{11}[\{\alpha\text{-SiW}_{11}O_{39}Ru^{IV/III}\}_2O]$.

In 1993 Finke *et al.* reported the crystal structure of $KLi_{15}[\{\alpha_2\text{-}P_2W_{17}O_{61}Ru^{IV}\text{-}Cl\}_2O]$, an oxo-bridged Ru-POM supported by two mono-vacant Wells–Dawson fragments [10a]. In this polyanion, the Ru–O–Ru angle was 180° and the molecule had C_s symmetry. However, the Ru–O–Ru bridging angle of our dimeric POM is

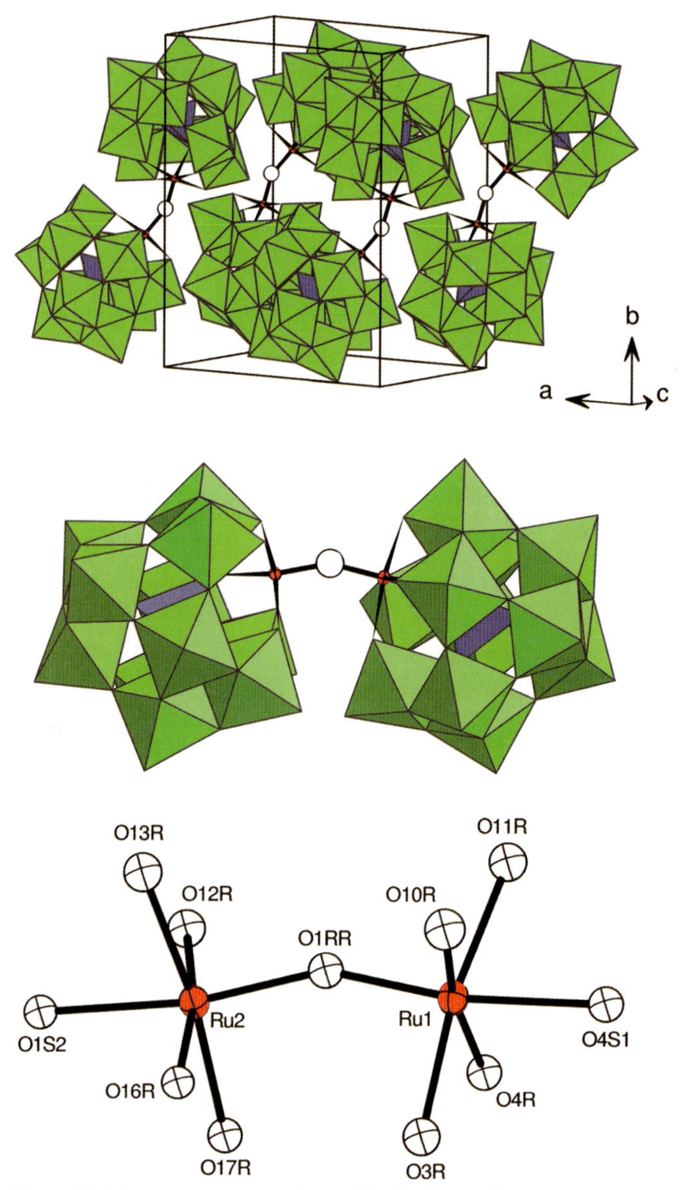

Figure 13.4 Structure representations of **3c**. Packing in the unit cell (top); molecular structure (middle); coordination sphere of Ru1 and Ru2 (bottom).

154° and the two Keggin units are tilted. The short $Ru^{IV}-O-Ru^{IV}$ average bond distance of 1.816 Å compared to the sum of the single-bond radii, 1.98 Å, suggests Ru–O multiple bonding in the Ru–O–Ru core of **3c**.

The ^{183}W-NMR spectrum of **3c** shows six resonances (integration ratio 2 : 2 : 1 : 2 : 2 : 2) (see Figure 13.5(a)). This is consistent with the C_s symmetry for a mono-substituted α-Keggin unit [12a–e, 27], indicating (1) intramolecular rotation of the two Keggin units in **3c** around the Ru–O–Ru bridge or (2) **3c** has C_{2v} symmetry in aqueous medium. If **3c** has indeed a tilted structure as indicated by XRD and if the intramolecular rotation is slower than the NMR timescale, then 22 lines are expected. The observed six lines appear between −80 and −150 ppm, which is a typical chemical shift range for tungstosilicates without a paramagnetic metal. This indicates that (1) **3c** is most likely diamagnetic (due to antiferromagnetic coupling of the two Ru^{IV} centers [11]) and (2) the Ru^{IV}–O–Ru^{IV} linkage in **3c** is hydrolytically stable.

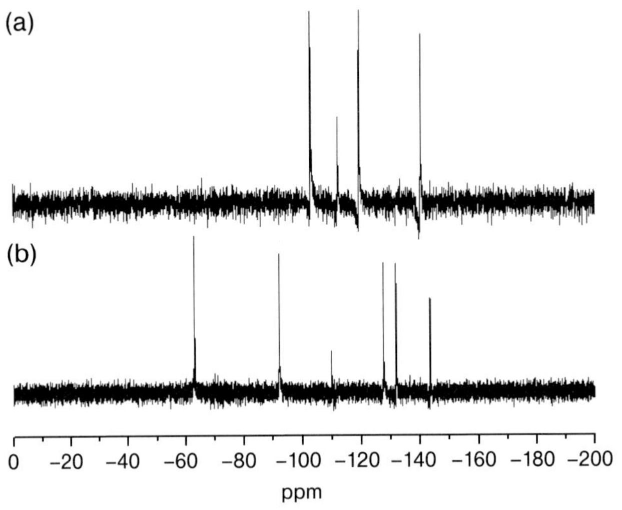

Figure 13.5 ^{183}W NMR spectra of **3c** (a) and **3b** (b).

The ^{183}W-NMR spectrum of **3b** shows only 4 resonances (integration ratio 2 : 1 : 2 : 2) between −100 and −150 ppm, see Figure 13.5 (b). The two lines assigned to the tungsten atoms adjacent to Ru are missing due to the paramagnetic nature of the Ru^{IV}–O–Ru^{III} moiety [27a]. In analogy to **3c** (see above), we can conclude that **3b** either shows fast intramolecular rotation around the Ru–O–Ru bridge or it has C_{2v} symmetry in aqueous solution.

The effects of pH, temperature, atmosphere and reaction time on the dimerization of **2** are summarized in Table 13.1. In a solution of **2** at pH 1.0, no dimeric species **3a**, **3b** or **3c** was detected even after 70 h (entry 1). The dimeric species **3b** was detected after increasing the pH of the solution, respectively in 27% (pH 6) and > 95% (pH 8) yields (entries 2 and 3). At room temperature, no dimeric species was observed (entry 4). The dimerization rate increased with increasing temperature and monomer concentration (entry 5). The dimer **3a** was only observed when the reaction was performed under Ar atmosphere (entries 6 and 7). The reaction under O_2 atmosphere produced the dimer **3c** (entries 8–10).

Dimerization of $[PW_{11}O_{39}Ti^{IV}(=O)]^{5-}$ has been reported to occur in acidic solution, because a protonated hydroxy species is the first intermediate [28]. In the case of the aqua-ruthenium species **2**, basic conditions are necessary to produce a deprotonated hydroxy-ruthenium species. Formation of the hydroxy-ruthenium species (Eq. (13.1), below) is the initial step, and hence dimerization is favored in basic solution (entries 1 – 3). The first dimer formed is **3a** (entry 6) (Eq. (13.2), below), which is easily oxidized by dioxygen in air (Eq. (13.3), below). This oxidation can be monitored by UV-Vis spectroscopy. The UV-Vis spectrum of **3a** changes to that of **3b** within 30 min in aqueous solution, even when just exposed to air at room temperature.

$$[\alpha\text{-SiW}_{11}\text{O}_{39}\text{Ru}^{\text{III}}(\text{H}_2\text{O})]^{5-} \rightleftharpoons [\alpha\text{-SiW}_{11}\text{O}_{39}\text{Ru}^{\text{III}}(\text{OH})]^{6-} + \text{H}^+ \quad \text{p}K\text{a} \approx 6.6 \tag{13.1}$$

$$2\,[\alpha\text{-SiW}_{11}\text{O}_{39}\text{Ru}^{\text{III}}(\text{OH})]^{6-} \rightleftharpoons [\{\alpha\text{-SiW}_{11}\text{O}_{39}\text{Ru}^{\text{III}}\}_2\text{O}]^{12-} + \text{H}_2\text{O} \tag{13.2}$$

$$[\{\alpha\text{-SiW}_{11}\text{O}_{39}\text{Ru}^{\text{III}}\}_2\text{O}]^{12-} \rightarrow [\{\alpha\text{-SiW}_{11}\text{O}_{39}\text{Ru}^{\text{IV-III}}\}_2\text{O}]^{11-} + \text{e}^- \tag{13.3}$$

$$[\{\alpha\text{-SiW}_{11}\text{O}_{39}\text{Ru}^{\text{IV-III}}\}_2\text{O}]^{11-} \rightarrow [\{\alpha\text{-SiW}_{11}\text{O}_{39}\text{Ru}^{\text{IV}}\}_2\text{O}]^{10-} + \text{e}^- \tag{13.4}$$

$$[\alpha\text{-SiW}_{11}\text{O}_{39}\text{Ru}^{\text{III}}(\text{H}_2\text{O})]^{5-} + \text{e}^- \rightleftharpoons [\alpha\text{-SiW}_{11}\text{O}_{39}\text{Ru}^{\text{II}}(\text{H}_2\text{O})]^{6-} \quad \text{pH} < \sim 6.6 \tag{13.5}$$

$$[\alpha\text{-SiW}_{11}\text{O}_{39}\text{Ru}^{\text{III}}(\text{OH})]^{6-} + \text{e}^- + \text{H}^+ \rightleftharpoons [\alpha\text{-SiW}_{11}\text{O}_{39}\text{Ru}^{\text{II}}(\text{H}_2\text{O})]^{6-} \quad \text{pH} > \sim 6.6 \tag{13.6}$$

Table 13.1 Effect of pH and temperature on the dimerization of **2** after redissolution of $K_5[\alpha\text{-SiW}_{11}O_{39}Ru^{III}(H_2O)]$ in H_2O (3 mL).

Entry	pH	Conc./mM	Temp./°C	Atmosphere	Time/h	Dimer and yield[e]/%
1	1[a]	10.7	100	air	70	0
2	6	10.7	100	air	70	Ru^{III}–O–Ru^{IV}, 27
3	8[b]	10.7	100	air	70	Ru^{III}–O–Ru^{IV}, > 95
4	8[b]	10.7	r. t.	air	70	0
5	8[b]	33.3	185[c]	air[d]	5	Ru^{III}–O–Ru^{IV}, > 95
6	8[b]	33.3	185[c]	Ar (1.0 MPa)[d]	5	Ru^{III}–O–Ru^{III}, 52
7	8[b]	33.3	185[c]	Ar (1.0 MPa)[d]	170	Ru^{III}–O–Ru^{III}, 57
8	8[b]	33.3	185[c]	O_2 (0.5 MPa)[d]	5	Ru^{III}–O–Ru^{IV} and Ru^{IV}–O–Ru^{IV}, 65
9	8[b]	33.3	185[c]	O_2 (0.5 MPa)[d]	20	Ru^{IV}–O–Ru^{IV}, 63
10	8[b]	33.3	185[c]	O_2 (0.5 MPa)[d]	170	Ru^{IV}–O–Ru^{IV}, 62

[a] pH adjusted using 0.1 M HCl.
[b] pH adjusted using 0.1 M KOH.
[c] T of oil bath.
[d] Reaction performed in 8 mL stainless autoclave.
[e] Yield based on CV and UV-Vis.

The p*Ka* value of **2** is estimated to be ca. 6.6 based on acid–base titration and the pH dependence of the $Ru^{II/III}(H_2O)$ redox potential. Figure 13.6 (a) shows titration curves obtained for a 10 mM solution of **2** in $HClO_4$.

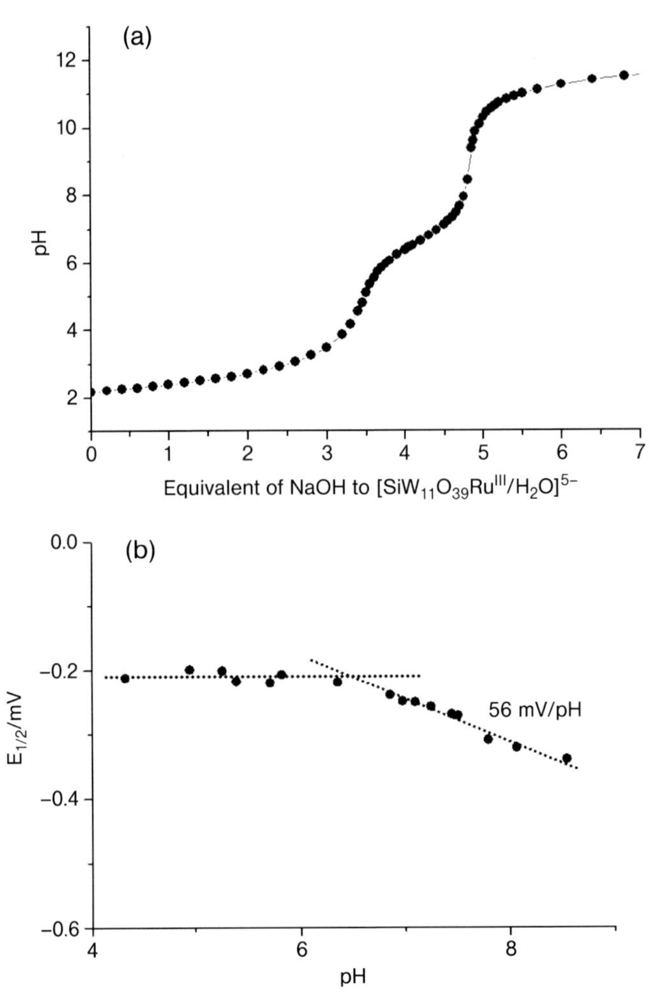

Figure 13.6 (a) Titration of **2** (10 mM) in 0.1 M $NaClO_4$ + 0.01 M $HClO_4$ by addition of NaOH. (b) pH dependence of the redox potential $E_{1/2}$ for $[\alpha\text{-}SiW_{11}O_{39}Ru^{III/II}(H_2O)]^{5/6-}$.

From the plateau observed at pH of ca. 6.6, the p*Ka* value is estimated to ca. 6.6. The redox potential of the $Ru^{III/II}(H_2O)$ was plotted against pH of the solution (Figure 13.6 b). The redox potential is independent of the solution pH value (< pH 6.6), indicating that the redox is not accompanied by deprotonation and protonation [Eq. (13.5)]. If the pH of the solution is above 6.6, the redox potential is dependent on pH with ca.

57 mV/pH, indicating that the oxidation of Ru^{II} to Ru^{III} is accompanied by mono-deprotonation (Eq. (13.6)) [3c, 29]. This result is consistent with a p*Ka* value of ca. 6.6. The p*Ka* value of $[PW_{11}O_{39}Ru^{III}(H_2O)]^{4-}$ was reported to be ca. 5.1 [30]. This difference in the p*Ka* can be explained by the different negative charges of tungstosilicates vs tungstophosphates; deprotonation increases the negative charge, and a smaller negative charge favors deprotonation. A similar difference in p*Ka* values was reported for $[SiW_{11}O_{39}Fe^{III}(H_2O)]^{5-}$ (p*Ka* ca. 6.8) and $[PW_{11}O_{39}Fe^{III}(H_2O)]^{4-}$ (p*Ka* ca. 5.3) [31].

The dimer **3b** can be further oxidized by molecular oxygen (Eq. (13.4)) resulting in **3c**, and the dimerization reaction under an oxygen atmosphere produced **3c** also (entries 8–10).

13.4 Conclusions

Reaction of $[Ru(p\text{-cymene})Cl_2]_2$ with the cyclic tungstophosphate $\mathbf{P_8W_{48}}$ in aqueous acidic medium results in the organo-ruthenium(II) derivative **1**. In addition to the four $\{Ru(p\text{-cymene})(H_2O)\}$ units, an unusual WO_6 group with four equatorial, terminal ligands is also grafted to the central cavity of the molecular $\mathbf{P_8W_{48}}$ precursor. Polyanion **1** represents the first organometallic derivative of $\mathbf{P_8W_{48}}$. We plan to perform homogeneous and heterogeneous oxidation catalysis, electrochemistry and electrocatalysis studies on **1**.

The ruthenium(IV)-containing Keggin-type tungstosilicate dimer **3c** has been prepared and structurally characterized in solution and in the solid state. The two Keggin units are connected by a μ-oxo group resulting in an Ru–O–Ru bridge. The diamagnetic nature of **3c** indicates that the μ-oxo moiety is stable in aqueous solution. Solution ^{183}W NMR on **3c** shows that either the two Keggin units rotate around the Ru–O–Ru bond or the dimeric molecule has indeed C_{2v} symmetry. The first step in the dimerization mechanism of **2** is probably the deprotonation of the aqua-ruthenium species $[\alpha\text{-}SiW_{11}O_{39}Ru^{III}(H_2O)]^{5-}$ to the hydroxy-ruthenium derivative $[\alpha\text{-}SiW_{11}O_{39}Ru^{III}(OH)]^{6-}$. Two molecules of the latter can dimerize in a condensation reaction resulting in **3a**. This in turn can be oxidized to the mixed-valence species **3b**, and further oxidation results in **3c**, depending on the concentration of dioxygen. Such versatile redox activity is attractive for designing oxidation catalysts and hence further studies on **2**, **3a**, **3b** and **3c** in this area are currently in progress.

Acknowledgments

This work was supported by the German Science Foundation (DFG, SPP 1118) and Jacobs University Bremen. I thank all my coworkers and collaborators who have contributed to the results presented here.

References

1 (a) Pope, M.T. (1983) *Heteropoly and Isopoly Oxometalates*, Springer, Berlin; (b) Souchay, P. (1963) *Polyanions et Polycations;* Gauthier-Villars; Paris.

2 Müller, A., Peters, F., Pope, M.T. and Gatteschi, D. (1998) *Chemical Reviews*, **98**, 239.

3 (a) Kozhevnikov, I.V. (1998) *Chemical Reviews*, **98**, 171; (b) Mizuno, N., and Misono, M. (1998) *Chemical Reviews*, **98**, 199; (c) Sadakane, M., Steckhan, E. (1998) *Chemical Reviews*, **98**, 219; (d) Okun, N., Anderson, T. and Hill, C.H. (2003) *Journal of the American Chemical Society*, **125**, 3194.

4 Rhule, J.T., Hill, C.L., Judd, D.A. and Schinazi, R.F. (1998) *Chemical Reviews*, **98**, 327.

5 Coronado, E. and Gómez-García, C. (1998) *Chemical Reviews*, **98**, 273.

6 (a) Hill, C.L. and Weinstock, I.A. (1997) *Nature*, **388**, 332; (b) Nishimura, T., Onoue, T., Ohe, K. and Uemura, S. (1998) *Tetrahedron Letters*, **39**, 6011; (c) Ten Brink, G.-J., Arends, I.W.C.E. and Sheldon, R.A. (2000) *Science*, **287**, 1639; (d) Sheldon, R.A., Arends, I.W.C.E. and Dijksman, A. (2000) *Catalysis Today*, **57**, 157; (e) d'Alessandro, N., Liberatore, L., Tonucci, L., Morbillo, A. and Bressan, M. (2001) *Journal of Molecular Catalysis A-Chemical*, **175**, 85.

7 Naota, T., Takaya, H. and Murahashi, S.I. (1998) *Chemical Reviews*, **98**, 2599.

8 (a) Finke, R.G. and Yin, C.X. (2005) *Inorganic Chemistry*, **44**, 4175; (b) Matsumoto, Y., Asami, M., Hashimoto, M. and Misono, M. (1996) *Journal of Molecular Catalysis A-Chemical*, **114**, 161; (c) Hill, C.J. and Prosser-McCartha, C.M. (1995) *Coordination Chemistry Reviews*, **143**, 407; (d) Bart, J.C. and Anson, F.C. (1995) *Journal of Electroanalytical Chemistry*, **390**, 11.

9 (a) Neumann, R. and Dahan, M. (1998) *Journal of the American Chemical Society*, **120**, 11969; (b) Neumann, R. and Dahan, M. (1998) *Polyhedron*, **17**, 3557; (c) Neumann, R. and Dahan, M. (1997) *Nature*, **388**, 353; (d) Neumann, R., Khenkin, A.M. and Dahan, M. (1995) *Angewandte Chemie–International Edition*, **34**, 1587.

10 (a) Randall, W.J., Weakley, T.J.R. and Finke, R.G. (1993) *Inorganic Chemistry*, **32**, 1068. (b) Neumann, R. and Khenkin, A.M. (1995) *Inorganic Chemistry*, **34**, 5753.

11 (a) Artero, V., Proust, A., Herson, P., Thouvenot, R. and Gouzerh, P. (2000) *Chemical Communications*, 883; (b) Artero, V., Proust, A., Herson, P. and Gouzerh, P. (2001) *Chemistry - A European Journal*, **7**, 3901; (c) Villanneau, R., Artero, V., Laurencin, D., Herson, P., Proust, A. and Gouzerh, P. (2003) *Journal of Molecular Structure*, **656**, 67; (d) Laurencin, D.E., Fidalgo, G., Villanneau, R., Villain, F., Herson, P., Pacifico, J., Stoeckli-Evans, H., Bènard, M., Rohmer, M.-M., Süss-Fink, G. and Proust, A. (2004) *Chemistry - A European Journal*, **10**, 208.

12 (a) Bi, L.-H., Hussain, F., Kortz, U., Sadakane, M. and Dickman, M.H. (2004) *Chemical Communications*, 1420; (b) Bi, L.-H., Kortz, U., Keita, B. and Nadjo, L. (2004) *Dalton Transactions*, 3184; (c) Bi, L.-H., Dickman, M.H., Kortz, U. and Dix, I. (2005) *Chemical Communications*, 3962; (d) Bi, L.-H., Kortz, U., Dickman, M.H., Keita, B. and Nadjo, L. (2005) *Inorganic Chemistry*, **44**, 7485; (e) Bi, L.-H., Chubarova, E.V., Nsouli, N.H., Dickman, M.H., Kortz, U., Keita, B. and Nadjo, L. (2006) *Inorganic Chemistry*, **45**, 8575.

13 (a) Nomiya, K., Torii, H., Nomura, K. and Sato, Y. (2001) *Journal of The Chemical Society-Dalton Transactions*, 1506; (b) Sakai, Y., Shinohara, A., Hayashi, K. and Nomiya, K. (2006) *European Journal of Inorganic Chemistry*, 163; (c) Kato, C.N., Shinohara, A., Moriya, N. and Nomiya, K. (2006) *Catalysis Communications*, **7**, 413.

14 Laurencin, D., Villanneau, R., Herson, P., Thouvenot, R., Jeannin, Y. and Proust, A. (2005) *Chemical Communications*, 5524.

15 Contant, R. and Tézé, A. (1985) *Inorganic Chemistry*, **24**, 4610.

16 (a) Mal, S.S. and Kortz, U. (2005) *Angewandte Chemie*, **117**, 3843; (2005) *Angewandte Chemie – International Edition*, **44**, 3777; (b) Jabbour, D., Keita, B., Nadjo, L., Kortz, U. and Mal, S.S. (2005) *Electrochemistry Communications*, **7**, 841; (c) Alam, M.S., Dremov, V., Müller, P., Postnikov, A.V., Mal, S.S., Hussain, F. and Kortz, U. (2006) *Inorganic Chemistry*, **45**, 2866; (d) Liu, G., Liu, T., Mal, S.S. and Kortz, U. (2006) *Journal of the American Chemical Society*, **128**, 10103; (e) Liu, G., Liu, T., Mal, S.S. and Kortz, U. (2007) *Journal of the American Chemical Society* (Addition/Correction), **129**, 2408.

17 Zimmermann, M., Belai, N., Butcher, R.J., Pope, M.T., Chubarova, E.V., Dickman, M.H. and Kortz, U. (2007) *Inorganic Chemistry*, **46**, 1737.

18 Pichon, C., Mialane, P., Dolbecq, A., Marrot, J., Riviere, E., Keita, B., Nadjo, L. and Secheresse, F. (2007) *Inorganic Chemistry*, **46**, 5292.

19 Müller, A., Pope, M.T., Todea, A.M., Bögge, H., van Slageren, J., Dressel, M., Gouzerh, P., Thouvenot, R., Tsukerblat, B. and Bell, A. (2007) *Angewandte Chemie*, **119**, 4561; (2007) *Angewandte Chemie – International Edition*, **46**, 4477.

20 Mal, S.S., Dickman, M.H., Kortz, U., Todea, A.M., Merca, A., Bögge, H., Glaser, T., Müller, A., Nellutla, S., Kaur, N., van Tol, J., Dalal, N.S., Keita, B. and Nadjo, L. (2008) *Chemistry - A European Journal*, **14**, 1186.

21 Mal, S.S., Nsouli, N.H., Dickman, M.H. and Kortz, U. (2007) *Dalton Transactions*, 2627.

22 Robert, F., Leyrie, M., Hervé, G., Tézé, A. and Jeannin, Y. (1980) *Inorganic Chemistry*, **19**, 1746.

23 Kortz, *et al.* unpublished results.

24 Sadakane, M., Tsukuma, D., Dickman, M.H., Bassil, B.S., Kortz, U., Capron, M. and Ueda, W. (2007) *Dalton Transactions*, 2833.

25 Brown, I.D. and Altermatt, D. (1985) *Acta Crystallographica. Section B, Structural Science*, **41**, 244.

26 Collman, J.P., Barnes, C.B., Brothers, P.J., Collins, T.J., Ozawa, T., Gallucci, J.C. and Ibers, J.A. (1984) *Journal of the American Chemical Society*, **106**, 5151.

27 (a) Rong, C. and Pope, M.T. (1992) *Journal of the American Chemical Society*, **114**, 2932 (b) Hayashi, Y., Ozawa, Y. and Isobe, K. (1991) *Inorganic Chemistry*, **30**, 1025.

28 Kholdeeva, O.A., Maksimov, G.M., Maksimovskaya, R.I., Kovaleva, L.A., Fedotov, M.A., Grigoriev, V.A. and Hill, C.L. (2000) *Inorganic Chemistry*, **39**, 3828.

29 (a) Sadakane, M. and Steckhan, E. (1996) *Journal of Molecular Catalysis A-Chemical*, **114**, 221; (b) Bard, A.J. and Faulkner, L.R. (1980) *in Electrochemical Methods*, Wiley, New York.

30 Higashijima, M. (1999) *Chemistry Letters*, 1093.

31 Toth, J.E. and Anson, F.C. (1988) *Journal of Electroanalytical Chemistry*, **256**, 361.

14
From NO to Peroxide Activation by Model Iron(III) Complexes

Alicja Franke, Natalya Hessenauer-Ilicheva, Joo-Eun Jee, and Rudi van Eldik

14.1
Introduction

The goal of our project within SPP 1118 was to apply low-temperature and high-pressure kinetic and thermodynamic techniques [1] to study particular reaction steps involved in oxygen activation and alkane hydroxylation in more detail in order to contribute to the mechanistic understanding of the catalytic activity of cytochrome P450 and related model complexes. The reaction steps to be studied included the lability of coordinated water, the binding of substrate molecules to the active center, the formation of the iron oxo complex, $(P^{+})Fe^{IV}{=}O$ via the activation of peroxide, and the subsequent oxygen transfer step to reach a directed oxidation/hydroxylation/epoxidation of the inert substrate. Kinetic studies using a variety of experimental techniques were to be performed on a series of model complexes for P450, and where possible high-pressure kinetic measurements would be employed to construct volume profiles for particular reaction steps [1]. In this way an improved understanding of the substrate oxidation mechanisms should be reached, which could lead to eventual industrial applications. The increasing interest in the elucidation of inorganic and bioinorganic mechanisms is geared in this direction [2].

In order to develop an understanding of the basic mechanistic behavior of the selected Fe(III) complexes, we first focused on the activation of NO by cytochrome P450 and functional model Fe(III) complexes since these reactions do not involve subsequent substrate oxidation, etc. The mechanistic insight gained from these reactions formed the basis for studies on the activation of peroxides by model Fe(III) complexes, which in general involves a much more complex reaction mechanism. Here we started with a functional model complex for cytochrome P450 and in parallel studied various Fe(III) porphyrin systems. As a result of these studies, we were finally able to observe with low-temperature rapid-scan spectroscopy the complete catalytic cycle for an epoxidation process catalyzed by the model P450 complex. Thus, mechanistic studies on the activation of NO led to a direct mechanistic clarification of the catalytic cycle involved in substrate oxidation via the activation of peroxides by such iron(III) complexes. Our findings are reported in this sequence in the present report.

Activating Unreactive Substrates: The Role of Secondary Interactions.
Edited by Carsten Bolm and F. Ekkehardt Hahn

ISBN: 978-3-527-31823-0

14.2
NO Activation by Fe(III) Complexes

Our work performed on the activation of NO by Fe(III) porphyrins, cytochrome P450 and functional models is summarized in the following sections.

14.2.1
Fe(III)-Porphyrins

The lability of coordinated solvent molecules (in many cases water) controls the substitution rate and mechanism of solvated metal ions, which in turn controls the activation of small molecules such as NO by such metal ions and complexes, since the coordination of the small molecule to the metal center is an essential part of the activation process. The mechanistic understanding of solvent exchange reactions is therefore of fundamental importance and has benefited greatly from the application of high-pressure kinetic techniques [3]. In the case of Fe(III)-porphyrins, the introduction of a porphyrin chelate induces electron density on the metal center and labilizes the coordinated water molecules in the axial position in aqueous medium. This labilization was studied by O^{17}-NMR for the different Fe(III)-porphyrins shown in Scheme 14.1 [4]. The water exchange rate constants and the corresponding activation enthalpy and entropy values were determined from the temperature dependence of the NMR line broadening. The water exchange rate constants were all found to be around $10^6\ s^{-1}$ at 25 °C for the studied complexes, i.e. ca. 10^4 times faster than water exchange on hexa-aqua Fe(III). The activation entropies are all in the region of $+100\ J\ K^{-1}\ mol^{-1}$ and suggest a dissociative water exchange mechanism. The pressure dependence of the exchange rate constants was

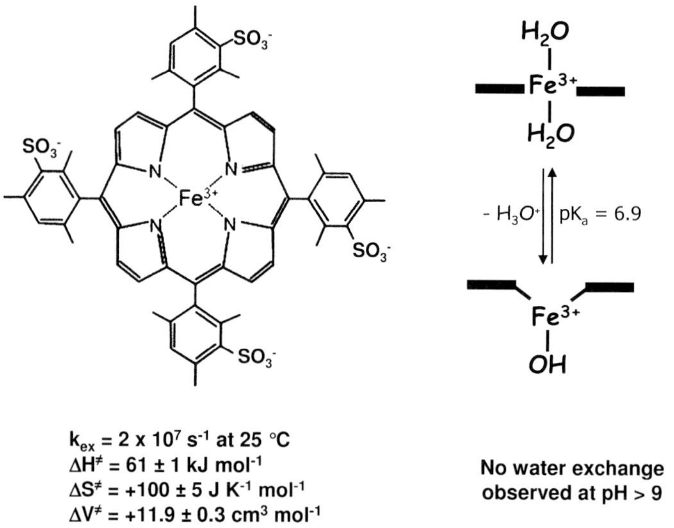

H_2O – Fe^{3+} – H_2O

$- H_3O^+$ ⇅ $pK_a = 6.9$

Fe^{3+} – OH

$k_{ex} = 2 \times 10^7\ s^{-1}$ at 25 °C
$\Delta H^{\neq} = 61 \pm 1\ kJ\ mol^{-1}$
$\Delta S^{\neq} = +100 \pm 5\ J\ K^{-1}\ mol^{-1}$
$\Delta V^{\neq} = +11.9 \pm 0.3\ cm^3\ mol^{-1}$

No water exchange observed at pH > 9

Scheme 14.1

studied at an appropriate temperature, as shown in Figure 14.1. The reported activation volumes are all significantly positive and underline the operation of a dissociative interchange (I_d) or even a limiting dissociative (D) mechanism in the case of the TMPS complex [4].

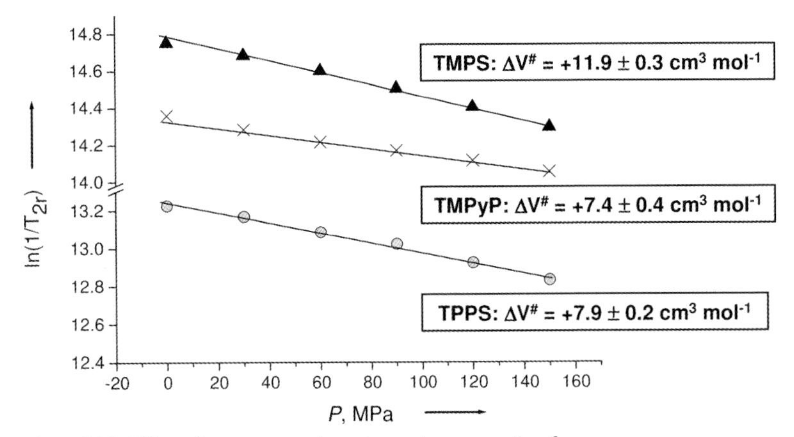

Figure 14.1 Effect of pressure on the water exchange reaction for the different Fe(III) pophyrins shown in Scheme 14.13 [4].

In the case of the $[Fe^{III}(TMPS^{4-})(H_2O)_2]$ complex, further studies indicated that the diaqua complex has a pK_a value of 6.9 and undergoes a change in spin state from a spin-admixed (S = 5/2, 3/2) state at pH 5 to a high-spin (S = 5/2) state at pH 9. Furthermore, no water exchange reaction could be detected at pH 9, from which it was concluded that a five-coordinate hydroxo complex is formed at higher pH (see Scheme 14.2) [5]. Both the diaqua and monohydroxo complexes reversibly bind NO to form diamagnetic Fe^{II}–NO^+ complexes as shown in Scheme 14.2, indicating that coordination of NO is accompanied by activation of NO through the formal oxidation

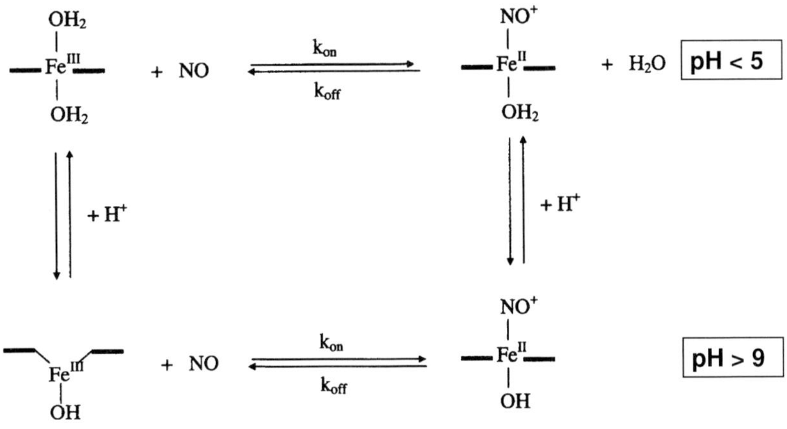

$$(OH_2)_2Fe^{III} + NO \underset{k_{off}}{\overset{k_{on}}{\rightleftharpoons}} (NO^+)(OH_2)Fe^{II} + H_2O \quad \text{pH < 5}$$

$$\updownarrow +H^+ \qquad\qquad \updownarrow +H^+$$

$$(OH)Fe^{III} + NO \underset{k_{off}}{\overset{k_{on}}{\rightleftharpoons}} (NO^+)(OH)Fe^{II} \quad \text{pH > 9}$$

Scheme 14.2

by the metal center. A detailed study of the kinetic data and activation parameters for both the 'on' and 'off' reactions as a function of pH indicated a changeover in binding mechanism from dissociative at pH 5 (diaqua complex) to associative at pH 9 (monohydroxo complex). The corresponding volume profiles reported in Figure 14.2 nicely illustrate the changeover in mechanism as a function of pH. The reaction at high pH involves a high-spin to low-spin (diamagnetic) changeover that contributes to the overall volume collapse observed for this process [5]. The activation volumes found for the water exchange reaction and the binding of NO to the diaqua complex at pH 5 are indeed very similar and suggest that the rate and mechanism of the binding of NO are controlled by the water exchange process. In the case of the coordination of NO to the vacant coordination site of the five-coordinate high-spin hydroxo complex, the activation barrier was suggested to be the spin change barrier that must be overcome during the binding process.

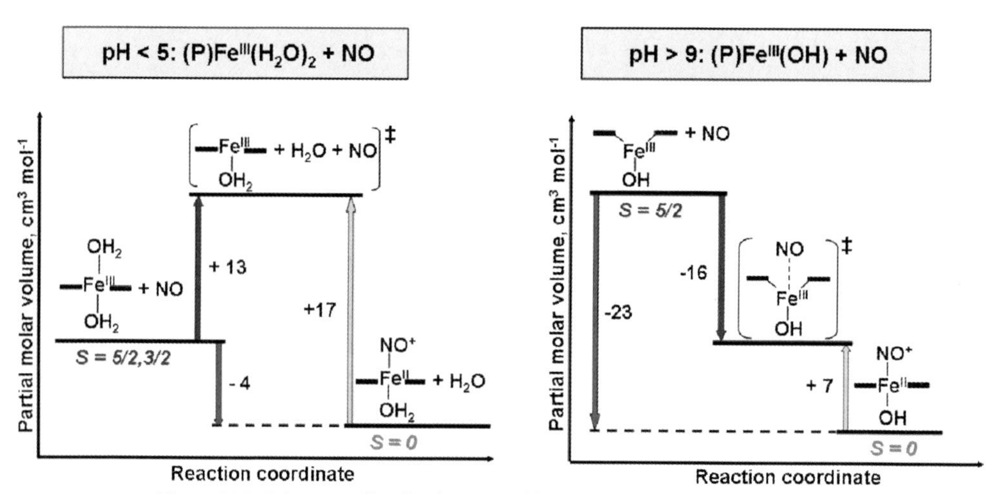

Figure 14.2 Volume profiles for the reversible binding of NO to [Fe^{III}(TMPS^{4-})(H$_2$O)$_2$] (left) and [Fe^{III}(TMPS^{4-})OH] (right) [5].

Following this study, a systematic variation of the overall charge on the porphyrin from -16 to $+8$ was undertaken by introducing different substituents on the porphyrin ring, with the result that the pK_a values of the diaqua complexes varied from 9.9 to 5.0 (see Table 14.1), respectively [6–8]. The water exchange rate constants and activation parameters for the diaqua complexes are summarized in Table 14.2. The rates of water exchange correlate with the contribution of the intermediate spin state S = 3/2 (Int%) in the spin admixed state (S = 3/2, 5/2) of the Fe(III) porphyrins, where the contribution of S = 3/2 was estimated from the proton β-pyrrole signal. The results show that the calculated Int% value increases with increasing lability of the water molecule coordinated to the iron(III) center, which is consistent with an increasing water exchange rate constant due to the lengthening of the Fe^{III}–OH$_2$ bond. The reported activation parameters, especially the values of $\Delta S^{\#}$ and $\Delta V^{\#}$,

Table 14.1 Values of pK_{a1} and structures of the synthetic water-soluble iron(III) porphyrins [8].

Iron(III) porphyrin	**(P^{16-})Fe**	**(P^{8-})Fe**	**($TMPS^{4-}$)Fe**	**($TPPS^{4-}$)Fe**	**(4-$TMPyP^{4+}$)Fe**	**(P^{8+})Fe**
Meso phenyl substituents	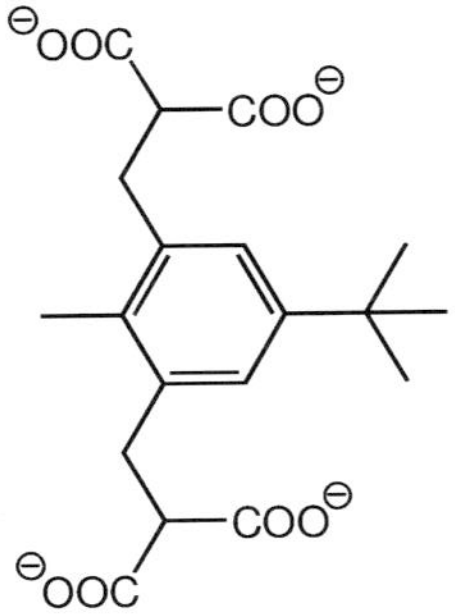 $^{\ominus}OOC$, $COO^{\ominus}$, $COO^{\ominus}$, $^{\ominus}OOC$	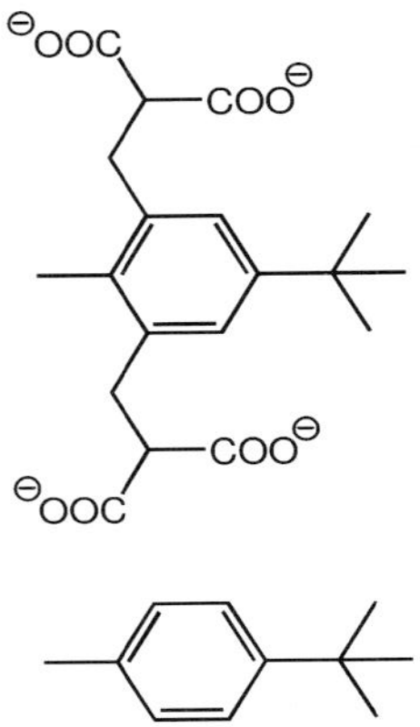 $^{\ominus}OOC$, $COO^{\ominus}$, $COO^{\ominus}$, $^{\ominus}OOC$	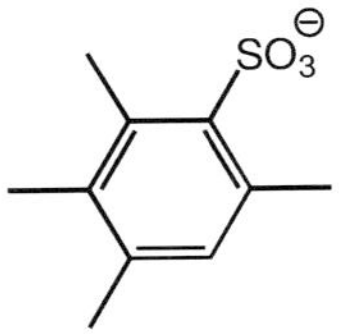 $SO_3^{\ominus}$	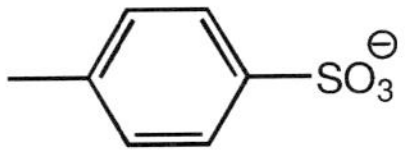 $SO_3^{\ominus}$	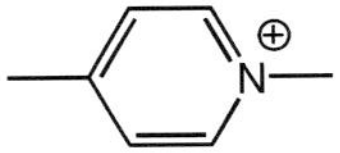 $N^{\oplus}$	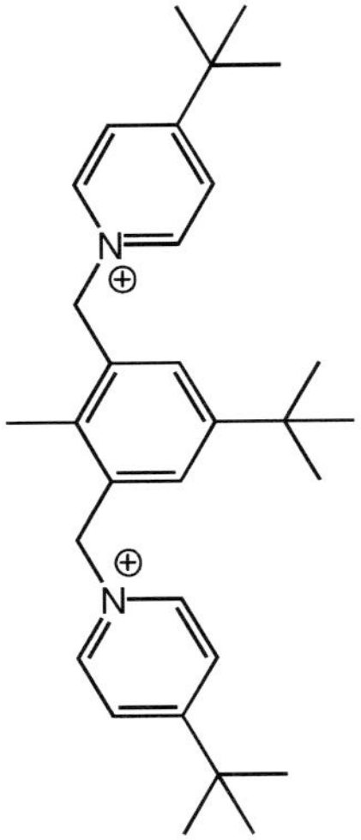 $N^{\oplus}$, $N^{\oplus}$
pK_{a1}	9.9	9.3	6.9	7.0	5.5	5.0

Table 14.2 Rate constants (at 298 K) and activation parameters for water exchange reactions of $(P)Fe^{III}(H_2O)_2$ complexes [8].

Iron(III) porphyrin	Int%[a)]	k_{ex} s^{-1}	$\Delta H^{\ddagger}_{ex}$ kJ mol^{-1}	$\Delta S^{\ddagger}_{ex}$ J mol^{-1} K^{-1}	$\Delta V^{\ddagger}_{ex}$ cm^3 mol^{-1}
$(P^{16-})Fe$	24.5	$(42 \pm 5) \times 10^5$	55 ± 3	$+66 \pm 10$	$+6.5 \pm 0.3$
$(P^{8-})Fe$	24	$(77 \pm 1) \times 10^5$	61 ± 6	$+91 \pm 23$	$+7.4 \pm 0.4$
$(TPPS^{4-})Fe$	20	$(20 \pm 1) \times 10^5$	67 ± 2	$+99 \pm 10$	$+7.9 \pm 0.2$
$(TMPS^{4-})Fe$	26	$(210 \pm 10) \times 10^5$	61 ± 1	$+100 \pm 5$	$+11.9 \pm 0.3$
$(TMPyP^{4+})Fe$	7	$(4.5 \pm 0.1) \times 10^5$	71 ± 2	$+100 \pm 6$	$+7.4 \pm 0.4$
$(P^{8+})Fe$	10	$(5.5 \pm 0.1) \times 10^4$	53 ± 3	$+28 \pm 9$	$+1.5 \pm 0.2$

a) Contribution of the intermediate spin state ($S = 3/2$) in the spin-admixed $(P)Fe^{III}(H_2O)_2$ porphyrins.

support the dissociative nature of the water exchange process. The reversible binding of NO was studied as a function of pH for the series of complexes for which the rate and activation parameters are summarized in Tables 14.3 and 14.4.

By way of example, mechanistic information on the 'on' and 'off' reactions of NO with $(P^{16-})Fe^{III}(H_2O)_2$ can best be obtained from the volume profile presented in Figure 14.3. The positive activation volumes for the binding and release of NO favor a dissociative mechanism which is analogous to that reported for the series of diaqua-ligated porphyrins summarized in Table 14.3. The overall volume change observed for the reaction is not only due to the displacement of water by NO, but also due to a change in spin state from $S = 3/2$, $3/2$ for the diaqua complex to $S = 0$ for the Fe^{II}–NO^+ complex. The data in Table 14.3 show that the rates for NO binding correlate to some extent with the contribution of the intermediate spin state (Int%) in the spin-admixed state ($S = 3/2$, $3/2$). As mentioned above, increasing Int% correlates with increasing lability of the ligand on the metal center as influenced by the increasing electron-donating properties of the meso substituents on the porphyrin. The trend in $\Delta V^{\ddagger}_{on}$ along the series of complexes suggests a gradual changeover from a dissociative mechanism for the anionic complexes to a dissociative interchange mechanism for the cationic complexes. Almost similar trends are observed for the water exchange reactions for the series of $(P)Fe^{III}(H_2O)_2$ complexes summarized in Table 14.2. The magnitude of the water exchange rate constant and the close agreement between the volumes of activation for the water exchange process and the binding of NO, suggest that the rate and mechanism of the latter process is controlled by the water exchange process of the diaqua complexes. Thus electron-donating substituents on the porphyrin induce a dissociative mechanism by weakening the $Fe-OH_2$, whereas electron-withdrawing substituents strengthen the $Fe-OH_2$ bond and tend to favor a dissociative interchange mechanism.

The rate constants for the dissociation of NO from the nitrosyl complex $(P)Fe^{II}(H_2O)(NO^+)$ summarized in Table 14.3 show for some of the complexes the trend that the nitrosyl complex is stabilized by positively charged as compared to negatively charged substituents on the porphyrin. In general, the value of $\Delta V^{\ddagger}_{off}$ is more positive than that of $\Delta V^{\ddagger}_{on}$ since the 'off' reaction involves bond cleavage, formal

Table 14.3 Rate constants (at 298 K) and activation parameters for reversible binding of NO to a series of diaqua-ligated water-soluble iron(III) porphyrins [8].

			NO binding				NO release			
Iron(III) porphyrin	**pK_{a1}**	**Int%**	$k_{on}/10^4$ M^{-1} s^{-1}	$\Delta H^{\ddagger}_{on}$ kJ mol^{-1}	$\Delta S^{\ddagger}_{on}$ J mol^{-1} K^{-1}	$\Delta V^{\ddagger}_{on}$ cm^3 mol^{-1}	k_{off} s^{-1}	$\Delta H^{\ddagger}_{off}$ kJ mol^{-1}	$\Delta S^{\ddagger}_{off}$ J mol^{-1} K^{-1}	$\Delta V^{\ddagger}_{off}$ cm^3 mol^{-1}
$(P^{16-})Fe$	9.8	24.5	113 ± 5	80 ± 1	$+138 \pm 4$	$+10.8 \pm 0.2$	22 ± 3	117 ± 13	$+173 \pm 24$	
$(P^{8-})Fe$	9.3	24	82 ± 1	51 ± 1	$+40 \pm 2$	$+6.1 \pm 0.1$	220 ± 2	101 ± 2	$+140 \pm 7$	$+16.8 \pm 0.4$
$(TMPS^{4-})Fe$	6.9	26	280 ± 20	57 ± 3	$+69 \pm 11$	$+13 \pm 1$	900 ± 200	84 ± 3	$+94 \pm 10$	$+17 \pm 3$
$(TPPS^{4-})Fe$	7.0	20	50 ± 3	69 ± 3	$+95 \pm 10$	$+9 \pm 1$	500 ± 400	76 ± 6	$+60 \pm 11$	$+18 \pm 2$
$(TMPyP^{4+})Fe$	5.5	7	2.9 ± 0.2	67 ± 4	$+67 \pm 13$	$+3.9 \pm 1.0$	59 ± 4	108 ± 5	$+150 \pm 12$	$+16.6 \pm 0.2$
$(P^{8+})Fe$	5.0	10	1.5 ± 0.1	77 ± 3	$+94 \pm 12$	$+1.5 \pm 0.3$	26.3 ± 0.5	83 ± 4	$+61 \pm 14$	$+9.3 \pm 0.5$

Table 14.4 Rate constants (at 298 K) and activation parameters for the reversible binding of NO to monohydroxo-ligated iron(III) porphyrins [8].

	NO binding				NO release			
Iron(III) porphyrin	$k_{on}/10^4$ M^{-1} s^{-1}	$\Delta H^{\ddagger}_{on}$ kJ mol^{-1}	$\Delta S^{\ddagger}_{on}$ J mol^{-1} K^{-1}	$\Delta V^{\ddagger}_{on}$ cm^3 mol^{-1}	k_{off} s^{-1}	$\Delta H^{\ddagger}_{off}$ kJ mol^{-1}	$\Delta S^{\ddagger}_{off}$ J mol^{-1} K^{-1}	$\Delta V^{\ddagger}_{off}$ cm^3 mol^{-1}
$(P^{16-})Fe$	3.1 ± 0.4	23 ± 1	-82 ± 4	-9.4 ± 0.2	8.0 ± 0.1	108 ± 7	$+136 \pm 19$	
$(P^{8-})Fe$	5.1 ± 0.2	34.6 ± 0.4	-39 ± 1	-6.1 ± 0.2	11.4 ± 0.3	107 ± 2	$+136 \pm 7$	$+17 \pm 3$
$(TMPS^{4-})Fe$	1.46 ± 0.02	28.1 ± 0.6	-128 ± 2	-16.2 ± 0.4	10.5 ± 0.2	90 ± 1	$+77 \pm 3$	$+7.4 \pm 1.0$
$(TMPyP^{4+})Fe$	0.36 ± 0.01	41.4 ± 0.5	-38 ± 5	-13.7 ± 0.6	3.2 ± 0.1	78 ± 2	$+25 \pm 7$	$+9.5 \pm 0.8$
$(P^{8+})Fe$	0.16 ± 0.01	41 ± 1	-45 ± 2	-13.8 ± 0.4	6.2 ± 0.1	72 ± 2	$+12 \pm 5$	$+2.6 \pm 0.2$

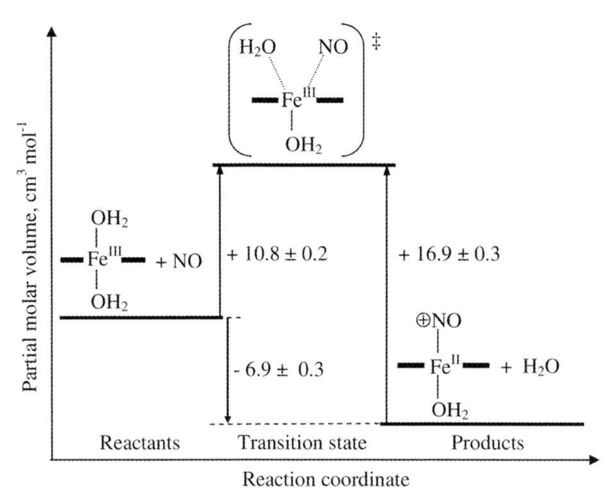

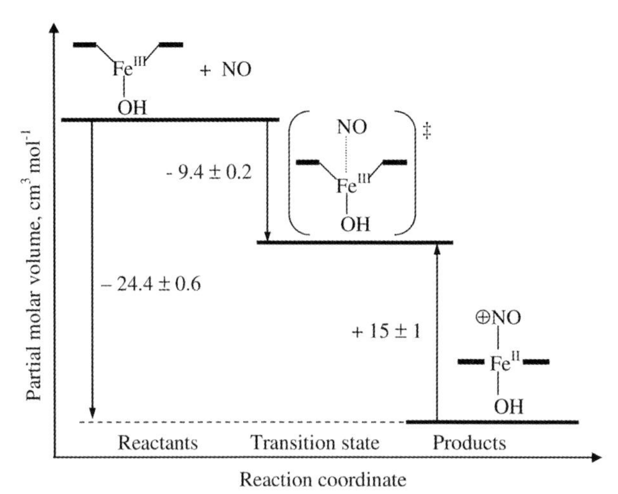

Figure 14.3 Volume profiles for the reversible binding of NO to $[Fe^{III}(P^{16-})(H_2O)_2]$ (top) and $[Fe^{III}(P^{16-})(OH)]$ (bottom) [8].

oxidation of Fe^{II} to Fe^{III} accompanied by a spin-state change ($S = 0$ to $S = 3/2$, $3/2$) and solvent reorganization due to charge neutralization during $Fe^{II}{-}NO^{+}$ bond cleavage.

The reversible binding of NO at higher pH revealed by way of example a different kinetic behavior for the monohydroxo-ligated $(P^{16-})Fe^{III}(OH)$ complex as compared to that of the $(P^{16-})Fe^{III}(H_2O)_2$ complex, as indicated by the negative activation parameters, namely $\Delta S^{\ddagger}_{on} = -82 \pm 4\,J\,mol^{-1}\,K^{-1}$ and $\Delta V^{\ddagger}_{on} = -9.4 \pm 0.2\,cm^3\,mol^{-1}$. It was in general concluded that the slow rate constant for NO binding to complexes of the type $(P)Fe^{III}(OH)$ is mainly controlled by the $Fe^{II}{-}NO^{+}$ bond formation step and the accompanying electronic and structural changes, rather than the lability of the metal center.[4,5] Electronic changes for the formation of the $Fe^{II}-NO^{+}$ bond are accompanied by reorganization of the spin arrangement in the iron(III) center from high spin (S = 3/2) to low spin (diamagnetic, S = 0). In addition, structural changes as a result of the iron center that moves from out-of-plane to in-plane during bond formation are also expected. These conclusions suggest that larger spin and structural changes for the monohydroxo-ligated species $(P)Fe^{III}(OH)$ demand a higher activation barrier for NO binding and release, such that lower reaction rates were observed. The observed results are consistent with the findings for related complexes summarized in Table 14.4. The rate-determining step for NO binding to $(P^{16-})Fe^{III}(OH)$ via an associative addition mechanism is mainly controlled by electronic changes on the Fe^{III} center and the accompanying structural rearrangement. Larger spin and structural changes during the formation and breakage of the $Fe^{II}-NO^{+}$ bond accompanied by higher activation barriers account for the slower reactions in the case of the monohydroxo-ligated species (see rate data in Table 14.4).

On the basis of results summarized in Tables 14.3 and 14.4, $(P)Fe^{III}(OH)$ complexes follow an associatively activated addition mechanism for NO binding and show slower binding rates in contrast to a dissociatively activated mode observed for $(P)Fe^{III}(OH_2)_2$ complexes. Furthermore, the reported values for k_{on} vary by a factor of approximately 10^2 for the diaqua-ligated porphyrins and appear to correlate with the contribution of the intermediate spin state, S = 3/2. In comparison, however, differences in the NO binding rate constants are rather small for all studied (P) $Fe^{III}(OH)$ complexes. This difference is due to the fact that the lability of the Fe^{III} complex controls the rate of NO binding to the six-coordinate $(P)Fe^{III}(OH_2)_2$ complexes, which does not play a role in the case of the five-coordinate $(P)Fe^{III}(OH)$ complexes.

The volume profile for the binding of NO to $(P^{16-})Fe^{III}(OH)$ is shown in Figure 14.3, from which it can clearly be seen that the overall reaction volume ($\Delta V = \Delta V^{\ddagger}_{on} - \Delta V^{\ddagger}_{off}$) has a value of $-24.4 \pm 0.6\ cm^3\ mol^{-1}$, which is close to a value of ca. $-23\ cm^3\ mol^{-1}$ found for several other hydroxo complexes (except for $(P^{8+})Fe^{III}$, see data in Table 14.4). This large overall volume collapse for the binding of NO is partially due to the formation of the Fe–NO bond and the change in spin state from a five-coordinate, high spin hydroxo reactant to a six-coordinate, diamagnetic (low spin) nitrosyl product. The activation volumes reported for $(P^{16-})Fe^{III}(OH)$ and (P^{8-}) $Fe^{III}(OH)$ suggest an 'early' transition state for the 'on' reaction and a 'late' transition state for the 'off' reaction, compared to the opposite trend for the other complexes cited in Table 14.4. This is also seen in the value of K_{NO} in terms of the more effective binding of NO.

14.2.2
Cytochrome P450 and Model Complexes

The results reported above demonstrate how a change in pH can drastically affect the binding of NO to Fe(III)-porphyrins because of a change from a six-coordinate, spin-admixed diaqua complex to a five-coordinate, high-spin hydroxo complex by increasing the pH. A rather similar situation can occur in cytochrome P450, however not depending on pH but on the absence or presence of a substrate. The resting state of cytochrome P450 is a six-coordinate, low-spin Fe(III) aqua complex, whereas in the presence of a substrate such as camphor the resting state is converted to a five-coordinate, high-spin complex with the substrate present in the active site as illustrated schematically in Scheme 14.3. It is known that NO inhibits the catalytic activity of cytochrome P450 by coordinating to the Fe(III) center to form a diamagnetic (low-spin), six-coordinate $Fe^{II}-NO^+$ complex in the absence and presence of substrate (see Ref. [9] and references cited therein). This means that coordination of NO to P450 will involve binding to a low-spin, six-coordinate complex in the absence of a substrate, but to a high-spin, five-coordinate complex in the presence of a substrate, similarly to the situation encountered for the Fe(III) porphyrins as a function of pH. In order to clarify the mechanism of NO binding to cytochrome P450, detailed kinetic studies were performed in the absence and presence of camphor as substrate (see Scheme 14.3).

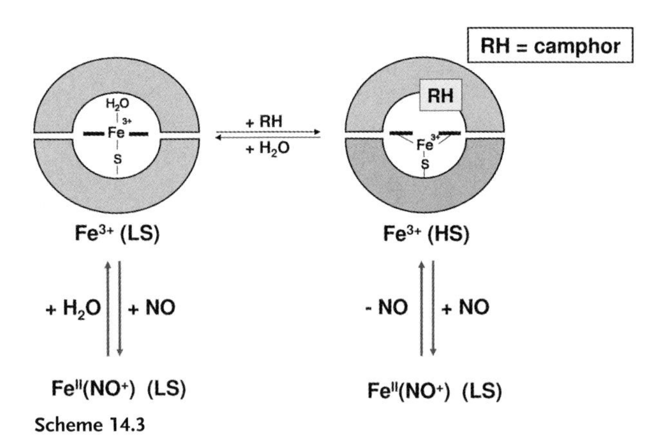

Scheme 14.3

Rate and activation parameters for the reversible binding of NO to cytochrome P450 in the absence and presence of camphor are summarized in Table 14.5 along with data for three functional models for P450 (see further Discussion in this Section). The overall binding constants, K_{NO}, are very similar for the reaction of NO with the six-coordinate aqua complex and the five-coordinate substrate complex. However, the kinetic data and activation parameters indicate that these forms of cytochrome P450 show totally different binding mechanisms, similar to that observed for the Fe(III)-porphyrins. Where the activation entropy and activation volume for the reversible binding of NO to the resting state of P450 clearly support the operation of a dissociative mechanism, these parameters for the reversible binding of NO to P450 in the presence of the substrate support the operation of an addition and an elimination process, respectively. This is nicely demonstrated by the volume profiles presented in Figure 14.4 [9]. The large volume increase

Table 14.5 Comparison of the rate and equilibrium constants, and thermodynamic and kinetic parameters for NO binding to the iron (III) center in $P450_{cam}$ (in the absence and presence of camphor) and model complexes.

	$P450_{cam}$ resting state [9]	**$P450_{cam}$ + camphor (E·S-$P450_{cam}$) [9]**	**SR complex in methanol [13]**	**Complex 3 in toluene [14]**	**Complex 4 in methanol [14]**
k_{on} ($M^{-1}\,s^{-1}$) at 25 °C	$(3.20 \pm 0.02) \times 10^5$	$(3.2 \pm 0.5) \times 10^6$	$(2.7 \pm 0.2) \times 10^6$	$(1.80 \pm 0.05) \times 10^6$	$(0.6 \pm 0.05) \times 10^5$
$\Delta H^{\#}_{on}$ (kJ mol^{-1})	92 ± 1	14.1 ± 0.1	75 ± 3	4 ± 2	14 ± 1
$\Delta S^{\#}_{on}$ (J $mol^{-1}\,K^{-1}$)	$+169 \pm 4$	-73.1 ± 0.4	$+130 \pm 11$	-111 ± 6	-107 ± 3
$\Delta G^{\#}_{on}$ (kJ mol^{-1}) at 25 °C	42 ± 1	35.9 ± 0.1	36 ± 3	37 ± 2	46 ± 1
$\Delta V^{\#}_{on}$ ($cm^3\,mol^{-1}$)	$+28 \pm 2$	-7.3 ± 0.2	$+6.4 \pm 1$	-25 ± 1	-21 ± 4
k_{off} (s^{-1}) at 25 °C	0.35 ± 0.02	1.93 ± 0.02	1.8 ± 2.1	12470 ± 120	2249 ± 167
$\Delta H^{\#}_{off}$ (kJ mol^{-1})	122 ± 4	83.8 ± 0.7		58 ± 1	44 ± 5
$\Delta S^{\#}_{off}$ (J $mol^{-1}\,K^{-1}$)	$+155 \pm 15$	$+41 \pm 2$		$+29 \pm 5$	-34 ± 22
$\Delta G^{\#}_{off}$ (kJ mol^{-1}) at 25 °C	76 ± 4	71.6 ± 0.7		50 ± 1	54 ± 5
$\Delta V^{\#}_{off}$ ($cm^3\,mol^{-1}$)	$+31 \pm 1$	$+24 \pm 1$		$+7 \pm 3$	$+7 \pm 3$
K_{NO} (M^{-1}) at 25 °C	$(9.0 \pm 0.2) \times 10^5$	$(1.2 \pm 0.4) \times 10^6$		122 ± 10	26.9 ± 2.9
ΔH° (kJ mol^{-1})	-30 ± 5	-69.7 ± 0.8		-71 ± 3	-59 ± 4
ΔS° (J $mol^{-1}\,K^{-1}$)	$+14 \pm 19$	-114 ± 2		-197 ± 10	-169 ± 13
ΔG° (kJ mol^{-1}) at 25 °C	-34 ± 4	-35.7 ± 0.7		-12 ± 3	-9 ± 4
ΔV° ($cm^3\,mol^{-1}$)	$+3 \pm 3$	-31.3 ± 1.2		-39 ± 2	-28 ± 1

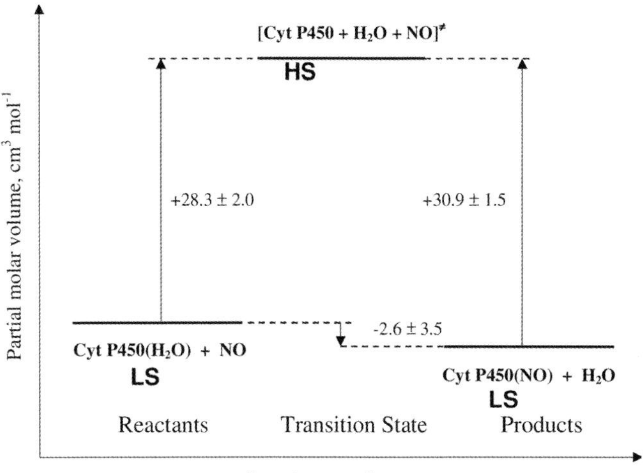

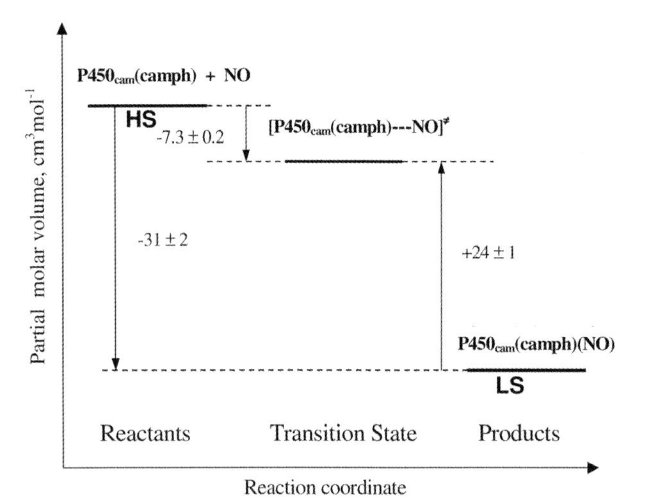

Figure 14.4 Volume profiles for the reversible binding of NO to P450 in the absence (top) and presenc (bottom) of camphor as substrate [9].

observed for the binding and release of NO in the case of substrate-free P450 is partly ascribed to the dissociative nature of the process that involves the dissociation of the coordinated water molecule and partly to a change from low-spin to high-spin Fe(III), which is typically accompanied by a volume increase of between 12 and 15 $cm^3\,mol^{-1}$ [10]. The 'off' reaction under these conditions is also accompanied by a large volume increase due to dissociation of NO and a low-spin to high-spin changeover. In the presence of the substrate, the binding of NO is accompanied by a decrease in volume typical of a bond-formation process, followed by a further volume decrease associated with a high-spin to low-spin changeover. The 'off' reaction under these conditions involves NO elimination and a spin change from low-spin to high-spin. The volume profiles in Figure 14.4 are in very good agreement with those reported for the studied Fe(III)-porphyrins as a function of pH in Figures 14.2 and 14.3.

This work was extended to the binding of NO by two functional models for cytochrome P450 developed by the research teams of Higuchi in Nagoya [11] and Woggon in Basel [12]. Figure 14.5 presents a summary of the Fe(III) active-site coordination spheres of the functional models and that of cytochrome P450. Higuchi and coworkers prepared a stable low-spin Fe^{III}–porphyrin alkanethiolate complex (**SR** complex, **1** in Figure 14.5), in which the thiolate ligand is sterically protected by bulky pivaloyl groups. The **SR** complex displays a similar catalytic activity to that of cytochrome P450, and the bulky protected axial thiolate ligand remains stable during the catalytic cycle.

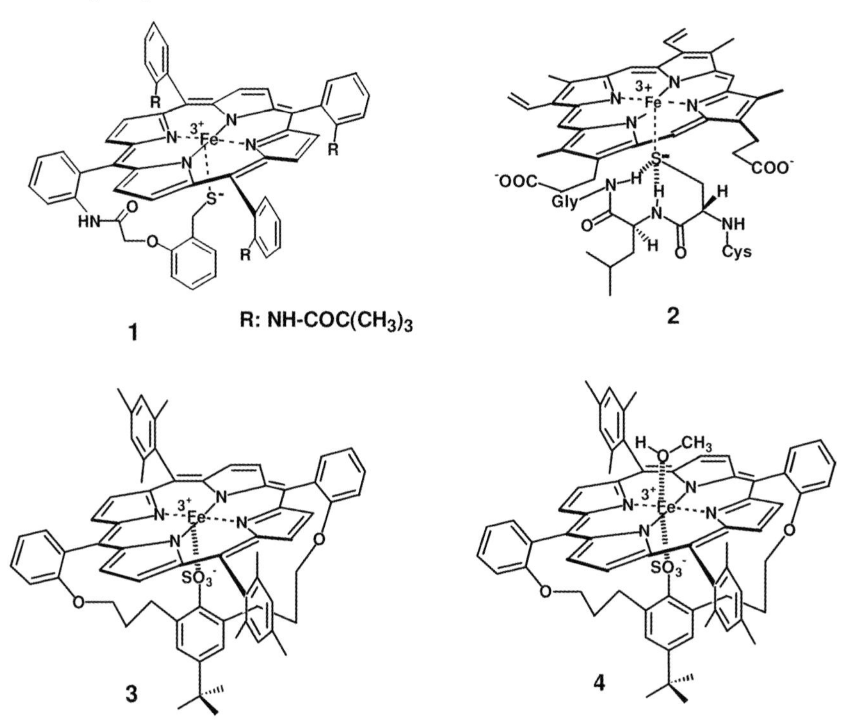

Figure 14.5 The active site coordination sphere of cytochrome $P450_{cam}$ **(2)** and functional model systems (**1** (**SR** complex), **3** and **4**) [11, 12].

We studied the mechanism of the reversible binding of NO to the **SR** complex in a coordinating (methanol) and a noncoordinating (toluene) solvent [13]. The kinetic studies revealed that the interaction of the **SR** complex with NO cannot be described as a simple reversible binding of NO to the iron(III) center as was found for the $P450_{cam}$/NO system. A rather complex reactivity pattern with several reaction steps was observed. For low NO concentrations, NO reversibly coordinates to the **SR** complex in methanol with an 'on' rate constant similar to that found for the corresponding reaction with $P450_{cam}$ (see Table 14.5). The NO dissociation rate constant is also of the same order of magnitude as that reported for the $P450_{cam}$/NO system. The activation parameters for the formation of the **SR**–NO complex in methanol suggest a limiting dissociative mechanism that is dominated by the dissociation of a coordinated methanol molecule. The three orders of magnitude higher value for the NO binding rate constant determined for this reaction in a noncoordinating solvent like toluene suggests that the NO binding dynamics is governed by the coordination mode of the iron(III) center.

For an excess of NO, rapid **SR–NO** complex formation is followed by subsequent slower processes. The second observed reaction can be accounted for in terms of direct attack of a second NO molecule on the sulfur atom of the thiolate ligand in the initially formed **SR**–NO complex as indicated in Scheme 14.4. This leads to the formation of the five-coordinate **SR**(Fe^{II}) nitrosyl complex, which was characterized by EPR and IR spectroscopy. Although the thiolate ligand in the **SR** complex is sterically protected, formation of the S-nitrosylated derivative does occur in the presence of an excess of NO. The two subsequent reactions are strongly accelerated by a large excess of NO and can be accounted for in terms of the dynamic equilibria between higher nitrogen oxides and reactive **SR** species, which result in the formation of a nitrosyl–nitrite complex of SR(Fe^{II}) as the final product [13].

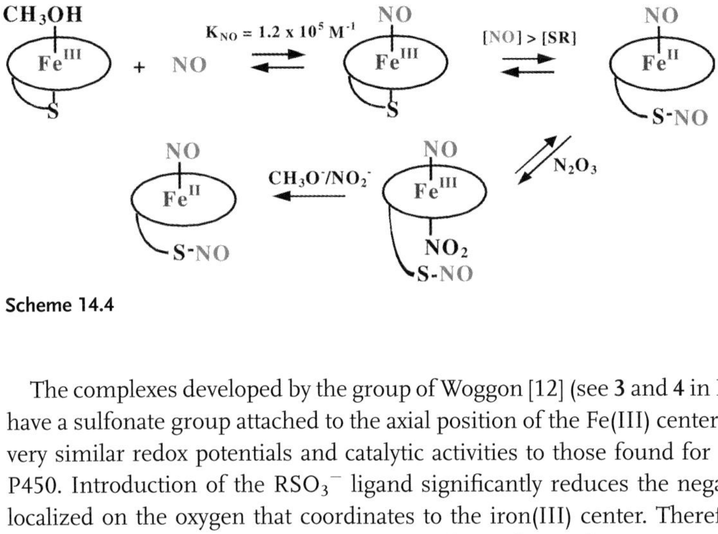

Scheme 14.4

The complexes developed by the group of Woggon [12] (see **3** and **4** in Figure 14.5) have a sulfonate group attached to the axial position of the Fe(III) center and exhibit very similar redox potentials and catalytic activities to those found for cytochrome P450. Introduction of the RSO_3^- ligand significantly reduces the negative charge localized on the oxygen that coordinates to the iron(III) center. Therefore, such a system can better mimic the coordination sphere of several native P450 enzymes in which electron donation from the sulfur atom is also reduced because of the presence of hydrogen bonding between the thiolate ligand and residues of the amino acids in the protein pocket. The influence of the proximal and distal ligands on the dynamics of NO binding to this new synthetic model was investigated by performing thermodynamic and kinetic studies on NO coordination to complexes **3** and **4** in non-coordinating toluene and in coordinating methanol, respectively [14]. In toluene, the five-coordinate, high-spin complex **3**, reminiscent of the camphor-bound $P450_{cam}$, is formed, whereas in methanol the six-coordinate complex **4**, resembling the resting state of $P450_{cam}$, is produced. The data summarized in Table 14.5 indicate that the model complex **3** displays an NO binding rate constant in a noncoordinating solvent similar to that for NO binding to the camphor-bound $P450_{cam}$. However, the rate constants for the NO release from the nitrosyl complexes of **3** and camphor-bound $P450_{cam}$ differ by almost four orders of magnitude, which results in a significantly lower value of the NO binding constant for complex **3** at room temperature. The binding constant increases significantly on going to lower temperatures and higher pressures, from which the following thermodynamic data were estimated: $K_{NO} = 40$ M^{-1} at 25 °C, $\Delta H° = -72\,kJ\,mol^{-1}$, $\Delta S° = -210\,J\,K^{-1}\,mol^{-1}$ and $\Delta V° = -39\,cm^3\,mol^{-1}$. The significantly negative reaction entropy and volume values suggest that NO binding to the high-spin iron(III) center is accompanied by a changeover to the low-spin (diamagnetic) $Fe^{II}{-}NO^+$ product, which was confirmed by spectroscopic observations [14].

Activation parameters reported for NO binding to complex **3** and native camphor-bound $P450_{cam}$ (Table 14.5) support the same mechanism for both reactions, this being dominated by $Fe^{III}{-}NO$ bond formation accompanied by a change in the iron (III) spin state (from S = 5/2 to S = 0). Despite this similarity in the reaction volume and entropy, the volume profiles for NO binding to these complexes (Figure 14.6) differ substantially in terms of the position of the transition state. In contrast to the volume profile constructed for the reaction between NO and camphor-bound $P450_{cam}$, reversible binding of NO to **3** is described as a 'late' transition state for the association reaction and an 'early' transition state for the dissociation reaction. This feature is in agreement with relatively slow 'on' and fast 'off' reactions and results in the much lower binding constant for NO coordination to **3** than for native $P450_{cam}$ [14]. The NO binding constants found for the enzyme models **3** and **4** are about 4 orders of magnitude smaller than those found for the native $P450_{cam}$/NO system, which is because of the rapid NO dissociation from the nitrosyl complexes of **3** and **4**. This feature clearly highlights the important role of the enzyme pocket of $P450_{cam}$ in stabilizing the NO^+ state and formation of more stable nitrosyl complexes at ambient temperature as compared to those observed for many protein-free model complexes.

14.3 Peroxide Activation by Fe(III) Complexes

The Fe(III) porphyrin complexes, cytochrome P450 and the functional models for P450 presented in the previous section, all include an iron center of which the coordination number and spin state can be tuned either through the pH, the presence of a substrate or the selected solvent, respectively. The mechanistic understanding of the interaction of NO with these Fe(III) systems under the

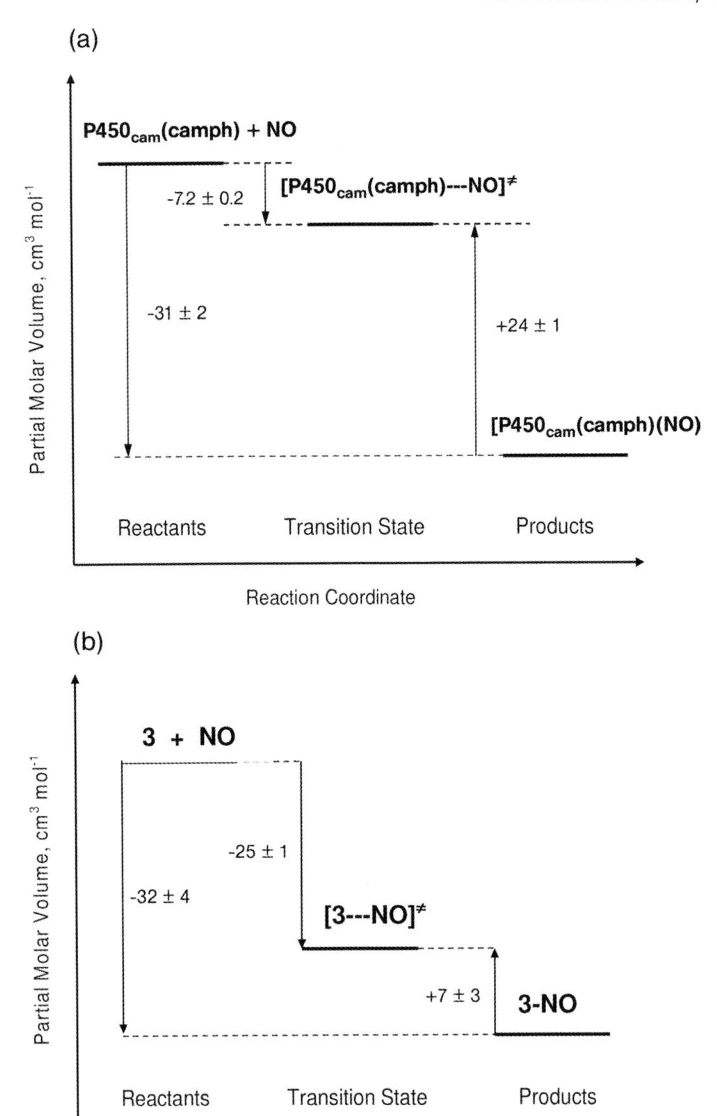

Figure 14.6 Comparison of volume profiles for NO binding to (a) camphor-bound $P450_{cam}$ [9], and (b) complex **3** in toluene [14].

selected conditions to tune the coordination number and the spin state of the metal center provides an excellent basis for studying the activation of peroxides by such Fe (III) systems. Where the activation of NO only involves the bonding of NO and a subsequent charge transfer process, the activation of peroxide involves coordination of peroxide to the Fe(III) center, subsequent homolysis or heterolysis of the peroxo

O–O bond, and formation of high oxidation state iron oxo complexes that are able to transfer oxygen to inert substrate molecules. This last step is of fundamental interest to many catalytic oxidation processes [15] and calls for detailed mechanistic understanding.

14.3.1
Cytochrome P450

The overall catalytic oxidation cycle of cytochrome P450 consists of the reaction steps summarized in Scheme 14.5 [16]. A number of reaction steps can be cut short along the 'peroxide shunt' pathway by combining O_2, $2H^+$ and $2e^-$ in the form of H_2O_2 (or as organic peroxides and peroxo acids) as shown schematically in Scheme 14.5.

$$RH + O_2 + 2H^+ + 2e^- \rightarrow ROH + H_2O$$

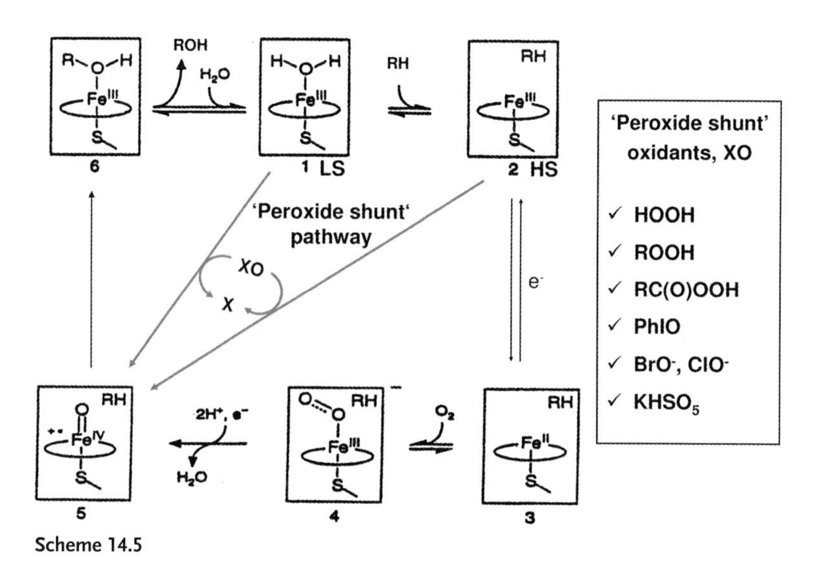

Scheme 14.5

Important to note is that the resting state of P450 in the absence and presence of substrate (species **1** and **2** in Scheme 14.5) has the properties discussed above for the interaction with NO as outlined in Scheme 14.3. Complex **5** in Scheme 14.5 is the actual important species since it accounts for the subsequent oxygen transfer process. The interaction of peroxo species with the reactant species **1** and **2** (in Scheme 14.5) will first of all involve rapid binding of peroxide to the Fe(III) porphyrin center (as observed for the reaction with NO) followed by homolytic or heterolytic cleavage of the O–O peroxo bond (see Scheme 14.6). In terms of mechanistic studies, it is essential to distinguish between these O–O bond cleavage modes, since they will control the chemistry of the subsequent oxidation process. Recent work in our laboratories demonstrated that low-temperature rapid-scan spectroscopy can be used very successfully to distinguish between heterolytic and homolytic cleavage of the peroxo bond in the absence of substrate molecules [15]. We were surprised to find that the selection between an overall two-electron oxidation process (heterolysis) and a one-electron oxidation process (homolysis) can be tuned by pH as the following example demonstrates.

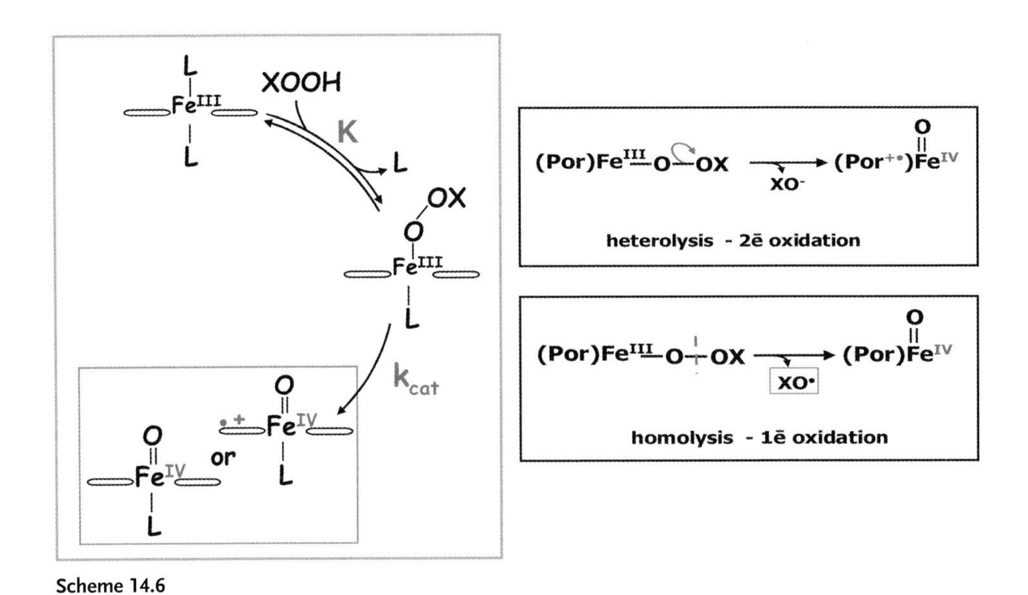

Scheme 14.6

14.3.2
Fe(III) Porphyrins

The $Fe^{III}(TMPS^{4-})(H_2O)_2$ has a pK_a value of 6.9 in aqueous solution.

$$(H_2O)_2Fe^{III} \rightleftharpoons Fe^{III}(OH) + H_3O^+ \qquad pK_a = 6.9$$

In order to follow reactions with the selected oxidants, the use of low temperatures (−30 to −40 °C) was necessary to stabilize and characterize the oxidized porphyrin products formed at various values of the pH. A 4 : 1 (v/v) MeOH/H_2O mixture was used as cryosolvent in the low-temperature spectroscopic studies. This caused a change in the pK_a value to 6.3 [15].

$$\text{(HOR)}_2\text{Fe}^{III} \rightleftharpoons \text{Fe}^{III}\text{–OR} + \text{HOR} + \text{H}^+ \qquad \text{p}K_a = 6.3$$

R = H or CH_3

The reaction of Fe^{III}(TMPS) with *m*-chloroperoxybenzoic acid (*m*-CPBA) was followed by low-temperature stopped-flow at −35 °C. These studies clearly revealed a pH-tuned change in the nature of the products formed at different values of the pH, as reported in Figure 14.7. At pH < 5, formation of a product with a Soret band of decreased intensity at 404 nm and a broad low-intensity band with a maximum at ca. 670 nm clearly indicated conversion to the oxoiron(IV) porphyrin cation radical $(TMPS^{+\bullet})Fe^{IV}(O)$, a 2ē-oxidation process. In contrast, the reaction at pH > 7.5 gave a product with Soret band at 425 nm and a low-intensity band at 545 nm assigned as $(TMPS)Fe^{IV}(O)$, a 1ē-oxidation process. Clean isosbestic points indicated that at pH < 5.5 and pH > 7.5 the reaction involves a single spectroscopically observable step at −35 °C.

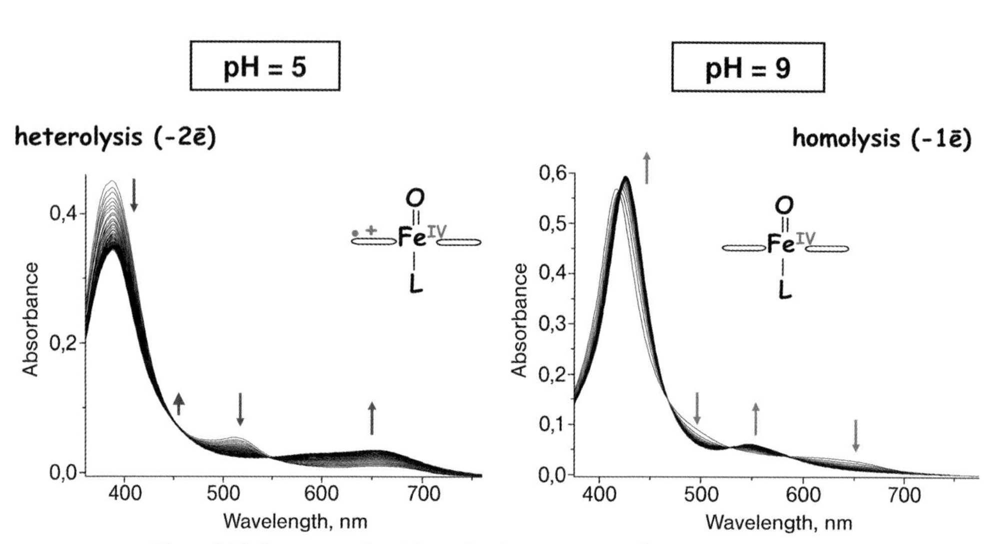

Figure 14.7 Spectroscopic evidence for the occurrence of heterolytic and homolytic bond cleavage during the reaction of $[Fe^{III}(TMPS^{4-})]$ with *m*-CPBA as a function of pH [15].

We proposed that the reactivity pattern observed for the reaction with *m*-CPBA, PhIO and H_2O_2 resulted from a pH-tuned change in the redox properties of the complex rather than its different ligation at low and high pH. The rationale for this proposal is based on consideration of the $E^{\circ\prime}$ vs pH diagram for electrochemical oxidation of this complex in aqueous solution depicted in Figure 14.8. As shown in the diagram, oxidation may be porphyrin-centered ($P \rightarrow P^{+\bullet}$) and/or metal-centered ($Fe^{III} \rightarrow Fe^{IV}$). Since the former does not involve release or uptake of protons, $E^{\circ\prime}_{P^{+\bullet}/P}$ is relatively pH-insensitive. This contrasts with the pH-dependent $Fe^{III} \rightarrow Fe^{IV}$ oxidation, a process coupled to acid–base equilibria involving formation of the oxo ligand. Because of the pH dependence of $E^{\circ\prime}_{Fe^{IV}/Fe^{III}}$, the relative positions of

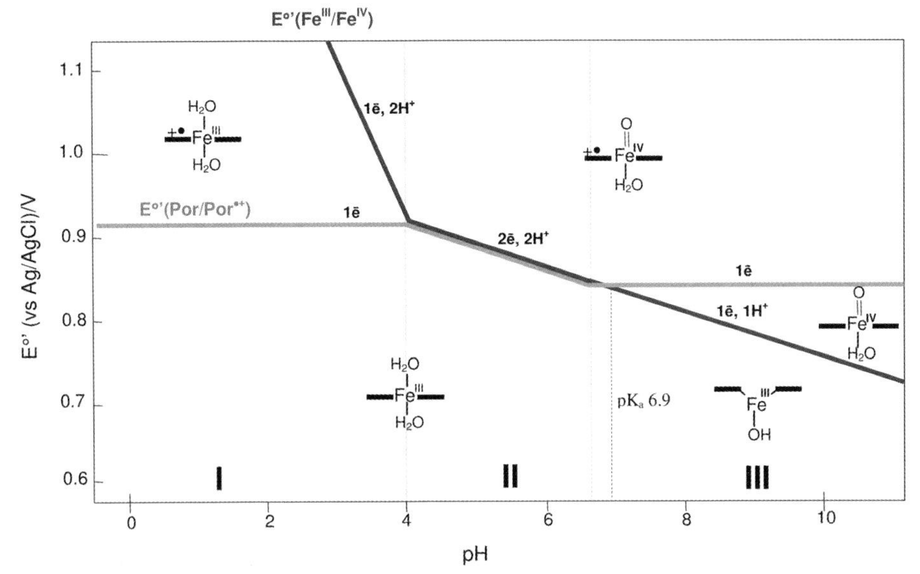

Figure 14.8 Schematic presentation of the pH dependence of $E^{\circ\prime}$ for electrochemical oxidations of Fe^{III}(TMPS) in aqueous solution [15].

$E^{\circ\prime}_{Fe^{IV}/Fe^{III}}$ and $E^{\circ\prime}_{P^{+\bullet}/P}$, and the nature of the most thermodynamically stabilized oxidized form of **1** change with pH. The diagram in Figure 14.8 can thus be divided into three sections, in which the first (thermodynamically most favored) oxidation reaction is given.

$$\text{Section } I \quad pH<3.5 \quad (P)Fe^{III}(H_2O)_2 \rightarrow (P^{+\bullet})Fe^{III}(H_2O)_2 + \bar{e}$$

$$\text{Section II} \quad 3.5<pH<6.5 (P)Fe^{III}(H_2O)_2 \rightarrow (P^{+\bullet})Fe^{IV}(O)(H_2O) + 2\bar{e} + 2H^+$$

$$\text{Section III} \quad pH \geq 7 \quad (P)Fe^{III}(OH) + H_2O \rightarrow (P)Fe^{IV}(O)(H_2O) + \bar{e} + H^+$$

In the context of chemical oxidation involving O–O bond cleavage of coordinated peroxide, Figure 14.8 suggests that Fe^{III}(TMPS) as a redox partner favors O–O bond homolysis ($1\bar{e}$ oxidation) at $pH<3.5$ and $pH>7$, since in these pH ranges its $1\bar{e}$ oxidation is thermodynamically most facile. In contrast, heterolysis of the O–O bond ($2\bar{e}$ oxidation) should be promoted in the pH range 4–7, where $E^{\circ\prime}_{P^{+\bullet}/P} \approx E^{\circ\prime}_{Fe^{IV}/Fe^{III}}$. The mode of O–O bond cleavage is, however, expected to depend not only on the redox properties of the complex at a given pH, but also on the oxidative power of the oxidant. A strong oxidant (like *m*-CPBA) with $E^{\circ\prime}$ exceeding both $E^{\circ\prime}_{Fe^{IV}/Fe^{III}}$ and $E^{\circ\prime}_{P^{+\bullet}/P}$ may thus effect a $2\bar{e}$ oxidation (O–O bond heterolysis) at pH > 7, while O–O bond homolysis ($1\bar{e}$ oxidation) should occur at this pH range for weaker oxidants with $E^{\circ\prime}_{Fe^{IV}/Fe^{III}} < E^{\circ\prime} < E^{\circ\prime}_{P^{+\bullet}/P}$.

A systematic variation of the m-CPBA concentration revealed that the observed first-order rate constant exhibits a saturation behavior on increasing the *m*-CPBA concentration. This is typical for a process that consists of a rapid pre-equilibrium (peroxide binding to the Fe(III) center in Scheme 14.6) followed by rate-determining cleavage of the O–O bond. In addition, a trap such as ABTS can be used to trap the higher oxidation state oxo complexes as shown in Scheme 14.7 by following the characteristic absorbance increase at 660 nm for the formation of the $ABTS^{\bullet+}$ radical cation. In this way the overall second-order rate constant for the catalyzed oxidation of ABTS by *m*-CPBA in the presence of $Fe(TMPS^{4-})$ could be determined as a function of temperature and pressure, such that mechanistic information on the oxygen transfer process can be extracted [17].

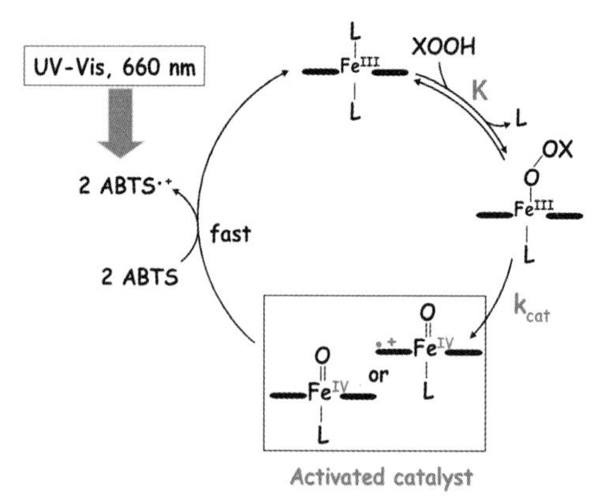

Scheme 14.7

It is clear from the presented information that only certain steps of the overall catalytic cycle in Scheme 14.7 can be observed directly under the selected conditions. In order to reach our goal to observe all reaction steps of the catalytic cycle, we turned to investigations at low temperature in a nonaqueous medium, in which rapid-scan techniques could be used to observe each reaction step separately as illustrated in the following section.

14.3.3 Catalytic Oxidation Cycle

For this purpose we used the high-spin, five-coordinate complex prepared by the Woggon group [12] shown in Scheme 14.8 (complex **1** in Scheme 14.9) that exhibits

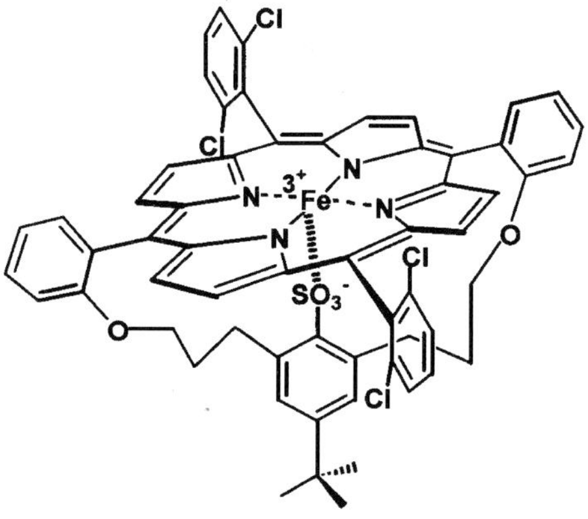

Scheme 14.8

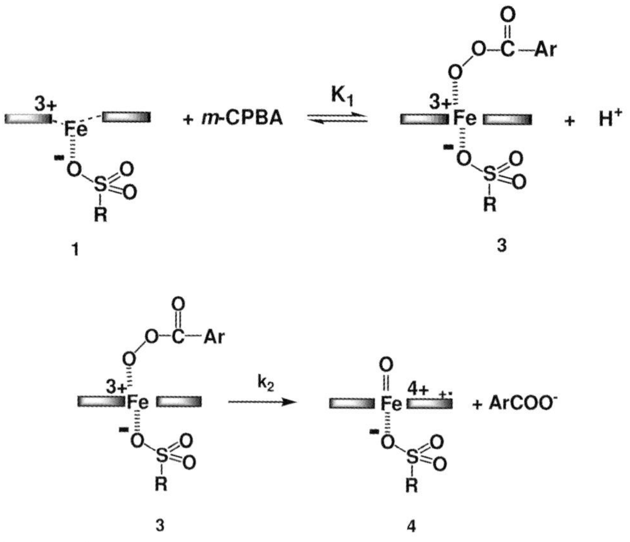

Scheme 14.9

excellent catalytic behavior in epoxidation reactions. We studied the mechanistic behavior of this complex with *m*-CPBA as oxidant in MeCN at −35 °C using a low-temperature stopped-flow system with a rapid-scan detector [18]. In the absence of substrate, rapid, reversible coordination of *m*-CPBA to the high-spin Fe(III) center (**1** and **3** in Scheme 14.9) is followed by the rate-determining acid-catalyzed heterolysis of the O−O bond to form the Fe(IV) porphyrin radical cation (**4** in Scheme 14.10), as evidenced by the repetitive scan spectra reported in Figure 14.9. The kinetic data for these combined steps show saturation kinetics as presented in Figure 14.10 from

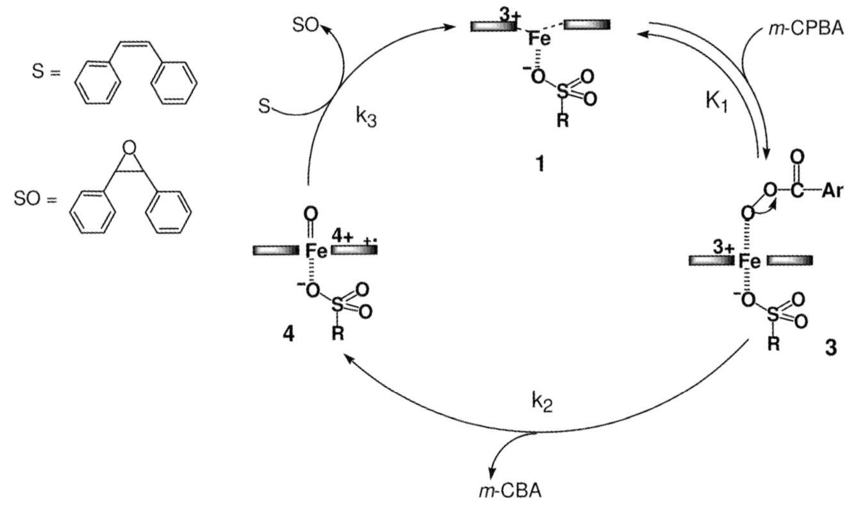

Scheme 14.10

which $K_1 = 4.4 \times 10^3\ \mathrm{M}^{-1}$ and $k_2 = 2.4\ \mathrm{s}^{-1}$ at $-35\ ^\circ\mathrm{C}$ were calculated using the rate law:

$$k_{obs} = \frac{k_2 K_1 [m\text{-CPBA}]}{1 + K_1 [m\text{-CPBA}]}$$

In the absence of reactive substrates, the oxo-iron porphyrin radical cation **4** is stable for almost 600 seconds at $-35\ ^\circ\mathrm{C}$ and $[m\text{-CPBA}] = 7.3 \times 10^{-4}$ M in acetonitrile

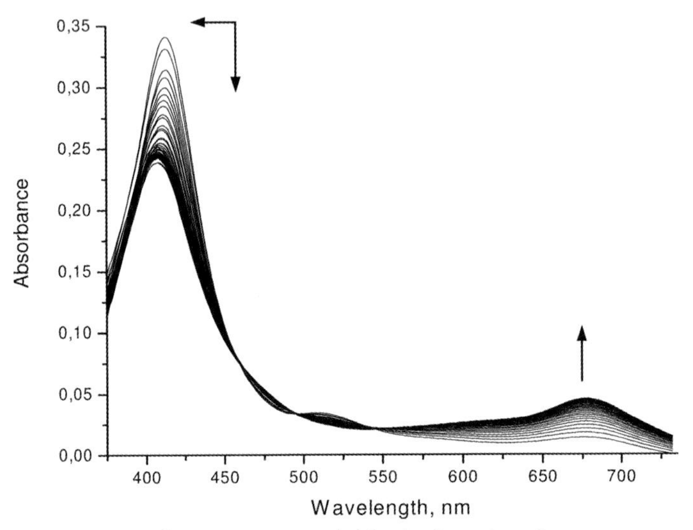

Figure 14.9 Rapid-scan spectra recorded for the formation of $(1^{+\bullet})Fe^{IV}{=}O$ in the reaction of **1** with *m*-CPBA. Experimental conditions: $[\mathbf{1}] = 4.3 \times 10^{-6}$ M, $[m\text{-CPBA}] = 5.4 \times 10^{-5}$ M in acetonitrile at $-35\ ^\circ\mathrm{C}$ [18].

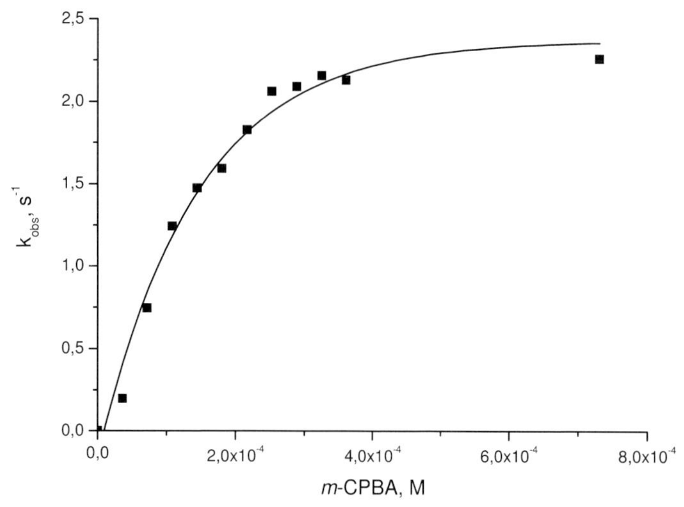

Figure 14.10 *m*-CPBA concentration dependence of k_{obs}. Experimental conditions: [**1**] = 3.6×10^{-6} M, *m*-CPBA = $(0.36–7.2) \times 10^{-4}$ M in acetonitrile at −35 °C [18].

(Figure 14.11a). The addition of *cis*-stilbene to the green solution of **4** results in a fast decomposition of **4** with concomitant 95% regeneration of the starting complex **1** (Figure 14.11b). In order to evaluate the effect of substrate concentration on the reaction between **4** and *cis*-stilbene, kinetic studies were performed in which the solutions of **1** containing various amounts of *cis*-stilbene were mixed with a solution of *m*-CPBA, and the re-formation of **1** (the product of reaction between **4** and substrate) was monitored at 412 nm in acetonitrile at −35 °C. The reaction profile in Figure 14.12 is characterized by a rapid decrease in the concentration of **1** (due to formation of **4**) followed by an apparent induction period which depends on the excess of substrate employed. If the *cis*-stilbene concentration is kept below a 500-fold excess, **4** is always sufficiently and rapidly regenerated by the 10-fold excess of *m*-CPBA (~20 s in the absence of *cis*-stilbene) such that the Soret band of **1** at 412 nm is not observed. This leads to a delay in the re-formation of **1** during which most of the *m*-CPBA is consumed and which can last hundreds of seconds when the excess of *cis*-stilbene is small under these conditions. The overall catalytic cycle is shown in Scheme 14.10. The value of k_3 was determined from the *cis*-stilbene concentration dependence of the observed rate constant for the re-formation of the starting complex **1** following the apparent induction period observed in Figure 14.12 and found to be $7\,M^{-1}\,s^{-1}$ at −35 °C. The reported values of K_1, k_2 and k_3, along with the concentrations employed, were used to simulate the kinetic traces in Figure 14.12 on the basis of the overall catalytic cycle in Scheme 14.10. The simulated kinetic traces display a good and satisfactory agreement with the experimental traces and confirm the credibility of the proposed mechanistic scheme. Thus, all the reaction steps, schematically shown in Scheme 14.10, could be observed by UV-vis spectroscopy under the selected low-temperature conditions [18].

(A)

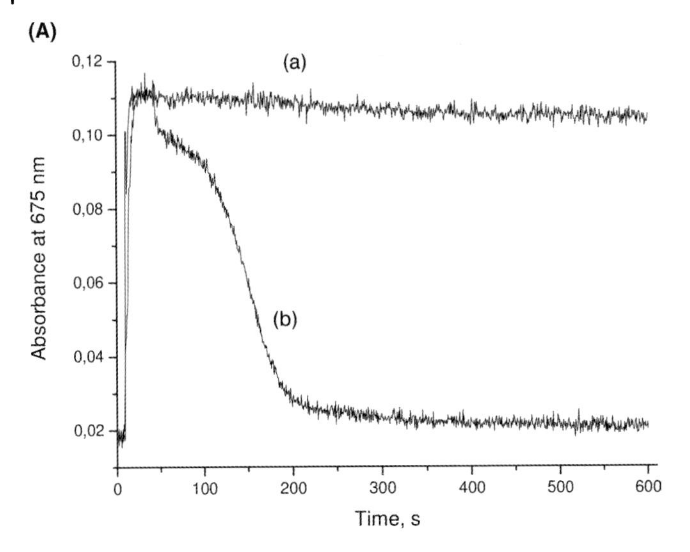

(B)

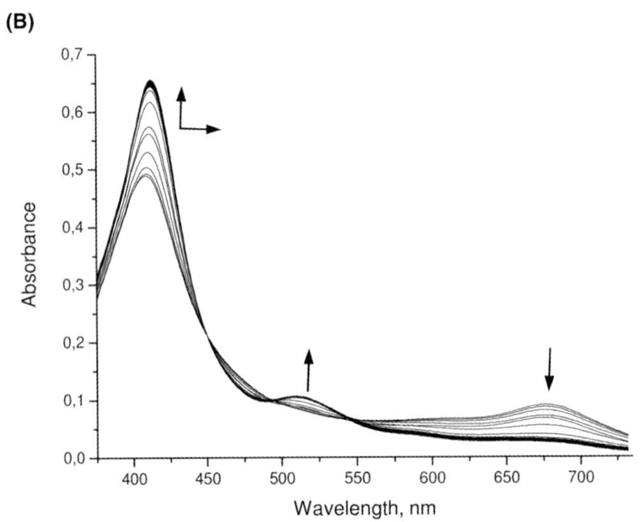

Figure 14.11 (A) Kinetic traces recorded with the dip-in detector at 675 nm for the formation and decay of **4**: (a) in the absence of the substrate; (b) after addition of 2.07×10^{-3} M *cis*-stilbene to the solution of **4**. Experimental conditions: [**1**] = 8.4×10^{-6} M, [*m*-CPBA] = 8.4×10^{-5} M in acetonitrile at −35 °C. (B) Spectral changes recorded during re-formation of the starting complex **1** in the presence of the substrate. Experimental conditions: [**1**] = 7.9×10^{-6} M, [*m*-CPBA] = 7.9×10^{-5} M, [*cis*-stilbene] = 1.6×10^{-2} M in acetonitrile at −35 °C [18].

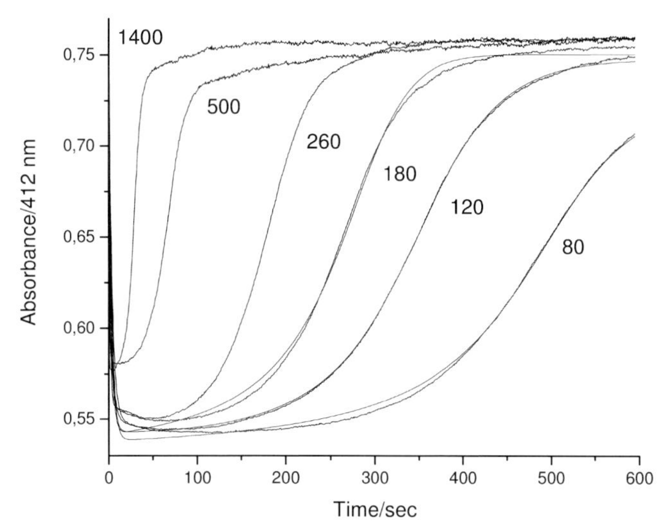

Figure 14.12 Typical kinetic traces recorded at 412 nm and −35 °C in acetonitrile for the reaction of **1** (7.9×10^{-6} M) with *m*-CPBA (7.9×10^{-5} M) in the presence of 6.35×10^{-4} (80), 9.53×10^{-4} (120), 1.43×10^{-3} (180), 2.06×10^{-3} (260), 3.95×10^{-3} (500) and 1.11×10^{-2} (1400) M *cis*-stilbene; the substrate excess in parentheses refers to **1**. Kinetic traces as smooth curves represent results of simulations obtained after introduction of the appropriate kinetic/thermodynamic constants and oxidant/catalyst/substrate concentrations for the reaction cycle proposed in Scheme 14.22 [18].

14.4 Conclusions

Our studies show how we could develop mechanistic concepts for the activation of NO and on the basis of the gained mechanistic insight extend the work to the activation of different peroxides by functional models for cytochrome P450. With the developed ability to spectroscopically observe all the reaction steps in the catalytic cycle, it will now be possible to perform systematic temperature- and pressure-dependent measurements for the activation of peroxide and the subsequent oxygen transfer process, from which further detailed mechanistic insight will be gained. In addition, with the vast possibilities to tune the catalytic activity of high-valent iron-oxo complexes via a systematic modification of the porphyrin macrocycle, pH and solvent composition, the general mechanistic understanding will be applied to the development of suitable catalysts for the oxidation of inert substrates, including organic dyes that are of environmental concern.

Acknowledgments

The authors gratefully acknowledge financial support from the Deutsche Forschungsgemeinschaft through SPP 1118 and the stimulating collaborations with

Grazyna Stochel (Jagiellonian University, Poland), Tsunehiko Higuchi (Nagoya City University, Japan), Norbert Jux (University of Erlangen-Nürnberg, Germany) and Wolf-D. Woggon (University of Basel, Switzerland).

References

1 (a) van Eldik, R. Dücker-Benfer, C. and Thaler, F. (2000) *Advances in Organometallic Chemistry*, **49**, 1–58. (b) Hubbard, C.D. and van Eldik, R. (2005) in *Chemistry at Extreme Conditions* (ed. M. Riad Manaa), Elsevier, Amsterdam, Chapter 4, pp. 109–164; (c) Hubbard, C.D. and van Eldik, R. (2007) *Journal of Coordination Chemistry* (invited review), **60**, 1–51.

2 Special issue on Inorganic and Bioinorganic Mechanisms, (2005) *Chemical Reviews*, **105**, 1917–2722.

3 Helm, L. and Merbach, A.E. (2005) *Chemical Reviews*, **105**, 1923–1959.

4 Schneppensieper, T., Zahl, A. and van Eldik, R. (2001) *Angewandte Chemie – International Edition*, **40**, 1678–1680.

5 Wolak, M. and van Eldik, R. (2005) *Journal of the American Chemical Society*, **127**, 13312–13315.

6 Jee, J.-E., Eigler, S., Hampel, F., Jux, N., Wolak, M., Zahl, A., Stochel, G. and van Eldik, R. (2005) *Inorganic Chemistry*, **44**, 7717–7731.

7 Jee, J.-E., Wolak, M., Balbinot, D., Jux, N., Zahl, A. and van Eldik, R. (2006) *Inorganic Chemistry*, **45**, 1326–1337.

8 Jee, J.-E., Eigler, S., Jux, N., Zahl, A. and van Eldik, R. (2007) *Inorganic Chemistry*, **46**, 3336–3352.

9 Franke, A., Stochel, G., Jung, C. and van Eldik, R. (2004) *Journal of the American Chemical Society*, **126**, 4181–4191.

10 Constable, E.C., Braun, G., Bill, E., Dyson, R., van Eldik, R., Fenske, D., Kaderli, S., Morris, D., Neubrand, A., Neuburger, M., Smith, D.R., Wieghardt, K., Zehnder, M. and Zuberbühler, A.D. (1999) *Chemistry - A European Journal*, **5**, 498–508, and references cited therein.

11 Suzuki, N., Higuchi, T., Urano, Y., Kikuchi, K., Uchida, T., Mukai, M., Kitagawa, T. and Nagano, T. (2000) *Journal of the American Chemical Society*, **122**, 12059–12060.

12 Woggon, W.-D. (2005) *Accounts of Chemical Research*, **38**, 127–136.

13 Franke, A., Stochel, G., Suzuki, N., Higuchi, T., Okuzono, K. and van Eldik, R. (2005) *Journal of the American Chemical Society*, **127**, 5360–5375.

14 Franke, A., Hessenauer-Ilicheva, N., Meyer, D., Stochel, G., Woggon, W.-D. and van Eldik, R. (2006) *Journal of the American Chemical Society*, **128**, 13611–13624.

15 Wolak, M. and van Eldik, R. (2007) *Chemistry - A European Journal*, **13**, 4873–4883, and literature cited therein.

16 Denisov, L.G., Makris, T.M., Sligar, S.G. and Schlichting, I. (2005) *Chemical Reviews*, **105**, 2253–2277.

17 Wolak, M., Franke, A., Hessenauer-Ilicheva, N., Theodoridis, A. and van Eldik, R., work in progress.

18 Hessenauer-Ilicheva, N., Franke, A., Meyer, D., Woggon, W.-D. and van Eldik, R. (2007) *Journal of the American Chemical Society*, **129**, 12473–12479.

15
Synthetic Nitrogen Fixation with Molybdenum and Tungsten Phosphine Complexes: New Developments

Gerald Stephan and Felix Tuczek

15.1
Introduction

The conversion of dinitrogen to ammonia under ambient conditions is a challenging problem which has been of interest in biological, inorganic and organometallic chemistry for many years [1]. In nature, this process is mediated by the enzyme nitrogenase according to Eq. (15.1), which indicates both the high energy consumption of this process and the fact that H_2 evolution accompanies dinitrogen reduction.

$$N_2 + 8H^+ + 8e^- + 16MgATP \rightarrow 2NH_3 + H_2 + 16(ADP + P_i) \qquad (15.1)$$

Nitrogenase consists of two proteins (Figure 15.1). The larger one, called molybdenum-iron protein (MoFe), is an $\alpha_2\beta_2$ tetramer which contains two unique iron-sulfur clusters, the P-Cluster and the iron-molybdenum cofactor (FeMoco) [2]. The electrons necessary for the reduction of N_2 are provided by the iron protein, which contains one iron-sulfur cluster located between its two subunits. During turnover the iron protein forms a complex with the MoFe protein and transfers one electron to the MoFe protein. This electron is transferred to the FeMoco via the P-cluster. After this process the complex dissociates and the Fe protein is recharged with one electron, being able to reduce the MoFe protein again. This process has to occur 8 times for a full catalytic cycle [3].

Despite many experimental attempts to elucidate the reaction mechanism of nitrogenase on a molecular level, details of this mechanism are still unclear. A number of theoretical studies have therefore been performed in order to obtain insight into possible reaction pathways [4]. So far, however, no commonly accepted reaction mechanism of nitrogenase has emerged from these studies. Model chemistry approaches have focused on binding and activating N_2 at mono-, di- or polynuclear metal centers. Importantly, some of these systems also allowed us to get information on further elementary reactions necessary for synthetic nitrogen fixation. In this context, two reaction schemes are exceptional in that they exhibit the full set of elementary steps for the conversion of N_2 to ammonia; these are the Schrock and the Chatt cycles.

Activating Unreactive Substrates: The Role of Secondary Interactions.
Edited by Carsten Bolm and F. Ekkehardt Hahn

ISBN: 978-3-527-31823-0

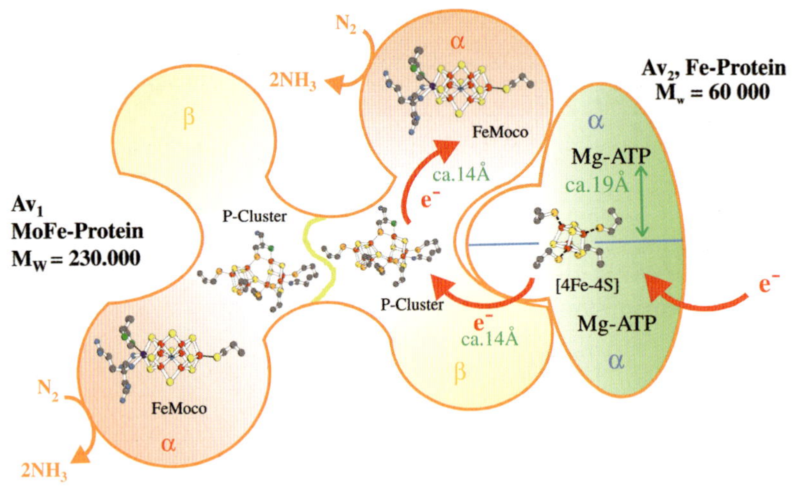

Figure 15.1

Schrock *et al.* were the first to realize that a truly catalytic ammonia synthesis proceeds under ambient conditions and through a series of well-defined intermediates [5]. The corresponding reactive system is based on the molybdenum(III) complex [Mo(HIPTN$_3$N)] containing the triamidoamine ligand HIPTN$_3$N (hexaisopropylterphenyl-triamidoamine), which provides a sterically shielded site for dinitrogen bonding and transformation to ammonia (Figure 15.2). Using lutidinium

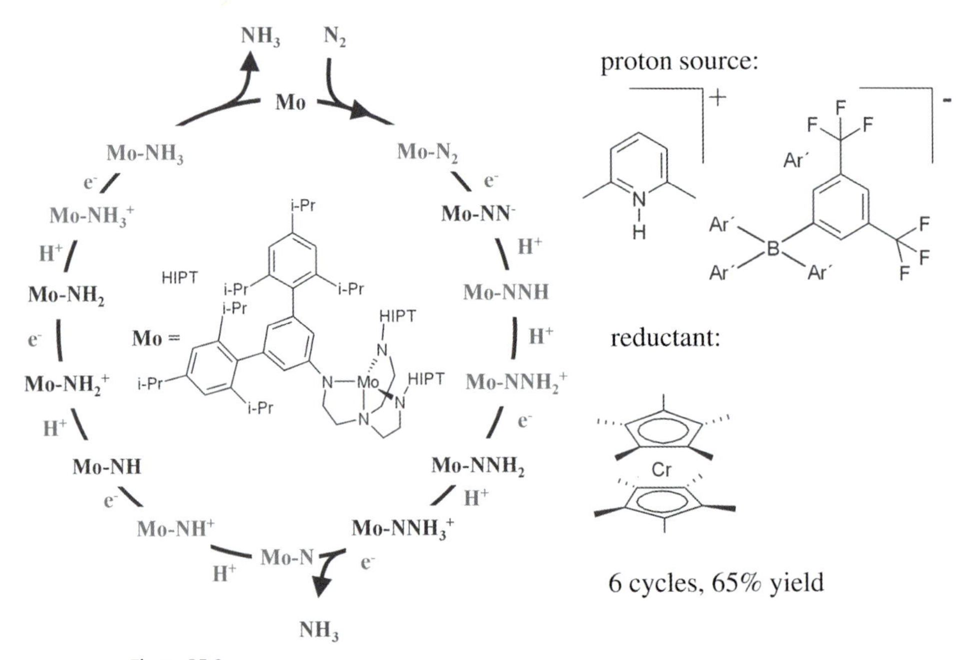

Figure 15.2

tetrakis(3,5-bis-trifluoromethyl-phenyl)-borate ($LutHBAr'_4$) as a proton source and decamethyl chromocene (Cp^*_2Cr) as reductant, ammonia was formed from N_2 at room temperature and normal pressure, achieving 6 cycles with an overall yield of 65%. We performed DFT calculations on all possible intermediates occurring in the catalytic transformation of N_2 to ammonia, including those characterized and invoked by Schrock *et al.* [5]. Based on these calculations a detailed mechanism and a free enthalpy profile of the entire cycle were derived [6].

The calculations indicate that the catalytic cycle consists of a sequence of strictly alternating protonation and reduction steps. This agrees with the mechanism postulated by Schrock *et al.* [5], except for the transformation of the dinitrogen complex to the neutral NNH complex. (Whereas the calculation predicts a primary protonation of coordinated N_2 followed by reduction, Schrock *et al.* postulate a primary reduction of the Mo(III)-N_2 complex in order to achieve the protonation leading to the NNH complex.) All relevant intermediates are thus either neutral or singly positively charged, while formation of negatively charged species appears to be excluded. Furthermore, most of the reactions are exergonic. The cleavage of the N−N bond is by far the most exergonic step, which in addition proceeds spontaneously. Only three steps of the cycle are endergonic, the most energy-demanding step being the first protonation of the N_2 complex. The most difficult reduction process involves the final conversion of the Mo(IV)-NH_3 complex to its Mo(III) counterpart, which is able to exchange NH_3 against N_2, closing the cycle.

The alternative mechanistic scenario for the protonation and reduction of end-on terminally coordinated N_2 is represented by the Chatt cycle, which was discovered many years before the Schrock cycle [1]. This system is based on Mo(0) and W(0) dinitrogen complexes with phosphine coligands (Figure 15.3). The intermediates of

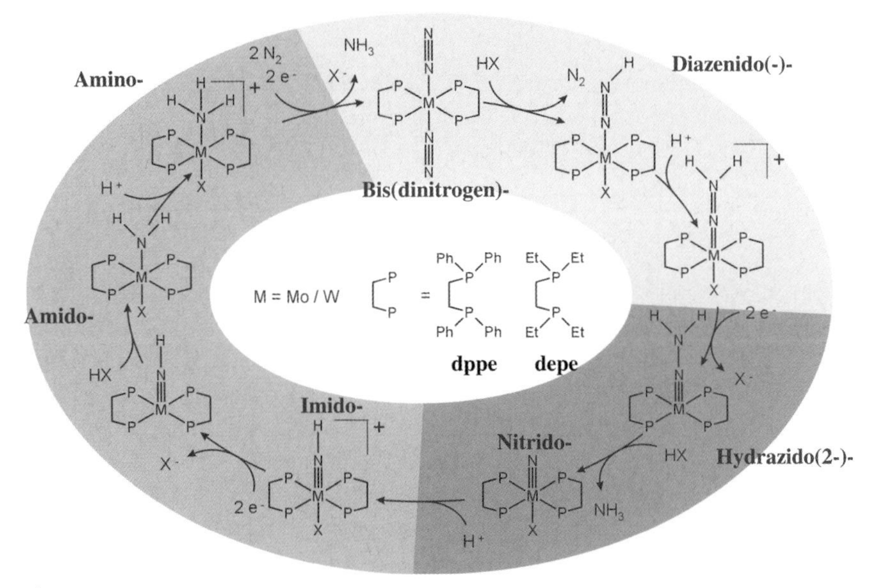

Figure 15.3

the dinitrogen reduction scheme are very similar to those of the Schrock cycle. Moreover, a cyclic generation of NH_3 from N_2 has been demonstrated on the basis of this system, though yields are very small [7, 8]. The goal of our research is to increase this number and make the system chemically more robust. New ligands developed in this context are presented in Section 15.3. In order to obtain general insight into the mechanism of the Chatt cycle we have studied most of the intermediates of Figure 15.2 by spectroscopy coupled to DFT calculations. The results of these studies are presented in the following section.

15.2 Mechanistic Investigation of the Chatt Cycle

For a detailed consideration of its mechanism it is meaningful to divide the Chatt cycle into three stages, (a) the protonation of bound N_2, (b) the cleavage of the N–N bond, generating the first molecule of NH_3, and (c) the reduction and protonation of nitrido complexes, leading to formation of the second equivalent of ammonia. These stages are treated separately in the following sections.

15.2.1 Protonation of N_2

In order to obtain insight into the factors determining the protonation of dinitrogen, the N_2, N_2H, and N_2H_2 complexes $[W(N_2)_2(dppe)_2]$ (**1**), $[WF(NNH)(dppe)_2]$ (**2**) and $[MF(NNH_2)(dppe)_2]^+$ (**3**) (dppe = 1,2-bis(diphenylphosphino)ethane) have been investigated by infrared and Raman spectroscopy coupled to DFT calculations [9]. More recently, these studies were complemented by investigation of the Mo and W hydrazidium complexes $[MF(NNH_3)(depe)_2](BF_4)_2$ (M = Mo (**4a**) and W (**4b**); depe 1,2-bis(diethylphosphino)ethane) [10]. The analogous NNH_3 compounds with dppe do not exist since protonation of the corresponding N_2 precursors stops at the NNH_2 stage; we therefore had to switch to depe in the latter investigation. The diazenido(−) and hydrazido(2−) complexes **2** and **3**, respectively, were prepared from complex **1** by treatment with HBF_4. The hydrazidium complexes **4** were obtained from the N_2 depe complexes $[M(N_2)_2(depe)_2]$, M=Mo and W, by protonation with HBF_4 as well. Protonation of the dinitrogen-depe complexes with HCl, on the other hand, led to the NNH_2 complexes $[MCl(NNH_2)(depe)_2]Cl$, M=Mo and W [11].

X-ray structures of Mo and W dinitrogen, hydrazido(2-) and hydrazidium complexes exist. In order to treat the Mo/W-N_2 and -N_2H_x complexes (x = 1,2,3) by DFT, they were simplified by approximating the phosphine ligands by PH_3 groups. Vibrational spectroscopic data were then evaluated by a Quantum Chemistry Assisted Normal Coordinate Analysis, which is based on the calculation of the f matrix by DFT and subsequent fitting of important force constants to match experimentally observed frequencies. Time-dependent (TD) DFT has been employed to calculate electronic transitions, which then were compared to experimental UV/Vis absorption spectra [10]. As a result, a close check of the quality of the quantum-chemical calculations has been obtained. This allowed us to employ these calculations as well to understand the chemical reactivity of the intermediates of N_2 fixation (see Section 15.2.4).

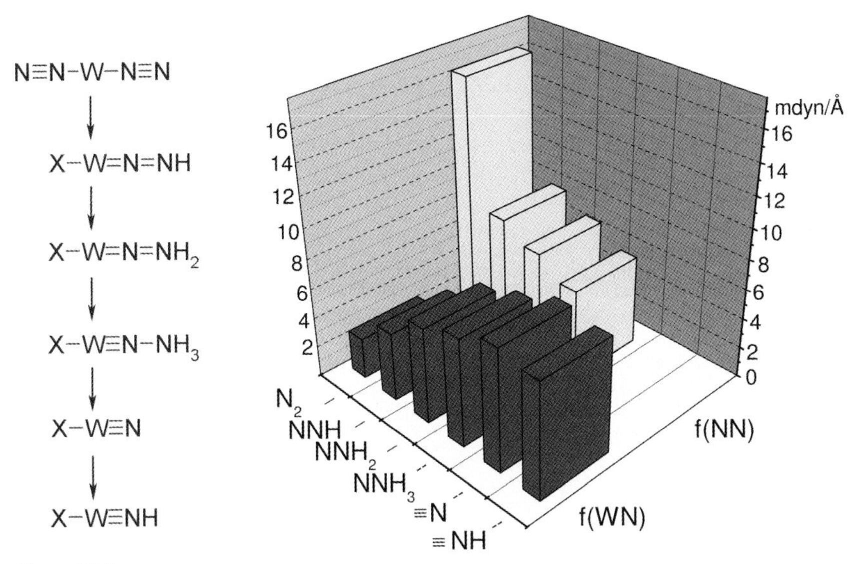

Figure 15.4

Figure 15.4 shows the N–N and metal–N force constants resulting from normal coordinate analysis of the tungsten N_2-, NNH-, NNH_2- and NNH_3-complexes **1**, **2**, **3** and **4b** [9, 11]. Upon protonation of $[W(N_2)_2(dppe)_2]$ to the NNH-complex [WF(NNH)(dppe)$_2$], the N–N force constant decreases from about 16 mdyn/Å to a value of 8.27 mdyn/Å and the M–N force constant increases from a value of about 2.5 to 4.5 mdyn/Å. This trend continues in the next protonation steps; i.e., for $[WF(NNH_2)(dppe)_2]^+$ the N–N force constant is further reduced to a value of 7.2 mdyn/Å and the metal–N force constant further increases to 6.3 mdyn/Å. In the NNH_3 complexes $[MF(NNH_3)(depe)_2]^{2+}$, M=Mo or W, the N–N force constant is found to be 6.03 mdyn/Å, close to the value for an N–N single bond, while the metal–N force constants (8.01 and 7.31 mdyn/Å, respectively) reach values typical for a metal–N triple bond (see below) [10]. In contrast to the N–N and metal-N force constants, the MNN bending force constants remain approximately constant upon successive protonation (around 0.7 mdyn/Å), with the exception of the in-plane bending force constants in the NNH_2 systems, which exhibit values of about 0.4 mdyn/Å.

The experimentally determined evolution of N–N force constants upon stepwise protonation indicates a successive decrease in N–N bond order, thus initiating bond cleavage, whereas the evolution of metal–N force constants reflects an increase of metal-ligand covalency, indicative of a successive strengthening of the metal–N bond. Besides providing an energetic driving force for the reduction of the N–N triple bond, this also acts to prevent loss of partly reduced NNH_x substrate, x = 1–3, in the course of the transformation of N_2 to ammonia.

In order to obtain an impression of the activation of N_2 in mono-dinitrogen complexes with alternative trans ligands, a series of tungsten nitrilo-dinitrogen complexes has been prepared. These complexes are of significant interest to synthetic nitrogen fixation as they can be protonated to the corresponding N_2H_2 complexes with retention of the trans ligand. Importantly, the N_2 ligand was found to be activated to a higher degree in the trans nitrilo as compared to the corresponding bis (dinitrogen) systems. On the other hand, the N_2H_2 ligand is less activated toward further protonation in the *trans*-nitrilo than in the analogous *trans*-fluoro complexes. Moreover, bonding of the nitrile group becomes labile at the N_2H_2 stage of N_2 reduction. The implications of these results with respect to synthetic nitrogen fixation were discussed [12].

15.2.2
N–N Cleavage

The ultimate stage of N_2 reduction and protonation at d^6 metal centers in the absence of external reductants is represented by Mo(IV)/W(IV) hydrazidium complexes. These systems exhibit a certain instability toward cleavage of the metal-P bonds [10]. However, if the phosphine coligands remain bound to the metal, the NNH_3 complexes are stable toward N-N splitting. This was supported by a DFT simulation of the heterolytic N–N cleavage of $[MoF(NNH_3)(PH_3)_4]^+$ leading to NH_3 and the Mo (VI) nitrido complex $[MoF(N)(PH_3)_4]^{2+}$. The calculation indicates that this process is endothermic ($\Delta H > +40$ kcal/mol, $\Delta G > +30\,\text{kcal mol}^{-1}$) [10]. Breaking the N–N single bond therefore requires the addition of electrons from an external source. As NNH_3 compounds are strongly acidic, this reduction is better performed at the NNH_2 stage. Moreover, mechanistic insight into the N–N cleavage process is favorably obtained using alkylated (NNR_2) derivatives [13, 14]. Thus, protonation of the five-coordinate complex $[Mo(dppe)_2\{NN(C_5H_{10})\}]$ (compound $\underline{B}^{Mo}$) with HBr in THF leads to N-N splitting under formation of HNC_5H_{10} (piperidine) and the imido complex $[Mo(dppe)_2Br(NH)]Br$ [14]. Compound $\underline{B}^{Mo}$ can be prepared by treatment of the Mo(IV) dialkylhydrazido(2-) complex $[MoBr\{NN(C_5H_{10})\}]Br$ (compound $\underline{A}^{Mo}$) with BuLi or electrochemically at –1.61 V vs SCE. Compound $\underline{A}^{Mo}$, in turn, is accessible through alkylation of $[M(N_2)_2(dppe)_2]$ with $Br\text{-}(CH_2)_5\text{-}Br$.

We investigated the spectroscopic properties and electronic structures of the W analogs of compounds $\underline{A}^{Mo}$ and $\underline{B}^{Mo}$, $\underline{A}^{W}$ and $\underline{B}^{W}$, and compared the results with those of the corresponding M(IV) NNH_2 complexes. A single-crystal X-ray structure determination of the five-coordinate W(II) complex $\underline{B}^{W}$ was reported and variable-temperature ^{31}P NMR spectra of this molecule were presented. Furthermore, the vibrational properties of compound $\underline{A}^{W}$ and $\underline{B}^{W}$ were evaluated. Infrared and Raman data were analyzed using the QCA-NCA procedure (see above). Finally, the electronic structures of compounds $\underline{A}^{M}$ and $\underline{B}^{M}$ (M = Mo,W) were evaluated using DFT, with special emphasis on the nature of the frontier orbitals.

Then the N-N cleavage reaction of the dialkylhydrazido complex $[W(dppe)_2(NNC_5H_{10})]$ ($\underline{B}^{W}$) upon treatment with acid, leading to the nitrido complex and piperidine, was investigated experimentally and theoretically [15]. In acetonitrile and propionitrile, $\underline{\mathrm{B}}^{\mathrm{W}}$ reacts orders of magnitude faster at room temperature with $HNEt_3BPh_4$ than its Mo analog, $[Mo(dppe)_2(N\text{-}NC_5H_{10})]$ ($\underline{\mathrm{B}}^{\mathrm{Mo}}$). A stopped-flow experiment was performed for the reaction of $\underline{\mathrm{B}}^{\mathrm{W}}$ with $HNEt_3BPh_4$ in propionitrile at $-70\,°C$. Evaluation of the kinetic data indicated that protonation of $\underline{\mathrm{B}}^{\mathrm{W}}$ is completed within the dead time of the stopped-flow apparatus, leading to the primary protonated intermediate $\underline{\mathrm{B^WH^+}}$ (Scheme 15.1). Propionitrile coordination to this species proceeds at a rate $k_{obs(1)}$ of $1.5 \pm 0.4\,s^{-1}$, generating intermediate $\underline{\mathrm{RCN\text{-}B^WH^+}}$ (R = Et) that subsequently mediates N-N bond splitting in a slower reaction ($k_{obs(2)} = 0.35 \pm 0.08\,s^{-1}$, 6 equivalents of acid). These findings are in contrast to the results obtained on the Mo analog $\underline{\mathrm{B}}^{\mathrm{Mo}}$, where only one phase corresponding to the N-N cleavage step had been observed. Moreover, $k_{obs(1)}$ and $k_{obs(2)}$ were found to be independent of the acid concentration, again in contrast to the Mo system where an acid-dependent rate constant had been observed. This suggests that the protonation equilibrium K_3 shown in Scheme 15.1 is saturated in the W system already at low acid concentrations.

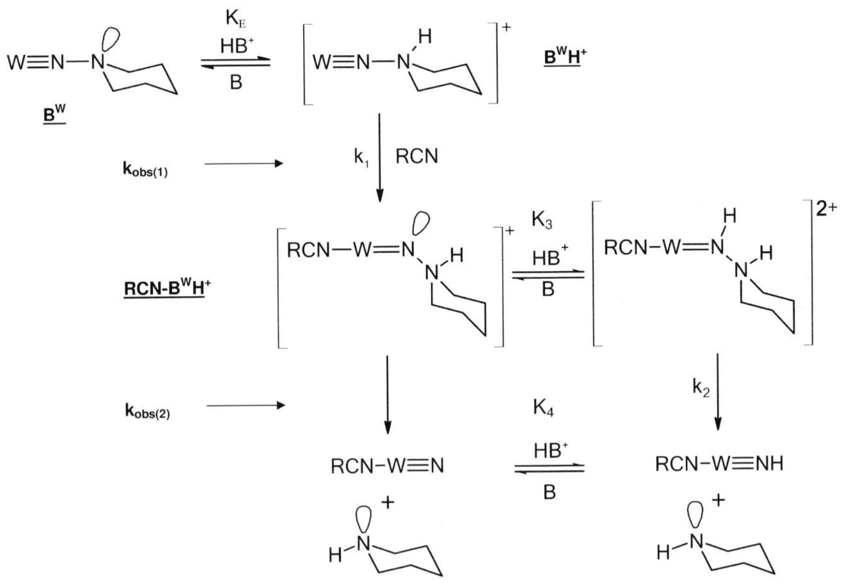

Scheme 15.1

All of these findings indicate that the W complex $\underline{\mathrm{B}}^{\mathrm{W}}$ exhibits a higher activation toward N–N cleavage than its Mo counterpart $\underline{\mathrm{B}}^{\mathrm{Mo}}$. In the meantime we have conducted temperature-dependent stopped-flow measurements on the latter complex and discovered that both stages of the reaction scheme of Scheme 15.1; i.e., solvent coordination and N–N cleavage, are thermally activated. Theoretical studies to interpret this result are in progress [16].

15.2.3
Reactivity of Nitrido and Imido Complexes

After cleavage of the N–N bond the parent dinitrogen complex has been converted to an Mo(IV) or W(IV) nitrido or imido complex, which after four-electron reduction and further protonation can be converted back to the dinitrogen complex, giving a second molecule of ammonia. Problems of this stage of the Chatt cycle are the strongly negative reduction potentials which are needed to regenerate the low-valent intermediate(s) capable of binding N_2.17 In a first step we wanted to get information on the reactivity of Mo and W nitrido and imido complexes toward acids and reductants. In this context we synthesized a number of such systems with various coligands and investigated their electronic structures and spectroscopic as well as electrochemical properties [18]. The starting compounds for these studies were the Mo azido-nitrido complexes $[Mo(N)(N_3)(diphos)_2]$ (diphos = depe (**5a**^{N3}) or dppe (**5b**^{N3})) which are accessible from the corresponding bis(dinitrogen) complexes by reaction with trimethylsilylazide (Scheme 15.2). These nitrido complexes can be protonated to the corresponding imido-azido systems **6a**^{X} and **6b**^{X} by treatment with acids HX. If a strong acid is employed, the trans-azido ligand is exchanged against the conjugated base X of that acid; otherwise protonation occurs with retention of the trans ligand (N_3^-). The former case is realized for protonation with HCl whereas the latter route applies to protonation with HLut BPh_4. The imido-chloro complexes **6a**^{Cl} and **6b**^{Cl} can be deprotonated with, e.g., MeLi, to their nitrido-chloro counterparts. If the chloro-imido Mo-dppe complex **6b**^{Cl} is deprotonated in acetonitrile, the corresponding cationic nitrido-acetonitrile complex **6b**^{MeCN} is formed.

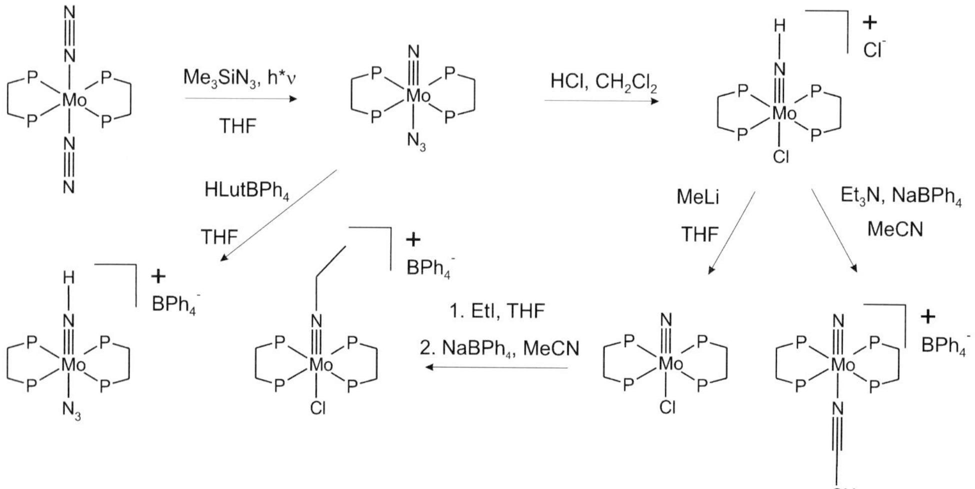

Scheme 15.2

The vibrational and optical spectroscopic properties of the various nitride and imido complexes were investigated. In the nitrido complexes with *trans*-azido and -chloro coligands, the metal-N stretch is found at about 980 cm^{-1}, shifting to about 920 cm^{-1} upon protonation. Force constants are: f(MoN) = 7.07 mdyn/Å, f(MoNH) = 6.70 mdyn/Å. The $^1A_1 \rightarrow {}^1E(n \rightarrow \pi^*)$ electronic transition is observed for complex **5a**^{N3} at 398 nm and shows a progression in the metal-N stretch of 810 cm^{-1}. The corresponding $^3E \rightarrow {}^1A(\pi^* \rightarrow n)$ emission band is observed at 542 nm, exhibiting a progression in the metal-N stretch of 980 cm^{-1}. In the imido systems the $n \rightarrow \pi^*$ transition is shifted to lower energy (518 nm) and markedly decreases in intensity.

Upon substitution of the anionic trans ligands by acetonitrile the metal-N(nitrido) stretching frequency increases to 1016 cm^{-1}. The $n \rightarrow \pi^*$ transition now is found at 450 nm, shifting to 525 nm upon protonation. Most importantly, the nitrido-nitrile complex can be reduced at −1.5 V vs Fc^+/Fc, indicating that reduction of this system is drastically facilitated compared to its counterparts with anionic trans ligands. On the other hand, the basicity of the nitrido group is decreased, as is evident from an increased acidity of $[Mo(NH)(NCCH_3)(dppe)_2](BPh_4)_2$ ($pK_a = 5$).

In order to study the following steps of the Chatt cycle, i.e. the conversion of Mo(IV) imido to Mo(II) amido complexes, it is useful to employ alkylated Mo-nitrido complexes, in analogy to the investigation of the two-electron reduction of the Mo (IV)-NNH_2 compounds (see above). We therefore prepared the Mo(IV) chloro-ethylimido complex [MoCl(NEt)$(dppe)_2$]Cl and reacted it with *n*-BuLi in order to prepare the corresponding five-coordinate Mo(II) ethylimido complex. Instead of the desired reaction product, however, the bis(dinitrogen) complex $[Mo(N_2)_2(dppe)_2]$ was obtained (Scheme 15.3). We ascribed this to a double deprotonation of the ethylimido group at the β-carbon atom, leading via the neutral Mo(II) azavinylidene complex

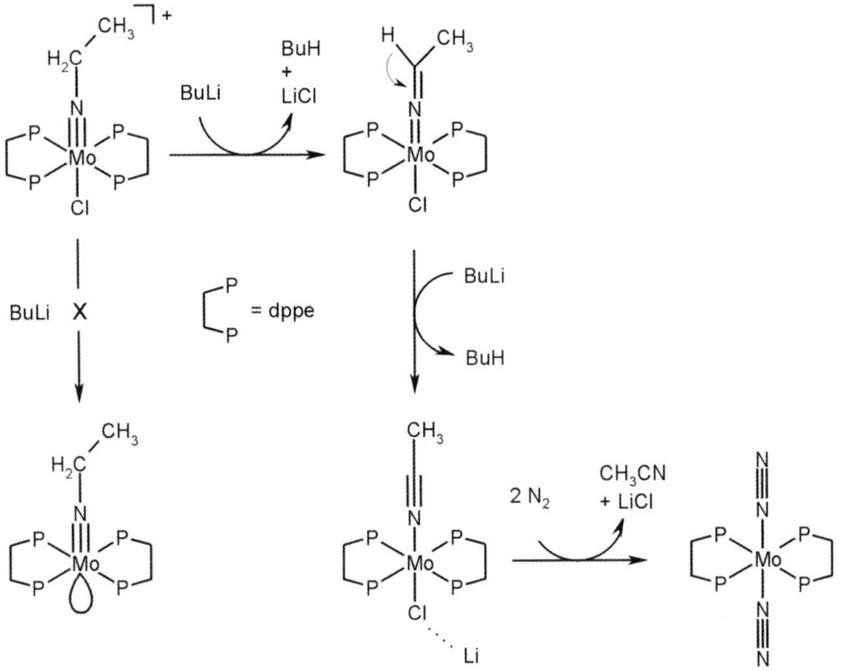

Scheme 15.3

$[MoCl(N=CH_2)(dppe)_2]$ to the anionic Mo(0) chloro-acetonitrile complex $[MoCl(CH_3CN)(dppe)_2]^-$ which under an N_2 atmosphere exchanges its two ligands against dinitrogen [19]. In order to avoid deprotonation we substituted the ethyl by a *tert*-butyl group; the corresponding *tert*-butylimido complex $[Mo(NtBu)(dppe)_2]$ is currently under investigation [20].

15.2.4 DFT Calculations of the Chatt Cycle

Having studied the elementary reactions of the Chatt cycle as described in the previous sections, it appeared of interest to also obtain a global picture of the energetics of this cycle in the same fashion as for the Schrock cycle. To this end, the intermediates of the Chatt cycle, the employed acid as well as the reductant were treated with DFT, and the free reaction enthalpies for all protonation and reduction reactions were determined. As for the Schrock cycle, decamethylchromocene was employed as reductant. For the protonation reactions two acids were considered, HBF_4/diethylether and lutidinium. For all protonation and reduction steps the corresponding free reaction enthalpy changes were calculated [21].

The derived energy profile and corresponding reaction mechanism bear strong similarities to the Schrock cycle [6]. In particular, the most endergonic reaction is the first protonation of the N_2 complex and the most exergonic reaction is the cleavage of the N−N bond. If lutidinium is employed as acid and $Cp^*{}_2Cr$ as reductant, the reaction course involves steps that are not thermally allowed. For HBF_4/diethylether as acid and $Cp^*{}_2Cr$ as reductant, however, a catalytic cycle consisting of thermally allowed reactions is predicted. This cycle involves an Mo(I) fluoro complex as dinitrogen intermediate. Reduction of Mo(I) fluoro to the $Mo(N_2)_2$ complex is not possible in this system. Although the Mo(I)-N_2 complex is able to activate dinitrogen toward protonation just like the Mo(0) bis(dinitrogen) complex, it is at significantly higher energy ($\sim$ 20 kcal mol^{-1}) than the latter intermediate. Bonding of N_2 to an Mo (I) fluoro complex thus is thermodynamically not particularly favorable, and the resulting intermediate is unstable toward disproportionation [22]. We theoretically determined a free-energy change of 30 kcal mol^{-1} for the reaction of two $[Mo^{(I)}F(N_2)(dpe)_2]$ molecules to form $[Mo^{(0)}(N_2)_2(dpe)_2]$ and $[Mo^{(II)}F_2(dpe)_2]$. The resulting Mo(0) bis(dinitrogen) complex gives another 2 molecules of NH_3 along the Chatt cycle, but the Mo(II) difluoro complex cannot be converted back to a Mo(I) or Mo(0) dinitrogen complex by the applied reductant (see above). Disproportionation at the Mo(I) level thus corresponds to 50% loss of catalyst per cycle.

If a truly catalytic action of the Chatt system is intended, strategies have be developed to avoid disproportionation of the Mo(I) dinitrogen complex to a Mo(II) complex carrying two anionic ligands. Important points in this respect are (1) avoiding the presence of strongly Lewis-basic species in solution (such as F^-) and (2) employing multidentate ligands that also occupy the trans-position of coordinated N_2, in contrast to the conventional Mo and W diphosphine systems. The synthesis of such systems is described in Sections 15.3.2 and 15.3.3. In order to obtain an impression of the corresponding energetics, the analogous calculations as described above were performed for an MoN_2 complex with a pentaphosphine coordination (Scheme 15.4);

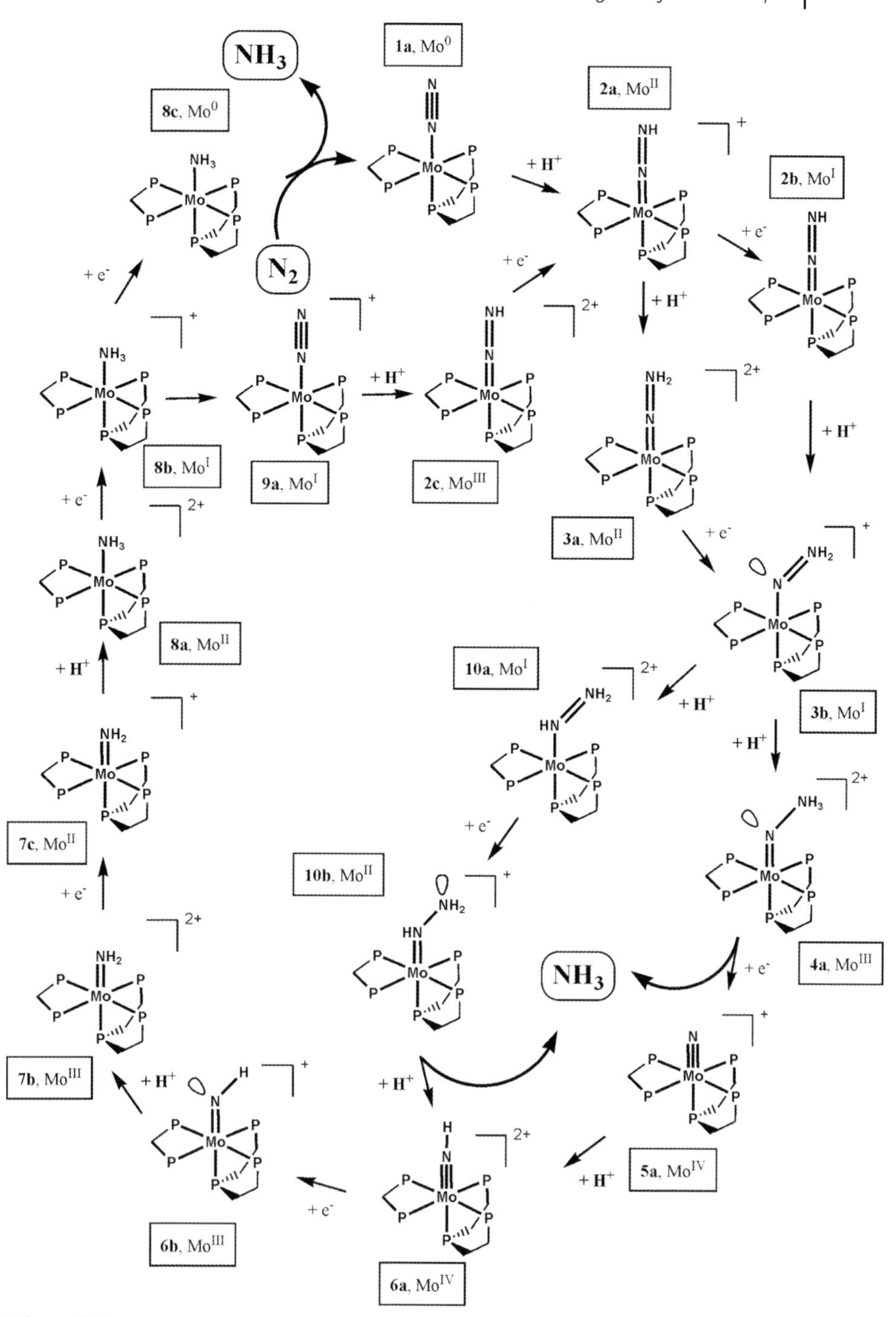
NH_3
1a, Mo^{0}
2a, Mo^{II}
2b, Mo^{I}
8c, Mo^{0}
N_2
+ H^+
+ e^-
8b, Mo^{I}
9a, Mo^{I}
2c, Mo^{III}
3a, Mo^{II}
8a, Mo^{II}
10a, Mo^{I}
3b, Mo^{I}
7c, Mo^{II}
10b, Mo^{II}
4a, Mo^{III}
NH_3
7b, Mo^{III}
5a, Mo^{IV}
6b, Mo^{III}
6a, Mo^{IV}

Scheme 15.4

i.e. the reduction steps were considered to be mediated by decamethylchromocene and the protonation reactions were assumed to be mediated by HBF_4 in ether, whereby a protonated ether effectively acts as acid (Figure 15.5).

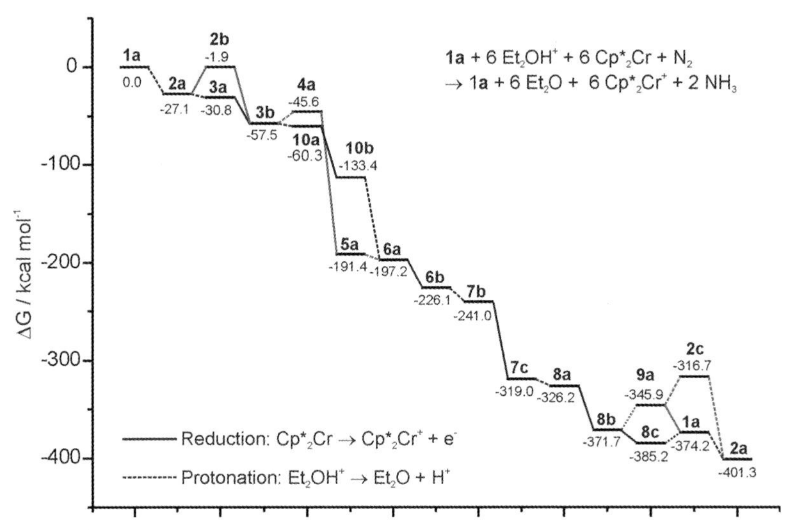

Figure 15.5

The reactive cycle starts with a double protonation of the dinitrogen complex **1a** via the molybdenum diazenido complex **2a**, leading to the hydrazido complex **3a**. One-electron reduction of this well-characterized intermediate leads to the Mo(I)-NNH_2 complex **3b**. The reaction of **1a** to **3b** via the reduced NNH species **2b** is thermodynamically unfavorable. In analogy to the fluoro system the reduced NNH_2 complex **3b** exhibits a bent Mo-NNH_2 unit and can be protonated at the terminal (β) or the coordinating (α) nitrogen. As a consequence there are two possible reaction pathways for the N-N splitting. The first alternative starts with the formation of the hydrazidium complex **4a** by protonation of **3b** at N_β. The reduction of this highly activated intermediate **4a** induces a spontaneous N-N cleavage, leading to the Mo(IV)-nitrido complex **5a** and the first equivalent of NH_3; the nitrido complex is subsequently protonated to the imido species **6a** in a weakly exergonic reaction. In the second alternative of the N-N splitting reaction the Mo(IV) imido complex **6a** is formed as well. Here the Mo(I)-NNH_2 intermediate **3b** is protonated at N_α, leading to the $NHNH_2$ complex **10a**. The reduction of **10a** to **10b** followed by a protonation at the terminal nitrogen results in a spontaneous N-N splitting with formation of NH_3 and the imido complex **6a**. Further reduction and protonation of the imido group leads to the formation of an Mo(I)-ammine complex (**8b**) in a mechanism similar to the fluoro system.

The Mo(I)-ammine complex **8b** is first reduced to the corresponding Mo(0) complex **8c**, at which stage the ammine ligand is exchanged against N_2, leading to the dinitrogen complex **1a**. The alternative scenario, exchange of NH_3 against N_2 at the Mo(I) level, is energetically unfavorable; moreover, the Mo(I)-dinitrogen complex does not allow protonation. This is in contrast to the Mo fluoro system where the Mo(I) fluoro-ammine complex cannot be reduced to the corresponding Mo(0) complex, and exchange of NH_3 against N_2 occurs at the level of the Mo(I) complex. The resulting Mo(I) fluoro-dinitrogen complex can be protonated with a strong acid, but is unstable with respect to disproportionation. These differences in reactivity reflect the influence of trans ligand (anionic fluoro vs neutral phosphine).

In conclusion, the calculations indicate that a catalytic cycle of ammonia synthesis on the basis of an Mo pentaphosphine complex should be feasible. Practically this concept is limited by the fact that the trans ligand of N_2 becomes labile at higher oxidation states of the metal and sooner or later will be replaced by another Lewis-basic ligand or a solvent molecule. Therefore the necessity persists to fix the trans ligand to the metal center in such a way that it cannot dissociate at any stage of the catalytic cycle. Concepts to achieve such an architecture are presented in the following sections.

15.3 New Phosphine and Mixed P/N Ligands for Synthetic Nitrogen Fixation

On the basis of detailed mechanistic insight into the elementary reactions of the Chatt cycle, it is possible to design alternative phosphine ligands or corresponding ligand systems which allow us to improve the performance of Mo and W complexes toward a catalytic reaction mode. To achieve this goal, three types of ligands were considered: tetraphos ligands, pentaphosphine ligand systems, and mixed phosphorus/nitrogen ligands. The results of our efforts toward synthesizing and coordinating these ligands to transition-metal centers are presented in the following.

15.3.1 Tetraphos Ligands

An obvious strategy to increase the thermal stability of Mo and W dinitrogen complexes and their protonated derivatives is to employ tetraphosphine instead of the diphosphine ligands which conventionally have been employed in this type of system. Against this background the tetradentate phosphine ligand prP_4 (1,1,4,8,11,11-hexaphenyl-1,4,8,11-tetraphosphaundecane) was prepared, and its coordination properties in mononuclear complexes were explored [23]. The free ligand was obtained as a mixture of meso and rac diastereomers from the reaction of dilithium-1,3-bis(phenylphosphido)propane tetrakis(tetrahydrofuran) (Lippp · 4 THF) with two equivalents of 1-chloro-2-diphenyl-phosphinoethane. For mononuclear hexacoordinated complexes of the type $[ML_2(prP_4)]$ five isomers are possible (Scheme 15.5); that is, the ligand can bind in its meso and rac forms in cis and trans configurations whereby two cis configurations (α and β) are possible. Before coordinating the prP_4 ligand to Mo(0) and W(0) dinitrogen complexes, corresponding iron(II) complexes with thiocyanato coligands were prepared. Reaction of the prP_4 with $FeCl_2$ and KSCN led to *cis*- and *trans*-$[Fe(NCS)_2(prP_4)]$ complexes of the meso and the rac ligand;

the composition of this mixture was fully analyzed. No hint of the formation of iron isothiocyanato complexes with cis-β geometry was found. The *cis*-α and *trans*-$[Fe(NCS)_2(rac\text{-}prP_4)]$ complexes were characterized by X-ray structure determination (Figure 15.6) and by infrared and Raman spectroscopy.

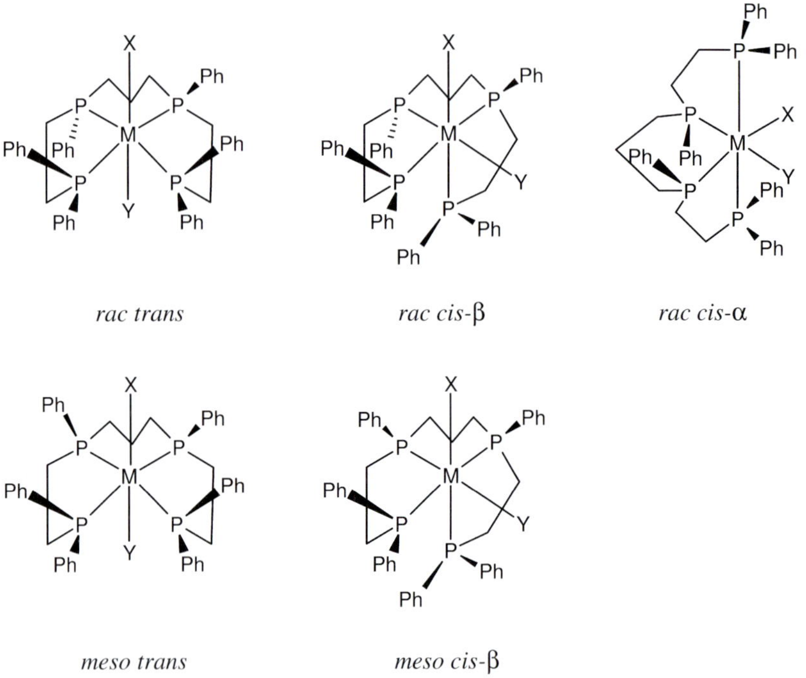

rac trans
rac cis-β
rac cis-α
meso trans
meso cis-β

Scheme 15.5

P1
P2
P3
P4
Fe1
N1
N2
C8
C9
S1
S2
C70
C80

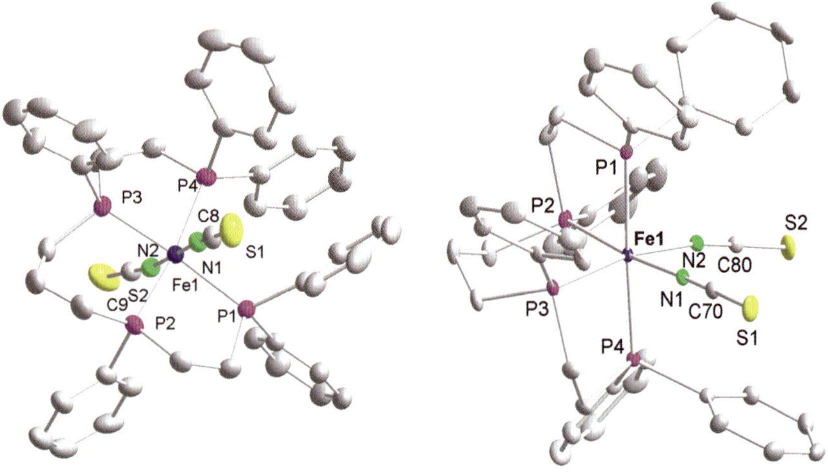

Figure 15.6

^{31}P NMR spectroscopy has been employed to obtain insight into the structure of these complexes in solution. The ^{31}P NMR spectrum of the free tetraphos ligand exhibits an AA'XX' coupling scheme with three groups of signals centered at −20.6, −20.7 ppm and −12.4 ppm which belong to the rac and meso isomers. ^{31}P NMR spectra of the *cis-* and *trans-* iron isothiocyanato complexes of *rac* prP_4 exhibit AA'XX' patterns as well which were reproduced by simulation. The *trans*-$[Fe(NCS)_2($*meso*-$prP_4)]$ complex was identified by its ^{31}P-NMR spectrum, which is qualitatively similar to that of its rac counterpart. Based on NMR spectroscopy it was found that the *cis*-α complex of the rac ligand converts to the trans rac complex with a half-life of approximate two days.

15.3.2
Pentaphosphine Complexes

In an effort to both enhance the stability of Mo diphos systems with respect to the cleavage of metal-phosphine bonds at higher metal oxidation states and to suppress the trans ligand exchange reactions occurring in the classic bis(dinitrogen) complexes, we prepared and characterized a number of Mo dinitrogen complexes containing a combination of a diphosphine and a triphosphine ligand [24]. As diphosphine ligands dppm, dppe, 1,2-dppp and depe were employed, and as the triphosphine component the ligand dpepp ($PhP(CH_2CH_2PPh_2)_2$) was chosen (Scheme 15.6) [25]. In analogy to the Mo triamidoamine complexes, these systems exhibit only one binding site for N_2 and can be protonated to give NNH_2 complexes. Moreover, these systems allow one to assess the influence of strongly σ-donating trans ligands on the bonding and reduction of N_2.

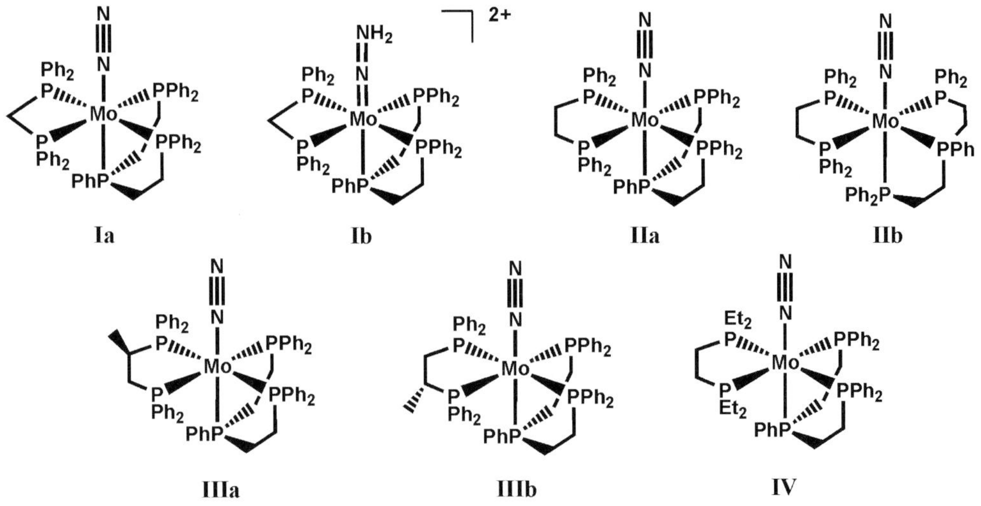

Scheme 15.6

First, the complex $[Mo(N_2)(dpepp)(dppm)]$ (**Ia**; dpepp = $PhP(CH_2CH_2PPh_2)_2$, dppm = $Ph_2PCH_2PPh_2$) and its hydrazido(2-) derivative **Ib** were prepared and characterized by ^{15}N- and ^{31}P-NMR spectroscopy. The spectroscopic results were interpreted with the help of DFT calculations. Moreover, infrared and Raman spectra were recorded and evaluated with the help of the quantum chemistry-assisted normal coordinate analysis (QCA_NCA) procedure (f(NN) = 16.3 mdyn/Å and f(MoN) = 2.8 mdyn/Å) (see above) [26]. These studies showed that the activation of the N_2 ligand of the parent N_2 complex is *lower* than in the *trans*-acetonitrile complex $[Mo(N_2)(NCCH_3)(dppe)_2]$, but *higher* than in the bis(dinitrogen) complex $[Mo(N_2)_2(dppe)_2]$. This was interpreted in terms of the relative σ/π donor and π-acceptor capabilities of the respective trans ligands. As a rule the mono-dinitrogen complexes exhibit a stronger activated ligand than the bis(dinitrogen) complex because of the lack of a second strong π-acceptor. Comparison of the two mono-N_2 complexes shows that the *trans*-nitrile system exhibits a stronger activation than the pentaphosphine system. This effect is due to the different σ-donor capabilities of nitrile and phosphine: in presence of a weak σ-donor in trans position like acetonitrile the N_2 ligand acts as a stronger σ-donor, which lowers the energy of its π^* orbitals. This enhances metal-ligand backbonding and overcompensates the effect of charge loss due to σ-donation, resulting in stronger activation. In the presence of a strong σ-donor in trans position like phosphine, this synergistic effect is absent; i.e., N_2 becomes a weaker σ-donor, its π^* orbitals are less lowered in energy, and as a consequence a weaker activation of N_2 is observed.

Another interesting result was the strong low-field shift of the ^{15}N signals in the complex $[Mo(N_2)(dpepp)(dppm)]$. The terminal N-atom N_β shows the strongest low-field shift (−16.7 ppm) for end-on-coordinated N_2 complexes. With the help of DFT GIAO calculations it could be shown that the origin of this unusual shift is the paramagnetic term. The magnitude of the paramagnetic shift is determined by the nitrogen contribution in the backbonding metal t_{2g} orbitals. This effect is stronger for N_β than for N_α as a larger negative charge resulting from backbonding is located on the former atom. In the case of the pentaphosphine complex, besides the comparatively strong backbonding interaction another effect contributes to the strong low-field shift of the ^{15}N signals, in particular of N_β: the d_{xy} orbital, which in the bis (dinitrogen) and in the *trans*-nitrile dinitrogen complex is nonbonding, is included in the backbonding interaction with the N_2 ligand in the P_5 system. For this reason *all* $t_{2g} \rightarrow e_g$ excitations contribute to the paramagnetic shift. This effect is basically a consequence of the symmetry reduction from (approximately) tetragonal in the bis (diphos) systems to mirror symmetry in the P_5 complex.

The results of the NMR and vibrational spectroscopic investigation of the hydrazido complex $[Mo(NNH_2)(dpepp)(dppm)](OTf)_2$ confirm the structure of a pentaphosphine complex with a linearly coordinated NNH_2 unit. On the basis of vibrational spectra and DFT calculations a normal coordinate analysis has been performed. The resulting force constants (f(NN) = 7.65 mdyn/Å, f(MoN) = 5.94 mdyn/Å) demonstrate the weakening of the N–N bond and the enhancement of the Mo–N bond strength as a consequence of the protonation (see Section 15.1.1). The force constants of the hydrazido ligand are comparable with those of other hydrazido complexes such as, for example, $[MoCl(NNH_2)(depe)_2]Cl$ (f(NN) = 7.16 mdyn/Å, f(MoN) = 5.52

mdyn/Å) [11]. However, no force constants exist for analogous Mo-dppe complexes with anionic or nitrile coligands, which precludes a direct comparison of the spectroscopic data of the NNH_2 complexes. With the help of DFT calculations on the model complexes $[Mo(NNH_2)(PH_3)_5]^{2+}$, $[MoF(NNH_2)(PH_3)_4]^+$ and $[Mo(NNH_2)(NCMe)(PH_3)_4]^+$ it could be shown, however, that the NPA charge of the NNH_2 ligand in the *trans*-fluoro complex (−0.03) is lower than that in the P_5 complex (+0.20). This reflects a lower activation of the NNH_2 ligand in the latter system toward further protonation. The reason for this finding is a lower charge transfer from the metal because of the exchange of a negative (F^-) against a neutral (phosphine) ligand.

In a second study we performed similar investigations on molybdenum dinitrogen complexes containing a combination of dpepp with diphos ligands exhibiting C_2 bridges, namely dppe, R-(+)-1,2-dppp ($Ph_2PC^*H(CH_3)CH_2PPh_2$) and depe.[27] Specifically, the three compounds $[Mo(N_2)(dpepp)(dppe)]$ (**II**), $[Mo(N_2)(dpepp)(1,2\text{-}dppp)]$ (**III**) and $[Mo(N_2)(dpepp)(depe)]$ (**IV**) were prepared and investigated by vibrational and ^{31}P-NMR spectroscopy. From compound **II** two isomers were identified, $[Mo(N_2)(dpepp)(dppe)]$ (**IIa**) and *iso*-$[Mo(N_2)\text{-}(dpepp)(dppe)]$ (**IIb**). The same applies to compound **III** containing the optically active ligand 1,2-dppp. Here, evidence for the existence of two diastereomers (**IIIa** and **IIIb**) was obtained. The methyl group of the 1,2-dppp ligand either points in the direction of the N_2 ligand (**IIIa**) or in the opposite direction (**IIIb**). Complex **IV**, in contrast, was found to exist in only one form. These results were found to be compatible with the corresponding vibrational (infrared and Raman) data, showing an unsplit N-N stretching vibration for **IV**, strongly split N−N stretching vibrations for compound **II** and weakly split N-N vibrations for compound **III**.

The N−N stretching frequencies (1950 cm^{-1} for **IV**, 1955 cm^{-1} for **IIa**, 2010 cm^{-1} for **IIb** and 1955 cm^{-1} for **III**) reflect the different degrees of activation imparted to the dinitrogen ligand by the phosphine coligands. It is well known that diphosphine ligands with terminal alkyl substituents are more strongly activating than those with terminal phenyl substituents [11]. As activation goes along with a lowering of the N-N stretching frequency, the following sequence of activation is obtained: $[Mo(N_2)(dpepp)(dppe)]$ (**IIa**) $\approx$ $[Mo(N_2)(dpepp)(1,2\text{-}dppp)]$ (**IIIa,b**) $<$ $[Mo(N_2)(dpepp)(depe)]$ (**IV**). In all of these systems, a dialkylphenylphosphine group exists in trans position to the N_2 ligand. In the iso form of **II** (**IIb**), this is changed to a configuration where a diphenylalkylphosphine ligand is located in trans position to N_2. This complex exhibits an N-N stretching frequency of 2010 cm^{-1}, markedly higher than in all other complexes. The reason for this observation is the fact that the *trans*-diphenylphosphine group is a stronger π-acceptor than a *trans*-monophenylphosphine group.

The ^{31}P-NMR spectra of compounds **II–IV** were fully analyzed, giving a complete set of chemical shifts and coupling constants. The ^{31}P-NMR trans coupling constants were found to be in the region of ~100 Hz. For cis coupling, two cases have to be distinguished. If the interaction occurs simultaneously over the metal center and the ethylene bridge, the coupling constant is very small (sometimes in the order of the linewidth), indicating that the corresponding coupling pathways correspond to coupling constants with almost identical absolute values but opposite signs. This is in agreement with literature data indicating that the ^2J coupling constant via a metal has a negative sign and the ^3J coupling constant via an ethylene bridge a positive sign; the trans coupling constant via the metal, in contrast, is large and positive [28]. The ^{31}P spectra also sensitively reflect the symmetry of the complexes. Quasi-tetragonal symmetry is exhibited by the dppe-dpepp complex **IIa** which has one phosphine group in the axial and four diphenylphosphine endgroups in the equatorial position. Here the signals deriving from the equatorial phosphines a,b,c,d appear almost unsplit. A similar geometry is present in the depe-dpepp complex **IV**. However, in this case the equatorial diethylphosphine and diphenylphosphine groups P_a/P_b and P_c/P_d, respectively, appear at different chemical shifts (48.80 and 62.12 ppm, respectively). Diethylphosphine is a stronger σ-donor than diphenylphosphine and the latter is a stronger π-acceptor than diethylphosphine; P_a and P_b are therefore more shielded than P_c and P_d. In complex **III**, finally, the symmetry of complexes **IIa** and **IV** is broken by an additional methyl group attached to the ethylene bridge of the equatorial diphos ligand. This leads to different environments for the phosphorus donors of the 1,2-dppp ligand, P_a and P_b. Moreover, the methyl group can point 'upwards' (i.e., in the direction of the N_2 ligand; isomer **IIIa**) or 'downwards' (i.e., in the opposite direction; isomer **IIIb**). In the downwards position the methyl group comes close to the phenyl ring of the axial phosphine, whereas in the upwards position no such interaction exists (there is only the N_2 ligand). We thus assume that the equatorial coordination of isomer **IIIb** is more distorted than in the case of **IIIa** and, correspondingly, a larger chemical shift difference between P_a and P_b results in the former than in the latter isomer.

For complex **II** an iso form (**IIb**) has been identified in which the central phosphorus atom of the dpepp ligand is in the equatorial position; i.e. cis to the dinitrogen group. From the relative intensities in the ^{31}P NMR spectrum, **IIa** and **IIb** occur in about equal amounts. Astonishingly, however, no iso forms could be evidenced for compounds **III** and **IV**. For the depe-dpepp complex **IV** this may have *electronic* reasons. That is to say, a disposition where the diethylphosphino units are in trans position to diphenylphosphino moieties may be particularly favourable, as strongly electron-donating groups are arranged in trans position to strongly electron-accepting groups (push-pull effect). For the 1,2-dppp complex **III**, on the other hand, *steric* reasons may account for the lack of the iso form. Specifically, for **IIIb** (with the methyl group pointing downwards) an iso form may be unfavourable, as in the trans position of the dinitrogen ligand there would be two phenyl groups instead of one, coming in close contact with the methyl group (see above). For **IIIa**, on the other hand, this argument would not apply, and thus it remains unclear why in this case no iso form has been observed.

To conclude, the vibrational and ^{31}P-NMR spectroscopic properties of four complexes **I–IV** with pentaphosphine coordination have been determined, and correlations between their geometries and their spectroscopic properties have been derived. It has been shown that the activation of N_2 in these complexes is higher than in corresponding bis(diphosphine) complexes with two dinitrogen ligands, but lower than in *trans*-acetonitrile dinitrogen complexes with two diphosphine coligands [24]. The latter systems, on the other hand, have the disadvantage that the nitrile coligands are labile. As dinitrogen complexes with a pentaphosphine ligation also can be protonated at the N_2 ligand, these systems appear to be superior with respect to a compromise between sufficient activation of the dinitrogen ligand on the one hand and thermodynamic/kinetic stability of the ancillary ligand sphere on the other.

15.3.3 Mixed P/N Ligands

In an effort to prepare Mo/W complexes with tetraphosphine ligands and donors other than phosphorus or acetonitrile in trans position to the N_2 ligand we also decided to synthesize a series of mixed phosphine/amine ligands which we wanted to coordinate to low-valent Mo centers in order to bind, activate and reduce dinitrogen. In the present case we additionally wanted to combine the advantage of phosphine ligands – stability vs acids – with the presence of Brønsted-basic groups which may assist the protonation of the N_2 ligand. The resulting target structure is shown in Scheme 15.7. An important side-effect of the presence of basic groups in the ligand framework would also be the fact that the 'secondary' (= ancillary ligand) protonation should shift the reduction potentials of these systems toward less negative values and thus make the NNH_x and NH_x intermediates of the Chatt cycle easier to reduce [18].

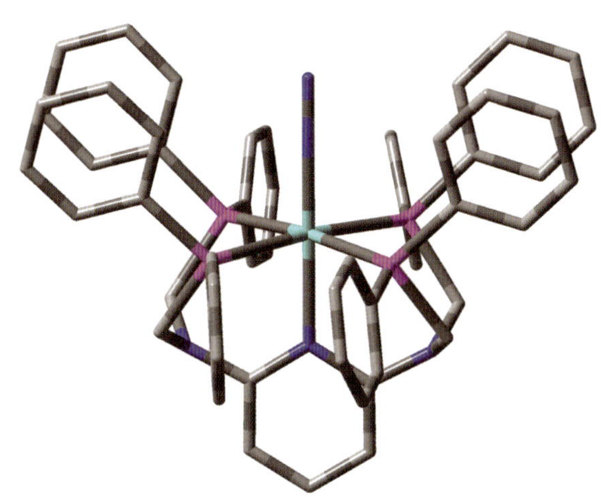

Scheme 15.7

Toward this goal a series of new mixed P/N ligands were prepared by the phosphorus-analogous Mannich reaction (Scheme 15.8). By means of this transformation primary or secondary amines can be substituted by methylenephosphine residues -CH_2PRR' (R, R' = alkyl, aryl) through reaction with a secondary phosphine and formaldehyde [29]. Based on the heterocyclic amines 2-aminopyridine, 2,6-diaminopyridine and 2-aminothiazole the P/N ligands N,N,N',N'-tetrakis-(diphenylphosphinomethyl)-2,6-diaminopyridine (pyN_2P_4) N,N-bis(diphenylphosphino-methyl)-2-aminopyridine ($pyNP_2$), N,N'-bis(diphenylphosphinomethyl)-2,6-diamino-pyridine (PpyP) and N-diphenylphosphinomethyl-2-aminothiazole (thiazNP) were prepared (Scheme 15.9) [30]. In contrast to aliphatic and aromatic amines, aminopyridines react sluggishly. Moreover, while aniline and its derivatives give bidentate PNP-ligands in high yields [31], aminopyridines prefer a monosubstitution of the amine, leading to pyNHCH$_2$PR$_2$ derivatives (Scheme 15.10) [32]. This may be due to the deactivating effect of the pyridine ring, considerably reducing the nucleophilicity of the amino group. For this reason the reaction under aprotic conditions leads to the monosubstituted amines while the disubstituted amines could be isolated under protic/acidic conditions.

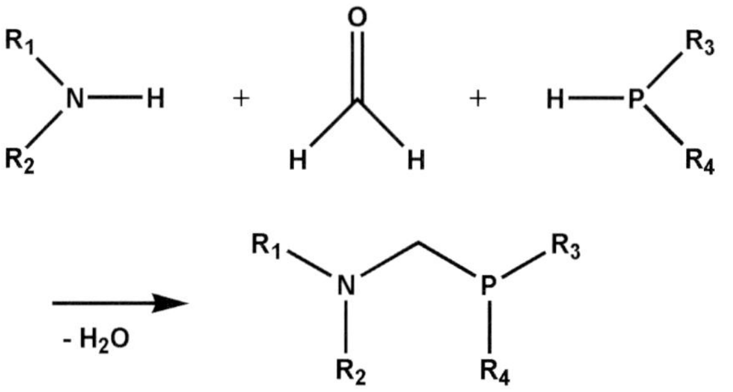

$$R_1R_2N{-}H + H_2C{=}O + H{-}PR_3R_4 \xrightarrow{-H_2O} R_1R_2N{-}CH_2{-}PR_3R_4$$

Scheme 15.8

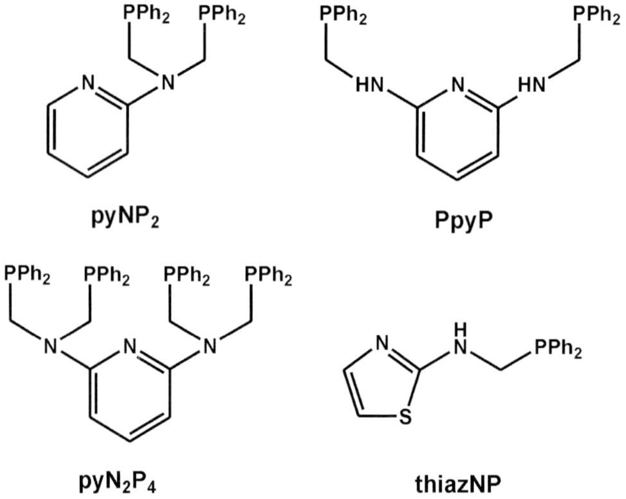

pyNP$_2$ PpyP pyN$_2$P$_4$ thiazNP

Scheme 15.9

$$\text{Py—NH}_2 \xrightarrow{\text{Ph}_2\text{PCH}_2\text{OH}} \text{Py—NHCH}_2\text{PPh}_2$$

$$\text{Py—NH}_2 \xrightarrow[\text{Et}_3\text{NHCl}]{\text{Ph}_2\text{PCH}_2\text{OH}} \text{Py—N(CH}_2\text{PPh}_2)_2$$

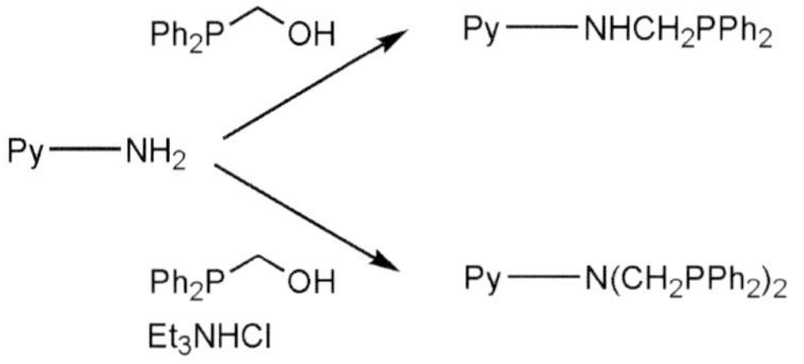

Scheme 15.10

The pyNP$_2$ ligand was used for the synthesis of a molybdenum(0) dinitrogen complex (Scheme 15.11) [30]. First the ligand was coordinated to a Mo(III) halide precursor to form [MoX$_3$(pyNP$_2$)(thf)] (X = Cl$^-$, Br$^-$). Reduction of the Mo(III) complex under N$_2$ with an additional diphosphine like 1,2-bis(diphenylphosphino)ethane (dppe) leads to the Mo(0) bis(dinitrogen) complex [Mo(N$_2$)$_2$(dppe)(pyNP$_2$)]. The formation of a mono-N$_2$ complex with pyNP$_2$ acting as a tridentante ligand was not observed. Obviously the *fac*-coordinating pyNP$_2$ cannot bind completely to the Mo(III) precursor, which has *mer*-geometry [33]. After the subsequent reduction step the free pyridine group then competes with a stronger backbonding ligand like N$_2$ for metal coordination. Formation of a mono-N$_2$ complex with pyNP$_2$ acting as a tridentate ligand thus probably requires a complete coordination of this ligand already at the level of the Mo(III)-precursor. Protonation of the Mo(0) pyNP$_2$ dinitrogen complex with HBF$_4$ and triflic acid leads to secondary protonated hydrazido complexes [MoF(NNH$_2$)(dppe)(pyHNP$_2$)](BF$_4$)$_2$ and [Mo(OTf)(NNH$_2$)

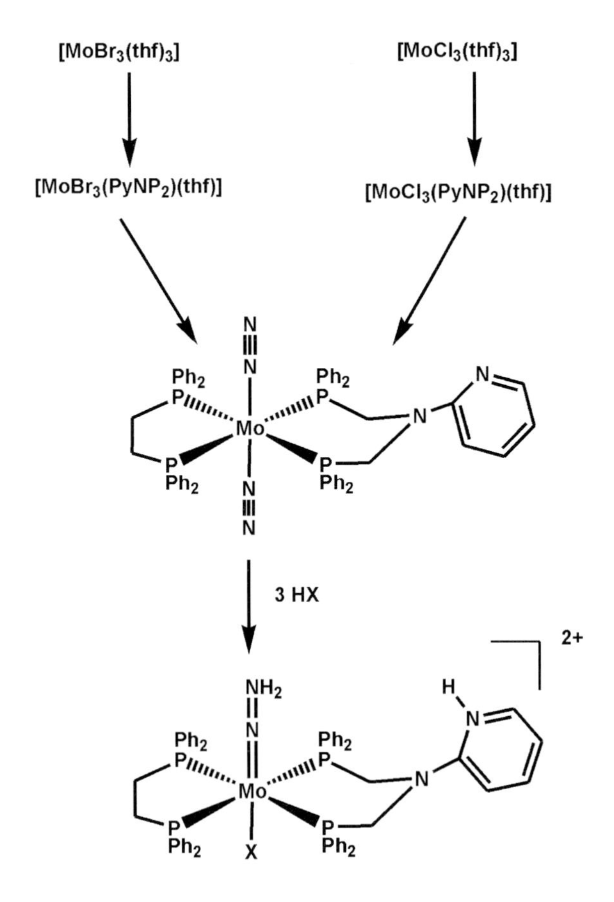

Scheme 15.11

(dppe)(pyHNP$_2$)](OTf)$_2$, respectively (Scheme 15.11). Besides the double protonation of the N_2 ligand the complex is also protonated at the pyridine nitrogen.

Although a dinitrogen complex with a protonatable ligand has been obtained, the final goal of these studies remains the synthesis of a dinitrogen complex with the pyN$_2$P$_4$ ligand (Scheme 15.5). So far all known methods for the synthesis of Mo (0)/W(0) dinitrogen complexes have failed in this regard. Therefore we are investigating the reactivities of low-valent metal compounds with respect to exchange reactions with phosphines. Moreover, we want to further evaluate the concept of protonatable ligands by careful stepwise protonation and deprotonation of the pyNP$_2$ dinitrogen and corresponding hydrazido complexes, respectively.

15.4 Summary and Conclusions

In the preceding sections our experimental and theoretical studies on the mechanism and various modifications of the Chatt cycle have been presented. Our investigations have started with the detailed spectroscopic and theoretical characterization of the intermediates of the Chatt cycle. These studies are not yet finished. That is to say we are currently investigating the reduction and protonation of Mo and W nitrido complexes to give NH_3 and the corresponding dinitrogen complexes. From the mechanistic insight obtained through these investigations new ligands can be designed which will allow us to improve the performance of the Chatt cycle with respect to a catalytic reaction mode. In this respect we have considered tetraphosphine ligands, pentaphosphine ligand systems and polydentate mixed P/N ligands. The coordination of these ligands, in particular of the last type, to Mo and W centers in order to prepare the corresponding dinitrogen complexes is, however, far from trivial and requires the development of new synthetic methods. Another possibility for an improved reaction mode of the Chatt cycle is to fix the Mo and W complexes to surfaces. Initial studies toward this goal have been reported elsewhere [34].

Acknowledgments

FT thanks all students and postdoctoral researchers who have been involved in this work for their valuable contributions and Deutsche Forschungsgemeinschaft (DFG) for financial support.

References

1 (a) MacKay, B.A. and Fryzuk, M.D. (2004) *Chemical Reviews*, **104**, 385; (b) Hidai, M. (1999) *Coordination Chemistry Reviews*, **185–186**, 99; (c) Kozak, C.M. and Mountford, P. (2004) *Angewandte Chemie*, **116**, 1206. (2004) *Angewandte Chemie – International Edition*, **43**, 1186; (d) Chatt, J., Dilworth, J.R. and Richards, R.L. (1978) *Chemical Reviews*, **78**, 589; (e) Hidai, M. and Mizobe, Y. (1995) *Chemical Reviews*,

95, 1115; (f) Leigh, G.J. (1992) *Accounts of Chemical Research*, **25**, 177; (g) Henderson, R.A., Leigh, G.J. and Pickett, C. (1983) *Advances in Inorganic Chemistry and Radiochemistry*, **27**, 197.

2 (a) Burgess, B.K. (1990) *Chemical Reviews*, **90**, 1377; (b) Kim, J. and Rees, D.C. (1992) *Nature*, **360**, 553.

3 Thorneley, R.N.F. and Lowe, D.J. (1985) *Molybdenum Enzymes* (ed. T.G. Spiro) John Wiley, New York.

4 (a) Studt, F. and Tuczek, F. (2006) *Journal of Computational Chemistry*, **27**, 1278; (b) Kästner, J. and Blöchl, P.E. (2007) *Journal of the American Chemical Society*, **129**, 2998; (c) Dance, I. (2007) *Journal of the American Chemical Society*, **129**, 1076.

5 (a) Schrock, R.R. (2005) *Accounts of Chemical Research*, **38**, 955; (b) Yandulov, D.V. and Schrock, R.R. (2002) *Journal of the American Chemical Society*, **124**, 6252; (c) Yandulov, D.V., Schrock, R.R., Rheingold, A.L., Ceccarelli, C. and Davis, W.M. (2003) *Inorganic Chemistry*, **42**, 796; (d) Ritleng, V., Yandulov, D.V., Weare, W.W., Schrock, R.R., Hock, A.S. and Davis, W.M. (2004) *Journal of the American Chemical Society*, **126**, 6150.

6 (a) Studt, F. and Tuczek, F. (2005) *Angewandte Chemie*, **117**, 5783. (2005) *Angewandte Chemie – International Edition*, **44**, 5639; (b) Neese, F. (2005) *Angewandte Chemie*, **118**, 202.(2005) *Angewandte Chemie – International Edition*, **45**, 196.

7 Pickett, C.J. and Talarmin, J. (1985) *Nature*, **317**, 652.

8 Pickett, C.J. (1996) *Journal of Biological Inorganic Chemistry: JBIC: a Publication of the Society of Biological Inorganic Chemistry*, 601.

9 (a) Lehnert, N. and Tuczek, F. (1999) *Inorganic Chemistry*, **38**, 1659; (b) Lehnert, N. and Tuczek, F. (1999) *Inorganic Chemistry*, **38**, 1671.

10 Horn, K.H., Lehnert, N. and Tuczek, F. (2003) *Inorganic Chemistry*, **42**, 1076.

11 Tuczek, F. Horn, K.H. and Lehnert, N. (2003) *Coordination Chemistry Reviews*, **245**, 107.

12 Habeck, C.M., Lehnert, N., Näther, C. and Tuczek, F. (2002) *Inorganica Chimica Acta*, **337**, 11.

13 Ishino, H., Tokunage, S., Seino, H., Ishii, Y. and Hidai, M. (1999) *Inorganic Chemistry*, **38**, 2489.

14 Henderson, R.A., Leigh, G.J. and Pickett, C.J. (1989) *Journal of The Chemical Society-Dalton Transactions*, 425.

15 Mersmann, K., Horn, K.H., Böres, N., Lehnert, N., Studt, F., Paulat, F., Peters, G., Ivanovic-Burmazovic, I., van Eldik, R. and Tuczek, F. (2005) *Inorganic Chemistry*, **44**, 3031.

16 Dreher, A. Mersmann, K. Näther, C. Ivanovic-Burmazovic, I., van Eldik, R. and Tuczek, F. *Inorganic Chemistry*, accepted.

17 Alias, Y., Ibrahim, S.K., Queiros, M.A., Fonseca, A., Talarmin, J., Volant, F. and Pickett, C.J. (1997) *Journal of The Chemical Society-Dalton Transactions*, 4807.

18 Mersmann, K., Hauser, A. and Tuczek, F. (2006) *Inorganic Chemistry*, **45**, 5044.

19 Sivasankar, C. and Tuczek, F. (2006) *Dalton Transactions*, **28**, 3396.

20 Barboza da Silva, C., Mersmann, K., Peters, G. and Tuczek, F.manuscript in preparation.

21 Stephan, G.C. Sivasankar, C. Studt, F. and Tuczek, F. (2008) *Chemistry-A European Journal*, **14**, 644.

22 Elson, C.M. (1976) *Inorganica Chimica Acta*, **18**, 209.

23 Habeck, C.M., Hoberg, C., Peters, G., Näther, C. and Tuczek, F. (2004) *Organometallics*, **23**, 3252.

24 Stephan, G.C., Peters, G., Lehnert, N., Habeck, C.M., Näther, C. and Tuczek, F. (2005) *Canadian Journal of Chemistry*, **83**, 385.

25 (a) George, T.A. and Tisdale, R.C. (1988) *Inorganic Chemistry*, **27**, 2909; (b) George, T.A. and Tisdale, R.C. (1985) *Journal of the American Chemical Society*, **107**, 5157; (c) George, T.A., Ma, L., Shailh, S.N., Tisdale, R.C. and Zubieta, J. (1990) *Inorganic Chemistry*, **29**, 4789.

26 Studt, F., MacKay, B.A., Fryzuk, M.D. and Tuczek, F. (2004) *Journal of the American Chemical Society*, **126**, 280.

27 Klatt, K., Stephan, G., Peters, G. and Tuczek, F. (2008), *Inorganic Chemistry*, **47**, 6541.

28 (a) Bertrand, R.D., Ogilie, F.B. and Verkade, J.G. (1970) *Journal of the American Chemical Society*, **92**, 1916; (b) Ogilie, F.B., Jenkins, J.M. and Verkade, J.G. (1970) *Journal of the American Chemical Society*, **92**, 1908; (c) Crumbliss, A.L. and Topping, R.J. (1987) *Phosphorus-31 NMR: Spectral Properties in Compound Characterization and Structural Analysis*, vol. 15, VCH; (d) Hughes, A.N. (1987) *Phosphorus-31 NMR: Spectral Properties in Compound Characterization and Structural Analysis*, vol. 19, VCH; (e) Clark, H.C., Kapoor, P.N. and McMahon, I.J. (1984) *Journal of Organometallic Chemistry*, **190**, C101; (f) Airey, A.A., Swiegers, G.F., Willis, A.C. and Wild, S.B. (1997) *Inorganic Chemistry*, **36**, 1588.

29 Keller, K. and Tzschach, A. (1984) *Zeitschrift fur Chemie*, **10**, 365.

30 Stephan, G.C., Näther, C., Sivasankar, C. and Tuczek, F. (2008) *Inorganica Chimica Acta*, **361**, 1008.

31 Märkl, G., Jin, G.Y. and Schoerner, C. (1980) *Tetrahedron Letters*, **21**, 1845.

32 (a) Durran, S.E., Smith, M.B., Slawin, A.M.Z. and Steed, J.W. (2000) *Journal of The Chemical Society-Dalton Transactions*, 2771; (b) Coles, S.J., Durran, S.E., Hursthouse, M.B., Slawin, A.M.Z. and Smith, M.B. (2001) *New Journal of Chemistry*, **25**, 416.

33 (a) Owens, B.E. Poli, R. and Rheingold, A.L. (1989) *Inorganic Chemistry*, **28**, 1456; (b) Poli, R. and Gordon, J.C. (1991) *Inorganic Chemistry*, **30**, 4550.

34 Hallmann, L., Bashir, A., Strunskus, R., Adelung, R., Staemmler, V., Wöll, H. and Tuczek, F. (2008) *Langmuir*, **24**, 5726.

16
Directed C–H Functionalizations

Carsten Bolm

16.1
Introduction

The selective conversion of a simple hydrocarbon (**1**) into a more complex molecule having a high degree of functionality (**2**–**6**) remains a dream of many chemists all over the world. Besides reactivity issues, selectivity aspects including chemo-, regio- and stereoselectivities come into play (Scheme 16.1).

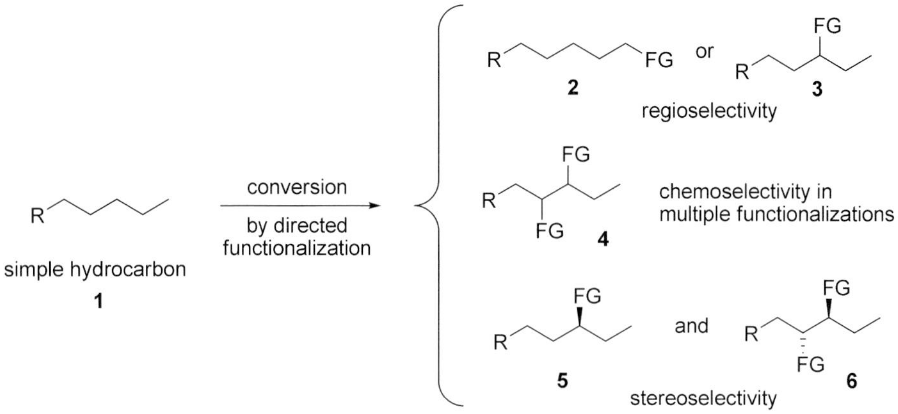

simple hydrocarbon **1**

conversion by directed functionalization

2 or **3** regioselectivity

4 chemoselectivity in multiple functionalizations

5 and **6** stereoselectivity

Scheme 16.1 Selectivity issues in directed functionalizations of simple hydrocarbons.

The recent literature indicates two major research directions. On one hand, the focus is on the very fundamental exploration of C–H bond activation. Gaining a basic understanding in this area could eventually allow us to convert alkanes and saturated hydrocarbons directly into valuable products needed as basic chemicals or synthetic intermediates [1–3]. An early example, which caught much attention, is shown in Scheme 16.2. As Shilov and coworkers discovered, simple alkanes (**7**) can be converted into alcohols (**8**) (and alkyl chlorides) by electrophilic activation using a platinum(IV) catalyst [4].

Activating Unreactive Substrates: The Role of Secondary Interactions.
Edited by Carsten Bolm and F. Ekkehardt Hahn

ISBN: 978-3-527-31823-0

$$\underset{\mathbf{7}}{\text{R-H}} + H_2O + PtCl_6^{2-} \xrightarrow[120\,°\text{C}]{PtCl_4^{2-}\ (\text{cat.})} \underset{\mathbf{8}}{\text{R-O-H}}\ (+\ \text{R-Cl}) + PtCl_4^{2-} + 2\ \text{HCl}$$

Scheme 16.2 Selective hydrocarbon oxidation according to Shilov and coworkers [4].

On the other hand, directed C–H bond functionalizations are more and more applied in complex organic synthesis [5]. A recent contribution by Chen and White, as summarized in Scheme 16.3, illustrates this approach [6]. Thus, using simple iron catalyst **11**, a selective C–H bond oxidation starting from the rather complex natural product (+)-artemisinin (**9**) can be achieved affording alcohol **10** in respectable 54% yield.

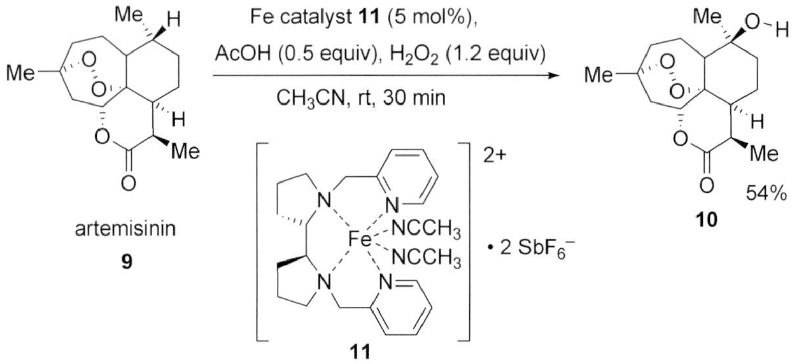

Fe catalyst **11** (5 mol%), AcOH (0.5 equiv), H_2O_2 (1.2 equiv), CH_3CN, rt, 30 min

artemisinin **9** → **10** (54%)

Scheme 16.3 Selectivity in the oxidative C–H functionalization of a complex natural product according to Chen and White [6].

As Chen and White showed, a variety of molecules can be selectively oxidized by the iron catalyst shown in Scheme 16.3 utilizing a combination of hydrogen peroxide and acetic acid as oxidant. Remarkable is the possibility to predict the position of the C–H bond activation, which appears to be solely based on the electronic and steric properties of the substrate. This distinguishes the approach from many previous ones, which rely on the use of directing groups. An example is shown in Scheme 16.4. There, Sames and coworkers used the chelating power of an imino anisole to coordinate palladium chloride, which activated a remote C–H bond of **12** upon binding. A subsequent Suzuki-type cross-coupling of **13** with a vinyl bronic acid (**14**) led to **15** by functionalization of a previously unactivated methyl group [7].

A fundamentally different tpye of C–H activation has been intensively studied by Doyle [8] and Davies [9]. Here, carbenes (derived from metal carbenoids) insert into C–H bonds, and by choosing appropriate chiral dirhodium complexes, excellent enantio- and diastereoselectivities have been achieved [10]. Two examples are shown in Scheme 16.5. Whereas the C–H insertion shown at the top (reported by Doyle and coworkers [8]) occurs *intra*molecularly (starting from **16** to give **17**), the other transformation shown (reported by Davies and coworkers [9]) proceeds in an

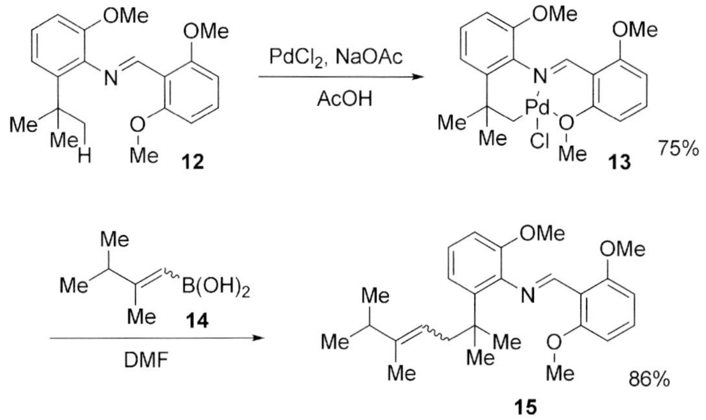

Scheme 16.4 Remote functionalization after activation through metal coordination [7].

*inter*molecular fashion (starting from **19** and **20** to give **22**). In both cases very high stereoselectivities were achieved.

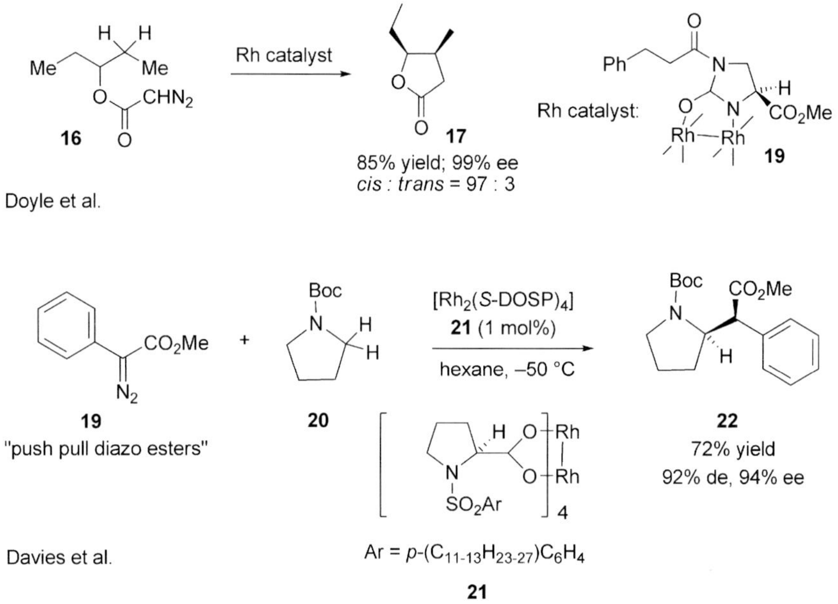

Scheme 16.5 Directed C–H functionalization through carbene insertion reactions as reported by Doyle and Davies [8–10].

Taking into account the structural variations of the dirhodium complexes **19** and **21** and considering the outstanding selectivies observed with these catalysts, the results by Davies and coworkers could be interpreted as affected by secondary effects imposed by the catalyst on the substrate during its activation and subsequent conversion to give the desired product.

16.2
Results and Discussion

Our own investigations were stimulated by the findings by Doyle and Davies (as exemplified in Scheme 16.5) [8–10], and studies that we ourselves had performed on the use of silylated diazo esters in the rhodium-catalyzed preparation of α-silyl-α-amino acids [11] and α-silyl-α-hydroxy acids [12] of type **25** and **26**, respectively (Scheme 16.6).

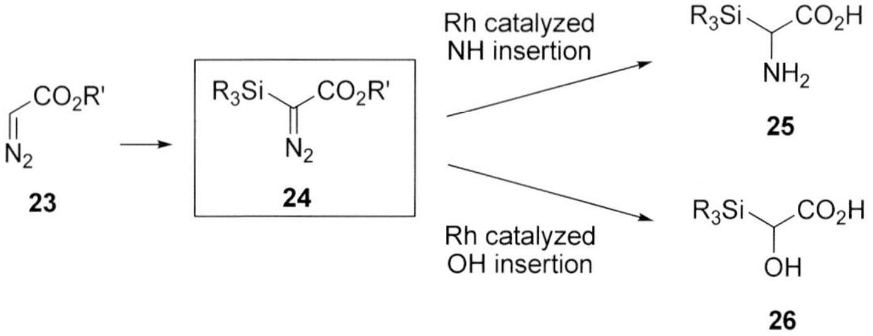

Scheme 16.6 Synthesis of α-silyl-α-amino acids and α-silyl-α-hydroxy acids by rhodium-catalyzed NH- and OH-insertions, respectively [11, 12].

According to Davies, push-pull diazo esters such as **19** work best in his rhodium catalyses, and thus we wondered if the use of silylated diazo esters (**24**) in directed C–H activations would allow us to overcome this limitation. The work by Maas [13] and Marsden [14] on intramolecular versions of such C–H functionalizations as well as early studies by Schöllkopf and coworkers [15] on photochemically activated diazo esters and their C–H insertions were also regarded as very significant for this project.

At the starting point of our investigations, it was decided to keep a strict analogy to the work by Davies and coworkers and to study conversions of substrates with site-directing heteroatoms such as cyclic ethers and acetals [16]. Unfortunately, all attempts to develop efficient transformations remained unsatisfactory, leading to products in only 22–50% yield. (Applying a chiral catalyst provided products in the range of 43–58% *ee* [16]). Scheme 16.7 summarizes the results.

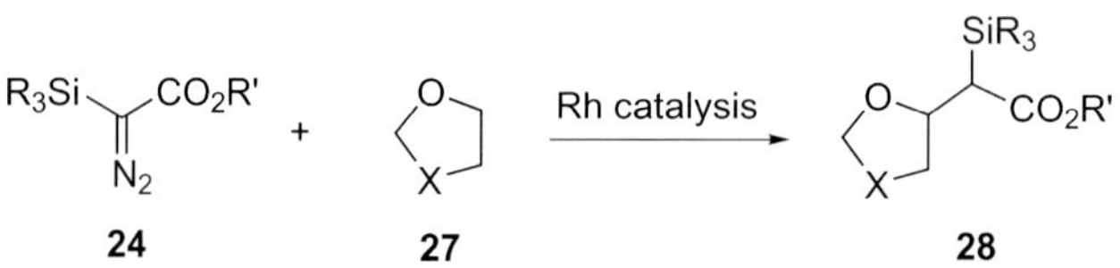

R = Alk, Ph, R' = Et, Bn; X = $(CH_2)_n$, O with n = 1 or 2; yield: 22-50% (ee up to 58%)

Scheme 16.7 Dirhodium-catalyzed reactions of N-protected α-silyl-α-amino esters with cyclic ethers or acetals [16].

Bearing in mind the activating and directing effects of carbonyl groups, C–H insertions into cyclic ketones were attempted next (Scheme 16.8). The well-known generation of carbonyl ylides resulting from reactions of diazo compounds with carbonyl substrates was hoped to be avoidable [17]. To our delight, ketones **29** reacted well with silylated diazo esters **24** under dirhodium catalysis. However, products **31** and **32** did not result from the desired C–H insertion reactions. Instead, the intermediacy of carbonyl ylide **30** could be proposed, which either reacted further by 1,4-proton shift (path A) or cyclized in a 1,5-manner followed by benzyl (cation) shift (path B).

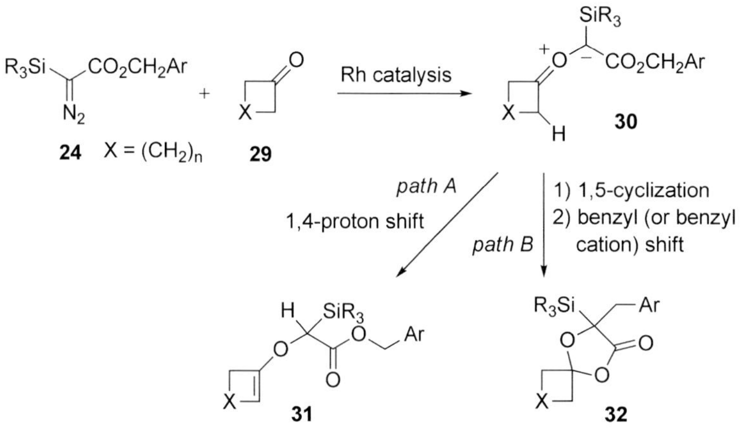

Scheme 16.8 Rhodium-catalyzed reactions of N-protected α-silyl-α-amino benzyl esters with cyclic ketones [18].

Subsequently, various carbonyl derivatives (**29**) were applied in this reaction, and the described rhodium catalysis led to a large number of variously substituted dioxolanes **32** [18, 19]. In no case, however, was a substantial C–H insertion reaction observed.

At this stage the research program was modified, and alternative C–H activation attempts became dominant (Figure 16.1) [20]. The starting point here was our discovery that 1,4,7-triazacyclonones (TACNs) of type **33** in combination with manganese salts were able to activate hydrogen peroxide, allowing an asymmetric

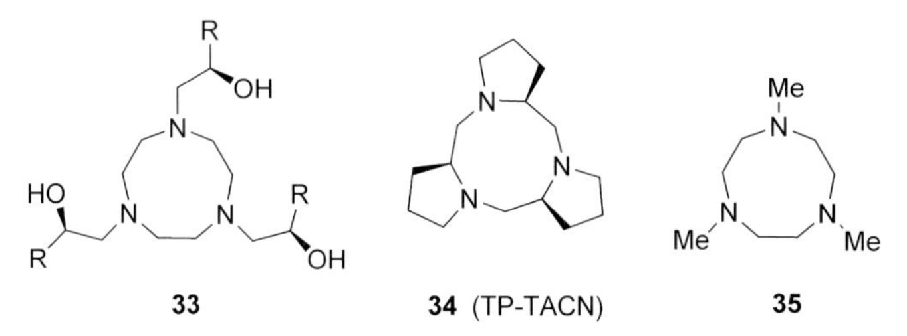

Figure 16.1 TACN derivatives applied as ligands in metal-catalyzed oxidation reactions [21–26].

olefin epoxidation reaction to occur [21–23]. Subsequently it was found that TP-TACN (**34** [24]) could also be applied and that this cyclotriamine was able to stabilize metals in higher oxidation states better than the parent Me_3-TACN (**35**) [24–26]. Furthermore, TP-TACN (**34**) showed an unusually high basicity, which even allowed us to study the doubly protonated form $\{[\mathbf{34}\cdot 2H]^{2+}(PF_6^-)_2\}$ by NMR spectroscopy and X-ray structure analysis [27].

At the beginning of our work on C–H activations with TACN/metal complexes, TACN derivatives had been applied not only in epoxidations [21–24] and bioinorganic chemistry [25, 26], but also in bleaching processes [28], alcohol [29] and sulfide oxidations [30], and olefin aziridinations [31]. Of particular interest for this project were the studies by Shul'pin, Süss-Fink and Lindsay Smith [32] as well as Che [33] on TACN/manganese and ruthenium systems, which had already resulted in selective hydrogen peroxide oxidations (and amidations) of (a limited range of) unfunctionalized alkanes. For our own investigations we considered oligomeric TACN derivatives **37** and **38**, which could readily be prepared by ROMP from **36** [22], as an adequate starting point (Figure 16.2).

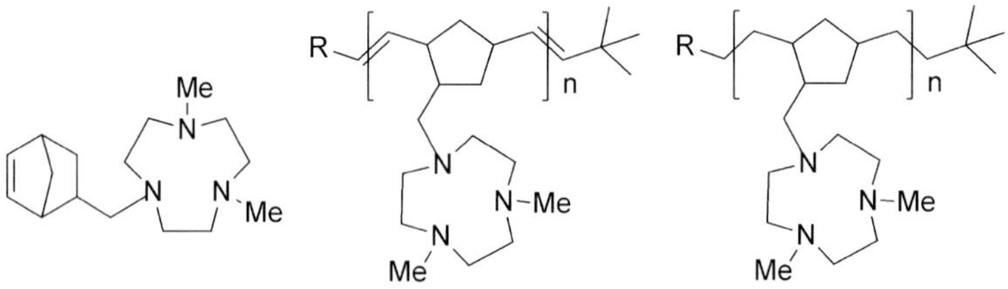

36 **37** (n = ~20) **38**: (n = ~20)

Figure 16.2 TACN derivatives applied in metal-catalyzed oxidation reactions [22, 34].

Thus, in collaboration with Shul'pin (from the Semenov Institute of Chemical Physics of the Russian Academy of Sciences) the use of **37** and **38** in manganese-catalyzed oxidations of simple alkanes with hydrogen peroxide as oxidant was studied [34]. Noteworthy findings were, for example, that in acetonitrile the presence of a small quantity of acetic acid was essential to obtain an active catalyst. Furthermore, both the kinetic isotope effect as well as the regioselectivity of the C–H activation differed significantly from those observed in reactions with Me_3-TACN (**35**). Thus, in the oxidation of 2,2,4-trimethylpentane (**39**) with a manganese catalyst prepared from **37** the main products were the two primary alcohols **40** and **41**. The corresponding secondary and tertiary alcohols (**42** and **43**, respectively) were only observed in minor quantities (Scheme 16.9) [34]. In contrast, analogous reactions with **35** as ligand provided a product ratio (primary: secondary: tertiary alcohols) of 1:5:55.

Furthermore, use of the oligomeric manganese complex in the oxidation of *cis*-decalin (**44**) led ('stereo*un*specifically') to a *trans/cis* product mixture (*trans*-**45** and *cis*-**45**) of 1.5:1 (Figure 16.3) [34], whereas the system derived from **35** gave a ratio of 0.12.

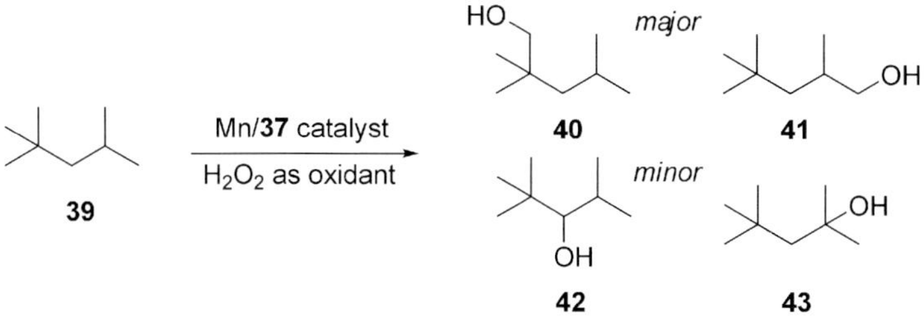

Scheme 16.9 C–H activation with hydrogen peroxide and a manganese complex obtained from oligomeric TACN derivative **37** [34].

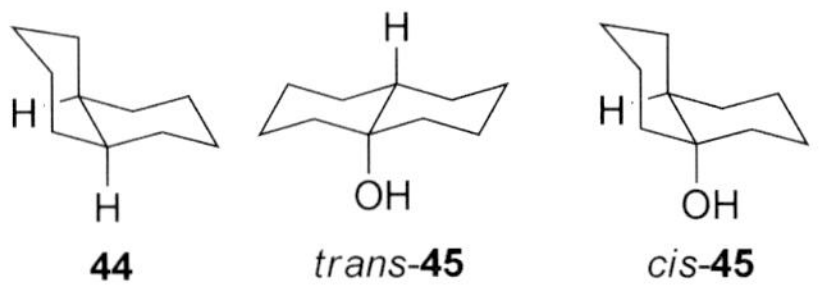

Figure 16.3 *Cis*-decalin (**44**) and alcohols **45** obtained by oxidative C–H functionalization [34].

A number of other TACN derivatives were prepared, and some of those are depicted in Figure 16.4. They included chiral compounds with stereogenic centers positioned at the nitrogen substituents such as **46–48** [35] as well as *N*-arylated TACN derivatives with multiple bonding sites such as **49–52** [36].

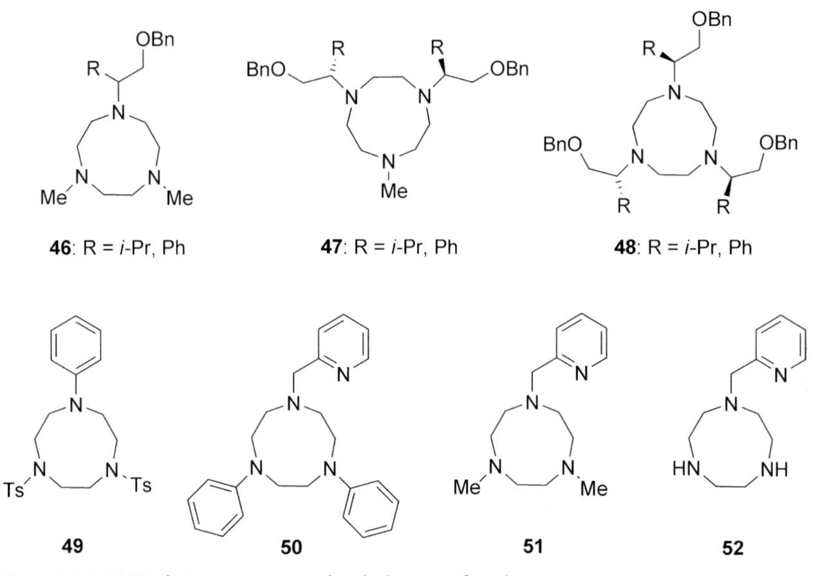

Figure 16.4 TACN derivatives prepared with the aim of applying them in metal-catalyzed C–H functionalizations [35, 36].

Of particular interest were the *N*-arylated TACN derivatives, and for their access new palladium catalysis protocols had to be developed [37]. Simple BINAP/palladium acetate combinations worked well for mono-arylations (cf. the synthesis of **54**, Scheme 16.10). For the preparation of di-arylated TACNs (such as **56**), however, a palladium(0) source together with Buchwald's phosphine **57** had to be applied (Scheme 16.10).

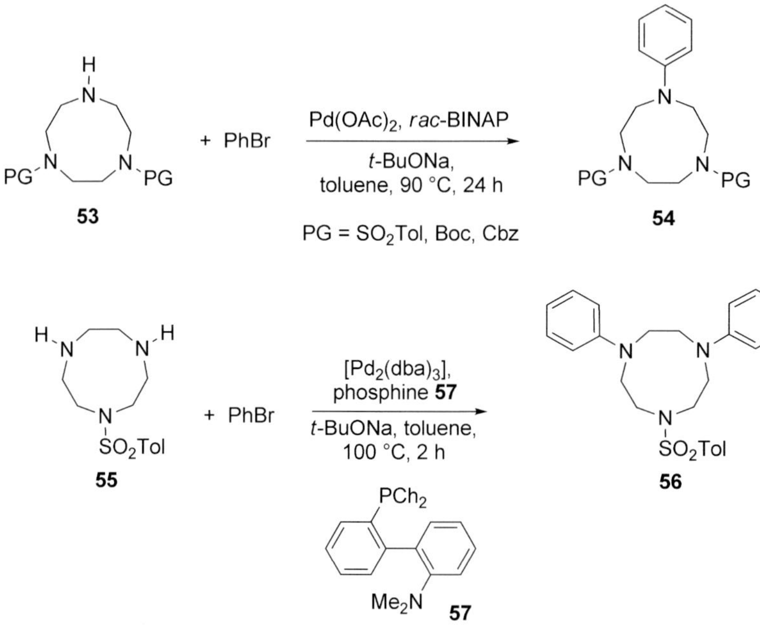

53 + PhBr → 54: $Pd(OAc)_2$, *rac*-BINAP; *t*-BuONa, toluene, 90 °C, 24 h

PG = SO_2Tol, Boc, Cbz

55 + PhBr → 56: $[Pd_2(dba)_3]$, phosphine **57**; *t*-BuONa, toluene, 100 °C, 2 h

Scheme 16.10 Palladium catalyses for selective arylations of TACN derivatives [37].

In the context of potential secondary effects imposed by TACN ligands on a given catalyst system, the palladium-catalyzed synthesis of highly fluorinated TACN **58** (Figure 16.5) was of interest. Its binding to copper chloride was proven by mass spectrometry [36].

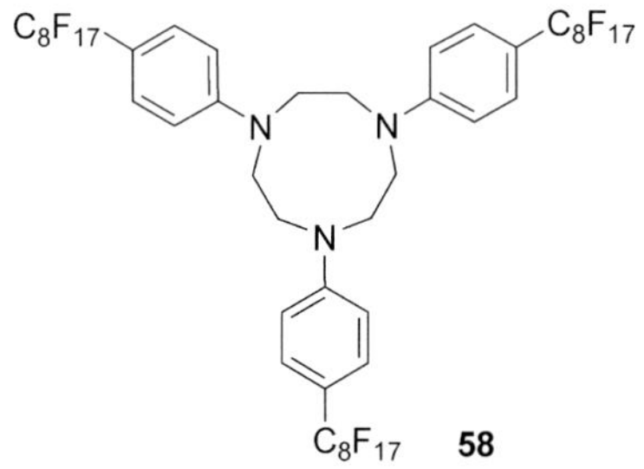

Figure 16.5 Highly fluorinated TACN derivative prepared by palladium-catalyzed *N*-arylation [36].

With respect to our emerging focus on iron catalyses [38] and taking into account the excellent results achieved by Que and coworkers providing a deep mechanistic understanding of iron-catalyzed oxidations [39], we decided to extend our studies towards pyridyl amines (and related compounds) such as those depicted in Figure 16.6 [36, 40, 41].

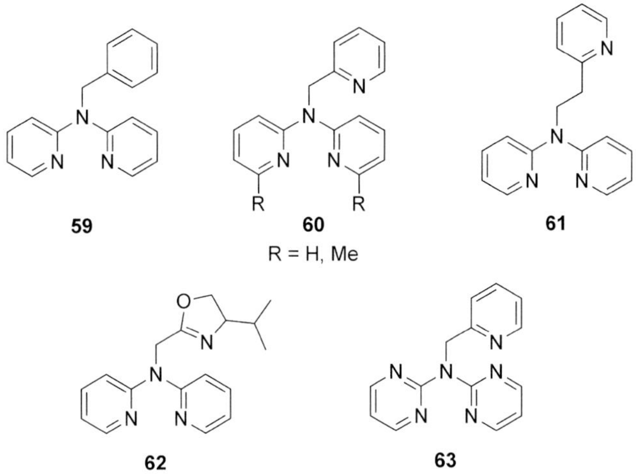

59 **60** R = H, Me **61** **62** **63**

Figure 16.6 Pyridyl amines (and related compounds) for use in iron-catalyzed oxidations [36].

To our disappointment, all attempts to obtain a concise picture in C–H functionalizations of unactivated substrates with iron salt/ligand combinations failed [42]. Despite the fact that the presence of the TACNs or the pyridyl amines affected reactions such as the one schematically shown below (Scheme 16.11), no conclusive results which could be taken as further guidelines were obtained. Furthermore, it was found that in some cases the presence of the ligand did not enhance the reactivity, but, in contrast, reduced it.

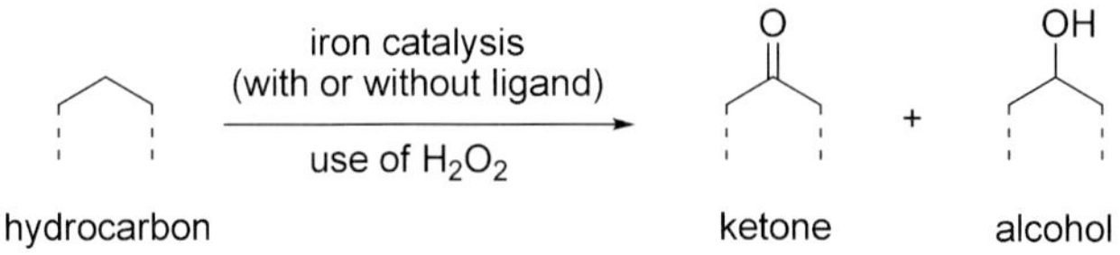

hydrocarbon → iron catalysis (with or without ligand), use of H_2O_2 → ketone + alcohol

Scheme 16.11 Attempted iron-catalyzed hydrocarbon oxidations.

These observations stimulated the search for iron-catalyzed C–H activations which proceeded *in the absence* of a ligand. After an intensive search and following a major optimization process it was found that iron(II) perchlorate (hexahydrate) was indeed able to catalyze C–H oxidations, providing carbonyl compounds in good yields. Despite the fact that only benzylic substrates worked well, the protocol proved interesting, since, starting from ethylbenzene (**64**), for example, the corresponding ketone **65** could be isolated as the major product in remarkable 50% yield (Scheme 16.12) [43]. This result compared well with those obtained by related protocols (e.g. GIF systems), which had first been developed by Barton and coworkers [44, 45].

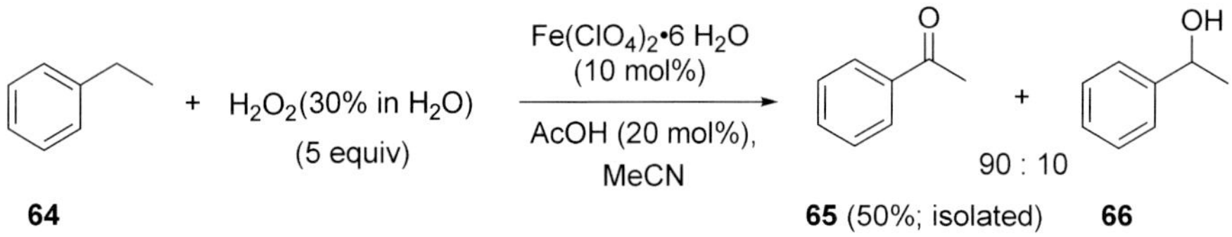

64 + H_2O_2(30% in H_2O) (5 equiv) → [$Fe(ClO_4)_2{\cdot}6\,H_2O$ (10 mol%); AcOH (20 mol%), MeCN] → 65 (50%; isolated) + 66 (90 : 10)

Scheme 16.12 Iron-catalyzed benzylic oxidation with hydrogen peroxide under ligand-free conditions [43].

Several features of this iron catalysis are noteworthy. First, besides iron(II) perchlorate, iron(II) chloride and iron(II) tetrafluoroborate could be used, leading to almost identical results. In contrast, iron(III) chloride, iron(II) acetate, and iron(II) acetylacetonate proved inapplicable, since they led to rapid decomposition of the oxidant, hydrogen peroxide (presumably by Fenton chemistry) [46]. Second, the use of acetonitrile was essential, and other solvents such as methanol, dichloromethane or ethyl acetate were incompatible (because of the loss of hydrogen peroxide). Third, the addition of an acid (such as acetic acid) enabled the exothermicity of the reaction to be controlled. A direct correlation of this behavior with the pK_a value of the acid did not exist [47].

Subsequently, an even more efficient iron-catalyzed oxidation of benzylic substrates was developed (Scheme 16.13). Here, simple iron(III) chloride served as the catalyst, which was used in combination with aqueous *tert*-butyl hydroperoxide as oxidant. In pyridine at elevated temperature (82 °C) a number of substrates (hydrocarbons and alcohols) could be converted, affording the corresponding carbonyl compounds in high yields [48, 49].

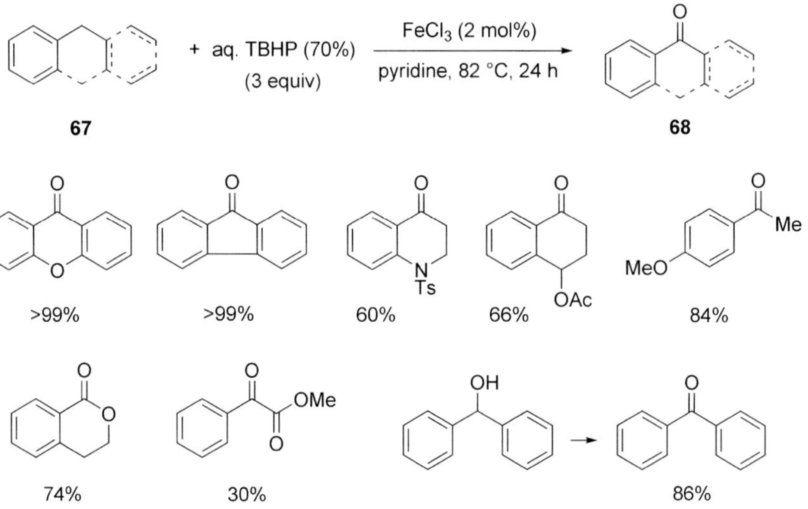

67 + aq. TBHP (70%) (3 equiv) → [$FeCl_3$ (2 mol%); pyridine, 82 °C, 24 h] → 68

>99% ; >99% ; 60% (N-Ts) ; 66% (OAc) ; 84% (MeO, Me) ; 74% ; 30% (OMe) ; 86% (OH → O)

Scheme 16.13 Iron-catalyzed benzylic oxidation with *tert*-butyl hydroperoxide as oxidant under ligand-free conditions [48].

Other oxidants, such as hydrogen peroxide, cumene hydroperoxide, NaOCl and dioxygen were less efficient (with $FeCl_3$ as catalyst), leading to lower product yields. Performing the reaction in air (open flask) was beneficial compared to performing them under argon. Other iron salts [such as $Fe(ClO_4)_2$, $Fe(ClO_4)_3$, $FeCl_2$, and $Fe(acac)_3$] proved applicable as well, but with the exception of iron(III) perchlorate they were less effective.

Currently, other iron-catalyzed oxidative transformations are under investigation [50], and it is most likely that alternative C–H activation protocols will be discovered within this program in due course.

16.3
Conclusions

Guided by the search for new C–H functionalizations of unreactive substrates (in particular hydrocarbons), we found ourselves in various exciting fields of modern organic chemistry. Most reactions involved metals, which we attempted to fine-tune in order to make them 'well-behaved' and useful as catalysts for reactions difficult to perform by other means. Starting with the chemistry of high-energy diazo compounds, which had to be activated by precious rhodium catalysts, we moved toward ligand design, trying to find systems for the reactivity control of metals such as manganese, copper, and iron. The latter search finally led us to reagent combinations which allowed hydrocarbon functionalizations with simple iron salts and hydrogen peroxide as the oxidant. In terms of their environmental impact, these compounds are most desirable, and, considering the operational simplicity of the developed processes, an industrially relevant scale-up could well be imagined. The need for alternative and even more efficient C–H functionalizations means that we cannot stop our search at this stage, and the future will show where nature is guiding us to next [51].

References

1 (a) Dyker, G. (ed.), (2005) *Handbook of C–H Transformations, Vols. 1 and 2*, Wiley-VCH, Weinheim; (b) Murai, S. (ed.), (1999) Activation of Unreactive Bonds and Organic Synthesis, Vol. 3 of *Topics in Organometallic Chemistry*, Springer, Berlin; (c) Davies, J.A., Watson, P.L., Greenberg, A. and Liebman, J.F. (eds), (1990) *Selective Hydrocarbon Oxidation and Functionalization*, VCH, New York; (d) C.L. Hill (ed.), (1989) *Activation and Functionalization of Alkanes*, Wiley, New York.

2 For leading references, see: (a) Crabtree, R.H. (1985) *Chemical Reviews*, **85**, 245; (b) Arnstsen, B.A., Bergman, R.G., Mobley, T.A. and Peterson, T.H. (1995) *Accounts of Chemical Research*, **28**, 154; (c) Crabtree, R.H. (1995) *Chemical Reviews*, **95**, 987; (d) Stahl, S.S., Labinger, J.A. and Bercaw, J.E. (1998) *Angewandte Chemie*, **110**, 2298; (1998) *Angewandte Chemie – International Edition*, **37**, 2181; (e) Guari, Y., Sabo-Etienne, S. and Chaudret, B. (1999) *European Journal of Inorganic*

Chemistry, 1047; (f) Sen, A. (1998) *Accounts of Chemical Research*, **31**, 559; (d) Crabtree, R.H. (2001) *Dalton Transactions*, 2437; (g) Jia, C.G., Kitamura, T. and Fujiwara, Y. (2001) *Accounts of Chemical Research*, **34**, 633; (h) Labinger, J.A. and Bercaw, J.E. (2002) *Nature*, **417**, 507; (i) Ritleng, V., Sirlin, C. and Pfeffer, M. (2002) *Chemical Reviews*, **102**, 1731; (j) Bergman, R.G. (2007) *Nature*, **446**, 391.

3 From these studies we could eventually also expect answers on how to convert methane into methanol; (a) Olah, G.A., Goeppert, A. and Surya Prakash, G.K. (2006) *Beyond Oil and Gas: The Methanol Economy*, Wiley-VCH, Weinheim; (b) Periana, R.A., Taube, D.J., Gamble, S., Taube, H., Satoh, T. and Fujii, H. (1998) *Science*, **280**, 560.

4 (a) Shilov, A.E. and Shulpin, G.B. (1987) *Russian Chemical Reviews*, **56**, 442; (b) Shilov, A.E. and Shulpin, G.B. (1997) *Chemical Reviews*, **97**, 2879; (c) Shilov, A.E. (1997) *Metal Complexes in Biomimetic Chemical Reactions*, CRC Press, New York. (d) Shilov, A.E. and Shteinman, A.A. (1999) *Accounts of Chemical Research*, **32**, 763; (e) See also: Siegbahn, P.E.M. and Crabtree, R.H. (1996) *Journal of the American Chemical Society*, **118**, 4442.

5 (a) Dyker, G. (1999) *Angewandte Chemie*, **111**, 1808; (1999) *Angewandte Chemie – International Edition*, **38**, 1698; (b) Kakiuchi, F. and Murai, S. (1999) *Topics in Current Chemistry*, **3**, 47; (c) Bringmann, G. and Menche, D. (2001) *Accounts of Chemical Research*, **34**, 615; (d) Murai, S. and Nomura, M. (2002) *Topics in Current Chemistry*, **219**, 211; (e) Kakiuchi, F. and Murai, S. (2002) *Accounts of Chemical Research*, **35**, 826; (f) Kakiuchi, F. and Chatani, N. (2003) *Advanced Synthesis and Catalysis*, **345**, 1077; (g) Campeau, L.-C. and Fagnou, K. (2006) *Chemical Communications*, 1253; (h) Dick, A.R. and Sanford, M.S. (2006) *Tetrahedron*, **62**, 2439; (i) Daugulis, O. Zaitsev, V.G. Shabashov, D. Phan, Q.-N. and Lazareva, A. (2006) *Synlett*, 3382; (j) Godula, K. and Sames, D. (2006) *Science*, **312**, 67; (k) Alberico, D., Scott, M.E. and Lautens, M. (2007) *Chemical Reviews*, **107**, 174; (l) Fairlamb, I.J.S. (2007) *Chemical Society Reviews*, **36**, 1036; (m) Seregin, I.V. and Gevorgyan, V. (2007) *Chemical Society Reviews*, **36**, 1173;

6 Chen, M.S. and White, M.C. (2007) *Science*, **318**, 783.

7 Dangel, B.D., Godula, K., Youn, S.W. Sezen, B. and Sames, D. (2002) *Journal of the American Chemical Society*, **124**, 11856.

8 (a) Doyle, M.P., Zhou, Q.-L., Dyatkin, A.B. and Ruppar, D.A. (1995) *Tetrahedron Letters*, **36**, 7579; (b) Doyle, M.P., Kalinin, A.V. and Ene, D.G. (1996) *Journal of the American Chemical Society*, **118**, 8837.

9 Davies, H.M.L., Venkataramani, C., Hansen, T. and Hopper, D.W. (2003) *Journal of the American Chemical Society*, **125**, 6462.

10 Review: Davies, H.M.L. and Beckwith, R.E.J. (2003) *Chemical Reviews*, **103**, 2861.

11 Bolm, C., Kasyan, A., Drauz, K., Günther, K. and Raabe, G. (2000) *Angewandte Chemie*, **112**, 2374; (2000) *Angewandte Chemie – International Edition*, **39**, 2288.

12 (a) Bolm, C., Kasyan, A., Heider, P., Saladin, S., Drauz, K., Günther, K. and Wagner, C. (2002) *Organic Letters*, **4**, 2265; (b) Bolm, C., Saladin, S., Claßen, A., Kasyan, A., Veri, E. and Raabe, G. (2005) *Synlett*, 461; (c) Mortensen, M., Husmann, R., Veri, E. and Bolm, C. *Chemical Society Reviews*, accepted for publication.

13 (a) Maas, G., Gimmy, M. and Alt, M. (1992) *Organometallics*, **11**, 3813; (b) Gettwert, V., Krebs, F. and Maas, G. (1999) *European Journal of Organic Chemistry*, 1213. Reviews: (c) Maas, G. (1998) *The chemistry of organic silicon compounds, Vol. 2* Part 1, (eds Z., Rappoport and Y., Apeloig), Wiley, Chichester, Chapter 13, p. 703; (d) Tomioka, H. (1989) *Methoden der Organischen Chemie (Houben Weyl)*, **E 19b** (ed. M., Regitz,), p. 1410.

14 (a) Kablean, S.N., Marsden, S.P. and Craig, A.M. (1998) *Tetrahedron Letters*, **39**, 5109; (b) Marsden, S.P., and Pang, W.-K. (1998) *Tetrahedron Letters*, **39**, 6077; (c) Marsden,

S.P. and Pang, E.-K. (1999) *Chemical Communications*, 1199.

15 (a) Schöllkopf, U., Hoppe, D., Rieber, N. and Jacobi, V. (1969) *Justus Liebigs Annalen der Chemie*, **730**, 1; (b) Schöllkopf, U. and Rieber, N. (1967) *Angewandte Chemie*, **79**, 906; See also: (1967) *Angewandte Chemie – International Edition*, **6**, 884.

16 Saladin, S. (2004) Dissertation at RWTH Aachen University.

17 (a) Padwa, A. and Weingarten, M.D. (1996) *Chemical Reviews*, **96**, 223; (b) Padwa, A., Brodney, M.A., Marino, J.P., Jr, and Sheehan, S.M. (1997) *The Journal of Organic Chemistry*, **62**, 78; (c) Padwa, A. and Price, A.T. (1998) *The Journal of Organic Chemistry*, **63**, 556.

18 Bolm, C. Saladin, S. and Kasyan, A. (2002) *Organic Letters*, **4**, 4631.

19 For analogous formations of dioxolanones in dirhodium perflourobutyrate-catalyzed reactions of allylic diazoacetates with various aldehydes, including acetaldehyde, crotonaldehyde and aromatic aldehydes, see: Maas, G. and Alt, M. (1994) *Chemische Berichte*, **127**, 1537.

20 For the extension of our research program focused on the application of silylated diazo esters, see: Bolm, C., Kasyan, A. and Saladin, S. (2005) *Tetrahedron Letters*, **46**, 4049.

21 Bolm, C., Kadereit, D. and Valacchi, M. (1997) *Synlett*, 687.

22 Grenz, A., Ceccarelli, S. and Bolm, C. (2001) *Chemical Communications*, 1726.

23 For the use of other TACN derivatives in transformations of olefins, see: (a) De Vos, D.E., Feijen, E.J.P., Schoonheydt, R.A. and Jacobs, P.A. (1994) *Journal of the American Chemical Society*, **116**, 4746; (b) De Vos, D.E. and Bein, T. (1996) *Journal of Organometallic Chemistry*, **520**, 195; (c) De Vos, D.E. and Bein, T. (1996) *Chemical Communications*, 917; (d) De Vos, D.E., Meinershagen, J.L. and Bein, T. (1996) *Angewandte Chemie*, **108**, 2355; (1996) *Angewandte Chemie – International Edition*, **35**, 2211; (e) Subba Rao, Y.V., De Vos, D.E., Bein, T. and Jacobs, P.A. (1997) *Journal of the Chemical Society. Chemical Communications*, 355; (f) De Vos, D.E. and Bein, T. (1997) *Journal of the American Chemical Society*, **119**, 9460; (g) De Vos, D.E., Sels, B.F., Reynaers, M., Subba Rao, Y.T. and Jacobs, P.A. (1998) *Tetrahedron Letters*, **39**, 3221; (h) Berkessel, A. and Sklorz, C.A. (1999) *Tetrahedron Letters*, **40**, 7965; (i) Vincent, J.M., Rabion, A., Yachandra, V.K. and Fish, R.H. (1997) *Angewandte Chemie*, **109**, 2438; (1997) *Angewandte Chemie – International Edition*, **36**, 2346; (j) Fish, R.H. (1999) *Chemistry - A European Journal*, **5**, 1677;
(k) Beller, M., Tafesh, A., Fischer, R. and Schabert, B., DE 195 23 890 C1; 30.06.95.
(l) Beller, M., Tafesh, A., Fischer, R. and Schabert, B. DE 195 23 891 C1; 30.06.95.
(m) Brinksma, J., Hage, R., Kerschner, J. and Feringa, B.L. (2000) *Chemical Communications*, 537; (n) *cis*-dihydroxylations: Brinksma, J., Schneider, L., van Vliet, G., Boaron, R., Hage, R., De Vos, D., Alster, P.L., Feringa, B.L. (2002) *Tetrahedron Letters*, **43**, 2619; (o) Muñiz-Fernandez, K. and Bolm, C. (1998) *Transition Metals for Organic Synthesis*, (eds. M., Beller and C. Bolm), Wiley-VCH, Weinheim, p. 271; (p) Muñiz, K. and Bolm, C. (2004) *Transition Metals for Organic Synthesis*, 2nd edn (eds. M. Beller and C. Bolm), Wiley-VCH, Weinheim, p. 344; For additional examples of TACN derivatives which have been prepared for the use as chiral ligands in asymmetric epoxidations, see: (q) Golding, S.W., Humbley, T.W., Lawrance, G.A., Luther, S.M., Maeder, M. and Turner, P. (1999) *Journal of The Chemical Society-Dalton Transactions*, 1975; (r) Pulacchini, S. and Watkinson, M. (2001) *European Journal of Organic Chemistry*, 4233; (s) Argoucharch, G., Gibson, C.L., Stones, G. and Sherrington, D.C. (2002) *Tetrahedron Letters*, **43**, 3795;
(t) Scheuermann, J.E.W., Ronketti, F., Motevalli, M., Griffiths, D.V. and Watkinson, M. (2002) *New Journal of Chemistry*, **26**, 1054; (u) Scheuermann, J.E.W., Ilashenko, G., Griffith, D.V. and

Watkinson, M. (2002) *Tetrahedron: Asymmetry*, **13**, 269; (v) Kim, B.M., So, S.M. and Choi, H.J. (2002) *Organic Letters*, **4**, 949.

24 (a) Bolm, C., Meyer, N., Raabe, G., Weyhermüller, T. and Bothe, E. (2000) *Chemical Communications*, 2435; (b) Meyer, N. (2000) Dissertation RWTH Aachen University.

25 For fundamental studies on the use of TACN derivatives in bioinorganic chemistry, see: (a) Chaudhuri, P. and Wieghardt, K. (1989) *Progress in Inorganic Chemistry*, Vol. **35** (ed. S.J., Lippard), Wiley, New York, p. 329; (b) Kimura, E. (1994) *Progress in Inorganic Chemistry*, **Vol. 41** (ed. K.D., Karlin), Wiley, New York, p. 443; (c) Halfen, J.A., Mahapatra, S., Wilkinson, E.C., Kaderli, S., Young, V.G., Jr., Que, L., Jr, Zuberbühler, A.D. and Tolman, W.B. (1996) *Science*, **271**, 1397; (d) Stockheim, C. Hoster, L., Weyhermüller, T., Wieghardt, K. and Nuber, B. (1996) *Journal of The Chemical Society-Dalton Transactions*, 4409; (e) Wainwright, K.P. (1997) *Coordination Chemistry Reviews*, **166**, 35; (f) Wieghardt, K., Bossek, U., Ventur, D. and Weiss, J. (1985) *Journal of the Chemical Society. Chemical Communications*, 347; (g) Wieghardt, K., Bossek, U., Nuber, B., Weiss, J., Bonvoisin, J., Corbella, M., Vitols, S.E. and Girerd, J.J. (1988) *Journal of the American Chemical Society*, **110**, 7398; (h) Bossek, U., Weyhermüller, T., Wieghardt, K., Nuber, B. and Weiss, J. (1990) *Journal of the American Chemical Society*, **112**, 6387; (i) Wieghardt, K., Pohl, J. and Gerbert, W. (1983) *Angewandte Chemie*, **95**, 739; (1983) *Angewandte Chemie – International Edition*, **22**, 727; (j) Armstrong, W.H. and Lippard, S.J. (1983) *Journal of the American Chemical Society*, **105**, 4837; (k) Hage, R., Gunnewegh, E.A., Niel, J., Tjan, F.S.B., Weyhermüller, T. and Wieghardt, K. (1998) *Inorganica Chimica Acta*, **268**, 43.

26 For excellent overviews on TACN chemistry, see: (a) Hage, R. (1996) *Recueil des Travaux Chimiques des Pays-Bas*, **115**, 385; (b) Sibbons, K.F., Shastri, K. and Watkinson, M. (2006) *Dalton Transactions*, 645.

27 Meyer, N.C., Bolm, C., Raabe, G. and Kölle, U. (2005) *Tetrahedron*, **61**, 12371.

28 (a) Hage, R., Iburg, J.E., Kerschner, J., Koek, J.H., Lempers, E.L.M., Martens, R.J., Racherla, U.S., Russell, S.W., Swarthoff, T., van Vllet, M.R.P., Warnaar, J.B., van der Wolf, L. and Krijnen, B. (1994) *Nature*, **369**, 637; (b) Koek, J.H., Kohlen, E.W.M.J., Russell, S.W., van der Wolf, L., ter Steeg, P.F. and Hellemons, J.C. (1999) *Inorganica Chimica Acta*, **295**, 189.

29 (a) Zondervan, C., Hage, R. and Feringa, B.L. (1997) *Journal of the Chemical Society. Chemical Communications*; 419; (b) Fung, W.-H., Yu, W.-Y. and Che, C.-M. (1998) *The Journal of Organic Chemistry*, **63**, 2873.

30 (a) Barton, D.H.R., Li, W. and Smith, J.A. (1998) *Tetrahedron Letters*, **39**, 7055; (b) Hay, R.W., Clifford, T. and Govan, N. (1998) *Transition Metal Chemistry*, **23**, 619.

31 Halfen, J.A., Hallman, J.K., Schultz, J.A. and Emerson, J.P. (1999) *Organometallics*, **18**, 5435.

32 (a) Lindsay Smith, J.R. and Shul'pin, G.B. (1998) *Tetrahedron Letters*, **39**, 4909; (b) Shul'pin, G.B. and Lindsay Smith, J.R. (1998) *Russian Chemical Bulletin*, **47**, 2379; (c) Shul'pin, G.B., Süss-Fink, G. and Lindsay Smith, J.R. (1999) *Tetrahedron*, **55**, 5345; (d) Shul'pin, G.B., Süss-Fink, G. and Shul'pina, L.S. (2001) *Journal of Molecular Catalysis A-Chemical*, **170**, 17; (e) Shul'pin, G.B., Süss-Fink, G. and Shilov, A.E. (2001) *Tetrahedron Letters*, **42**, 7253; (f) Review: Shul'pin, G.B. and (2002) *Journal of Molecular Catalysis A-Chemical*, **189**, 39.

33 (a) Cheng, W.-C., Yu, W.-Y., Cheung, K.-K. and Che, C.-M. (1063) *Journal of the Chemical Society. Chemical Communications*, **1994**, (b) Au, S.-M., Zhang, S.-B., Fung, W.-H., Yu, W.-Y., Che, C.-M. and Cheung, K.-K. (1998) *Chemical Communications*, 2677; (c) Au, S.-M., Huang, J.-S., Che, C.-M. and Yu, W.-Y. (2000) *The Journal of Organic Chemistry*, **65**, 7858.

34 Nizova, G.V., Bolm, C., Ceccarelli, S., Pavan, C. and Shul'pin, G.B. (2002) *Advanced Synthesis and Catalysis*, **344**, 899.

35 Pavan, C., Dissertation RWTH Aachen University, (2005).

36 Nakanishi, M., Dissertation RWTH Aachen University, (2007).

37 Nakanishi, M. and Bolm, C. (2007) *Advanced Synthesis and Catalysis*, **349**, 861.

38 For reviews on iron catalyses stemming from this group, see: (a) Bolm, C. Legros, J., Le Paih, J. and Zani, L. (2004) *Chemical Reviews*, **104**, 6217; (b) Correa, A., García Mancheño, O. and Bolm, C. (2008) *Chemical Society Reviews*, **37**, 1108.

39 (a) Halfen, J.A., Mahapatra, S., Wilkinson, E.C., Kaderli, S., Young, V.G., Jr, Que, L., Jr, Zuberbühler, A.D. and Tolman, W.B. (1996) *Science*, **271**, 1397; (b) Rohde, J.-U., In, J.H., Lim, M.H., Brennessel, W.W., Bukowski, M.R., Stubna, A., Münck, E., Nam, W. and Que, L. Jr, (2003) *Science*, **299**, 1037; (c) Costas, M., Chen, K. and Que, L. Jr, (2000) *Coordination Chemistry Reviews*, **200–202**, 517; (d) Costas, M., Mehn, M.P., Jensen, M.P. and Que, L. Jr, (2004) *Chemical Reviews*, **104**, 939; (e) Rohde, J.-U., Torelli, S., Shen, X., Lim, M.H., Klinker, E.J., Kaizer, J., Chen, K., Nam, W. and Que, L. Jr, (2004) *Journal of the American Chemical Society*, **126**, 16750; (f) Mas-Ballesté, R. and Que, L Jr, (2007) *Journal of the American Chemical Society*, **129**, 15964; and references therein.

40 (a) For the first synthesis of **60** (with R = H), see: Foxon, S.P., Walter, O. and Schindler, S. (2002) *European Journal of Inorganic Chemistry*, 111; For related compounds see also: (b) Schareina, T. Hildebrand, G., Fuhrmann, H. and Kempe, R. (2001) *European Journal of Inorganic Chemistry*, 2421; (c) Silberg, J. Schareina, T., Kempe, R., Wurst, K. and Buchmeiser, M.R. (2001) *Journal of Organometallic Chemistry*, **622**, 6; (d) Miao, D., Guo, Y.-M., Chen, S.T., Bu, X.-H. and Ribas, J. (2003) *Inorganica Chimica Acta*, **346**, 207; (e) de Hoog, P., Gamez, P., Roubeau, O., Lutz, M., Driessen, W.L., Spek, A.L. and Reedijk, J. (2003) *New Journal of Chemistry*, **27**, 18; (f) Youngme, S., van Albada, G.A., Roubeau, O., Pakawatchai, C., Chaichit, N. and Reedijk, J. (2003) *Inorganica Chimica Acta*, **342**, 48; (g) Carranza, J., Brennan, C., Sletten, J., Lloret, F. and Julve, M. (2002) *Journal of The Chemical Society-Dalton Transactions*, 3164.

41 For previous studies related to such compounds, see: (a) Bolm, C., Frison, J.-C., Le Paih, J. and Moessner, C. (2004) *Tetrahedron Letters*, **45**, 5019; (b) Bolm, C., Frison, J.-C., Le Paih, J., Moessner, C. and Raabe, G. (2004) *Journal of Organometallic Chemistry*, **689**, 3767.

42 For previous studies along these lines, see: (a) Kim, C., Dong, Y. and Que, L. Jr, (1997) *Journal of the American Chemical Society*, **119**, 3635; (b) Klopstra, M. Hage, R. Kellogg, R.M. and Feringa, B.L. (2003) *Tetrahedron Letters*, **44**, 4581.

43 Pavan, C., Legros, J. and Bolm, C. (2005) *Advanced Synthesis and Catalysis*, **347**, 703.

44 (a) Barton, D.H.R. and Doller, D. (1992) *Accounts of Chemical Research*, **25**, 504; (b) Barton, D.H.R. and Chavasiri, W. (1994) *Tetrahedron*, **50**, 19; (c) Barton, D.H.R. (1998) *Tetrahedron*, **54**, 5805; (d) Stavropoulos, P., Çelenligil-Çetin, R. and Tapper, A.E. (2001) *Accounts of Chemical Research*, **34**, 745; (e) For a summary of Gif reactions, see: Stavropoulos, P., Çelenligil-Çetin, R., Kiani, S., Tapper, A. Pinnapareddy, D., Paraskevopoulou, P. in Ref. [1], 497.

45 For a related system, see: Shul'pin, G.B., Nizova, G.V., Kozlov, Y.N., Gonzalez Cuerva, L. and Süss-Fink, G. (2004) *Advanced Synthesis and Catalysis*, **346**, 317.

46 (a) Sawyer, D.T., Sobkowiak, A. and Matsushita, T. (1996) *Accounts of Chemical Research*, **29**, 409; (b) Sawyer, D.T. (1997) *Coordination Chemistry Reviews*, **165**, 297; (c) Walling, C. (1998) *Accounts of Chemical Research*, **31**, 155; (d) MacFaul, P.A. Wayner, D.D.M. and Ingold, K.U. (1998) *Accounts of Chemical Research*, **31**, 159;

(e) Goldstein, S. and Meyerstein, D. (1999) *Accounts of Chemical Research*, **32**, 547.

47 For positive effects of acids in iron catalyses, see: (a) White, M.C., Doyle, A.G. and Jacobsen, E.N. (2001) *Journal of the American Chemical Society*, **123**, 7194; (b) Legros, J. and Bolm, C. (2004) *Angewandte Chemie*, **116**, 4321; (2004) *Angewandte Chemie – International Edition*, **43**, 4225; (c) Fujita, M. and Que, L., Jr, (2004) *Advanced Synthesis and Catalysis*, **346**, 190; (d) Legros, J. and Bolm, C. (2005) *Chemistry - A European Journal*, **11**, 1086.

48 Nakanishi, M. and Bolm, C. (2006) *Advanced Synthesis and Catalysis*, **348**, 1823.

49 For a related system., see: Kim, S.S., Sar, K.S. and Tamrakar, P. (2002) *Bulletin of the Korean Chemical Society*, **23**, 937.

50 (a) Nakanishi, M., Salit, A.-F. and Bolm, C. (2008) *Advanced Synthesis and Catalysis*, **350**, 1835; (b) Mayer, A.C., Salit, A.-F. and Bolm, C. (2008) *Chemical Communication*, **45**, 5975–5977.

51 For contributions of this group on alkane oxidations involving microorganisms, see, (a) Wilkes, H., Kühner, S., Bolm, C., Fischer, T., Classen, A., Widdel, F. and Rabus, R. (2003) *Organic Geochemistry*, **34**, 1313; (b) Kniemeyer, O., Musat, F., Sievert, S.M., Knittel, K. Wilkes, H., Blumenberg, M. Michaelis, W., Classen, A., Bolm, C., Joye, S.B. and Widdel, F. (2007) *Nature*, **449**, 898.

17
Development of Novel Ruthenium and Iron Catalysts for Epoxidation with Hydrogen Peroxide

Man Kin Tse, Bianca Bitterlich, and Matthias Beller

17.1
Introduction

Oxidation of olefins to obtain value-added products is of major importance in the chemical industry [1]. Among the various oxidation methods, epoxidation of olefins continues to be an interesting field of research in industry and academia. The formation of two C–O bonds in one reaction and the facile ring-opening reactions make epoxides versatile building blocks for materials, bulk and fine chemicals as well as agrochemicals and pharmaceuticals [2]. Despite all advances in catalytic oxidations, epoxides are still often synthesized by stoichiometric reaction of olefins with peroxoacids generated from hydrogen peroxide and acids or acid derivatives [3]. A drawback of this convenient method is the limited use for acid-labile olefins or epoxides and the generation of significant amounts of waste (salts). Obviously, in the context of more sustainable production processes, reduction of waste by-products is highly desired.

Among the different epoxides the industrially most important examples with respect to scale are ethylene oxide and propylene oxide. For example, nowadays, approximately 7 million tons of propylene oxide are produced annually and the demands grow steadily [4]. Three major routes are being used for this simple epoxide manufacture: chlorohydrin process, propylene oxide/styrene monomer process and propylene oxide/*tert*-butyl alcohol process (Scheme 17.1). These processes produce various by-products and must be incorporated with other downstream treatment or production plants. Even the by-products, such as methyl *tert*-butanol and styrene, are useful, but the price of the epoxide is adversely affected if a by-product is over produced. From the point of view of environmental and economical considerations, the applied oxidant hence determines the value of the system to a significant extent [5]. It is apparent that molecular oxygen is the most ideal oxidant for this transformation [6]. However, usually only one oxygen atom of molecular oxygen is used productively for oxidation (50% atom efficiency) [7], and thus at least stoichiometric amounts of unwanted by-products are generated during the reactions. In spite of this,

Activating Unreactive Substrates: The Role of Secondary Interactions.
Edited by Carsten Bolm and F. Ekkehardt Hahn

ISBN: 978-3-527-31823-0

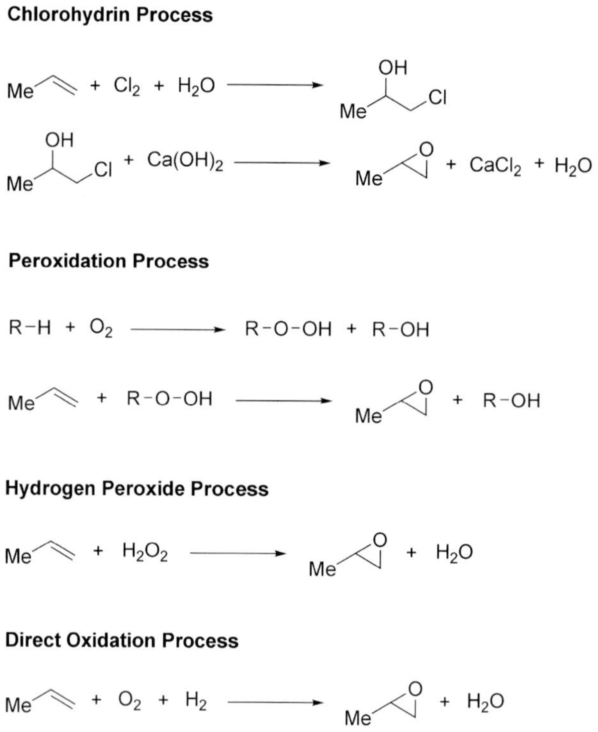

Chlorohydrin Process

$$MeCH{=}CH_2 + Cl_2 + H_2O \longrightarrow MeCH(OH)CH_2Cl$$

$$MeCH(OH)CH_2Cl + Ca(OH)_2 \longrightarrow \text{propylene oxide} + CaCl_2 + H_2O$$

Peroxidation Process

$$R{-}H + O_2 \longrightarrow R{-}O{-}OH + R{-}OH$$

$$MeCH{=}CH_2 + R{-}O{-}OH \longrightarrow \text{propylene oxide} + R{-}OH$$

Hydrogen Peroxide Process

$$MeCH{=}CH_2 + H_2O_2 \longrightarrow \text{propylene oxide} + H_2O$$

Direct Oxidation Process

$$MeCH{=}CH_2 + O_2 + H_2 \longrightarrow \text{propylene oxide} + H_2O$$

Scheme 17.1

development of catalysts for the direct epoxidation of propylene using molecular oxygen, with hydrogen as the co-reductant, is still one of the "Holy Grail" reactions in this field.

Apart from molecular oxygen, hydrogen peroxide, H_2O_2, is an environmentally benign oxidant, which theoretically generates only water as co-product [8]. BASF and Dow therefore launched the construction of the first propylene oxide production plant using H_2O_2 as the oxidant recently (Scheme 17.1) [4]. We believe that H_2O_2 will have many more applications in the synthesis of fine chemicals, pharmaceuticals, agrochemicals and electronic materials because of its characteristic physical properties. Hence, the discovery and advancement of new improved catalysts using H_2O_2 is an important and challenging goal in oxidation chemistry [9].

17.2 Development of Epoxidation Catalysts Using H_2O_2

Excellent reviews have appeared in recent years on the utilization of hydrogen peroxide as the epoxidation oxidant in the presence of various catalysts [9, 10].

The combination of environmentally friendly H_2O_2 with non-toxic and inexpensive metal catalysts is undoubtedly an ideal system for epoxidation reactions [11]. Among the various potential oxidation metals, iron-based catalysts are preferable, because iron is the most abundant metal in the earth's crust and is essential in nearly all organisms [12]. Many biological systems such as hemoglobin, myoglobin, cytochrome oxygenases, non-heme oxgenases and [FeFe] hydrogenase are iron-containing enzymes or co-enzymes [13, 14]. With nature as our guide, we should be capable of using biomimetic or bio-inspired approaches to develop new synthetically useful epoxidation protocols with iron catalysts [15].

In this report we mainly describe our efforts to develop novel ruthenium- and iron-based epoxidation catalysts. For comparison we also included recent work of other research groups in this area.

17.2.1 Ruthenium-catalyzed Epoxidation

Ruthenium porphyrins have long been used as models for cytochrome P-450 for the epoxidation of olefins [16]. Pyridine *N*-oxides and iodosobenzenes have usually been used as the terminal oxidant. Comparatively few examples showed epoxidation activity with H_2O_2 or oxygen in the presence of ruthenium catalysts. Most notably a sterically encumbered *trans*-dioxo ruthenium porphyrin complex showed catalytic activity toward epoxidation of olefins using both oxygen atoms of molecular oxygen [17, 18]. Though these types of high-valent ruthenium(VI) porphyrins are highly interesting epoxidation catalysts, their tedious synthesis has hindered their further application, both in asymmetric and bulk chemical epoxidation [13, 19].

Ruthenium(VIII) tetraoxide (RuO_4), which can be generated from $RuCl_3$ or RuO_2 with various oxidizing agents, is known to be a powerful oxidant. It has been applied for cleavage of olefins to aldehydes, ketones and carboxylic acids [16]. However, it is also possible to perform *cis*-dihydroxylation of olefins at low temperatures in a short time [20]. Under particular conditions, α-ketols can also be formed preferentially [21]. Before our work, relatively few examples of ruthenium-catalyzed epoxidations of olefins using H_2O_2 had been described [22].

During the re-investigation on the ruthenium-catalyzed asymmetric epoxidation system of Nishiyama and co-workers using various oxidants [23], we discovered that slow addition of alkyl peroxides or hydrogen peroxide significantly improved the yield of chiral epoxides (vide infra) [24]. Hence, the unproductive decomposition of H_2O_2 to O_2 is minimized. As model reaction the epoxidation of *trans*-β-methylstyrene was studied (Scheme 17.2). We were able to show that 2,6-pyridinedicarboxylic acid (H_2pydic) is an efficient and essential ligand for this reaction (Table 17.1). It is evident that the *O,N,O*-dianionic tridentate moiety is the origin of the catalyst activity. Removal of only one carboxylic group or the pyridine nitrogen gave inferior

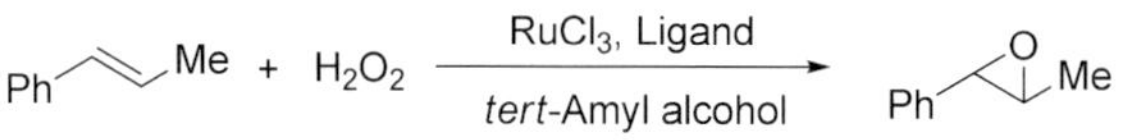

Ph–CH=CH–Me + H_2O_2 —[$RuCl_3$, Ligand / *tert*-Amyl alcohol]→ Ph–(epoxide)–Me

Scheme 17.2 Epoxidation of *trans*-β-methylstyrene with H_2O_2.

Table 17.1 Ligand effects of $RuCl_3 \cdot xH_2O$-catalyzed epoxidation of *trans*-β-methylstyrene with H_2O_2[a] [24].

Entry	Ligand	Conv. [%]	Yield [%]	Selec. [%]
1	–	34	0	0
2	HO_2C, N, CO_2H	100	>99	>99
3	Me, N, Me, O, O	26	0	0
4	CO_2H, N, CO_2H	30	25	83
5	N, CO_2H	5	0	0
6	HO_2C, CO_2H	29	0	0
7	HO_2C, N, H, CO_2H	56	46	82

[a] General conditions: 0.5 mmol *trans*-β-methylstyrene, 1 mol% $RuCl_3 \cdot xH_2O$ in 9 mL *tert*-amyl alcohol, 12 h slow addition of 3.0 equiv. 30% H_2O_2 in 1 mL *tert*-amyl alcohol, room temperature, 10 mol% ligand.

results. The developed reaction protocol is relatively general (Table 17.2). Both aliphatic and aromatic olefins can be oxidized to the corresponding epoxides in good to excellent yields. In most of the cases, 1 mol% of ruthenium is sufficient for high product yield. In some cases, as little as 0.01 mol% of ruthenium catalyzes the reaction efficiently. Hence, a maximum turnover number (TON) of 16 000 was determined. The major drawback for this reaction is the necessity for a relatively high concentration of ligand loading (10 mol%).

With respect to asymmetric epoxidation, a variety of chiral ruthenium catalysts have been developed (Scheme 17.3) [16]. Mono-, bi-, tri- and tetradentate ligands as well as macrocyclic porphyrins have been demonstrated as selective ligands for this reaction. Moreover, various coordinating atoms including N, O, S or P can all be beneficial to these Ru catalysts. In some cases a combination of ligands was advantageous. For us the coordinatively saturated ruthenium(II) complex, Ru(pyridine-2,6-bisoxazoline)(2,6-pyridinedicarboxylate) **1**, which is composed of two different meridional ligands, was of particular interest [23]. Indeed, **1** becomes a more practical epoxidation catalyst with $PhI(OAc)_2$ on adding a defined amount of water ito the reaction mixture [25]. Later, it was shown that alkyl peroxides [26] and hydrogen peroxide [27] can be used

Table 17.2 Ru-catalyzed epoxidation of different olefins with H_2O_2.[a] [24].

Entry	Substrate	$RuCl_3 \cdot xH_2O$ [mol%]	Conv. [%]	Yield [%]	Selec. [%]
1	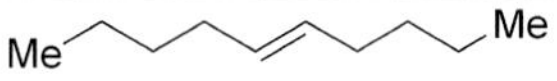 Me, Me	5	78	74	95
2		1[b]	100	90	90
3	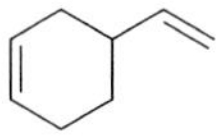	0.1	79	66	83[c]
4	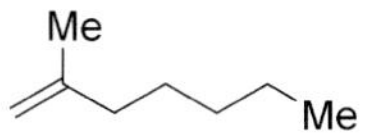 Me, Me	0.1	100	95	95
5	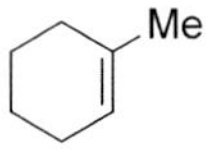 Me	0.01	100	>99	>99
6	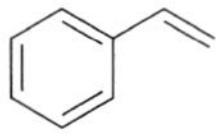	1	100	71	71
7	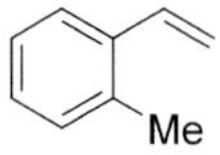 Me	1	100	70	70
8	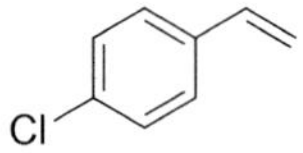 Cl	1[b]	100	>99	>99
9	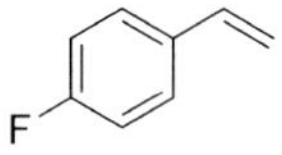 F	1	100	76	76
10	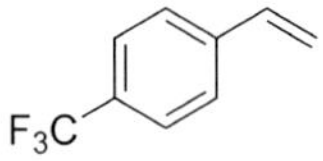 F_3C	1[c]	49	46	93
11	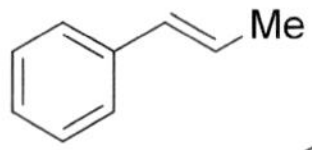 Me	0.01	100	96	96
12	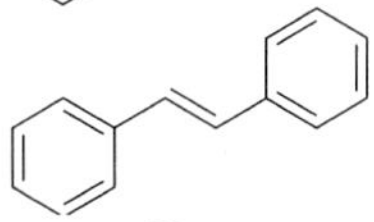	0.1	100	93	93
13	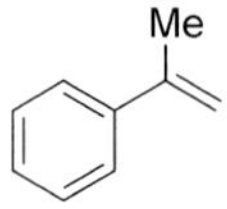 Me	0.1	100	90	90
14	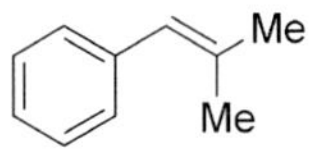 Me, Me	0.1	100	96	96
15	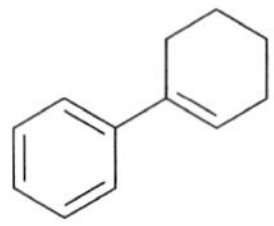	0.01	100	97	97[d]

(*Continued*)

Table 17.2 (*Continued*)

Entry	Substrate	$RuCl_3 \cdot xH_2O$ [mol%]	Conv. [%]	Yield [%]	Selec. [%]
16	(cinnamyl acetate, structure: Ph–CH=CH–CH₂–OAc) OAc	1	100	80	80

[a] General conditions: 0.5 mmol substrate, $RuCl_3 \cdot xH_2O$, 10 mol% pyridine-2,6-dicarboxylic acid in 9 mL *tert*-amyl alcohol, 12 h slow addition of 3 equiv. 30% H_2O_2 in 1 mL *tert*-amyl alcohol, room temperature.
[b] 20 mol% pyridine-2,6-dicarboxylic acid.
[c] Product: 3-Vinyl-7-oxabicyclo[4.1.0]heptane.
[d] Respective values for Ru-concentration of 0.001 mol%: conversion 16%, yield 16%, selectivity >99%.

as oxidants. Applying complexes of type **1**, it seemed possible to tune up the reactivity and (enantio)selectivity by modifying the chiral (pybox) and the achiral (pydic) ligands separately.

Based on this concept, several ruthenium(II) pre-catalyst types for epoxidation were synthesized by us (Scheme 17.4). These are all active catalysts toward various aromatic olefins and produce the corresponding epoxides in good to excellent yields (Table 17.3). With respect to the asymmetric induction, the complexes of pyridine-2,6-bisoxazoline (pybox) (**1**) and pyridine-2,6-bisimidazoline (pybim) (**12**) gave very similar results. The complex of pyridine-2,6-bisoxazine (pyboxazine) (**11**) gave the best stereoselective induction, and up to 84% *ee* was obtained. It was suggested that the 6-membered ring structure of the oxazine causes the chiral carbon center to be closer to the active metal center. This argument was supported by comparing the corresponding X-ray crystal structures of **1** and **11** [27a,b]. With spectroscopic methods, independent synthesis and DFT calculations, mechanistic studies revealed an unusual pyridine *N*-oxide intermediate [27b,37]. Particularly interesting is the ruthenium complex **13**. On the one hand, it is a very general epoxidation catalyst. Both aliphatic and aromatic olefins can be epoxidized smoothly, all the 6 classes of substituted olefins work nicely [27e], and a turnover number (TON) of 8800 with high conversion (93%) was also observed. On the other hand, **13** is also an excellent oxidation catalyst for naphthalene [38] and alcohol oxidation [39]. It is noteworthy that **13** reached a TON of 14 800 for alcohol oxidation with 30% aqueous H_2O_2 in solvent-free and base-free conditions. The usefulness of this type of ruthenium catalysts is summarized in Scheme 17.5

17.2.2
Biomimetic Iron-catalyzed Epoxidation

Iron is the most abundant element by mass on the earth [47]. Because of its low cost, environmentally friendly properties and biological relevance, there is an increasing interest in using iron complexes as catalysts for a wide range of reactions [48]. In this respect, iron catalysts for C–C [48a,49] and C–N [50] bond formation

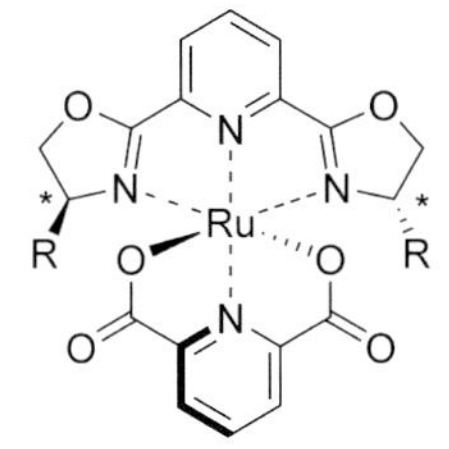

1 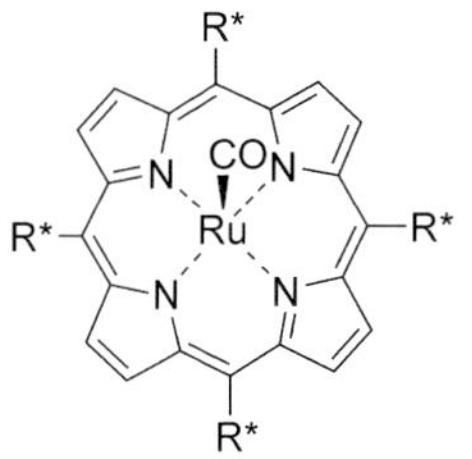 2 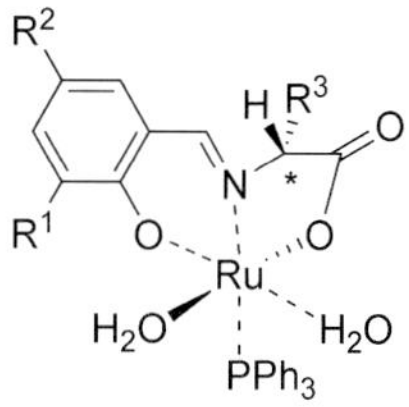 3

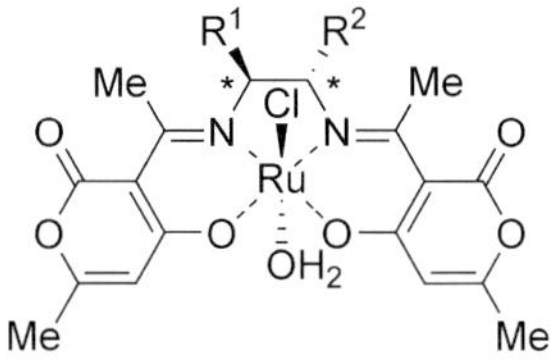

4  5 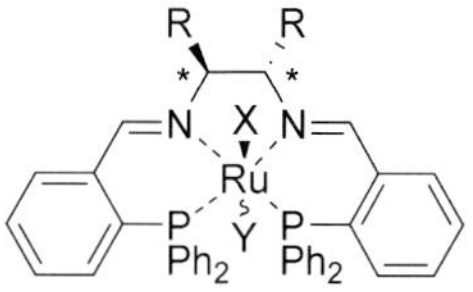 6

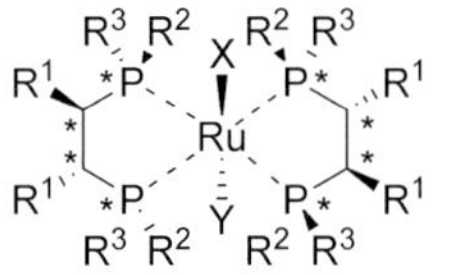

7 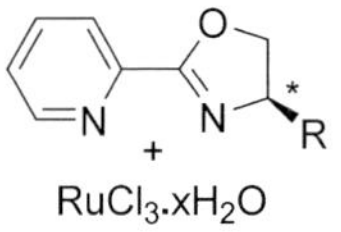 8

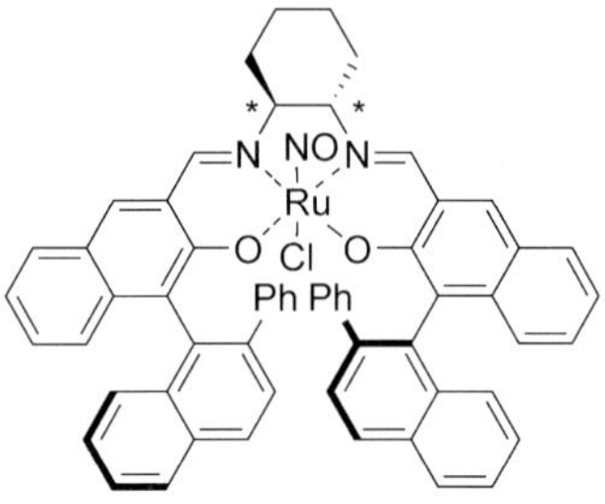

9 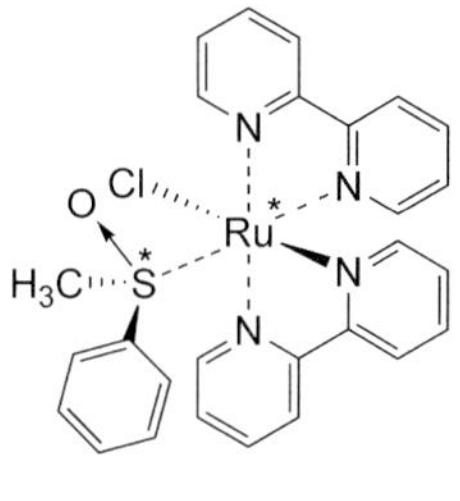 10

Scheme 17.3 Ruthenium catalysts for asymmetric epoxidation of olefins.

reactions, hydrogenations [51] and oxidations are attracting much attention [9, 10, 15]. However, the use of H_2O_2 in combination with simple iron complexes as oxidation catalysts is limited [52, 53]. This is partly because H_2O_2 decomposes vigorously in the presence of iron, which reduces the efficiency of the oxidant [54]. The generation of highly reactive free hydroxyl radical by Fenton or Gif chemistry induces the

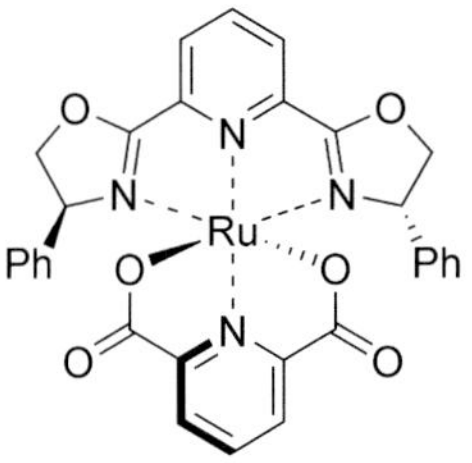

Scheme 17.4 Selected ruthenium(II) catalysts for epoxidation of aromatic olefins using 30% H_2O_2.

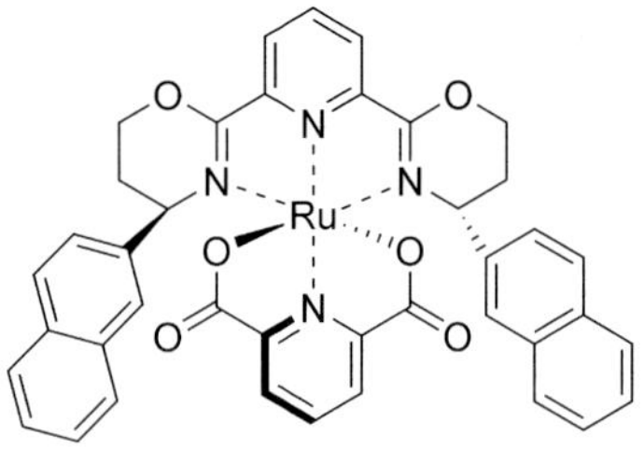

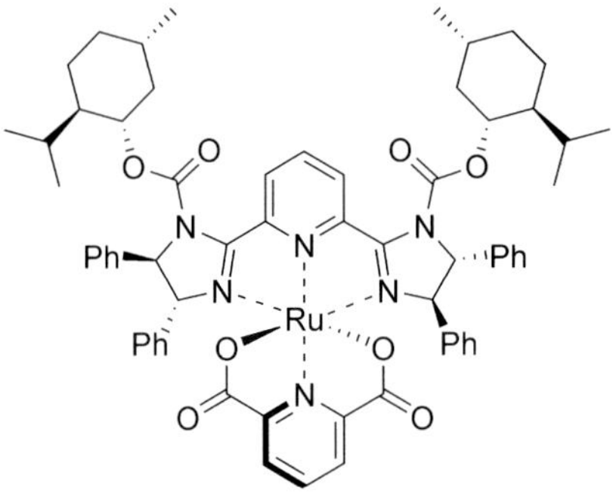

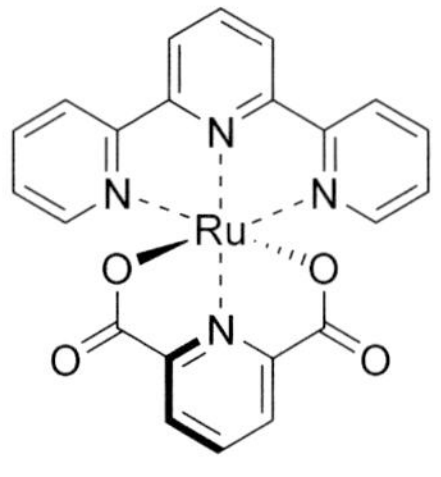

decomposition of the ligand and substrate as well as the product [52d–e]. Hence, a high ratio of substrate to hydrogen peroxide is usually employed to solve these problems. In fact, only a few examples of iron catalysts utilizing H_2O_2 as the oxidant are known (Scheme 17.6).

Olefins can be oxidized in the presence of anhydrous $FeCl_3$ to a mixture of dimer, epoxide, aldehydes and other products using 100% H_2O_2 in anhydrous CH_3CN as the oxidant [53]. The exceptional oxidizing power of anhydrous H_2O_2 and the low selectivity of the reaction restrict further applications of this system. Heme models such as **14** and **15** have been reported to catalyze the epoxidation of olefins to the corresponding epoxides in good yield. However, they have different mechanisms to improve the efficiency of H_2O_2. While **14** activates H_2O_2 with hydrogen bonding to the NH proton, **15** enhances the selectivity of epoxidation towards H_2O_2 oxidation. The major drawback for these heme-model systems is the tunability of the catalysts for different olefins. A number of non-heme models were reported recently. Of particular interest are complexes **16** and **17**, which were synthesized by Jacobsen's group and Que's group with the same parent ligand, respectively [57, 58]. The self-assembling MMO mimic **16** catalyzed the epoxidation of a range of aliphatic olefins. Even the relatively non-reactive substrate, 1-decene, can be oxidized

Table 17.3 Epoxidation of aromatic olefins with ruthenium complexes **1** and **11–13** with 30% H_2O_2.

$R^1R^2C{=}CR^3R^4$ + 30% H_2O_2 → (5 mol% **catalyst**, *tert*-AmOH, r.t., 12 h addition H_2O_2) → epoxide

Entry	Substrate	Catalyst 1	Catalyst 11	Catalyst 12	Catalyst 13
1	Ph–CH=CH$_2$	75% yield 42% *ee*	85% yield 59% *ee*	76% yield 42% *ee*	71% yield –
2	Ph–CH=CH–Me	82% yield 58% *ee*	95% yield 72% *ee*	100% 65% *ee*	>99% yield –
3	Ph–CH=CH–Ph	100% yield 67% *ee*	100% yield 54% *ee*	97% yield 71% *ee*	– –
4	Ph–CH=C(Me)–Me	91% yield 72% *ee*	91% yield 84% *ee*	82% yield 68% *ee*	>99% yield –

to the corresponding epoxide in 85% yield in 5 min. The presence of acetic acid not only affects the structure of the intermediate, but also the selectivity of the products (Table 17.4). A higher ratio of epoxide to *cis*-diol was observed in the presence of HOAc. This effect is more significant when **18** is employed as the catalyst. *In-situ* generation of peracetic acid was therefore suggested [59b]. Though these catalysts show interesting selectivity toward epoxides and diols for aliphatic olefins, they decompose aromatic olefins to unidentified products. This reduces the general application of these types of catalysts for synthetic purpose. Stack's catalyst derived from phenanthroline is synthetically more practical [60]. In spite of the fact that peracetic acid is used, a wide range of olefins can be oxidized to the corresponding epoxides with only 0.25 mol% of catalyst.

Despite this excellent work, Fe catalysts are still underdeveloped for general epoxidation using aqueous H_2O_2 under neutral conditions. Instead of synthesizing molecularly well-defined iron complexes as epoxidation catalysts, we tackled this problem in another way. Based on the *in-situ* generation method from our ruthenium system [25], we transferred the technology to simple iron systems using a combinatorial ligand strategy (Scheme 17.7) [61]. It was found that $FeCl_3{\cdot}6H_2O$ can substitute the ruthenium source in the model reaction with 2,6-bis(4-phenyl-4,5-dihydrooxazol-2-yl)pyridine (Ph_2-pybox) and H_2pydic as ligands (Table 17.5). However, the synthesized iron(II) complex with the *N,N,N*-tridentate Ph_2-pybox is less active than the *in-situ* generated catalyst. Noteworthy evidence, including mass spectrometric analysis and electronic absorption spectra, suggests that the pybox ligand was decomposed during the reaction and generated the active catalyst species. Also, combination of the fragments gave a comparable result to the pybox ligand alone. With this information and the screening method in hand, the effect of the base was demonstrated to be crucial in this system (Table 17.6).

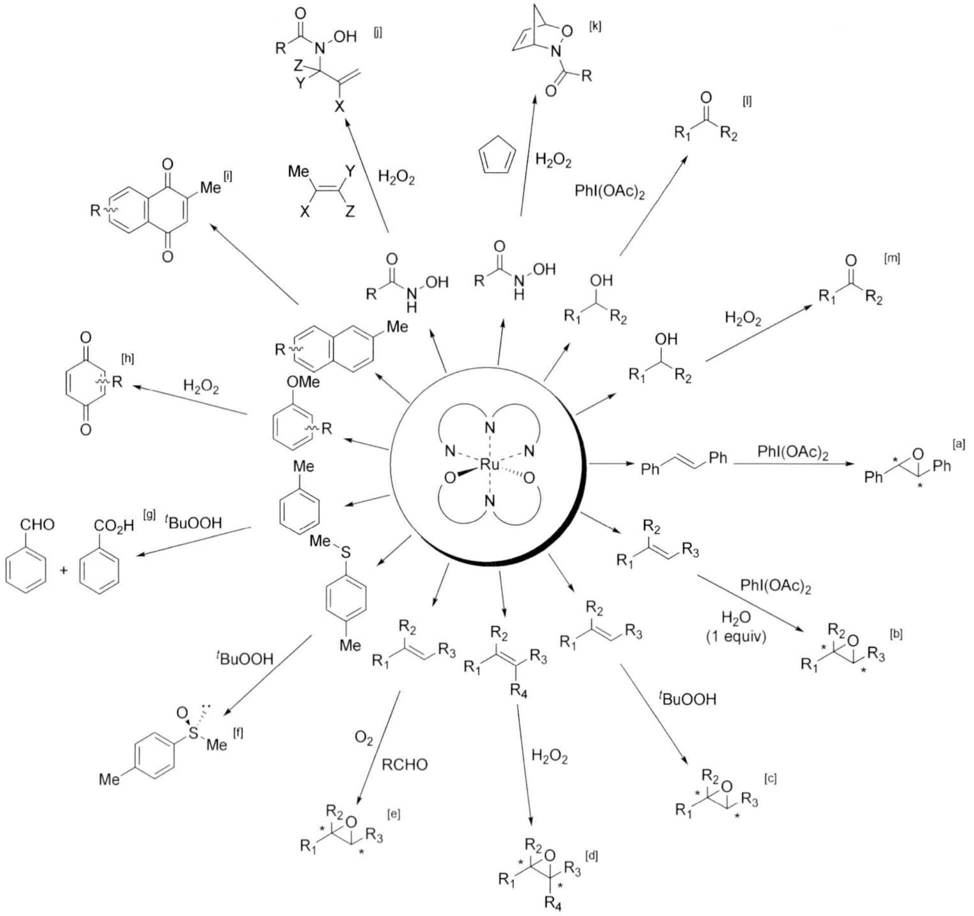

Scheme 17.5 [a] see Ref. [23]. [b] see Ref. [25]. [c] see Ref. [26]. [d] see Ref. [27]. [e] see Ref. [40]. [f] see Ref. [41]. [g] see Ref. [42]. [h] see Ref. [43]. [i] see Ref. [38]. [j] see Ref. [44]. [k] see Ref. [45]. [l] see Ref. [46]. [m] see Ref. [39].

It is apparent that all organic bases form active species with $FeCl_3{\cdot}6H_2O$ and H_2pydic. Notably, excellent activity was found with pyrrolidine, benzyl amine and 4-methylimidazole. For some aromatic olefins, the reaction is finished in 5 min and slow dosage of H_2O_2 is not necessary. However, the structure of the base has a dramatic effect on the system. Hence, it is evident that the organic base deprotonates the H_2pydic and acts as the ligand as well. As imidazole derivatives also play an essential role as ligands or base in enzymes [62], this combinatorial-activity approach provides a general platform for searching functional mimics for iron enzymes. Understanding the fundamental aspects of these systems should allow us to develop synthetically more useful biomimetic iron-catalyzed epoxidation systems. Hence, by fine tuning the imidazole ligand, a bio-inspired $FeCl_3$/imidazole-system

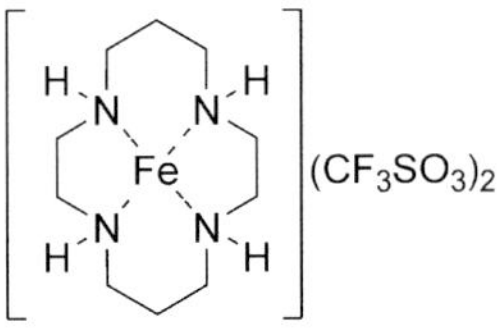

14

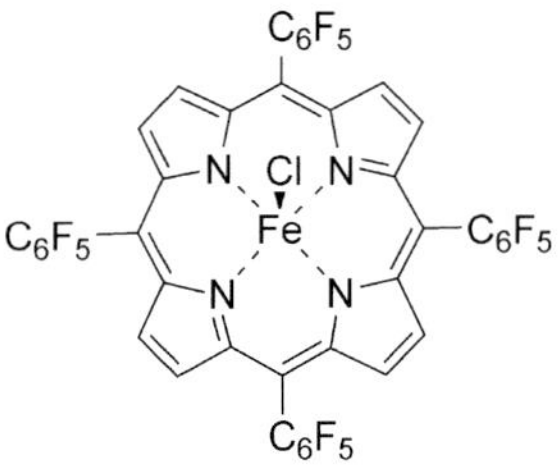

15

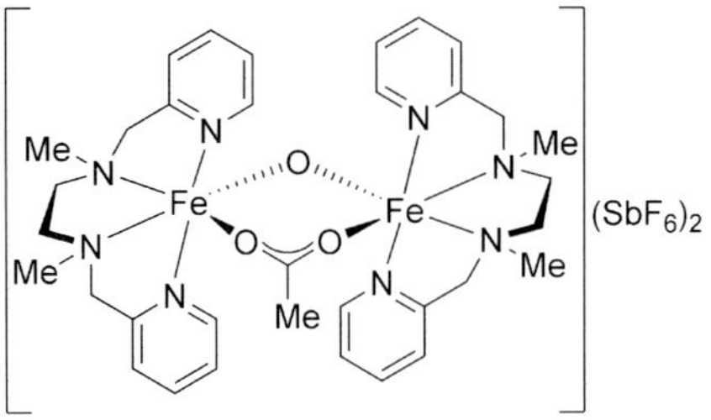

16

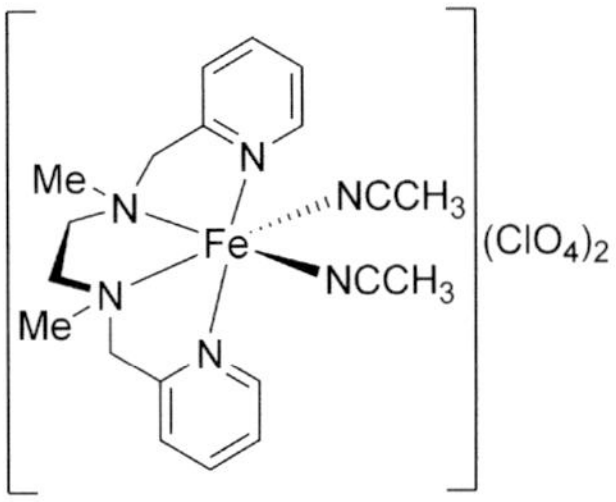

17

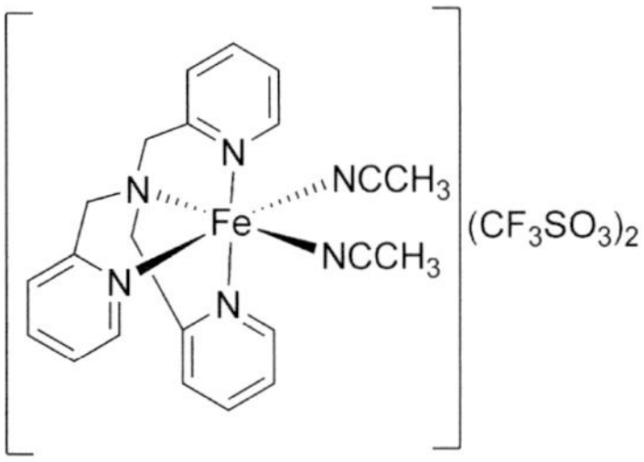

18

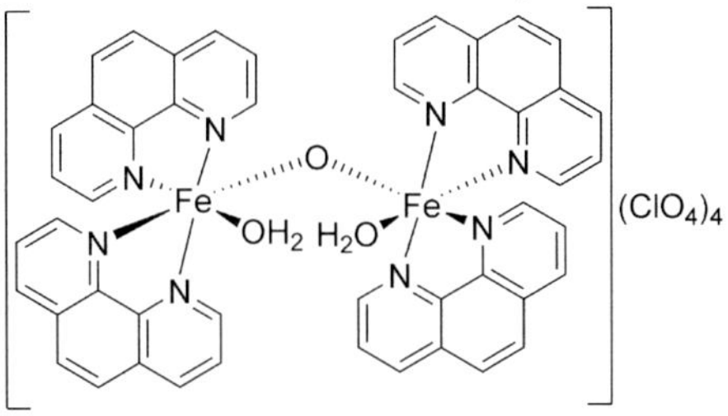

19

Scheme 17.6 Iron catalysts for epoxidation of olefins using H_2O_2.

could also be developed for epoxidations of olefins using aqueous hydrogen peroxide as the oxidant. A wide range of imidazoles displayed good selectivity for epoxidation of *trans*-stilbene. With 15 mol% of 5-chloro-1-methylimidazole and 5 mol% $FeCl_3 \cdot 6H_2O$, 87% yield of *trans*-stilbene oxide with 94% selectivity is achieved [63].

Table 17.4 Epoxidation of cyclooctene with **16** and **17** with 50% H_2O_2.

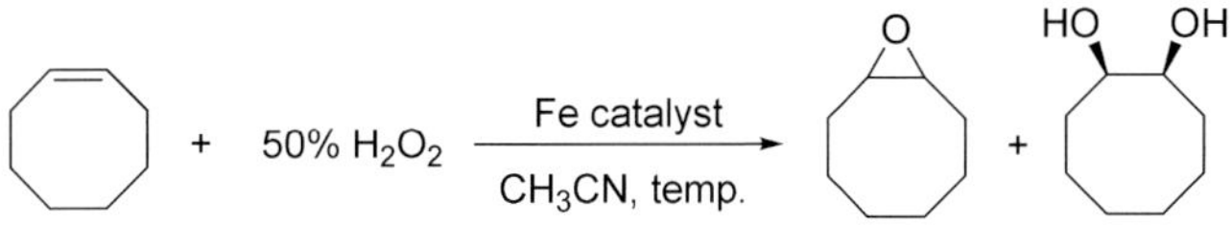

Entry	Catalyst	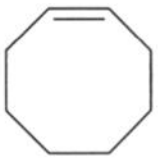 cyclooctene:H_2O_2	Temp. [°C]	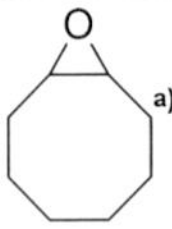 epoxide[a]	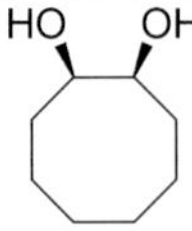 diol[b]	Additive	Ref.
1	**16** 3 mol%	1:1.5	4	85%	0%	30 mol% HOAc	[57]
2	**17** 10 mol%	100:1	30	75%	9%	nil	[58b]
3	**17**[b] 7 mol%	100:2.8	r.t.	91%	5%	nil	[59b]
4	**17**[c] 7 mol%	100:2.8	r.t.	102%[c]	3%	34 mol% HOAc	[59b]
5	**18** 7 mol%	100:2.8	r.t.	30%	41%	29 mol% HOAc	[59b]
6	**18** 7 mol%	100:2.8	r.t.	88%[c]	14%	29 mol% HOAc	[59b]
7	**19** 0.25 mol%	1:2 CH_3CO_3H	0	100%	0%	nil	[60]

[a] Yield based on the limiting reagent.
[b] Trifluoromethansulfonate was the anion of the iron complex instead of perchlorate.
[c] A total yield over 100% has been reported due to experimental error.

This system showed a more general activity and higher selectivity toward various olefins than that of the $FeCl_3{\cdot}6H_2O/H_2$pydic/pyrrolidine system (Table 17.7). However, the H_2pydic system has a higher activity for some substrates. Both reaction systems demonstrated that it is possible to perform epoxidations with hydrogen peroxide in the presence of easily manageable and simple Fe/Ligand catalysts. Reports of the efforts to examine the reaction mechanism and to realize an asymmetric version of these reactions are currently in preparation in our laboratory.

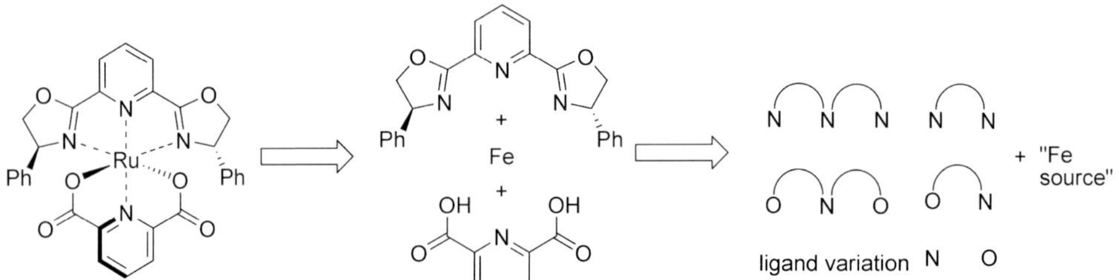

Scheme 17.7 Iron catalyst modification strategy.

Until very recently only a few examples of asymmetric epoxidations using iron-based catalysts had been reported in the literature (Scheme 17.8). This can be partly

Table 17.5 Epoxidation of *trans*-stilbene with *in-situ* generated Fe catalysts.

Ph–CH=CH–Ph → (5 mol% "Fe source", L_1, L_2, 1 h, 65 °C; 3 equiv. H_2O_2, *tert*-amyl alcohol, r.t., 12 h addition) → *trans*-stilbene oxide (Ph/Ph epoxide)

Entry	Fe source	L_1	L_2	Conv.[a),b)] [%]	Yield[b)] [%]	Selec.[c)] [%]
1	$FeCl_3 \cdot 6H_2O$	Ph_2-pybox[d)]	H_2pydic	55	48	87
2	$Fe(Ph_2\text{-pybox})Cl_2$	–	–	22	15	67
3	$FeCl_3 \cdot 6H_2O$	–	–	13	7	52
4	$FeCl_3 \cdot 6H_2O$	Ph_2-pybox	–	100	82	82
5	$FeCl_3 \cdot 6H_2O$	–	H_2pydic	69	53	77
6	$FeCl_3 \cdot 6H_2O$	HO–CH₂–CH(NH₂)–Ph (amino alcohol: HO, NH_2, Ph)	H_2pydic	100	84	84
7	$FeCl_3 \cdot 6H_2O$	HO–CH₂–CH(NH₂)–Ph (amino alcohol: HO, NH_2, Ph)	–	9	6	69

a) Reaction conditions: In a 25 mL Schlenk tube, iron trichloride hexahydrate (0.025 mmol), Ph_2-pybox (0.025 mmol) and H_2pydic (0.025 mmol) were dissolved in *tert*-amyl alcohol (4 mL) and heated for 1 h at 65 °C. Afterwards *trans*-stilbene (0.5 mmol), *tert*-amyl alcohol (5 mL) and dodecane (GC internal standard, 100 μL) were added in sequence at r.t. in air. To this mixture, a solution of 30% hydrogen peroxide (170 μL, 1.5 mmol) in *tert*-amyl alcohol (830 μL) was added over a period of 12 h at r.t. by a syringe pump.
b) Conversion and yield were determined by GC analysis.
c) Selectivity refers to the ratio of yield to conversion in percentage.
d) 2,6-bis(4-phenyl-4,5-dihydrooxazol-2-yl)pyridine = Ph_2-pybox.

attributed to the rapid decomposition of H_2O_2 with iron catalysts. The other reason possibly comes from inferior activity of iron porphyrins compared to their manganese counterparts in the early studies of heme models. Thus, researchers concentrated on the development of manganese catalysts in the 1990 s [64]. As a result, only a handful of asymmetric epoxidation systems using iron porphyrin heme mimics are known, and only iodosobenzene was successfully applied as an oxidant [65]. Electron-deficient polyfluorinated porphyrin was the catalyst for the catalyzed epoxidation of olefins with H_2O_2 [56]. However, the synthesis of chiral porphyrins with electron-withdrawing groups has still not yet been realized because of their notorious reputation for labor-intensive and expensive syntheses [13, 19]. Aerobic epoxidation of styrene derivatives with an aldehyde co-reductant catalyzed by tris(δ,δ-dicampholylmethanato) iron(III) complex **21** was also reported [66]. In spite of the encouraging results, more environmentally benign oxidant and solvent are still to be developed. Notably, in the dihydroxylation of of *trans*-2-heptene with H_2O_2 using biomimetic non-heme Fe catalysts, $[Fe(BPMCN)(CF_3SO_3)_2]$ **22**, 58% of the epoxide with 12% *ee* was obtained [58b]. By elaborating 5760 metal-ligand combinations, Francis and Jacobsen identified three Fe complexes with peptide-like ligands, which gave the epoxide in 15–20% *ee* in the asymmetric epoxidation of *trans*-β-methylstyrene

Table 17.6 Epoxidation of *trans*-stilbene with different bases.

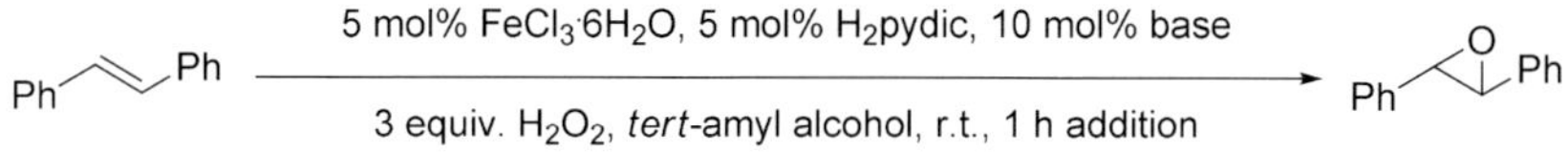

Ph–CH=CH–Ph → (5 mol% $FeCl_3 \cdot 6H_2O$, 5 mol% H_2pydic, 10 mol% base; 3 equiv. H_2O_2, *tert*-amyl alcohol, r.t., 1 h addition) → Ph–(O)–Ph

Entry	Base	Conv.[a), b)] [%]	Yield [b)] [%]	Selec.[c)] [%]
1	KOH	33[d)]	30	91
2	Et_3N	86	74	86
3	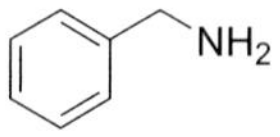 NH_2	100	97	97
4	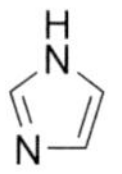 H N N	91	90	99
5	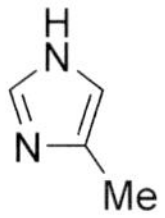 H N N Me	100[e)]	97	97
6	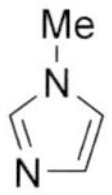 Me N N	78[e)]	72	92
7	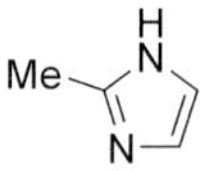 Me H N N	12[e)]	11	92
8	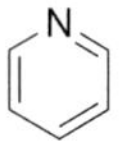 N	56	50	89
9	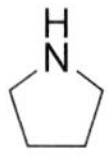 H N	100	97	97

a) Reaction conditions: In a 25 mL Schlenk tube, $FeCl_3 \cdot 6H_2O$ (0.025 mmol), H_2pydic (0.025 mmol), *tert*-amyl alcohol (9 mL), base (0.05 mmol), *trans*-stilbene (0.5 mmol) and dodecane (GC internal standard, 100 μL) were added in sequence at r.t. in air. To this mixture, a solution of 30% hydrogen peroxide (170 μL, 1.5 mmol) in *tert*-amyl alcohol (830 μL) was added over a period of 1 h at r.t. by a syringe pump.
b) Conversion and yield were determined by GC analysis.
c) Selectivity refers to the ratio of yield to conversion in percentage.
d) H_2O_2-addition over a period of 12 h.
e) Addition of 2 equiv. of H_2O_2.

utilizing 30% H_2O_2. The homogeneous catalyst **23** derived from this study gave 48% *ee* with 100% conversion of *trans*-β-methylstyrene. It is clear in this example that a combinatorial ligand approach can lead to promising new bio-inspired iron epoxidation catalysts.

The first breakthrough in nonporphyrin iron-catalyzed asymmetric epoxidation of aromatic alkenes using hydrogen peroxide has been reported very recently by us.

Table 17.7 Scope and limitations of $FeCl_3 \cdot 6H_2O/H_2pydic$/pyrrolidine and $FeCl_3 \cdot 6H_2O$/5-chloro-1-methylimidazole systems.

$$Ar^1\text{-}CH{=}C(R^3)\text{-}R^2 \xrightarrow[\text{3 equiv. } H_2O_2,\ tert\text{-amyl alcohol, r.t., 1 h addition}]{\text{5 mol\% "Fe catalyst"}} Ar^1\text{-epoxide}(R^2)(R^3)$$

Entry	Substrate	$FeCl_3 \cdot 6H_2O/H_2pydic$/pyrrolidine[61]			$FeCl_3 \cdot 6H_2O$/5-chloro-1-methylimidazole[63]		
		Conv.a),b) [%]	Yield b) [%]	Selec.c) [%]	Conv.a),b) [%]	Yield b) [%]	Selec.c) [%]
1	trans-stilbene	100	97	97	92	87	94
2	α-methyl-stilbene (Me)	40d)	21	53	46	41	88
3	benzylidenecyclohexane	25d)	11	44	39	25	64
4	cis-stilbene	22d)	8	36	34	24e)	71
5	1-phenylcyclohexene	85d)	21	25	52	36	68
6	styrene	94d)	93	99	74	70	95

a)Reaction conditions: In a 25 mL Schlenk tube, $FeCl_3 \cdot 6H_2O$ (0.025 mmol), H_2pydic (0.025 mmol), *tert*-amyl alcohol (9 mL), base (0.05 mmol), *trans*-stilbene (0.5 mmol) and dodecane (GC internal standard, 100 μL) were added in sequence at r.t. in air. To this mixture, a solution of 30% hydrogen peroxide (170 μL, 1.5 mmol) in *tert*-amyl alcohol (830 μL) was added over a period of 1 h at r.t. by a syringe pump.

b)Conversion and yield were determined by GC analysis;

c)Selectivity refers to the ratio of yield to conversion in percentage;

d)Addition of 2 equiv. of H_2O_2.

e)Additionally 3% *trans*-stilbene oxide and *trans*-stilbene were observed.

Here, good to excellent isolated yields of aromatic epoxides were obtained with *ee* values of up to 97% for stilbene derivatives.

In these investigations, *trans*-stilbene was used as the model substrate (Table 17.8) [68]. In earlier studies of nonasymmetric iron-catalyzed epoxidations

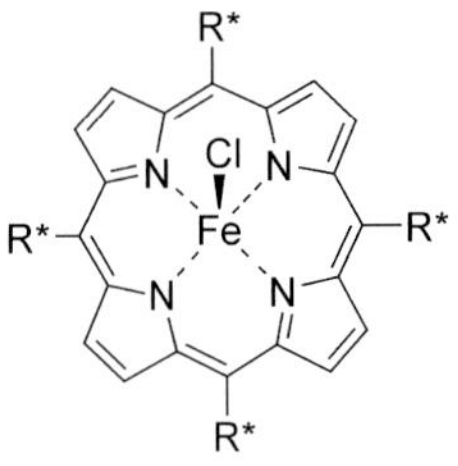

Scheme 17.8 Iron catalysts for asymmetric epoxidation of olefins.

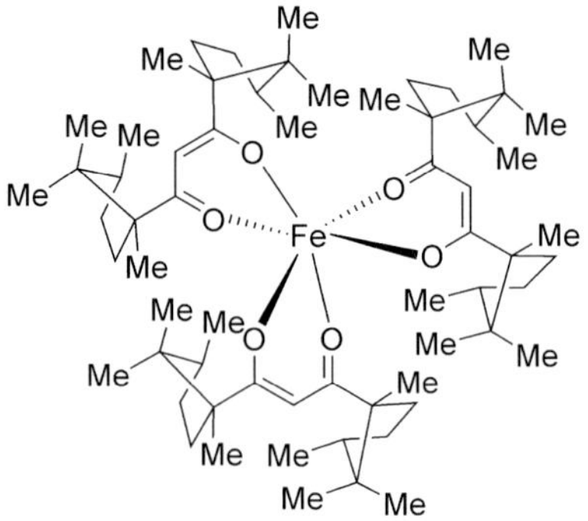

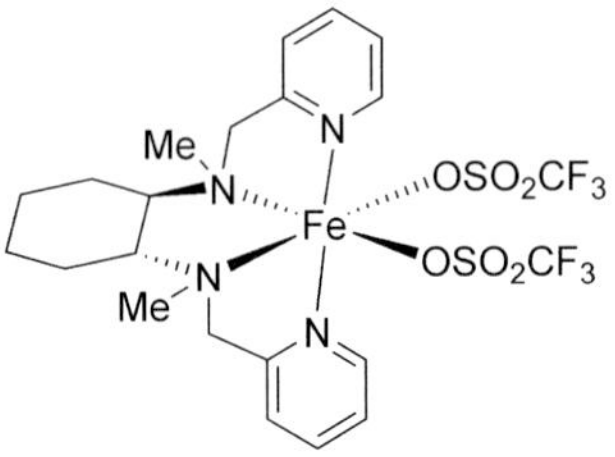

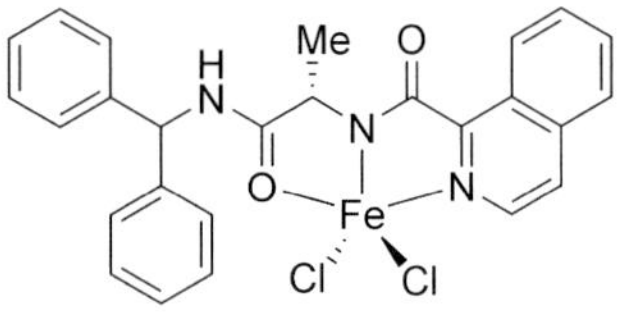

decomposition of the Ph_2-pybox ligand was observed [61]. Comparable activity is obtained with H_2pydic and phenylglycinol as ligand and base. However, less than 5% *ee* was observed with the enantiomerically pure aminoalcohol. Further investigation of other optically pure amines showed a high yield of the epoxide, albeit with low enantioselectivity. As ligands with a *p*-tolylsulfonamide substituent gave significantly higher *ee* values and good activity, formation of hydrogen bonding as the key chiral information-transferring step is suggested. Optimization by ligand modification showed that *N*-(*1R,2R*)-2-(benzylamino)-1,2-diphenylethyl)-4-methylbenzenesulfonamide (Table 17.8, entry 7) is the best ligand in terms of enantioselectivity (42% *ee*).

It has been demonstrated that *trans*-1,2-disubstituted aromatic olefins give excellent activity with moderate to excellent *ees* (Table 17.9). It should be mentioned that the activity of substituted *trans*-stilbenes is in the order: para > meta > ortho, presumably for steric reasons. The higher steric bulkiness of the 4,4′-dialkyl substituted compound, the higher is the enantiomeric excess (^{t}Bu>Me>H). The highest *ee* value was achieved with 2-(4-*tert*-butylstyryl)naphthalene as the substrate in the presence of 10 mol% of the iron catalyst (Table 17.9, entry 9). 91% *ee* and 46% isolated yield were obtained with the corresponding epoxide. By slightly lowering the reaction temperature to 10 °C, 97% *ee* was reached with complete substrate conversion within one hour (Table 17.9, entry 10).

Table 17.8 Iron-catalyzed asymmetric epoxidation of *trans*-stilbene.

Ph–CH=CH–Ph → (5 mol% $FeCl_3 \cdot 6H_2O$, 5 mol% H_2pydic; 12 mol% Ligand; H_2O_2, *tert*-amyl alcohol, r.t.) → Ph–*CH(O)CH*–Ph

Entry	Ligand	Conv.[a] [%]	Yield[a] [%]	Selec.[c] [%]	*ee*[b] [%]	Ref.
1	HO, NH_2, Ph	100	84	84	<5	[61]
2	N H, OH	95	73	77	0	[68]
3[d]	N H, NH_2	60	58	97	1	[68]
4[d]	Ph, Ph, N H, OH	78	53	68	10	[68]
5[e]	Ph, Ph, N H, F	100	98	98	17	[68]
6[f]	TsHN, NH_2, Ph, Ph	100	86	86	28	[68]
7[f]	TsHN, NHBn, Ph, Ph	100	87	87	42	[68]

[a] Reaction conditions: In a 25 mL Schlenk tube, iron trichloride hexahydrate (0.025 mmol), chiral ligand (0.060 mmol) and H_2pydic (0.025 mmol) and *trans*-stilbene (0.5 mmol) were dissolved in hot *tert*-amyl alcohol (9 mL). Dodecane (GC internal standard, 100 μL) were added in sequence at r.t. in air. To this mixture, a solution of 30% hydrogen peroxide (170 μL, 1.5 mmol) in *tert*-amyl alcohol (830 μL) was added over a period of 1 h at r.t. by a syringe pump.

[b] Conversion and yield were determined by GC analysis.

[c] Selectivity refers to the ratio of yield to conversion in percentage.

[d] 36 h.

[e] 0 °C, 14 h.

[f] 1.0 mmol H_2O_2.

In summary, we have demonstrated that high activity and chemoselectivity and even excellent enantioselectivity can be achieved in Fe-catalyzed epoxidations with hydrogen peroxide. This long-standing goal in oxidation catalysis is realized by combining $FeCl_3$ with appropriately chosen ligands. Chiral diamine ligands and pyridine-2,6-dicarboxylic acid as the combination provide an excellent platform for further improvement of this developing bio-inspired oxidation chemistry. It is apparent that an improvement in the generality of these catalytic systems is still awaited.

Table 17.9 Iron-catalyzed asymmetric epoxidation of various aromatic olefins [68].

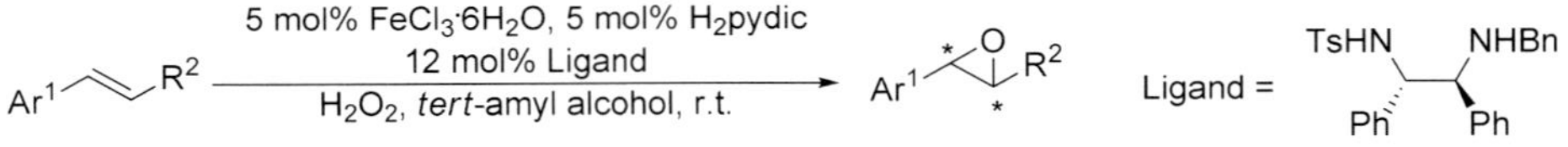

Ar^1–CH=CH–R^2 $\xrightarrow[H_2O_2,\ \textit{tert}\text{-amyl alcohol, r.t.}]{5\ mol\%\ FeCl_3{\cdot}6H_2O,\ 5\ mol\%\ H_2pydic,\ 12\ mol\%\ Ligand}$ Ar^1–*CH(O)*CH–R^2

Ligand = TsHN, NHBn, Ph, Ph

Entry	Substrate	Conv.[a] [%]	Yield[b] [%]	*ee*[c] [%], abs. Conf.
1	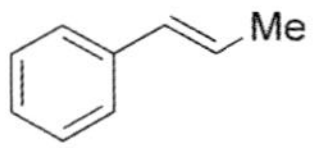 Me	100[a]	94[d]	28, (+)-(2*R*,3*R*)[e]
2	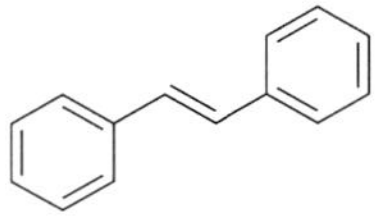 –	100	87	42, (+)-(2*R*,3*R*)
3	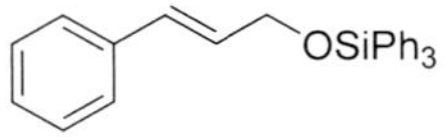 $OSiPh_3$	100	67	35, (+)-(2*R*,3*R*)[f]
4	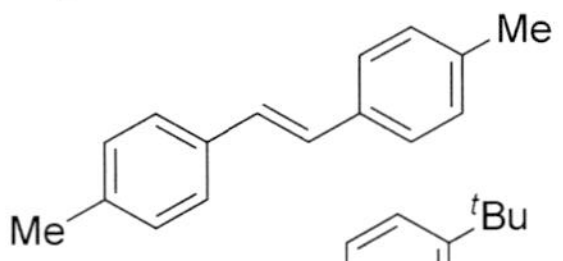 Me, Me	100	92	64, (+)-(2*R*,3*R*)[g]
5	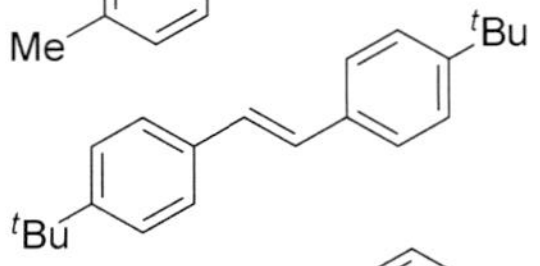 ^{t}Bu, ^{t}Bu	100	82	81, (+)-(2*R*,3*R*)[g]
6	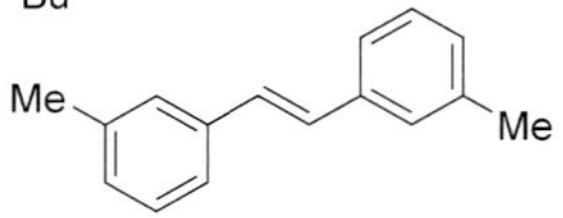 Me, Me	>95	88	27, (+)-(2*R*,3*R*)[h]

7	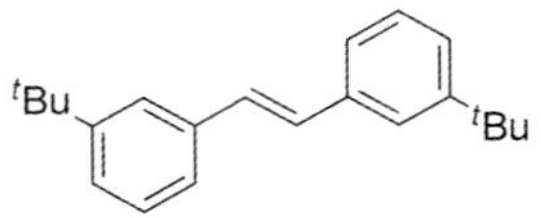 ^{t}Bu, ^{t}Bu	>95	66	10, (−)-(2*S*,3*S*)[h]
8	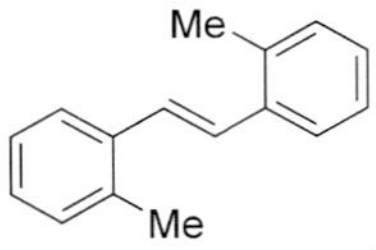 Me, Me	60[i]	57[i]	55, (−)-(2*S*,3*S*)[h]
9[i]	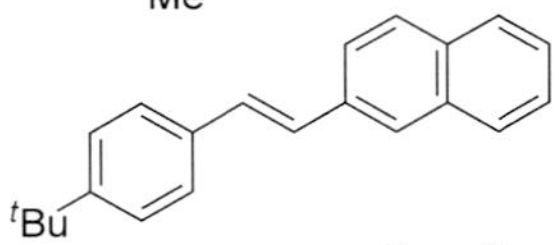 ^{t}Bu	100	46	91, (+)-(2*R*,3*R*)[h]
10[j,k]	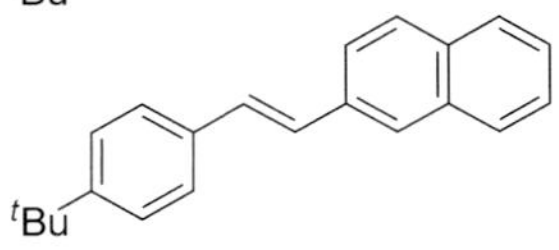 ^{t}Bu	100	40	97, (+)-(2*R*,3*R*)[h]

[a] Estimated by GC-MS and/or TLC which respectively showed absence of substrate peak and traces.
[b] Isolated yield of pure product.
[c] Determined by HPLC on chiral columns.
[d] Determined by GC.
[e] Assigned by comparing the retention times of the enantiomers on a chiral HPLC with that of an authentic sample of the (*S*,*S*)-enantiomer.
[f] Assigned by desilylation to the corresponding epoxy alcohol by analogy with literature protocol and comparing the sign of optical rotation of the resulting product with that of an authentic sample.[69]
[g] Determined by comparing the sign of the optical rotation of the major enantiomer on a chiral detector coupled with a chiral HPLC with known data; the CD spectra of these products are positive, in contrast to those reported for the (*S*,*S*)-enantiomers [70].
[h] Tentatively assigned by comparing the CD spectrum with those of substituted *trans*-stilbene oxides.
[i] Determined after 24 h by ^1H NMR of crude product using an internal standard.
[j] 4 equiv. H_2O_2, 10 mol% H_2-pydic, 10 mol% $FeCl_3 \cdot 6H_2O$, 24 mol% ligand;
[k] Reaction at 10 °C.

Further work to extend the range of substrates and to gain a mechanistic understanding of these new catalysts is under way in our laboratory.

Acknowledgments

We thank the State of Mecklenburg – Western Pomerania, the Federal Ministry of Education and Research (BMBF) and the Deutsche Forschungsgemeinschaft (SPP 1118 and Leibniz-prize) for financial support of our oxidation chemistry. Dr. Santosh Bhor, Dr. Christian Döbler, Dr. Feyissa Gadissa, Dr. Anilkumar Gopinathan, Dr. Markus Klawonn, Dr. Feng Shi, Kristin Schröder, and Dr. Xiaofeng Tong are thanked for their excellent work in the oxidation group of LIKAT.

References

1 (a) Weissermel, K. and Arpe, H.-J. (2004) *Industrial Organic Chemistry*, 4th Edn, Wiley-VCH, Weinheim; (b) Bäckvall, J-.E. (ed.) (2004) *Modern Oxidation Methods*, Wiley-VCH, Weinheim; (c) Beller, M. and Bolm, C. (eds), (2004) *Transition Metals for Organic Synthesis, Building Blocks and Fine Chemicals*, 2nd Edn, Vol. 1–2, Wiley-VCH, Weinheim.

2 (a) Nielsen, L.P.C. and Jacobsen, E.N. (2006) *in Aziridines and Epoxides in Organic Synthesis*, (ed. A.K. Yudin), Wiley-VCH, Weinheim, pp. 229–269; (b) Larrow, J.F. and Jacobsen, E.N. (2004) *Topics in Organometallic Chemistry*, **6**, 123–152; (c) Jacobsen, E.N. and Wu, M.H. (1999) *Ring opening of epoxides and related reactions Comprehensive Asymmetric Catalysis*, Vol. 3 (eds E.N. Jacobsen, A. Pfaltz and H. Yamamoto), pp. 1309–1326.

3 Examples using peracids for olefin epoxidation without metal catalyst: (a) Crawford, K., Rautenstrauch, V. and Uijttewaal, A. (2001) *Synlett*, 1127–1128; (b) Wahren, U., Sprung, I., Schulze, K., Findeisen, M. and Buchbauer, G. (1999) *Tetrahedron Letters*, **40**, 5991–5992; (c) Kelly, D.R. and Nally, J. (1999) *Tetrahedron Letters*, **40**, 3251–3254.

4 Tullo, A.H. and Short, P.L. (2006) *C&EN*, **84**, 22–23; and references therein.

5 For a list of common oxidants, their active oxygen contents and waste products, see: Adolfsson, H. (2004) *Modern Oxidation Methods* (ed. J.-E. Bäckvall), Wiley-VCH, Weinheim, pp. 22–50;

6 Sheldon, R.A. and Kochi, J.K. (1981) *Metal-Catalyzed Oxidations of Organic Compounds*, Academic Press, New York.

7 (a) Simándi, L.I. (ed.), (2003) *Advances in Catalytic Activation of Dioxygen by Metal Complexes*, Kluwer Academic, Dordrecht; (b) Barton, D.H.R., Bartell, A.E. and Sawyer, D.T. (eds), (1993) *The Activation of Dioxygen and Homogeneous Catalytic Oxidation*, Plenum, New York; (c) Simándi, L.I. (1992) *Catalytic Activation of Dioxygen by Metal Complexes*, Kluwer Academic, Dordrecht;

8 (a) Jones, C.W. (1999) *Applications of Hydrogen Peroxide and Derivatives*, Royal Society of Chemistry, Cambridge; (b) Strukul, G. (ed.) (1992) *Catalytic Oxidations with Hydrogen Peroxide as Oxidant*, Kluwer, Academic, Dordrecht;

9 For reviews of H_2O_2 as epoxidation oxidant see: (a) Grigoropoulou, G., Clark, J.H. and Elings, J.A. (2003) *Green Chemistry*, **5**, 1–7; (b) Lane, B.S. and Burgess, K. (2003) *Chemical Reviews*, **103**, 2457–2473; For a commentary see: (c) Beller, M. (2004) *Advanced Synthesis and Catalysis*, **346**, 107–108.

10 Adolfsson, H. and Balan D. (2006) *in Aziridines and Epoxides in Organic Synthesis* (ed. A.K. Yudin), Wiley-VCH, Weinheim, pp. 185–228.

11 For recent examples of transition metal catalyzed epoxidations using hydrogen peroxide see: (a) Colladon, M., Scarso, A., Sgarbossa, P., Michelin, R.A. and Strukul, G. (2007) *Journal of the American Chemical Society*, **129**, 7680–7689; (b) Colladon, M., Scarso, A., Sgarbossa, P., Michelin, R.A. and Strukul, G. (2006) *Journal of the American Chemical Society*, **128**, 14006–14007; (c) Sawada, Y., Matsumoto, Z., Kondo, S., Watanabe, H., Ozawa, T., Suzuki, K., Saito, B. and Katsuki, T. (2006) *Angewandte Chemie-International Edition*, **45**, 3478–3480; (d) Matsumoto, K., Sawada, Y., Saito, B., Sakai, K. and Katsuki, T. (2005) *Angewandte Chemie-International Edition*, **44**, 4935–4939; (e) Mahammed, A. and Gross, Z. (2005) *Journal of the American Chemical Society*, **127**, 2883–2887; (f) Kühn, F.E., Scherbaum, A. and Herrmann, W.A. (2004) *Journal of Organometallic Chemistry*, **689**, 4149–4164; (g) Kamata, K., Yamaguchi, K., Hikichi, S. and Mizuno, N. (2003) *Advanced Synthesis and Catalysis*, **345**, 1193–1196; (h) Adam, W., Alsters, P.L., Neumann, R., Saha-Möller, C.R., Sloboda-Rozner, D. and Zhang, R. (2003) *The Journal of Organic Chemistry*, **68**, 1721–1728; (i) Lane, B.S., Vogt, M., DeRose, V.J. and Burgess, K. (2002) *Journal of the American Chemical Society*, **124**, 11946–11954.

12 Lippard, S.J. and Berg, J.M. (1994) *Principles of Bioinorganic Chemistry*, University Science Books, Mill Valley, CA;

13 Sheldon, R.A. (ed.) (1994) *Metalloporphyrins in Catalytic Oxidations*, Marcel Dekker Ltd, New York;

14 (a) Trautheim, A.X. (ed.) (1997) *Bioinorganic Chemistry: Transition Metals in Biology and their Coordination Chemistry*, Wiley-VCH, Weinheim; (b) Ponka, P., Schulman, H.M. and Woodworth, R.C. (1990) *Iron Transport and Storage*, CRC Press. Inc, Boca Raton, Florida;

15 (a) van Eldik, R. (ed.) (2006) *Advances in Inorganic Chemistry Vol. 58: Homogeneous Biomimetic Oxidation Catalysis*, Academic Press, London; (b) Meunier, B. (ed.) (2000) *Biomimetic Oxidations Catalyzed by Transition Metals*, Imperial College Press, London;

16 Murahashi, S.-I. and Komiya N. (2004) *Ruthenium in Organic Synthesis* (ed. S.-I. Murahashi), Wiley-VCH, Weinheim;

17 For some examples of using oxygen or air see: (a) Christoffers, J. (1999) *The Journal of Organic Chemistry*, **64**, 7668–7669; (b) Coleman, K.S., Lorber, C.Y. and Osborn, J.A. (1998) *European Journal of Inorganic Chemistry*, 1673–1675; (c) Peterson, K.P. and Larock, R.C. (1998) *The Journal of Organic Chemistry*, **63**, 3185–3189; (d) Markó, I.E., Giles, P.R., Tsukazaki, M., Chellé-Regnant, I., Urch, C.J. and Brown, S.M. (1997) *Journal of the American Chemical Society*, **119**, 12661–12662; (e) Paeng, I.R. and Nakamoto, K. (1990) *Journal of the American Chemical Society*, **112**, 3289–3297; (f) Groves, J.T. and Quinn, R. (1985) *Journal of the American Chemical Society*, **107**, 5790–5792.

18 For osmium-catalyzed dihydroxylation using oxygen or air see: (a) Sundermeier, U., Döbler, C., Mehltretter, G.M., Baumann, W. and Beller, M. (2003) *Chirality*, **15**, 127–134; (b) Döbler, C., Mehltretter, G.M., Sundermeier, U. and Beller, M. (2001) *Journal of Organometallic Chemistry*, **621**, 70–76; (c) Mehltretter, G.M., Döbler, C., Sundermeier, U. and Beller, M. (2000) *Tetrahedron Letters*, **41**, 8083–8087; (d) Döbler, C., Mehltretter, G., Sundermeier, U. and Beller, M. (2000) *Journal of the American Chemical Society*, **122**, 10289–10297; (e) Döbler, C., Mehltretter, G. and Beller, M. (1999)

Angewandte Chemie-International Edition, **38**, 3026–3028.

19 Montanari, F. and Casella, L. (ed.) (1994) *Metalloporphyrins Catalyzed Oxidations*, Kluwer, Dordrecht;

20 (a) Shing, T.K.M. and Tam, E.K.W. (1999) *Tetrahedron Letters*, **40**, 2179–2180; (b) Shing, T.K.M., Tam, E.K.W., Tai, V.W.-F., Chung, I.H.F. and Jiang, Q. (1996) *Chemistry - A European Journal*, **2**, 50–57; (c) Shing, T.K.M., Tai, V.W.-F. and Tam, E.K.W. (1994) *Angewandte Chemie (International Edition in English)*, **33**, 2312–2313.

21 (a) Hotopp, T., Gutke, H.-J. and Murahashi, S.-I. (2001) *Tetrahedron Letters*, **42**, 412–415; (b) Beifuss, U. and Herde, A. (1998) *Tetrahedron Letters*, **39**, 7691–7692; (c) Murahashi, S.-I., Saito, T., Hanaoka, H., Murakami, Y., Naota, T., Kumobayashi, H. and Akutagawa, S. (1993) *The Journal of Organic Chemistry*, **58**, 2929–2930.

22 (a) Fisher, J.M., Fulford, A. and Bennett, P.S. (1992) *Journal of Molecular Catalysis*, **77**, 229–234; (b) Taqui Khan, M.M. and Shukla, R.S. (1991) *Journal of Molecular Catalysis*, **70**, 129–140.

23 Nishiyama, H., Shimada, T., Itoh, H., Sugiyama, H. and Motoyama, Y. (1997) *Chemical Communications*, 1863–1864.

24 Klawonn, M., Tse, M.K., Bhor, S., Döbler, C. and Beller, M. (2004) *Journal of Molecular Catalysis A-Chemical*, **218**, 13–19.

25 Tse, M.K., Bhor, S., Klawonn, M., Döbler, C. and Beller, M. (2003) *Tetrahedron Letters*, **44**, 7479–7483.

26 Bhor, S., Tse, M.K., Klawonn, M., Döbler, C., Mägerlein, W. and Beller, M. (2004) *Advanced Synthesis and Catalysis*, **346**, 263–267.

27 (a) Tse, M.K., Bhor, S., Klawonn, M., Anilkumar, G., Jiao, H.-j., Spannenberg, A., Döbler, C., Mägerlein, W., Hugl, H. and Beller, M. (2006) *Chemistry - A European Journal*, **12**, 1875–1888; (b) Tse, M.K., Bhor, S., Klawonn, M., Anilkumar, G., Jiao, H.-j., Döbler, C., Spannenberg, A., Mägerlein, W., Hugl, H. and Beller, M. (2006) *Chemistry - A European Journal*, **12**, 1855–1874; (c) Anilkumar, G., Bhor, S., Tse, M.K., Klawonn, M., Bitterlich, B. and Beller, M. (2005) *Tetrahedron: Asymmetry*, **16**, 3536–3561; (d) Bhor, S., Anilkumar, G., Tse, M.K., Klawonn, M., Bitterlich, B., Grotevendt, A. and Beller, M. (2005) *Organic Letters*, **7**, 3393–3396; (e) Tse, M.K., Klawonn, M., Bhor, S., Döbler, C., Anilkumar, G., Hugl, H., Mägerlein, W. and Beller, M. (2005) *Organic Letters*, **7**, 987–990; (f) Tse, M.K., Döbler, C., Bhor, S., Klawonn, M., Mägerlein, W., Hugl, H. and Beller, M. (2004) *Angewandte Chemie-International Edition*, **43**, 5255–5260.

28 For recent examples see: (a) Berkessel, A., Kaiser, P. and Lex, J. (2003) *Chemistry - A European Journal*. **9**, 4746–4756; (b) Gross, Z. and Ini, S. (1999) *Organic Letters*, **1**, 2077–2080; (c) Lai, T.-S., Kwong, H.-L., Zhang, R. and Che, C.-M. (1998) *Journal of The Chemical Society-Dalton Transactions*, 3559–3564; (d) Berkessel, A. and Frauenkron, M. (1997) *Journal of the Chemical Society-Perkin Transactions 1*, 2265–2266; (e) Gross, Z., Ini, S., Kapon, M. and Cohen, S. (1996) *Tetrahedron Letters*, **37**, 7325–7328; (f) Ini, S., Kapon, M., Cohen, S. and Gross, Z. (1996) *Tetrahedron: Asymmetry*, **7**, 659–662.

29 (a) Kureshy, R.I., Khan, N.H., Abdi, S.H.R., Patel, S.T. and Iyer, P. (1999) *Journal of Molecular Catalysis*, **150**, 175–183; (b) Kureshy, R.I., Khan, N.H., Abdi, S.H.R., Patel, S.T. and Iyer, P. (1999) *Journal of Molecular Catalysis*, **150**, 163–173; (c) Kureshy, R.I., Khan, N.H. and Abdi, S.H.R. (1995) *Journal of Molecular Catalysis*, **96**, 117–122; (d) Kureshy, R.I., Khan, N.-u., Abdi, S.H.R. and Bhatt, K.N. (1993) *Tetrahedron-Asymmetry*, **4**, 1693–1701.

30 Kureshy, R.I., Khan, N.H., Abdi, S.H.R. and Iyer, P. (1997) *Journal of Molecular Catalysis*, **124**, 91–97.

31 (a) End, N. and Pfaltz, A. (1998) *Chemical Communications*, 589–590; (b) End, N., Macko, L., Zehnder, M. and Pfaltz, A.

(1998) *Chemistry - A European Journal*, **4**, 818–824.

32 (a) Stoop, R.M., Bachmann, S., Valentini, M. and Mezzetti, A. (2000) *Organometallics*, **19**, 4117–4126; (b) Stoop, R.M. and Mezzetti, A. (1999) *Green Chemistry*, 39–41.

33 Stoop, R.M., Bauer, C., Setz, P., Wörle, M., Wong, T.Y.H. and Mezzetti, A. (1999) *Organometallics*, **18**, 5691–5700.

34 Augier, C., Malara, L., Lazzeri, V. and Waegell, B. (1995) *Tetrahedron Letters*, **36**, 8775–8778.

35 (a) Nakata, K., Takeda, T., Mihara, J., Hamada, T., Irie, R. and Katsuki, T. (2001) *Chemistry - A European Journal*, **7**, 3776–3782; (b) Takeda, T., Irie, R., Shinoda, Y. and Katsuki, T. (1999) *Chemistry Letters*, **7**, 1157–1159.

36 The authors withdrew the paper later due to problem of reproducibility. Pezet, F., Ait-Haddou, H., Daran, J.-C., Sadaki, I. and Balavoine, G.G.A. (2002) *Chemical Communications*, 510–511.

37 Tse, M.K., Jiao, H.-j., Anilkumar, G., Bitterlich, B., Gelalcha, F.G. and Beller, M. (2006) *Journal of Organometallic Chemistry*, **691**, 4419–4433.

38 (a) Shi, F., Tse, M.K. and Beller, M. (2007) *Journal of Molecular Catalysis A-Chemical*, **270**, 68–75; (b) Shi, F., Tse, M.K. and Beller, M. (2007) *Advanced Synthesis and Catalysis*, **349**, 303–308.

39 Shi, F., Tse, M.K. and Beller, M. (2007) *Chemistry, an Asian Journal*, **2**, 411–415.

40 Bhor, S., Tse, M.K., Klawonn, M., Anilkumar, G., Bitterlich, B. and Beller, M., unpublished results (2005).

41 Klawonn, M. (2005) University Rostock and Leibniz-Institut für Organische Katalyse an der Universität Rostock Dissertation.

42 Bhor, S., Tse, M.K., Klawonn, M., Döbler, D. and Beller, M. (2003) unpublished results.

43 Iwasa, S., Fakhruddin, A., Widagdo, H.S. and Nishiyama, H. (2005) *Advanced Synthesis and Catalysis*, **347**, 517–520.

44 Fakhruddin, A., Iwasa, S., Nishiyama, H. and Tsutsumi, K. (2004) *Tetrahedron Letters*, **45**, 9323–9326.

45 Iwasa, S., Tajima, K., Tsushima, S. and Nishiyama, H. (2001) *Tetrahedron Letters*, **42**, 5897–5899.

46 Iwasa, S., Morita, K., Tajima, K., Fakhruddin, A. and Nishiyama, H. (2002) *Chemistry Letters*, 284–285.

47 (a) Comell, R.M. and Schwertmann, U. (2003) *The Iron Oxide: Structures, Properties, Occurences and Uses*, Wiley-VCH, Weinheim; (b) Mielczarek, E.V., McGrayne, S.B., Mielczarek, E.V. and McGrayne, S.B. (2000) *Iron Nature's Universal Element: Why People Need Iron & Animals Make Magnets*, Rutgers University Press, New Brunswick, N. J;

48 (a) Fürstner, A. and Martin, R. (2005) *Chemistry Letters*, **34**, 624–629; (b) Zecchina, A., Bordiga, S., Spoto, G., Damin, A., Berlier, G., Bonino, F. and Lamberti, C.P.C. (2005) *Catalysis Reviews-Science and Engineering*, **47**, 125–172; (c) Bolm, C., Legros, J., Le Paih, J. and Zani, L. (2004) *Chemical Reviews*, **104**, 6217–6254; (d) Costas, M., Chen, K. and Que, L. Jr. (2000) *Coordination Chemistry Reviews*, **200–202**, 517–544; (e) Fontecave, M., Ménage, S. and Duboc-Toia, C. (1998) *Coordination Chemistry Reviews*, **178–180**, 1555–1572.

49 (a) Hatakeyama, T. and Nakamura, M. (2007) *Journal of the American Chemical Society*, **129**, 9844–9845; (b) Li, Z.-p., Cao, L. and Li, C.-J. (2007) *Angewandte Chemie-International Edition*, **46**, 6505–6507; (c) Cahiez, G., Habiak, V., Duplais, C. and Moyeux, A. (2007) *Angewandte Chemie-International Edition*, **46**, 4364–4366; (d) Kischel, J., Michalik, D., Zapf, A. and Beller, M. (2007) *Chemistry, an Asian Journal*, **6**, 865–870; (e) Kischel, J., Mertins, K., Michalik, D., Zapf, A. and Beller, M. (2007) *Advanced Synthesis and Catalysis*, **349**, 871–875; (f) Kischel, J., Jovel, I., Mertins, K., Zapf, A. and Beller, M. (2006) *Organic Letters*, **8**, 19–22; (g) Jovel, I., Mertins, K., Kischel, J., Zapf, A. and Beller, M. (2005) *Angewandte Chemie-International Edition*, **44**, 3913–3916; (h) Nakamura, M., Matsuo, K., Ito, S. and

Nakamura, E. (2004) *Journal of the American Chemical Society*, **126**, 3686–3687; (i) Fürstner, A., Leitner, A., Méndez, M. and Krause, H. (2002) *Journal of the American Chemical Society*, **124**, 13856–13863.

50 (a) Correa, A. and Bolm, C. (2007) *Angewandte Chemie-International Edition*, **46**, DOI: 10.1002/anie.200703299 (b) Plietker, B. (2006) *Angewandte Chemie-International Edition*, **45**, 6053–6056; (c) Komeyama, K., Morimoto, T. and Takaki, K. (2006) *Angewandte Chemie-International Edition*, **45**, 2938–2941; (d) Srivastava, R.S., Khan, M.A. and Nicholas, K.M. (1996) *Journal of the American Chemical Society*, **118**, 3311–3312.

51 (a) Casey, C.P. and Guan, H.-r. (2007) *Journal of the American Chemical Society*, **129**, 5816–5817; (b) Enthaler, S., Erre, G., Tse, M.K., Junge, K. and Beller, M. (2006) *Tetrahedron Letters*, **47**, 8095–8099; (c) Bart, S.C., Lobkovsky, E. and Chirik, P.J. (2004) *Journal of the American Chemical Society*, **126**, 13794–13807; (d) Daida, E.J. and Peters, J.C. (2004) *Inorganic Chemistry*, **43**, 7474–7485; (e) Radhi, M.A. and Marko, L. (1984) *Journal of Organometallic Chemistry*, **262**, 359–364; (f) Schroeder, M.A. and Wrighton, M.S. (1976) *Journal of the American Chemical Society*, **98**, 551–558.

52 For alkane oxygenation with H_2O_2 catalyzed by $FeCl_3$ see: (a) Shul'pin, G.B., Golfeto, C.C., Süss-Fink, G., Shul'pina, L.S. and Mandelli, D. (2005) *Tetrahedron Letters*, **46**, 4563–4567; (b) Klopstra, M., Hage, R., Kellogg, R.M. and Feringa, B.L. (2003) *Tetrahedron Letters*, **44**, 4581–4584; (c) Martín, S.E. and Garrone, A. (2003) *Tetrahedron Letters*, **54**, 549–552; (d) Barton, D.H.R. and Hu, B. (1997) *Pure and Applied Chemistry*, **69**, 1941–1950; (e) Barton, D.H.R. and Taylor, D.K. (1996) *Pure and Applied Chemistry*, **68**, 497–504.

53 For epoxidation with H_2O_2 catalyzed by $FeCl_3$ see: (a) Sugimoto, H. and Sawyer, D.T. (1985) *The Journal of Organic Chemistry*, **50**, 1784–1786; (b) Sugimoto, H., Spencer, L. and Sawyer, D.T. (1987) *Proceedings of the National Academy of Sciences of the United States of America*, **84**, 1731–1733.

54 (a) Wu, A.J., Penner-Hahn, J.E. and Pecoraro, V.L. (2004) *Chemical Reviews*, **104**, 903–938; (b) Yagi, M. and Kaneko, M. (2001) *Chemical Reviews*, **101**, 21–35; (c) Ortiz de Montellano, P.R. (1987) *Accounts of Chemical Research*, **20**, 289–294.

55 Nam, W., Ho, R. and Valentie, J.S. (1991) *Journal of the American Chemical Society*, **113**, 7052–7054.

56 Traylor, T.G., Tsuchiya, S., Byun, Y.S. and Kim, C. (1993) *Journal of the American Chemical Society*, **115**, 2775–2781 and references therein.

57 White, M.C., Doyle, A.G. and Jacobsen, E.N. (2001) *Journal of the American Chemical Society*, **123**, 7194–7195.

58 (a) Chen, K., Costas, M., Kim, J., Tipton, A.T. and Que, L. Jr. (2002) *Journal of the American Chemical Society*, **124**, 3026–3035; (b) Costas, M., Tipton, A.K., Chen, K., Jo, D.-H. and Que, L. Jr. (2001) *Journal of the American Chemical Society*, **123**, 6722–6723; (c) Chen, K. and Que, L. Jr. (1999) *Chemical Communications*, 1375–1376.

59 (a) Bukowski, M.R., Comba, P., Lienke, A., Limberg, C., Lopez de Laorden, C., Mas-Ballesté, R., Merz, M. and Que, L. Jr. (2006) *Angewandte Chemie-International Edition*, **45**, 3446–3449; (b) Fujita, M. and Que, L. Jr. (2004) *Advanced Synthesis and Catalysis*, **346**, 190–194.

60 Dubois, G., Murphy, A. and Stack, T.D.P. (2003) *Organic Letters*, **5**, 2469–2472.

61 (a) Anilkumar, G., Bitterlich, B., Gelalcha, F.G., Tse, M.K. and Beller, M. (2007) *Chemical Communications*, 289–291; (b) Bitterlich, B., Anilkumar, G., Gelalcha, F.G., Spilker, B., Grotevendt, A., Jackstell, R., Tse, M.K. and Beller, M. (2007) *Chemistry, an Asian Journal*, **2**, 521–529.

62 Costas, M., Mehn, M.P., Jensen, M.P. and Que, L. (2004) *Chemical Reviews*, **104**, 939–986.

63 Schröder, K., Tong, X.-f., Bitterlich, B., Tse, M.K., Gelalcha, F.G., Brückner, A. and Beller, M. (2007) *Tetrahedron Letters*, **48**, 6339–6342.

64 Reviews for Mn: (a) Katsuki, T. (2002) *Advanced Synthesis and Catalysis*, **344**, 131–147; (b) Katsuki T. (2000) *Catalytic Asymmetric Synthesis* (ed. I. Ojima), Wiley-VCH, New York, pp. 287–25; (c) Jacobsen, E.N. (1993) *Catalytic Asymmetric synthesis*, 2nd Edn (ed. I. Ojima), VCH, New York, Chapter 42.

65 (a) Rose, E., Ren, Q.-z. and Andrioletti, B. (2004) *Chemistry - A European Journal*, **10**, 224–230; (b) Lindsay Smith, J.R. and Reginato, G. (2003) *Organic and Biomolecular Chemistry*, **1**, 2543–2549; (c) Adam, W., Stegmann, V.R. and Saha-Möller, C.R. (1999) *Journal of the American Chemical Society*, **121**, 1879–1882; (d) Groves, J.T. and Myers, R.S. (1983) *Journal of the American Chemical Society*, **105**, 5791–5796.

66 Cheng, Q.F., Xu, X.Y., Ma, W.X., Yang, S.J. and You, T.P. (2005) *Chinese Chemical Letters*, **16**, 1467–1470.

67 (a) Jacobsen, E.N. (2001) *221st ACS National Meeting*, San Diego, CA, United States, April 1–5, ORGN-427. (b) Jacobsen, E.N. (2000) 219th ACS National Meeting, San Francisco, CA, United States, March 26–30, INOR-004. (c) Francis, M.B. and Jacobsen, E.N. (1999) *Angewandte Chemie-International Edition*, **38**, 937–941.

68 Gelalcha, F.G., Bitterlich, B., Anilkumar, G., Tse, M.K. and Beller, M. (2007) *Angewandte Chemie-International Edition*, **46**, 7293–7296.

69 Bayer, A. and Svendsen, J.S. (2001) *European Journal of Organic Chemistry*, 1769–1780.

70 Yang, D., Wong, M.-K., Yip, Y.-C., Wang, X.-C., Tang, M.-W., Zheng, J.-H. and Cheung, K.-K. (1998) *Journal of the American Chemical Society*, **120**, 5943–5952.

18
Pentacoordinating Bis(oxazoline) Ligands with Secondary Binding Sites

Caroline A. Schall, Michael Seitz, Anja Kaiser, and Oliver Reiser

Nature demonstrates most efficiently the use of iron complexes for the detoxification of organic compounds in living beings by oxidation [1, 2], the oxidized products being more water soluble and thus easily excretable. In order to achieve this goal, which is essential for life, nature had to overcome a number of obstacles: (1) Given the large amounts of organic material taken up by living organisms to meet their energy demands, combined with the irresponsible consumption of quite toxic compounds (especially by human beings), the detoxification processes require highly active catalysts that are able to perform oxidations with high turnover numbers and rates. (2) The oxidation must be possible for a broad range of substrates, including nonactivated ones, calling for their C−H activation. (3) Molecular oxygen, being taken up in large quantities by living beings, is best employed as the terminal oxidant. It shows, however, little tendency to trigger oxidations at ambient temperatures, and, moreover, its catalytic activation by metals to allow its transfer to organic substrates is difficult because of the stable metal-oxo complexes generally formed. (4) Finally, since nature has presented a solution to the challenges posed in (1)–(3) by use of iron complexes, the well-known tendency of iron(III) to form Fe-O-Fe dimers that would precipitate from solution has to be prevented.

Three general types of iron-containing enzymes are employed by nature for the oxidation of organic compounds, i.e. heme proteins [3–5], proteins containing iron-sulfur clusters [6–8], and nonheme proteins [9, 10]. This last, arguably most important class of iron-containing enzymes is predominantly based on oxygen- and nitrogen-containing ligands. These allow the oxidative utilization of dioxygen under ambient conditions, catalysis of oxidation reactions of nonactivated C−H bonds, detoxification of biologically harmful radicals and reversible O_2-binding reactions.

Despite a great number of elegant studies to devise small iron complexes that can mimic the action of these enzymes, the goal to achieve the efficient and selective oxidation of nonactivated compounds such as hydrocarbons, considered to be one of the great chemical challenges to be solved in this century, is still not met. In contrast to nature, which strives to achieve detoxification by creating more water soluble oxidation products, the goal of developing oxidations of hydrocarbons for fine

Activating Unreactive Substrates: The Role of Secondary Interactions.
Edited by Carsten Bolm and F. Ekkehardt Hahn

ISBN: 978-3-527-31823-0

chemicals includes the additional challenge of selectivity: first of all only partial oxidation is desired as exemplified by the transformation of methane to methanol, posing the difficulty that the products are often more reactive toward further oxidation than the starting material. Moreover, stereoselective hydroxylation reactions would be highly useful for creating building blocks that could be applied in pharmaceutical and agrochemical syntheses.

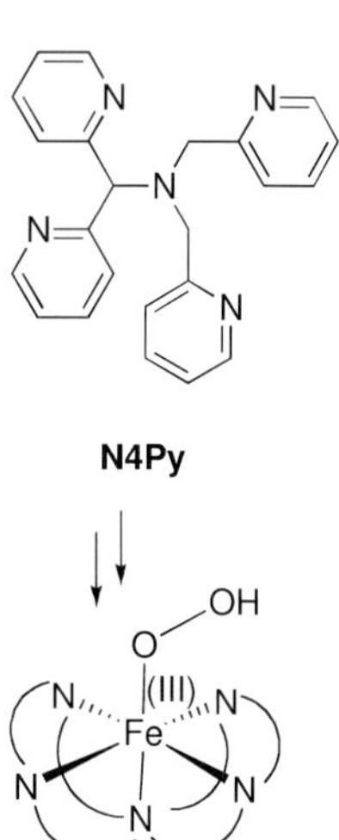

Figure 18.1 Examples of pentacoordinating ligands suitable for iron-catalyzed oxidation of hydrocarbons with hydrogen peroxide.

Keeping these goals in mind, we wanted to design pentacoordinating chiral ligands that would form octahedral iron and related complexes with one open coordination site for activating oxygen, following, for example, the lead from bleomycin, which is effective for the oxidative cleavage of DNA but also shows good activity for the hydroxylation of hydrocarbons (Figure 18.1) [11]. A simple but effective mimic of bleomycin was found, namely the iron complex of N4Py, which is able to oxidize cyclohexane, using hydrogen peroxide, to a mixture of cyclohexanol and cyclohexanone with 36 turnovers per hour [12]. Our choice to develop chiral, enantiopure ligands was motivated not only by the hope of achieving asymmetric transformations catalyzed by their corresponding metal complexes, but also by the possibility of having an additional tool to study the structure of the complexes in solution, i.e. by CD-spectroscopy.

Intrigued by the the pentacoordinating ligands **1** [13] and **2** [14] as well as by the excellent ability of oxazoline moieties to coordinate metals and be easily rendered chiral [15], we envisioned ligands of the general structure **3** for our investigation (Figure 18.2). In addition, we also wanted to incorporate hydrogen donors (e.g. X = OH) in **3** that might assist oxygen activation at a metal center coordinated to the ligand, a concept that was demonstrated by Borovik et al. for iron complex **4** [16].

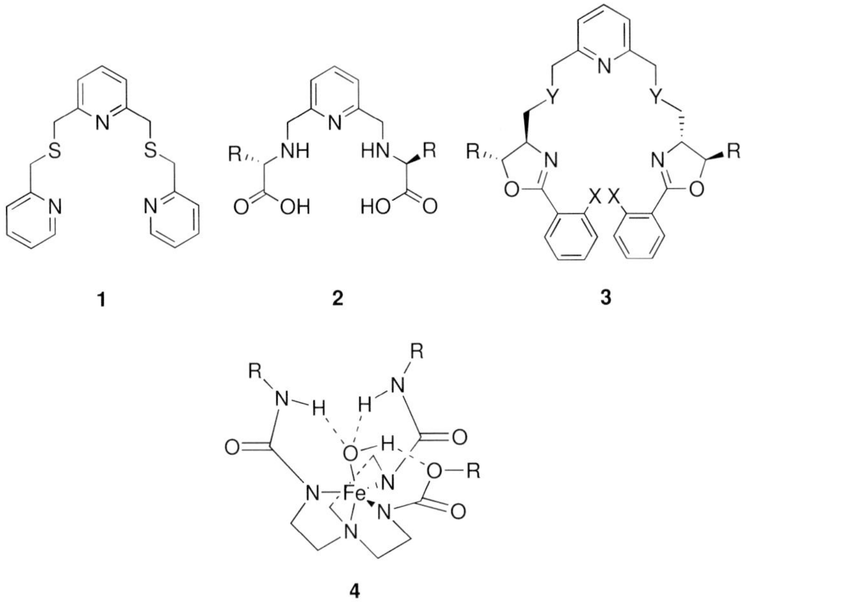

Figure 18.2 Ligands **3** developed in this study in comparison to known structures **1**, **2** and **4**.

Oxazoline moieties can be readily synthesized by coupling/cyclization sequences of carboxylic acid derivates and amino alcohols. However, we wanted to develop the synthesis of **3** in a modular approach, being able to transfer an already assembled oxazoline moiety to a given backbone. Thus, we were able to identify **8** as a nucleophilic and **10** as an electrophilic oxazoline building block that could be readily combined with an appropriately functionalized counterpart **5** and **7**, giving rise to the C_2-symmetrical ligands **3** with wide structural variety. Besides **3** (Scheme 18.1), which has a central pyridine unit, we were also able to synthesize ligands with central pyridine-*N*-oxide, phenol and anisole moieties by this approach [17].

Ligands **3a–c** proved to be especially suitable to study as metal complexes. Their ability to act as pentacoordinating ligands was proved by forming complexes with zinc(II) perchlorate, which could be characterized both in the solid state by X-ray structure analysis and in solution by NMR- and CD-spectroscopy [18]. Not surprisingly, trigonal bipyramidal zinc complexes were formed with the oxazoline moieties placed in the axial positions. Moreover, the chirality of the ligands induced only one configuration around the metal in the solid state, but, quite unexpectedly, with opposite sense for **3a, b** in comparison to **3c**: X-ray structure analyses revealed that **7a, b** was in the Λ_2-configuration, while the Δ_2-configuration was observed for **7c** (Figure 18.3).

The complexes **7b** and **7c** were also configurationally stable in solution, judging from a single set of signals in the ^{1}H and ^{13}C NMR spectra and particularly from the CD-data (Figure 18.3), reflecting the structures observed in the solid state. In contrast, the NMR-spectra of **7a** showed more than one signal set, and, moreover, its CD-spectrum was not very well defined, indicating that more than one zinc species

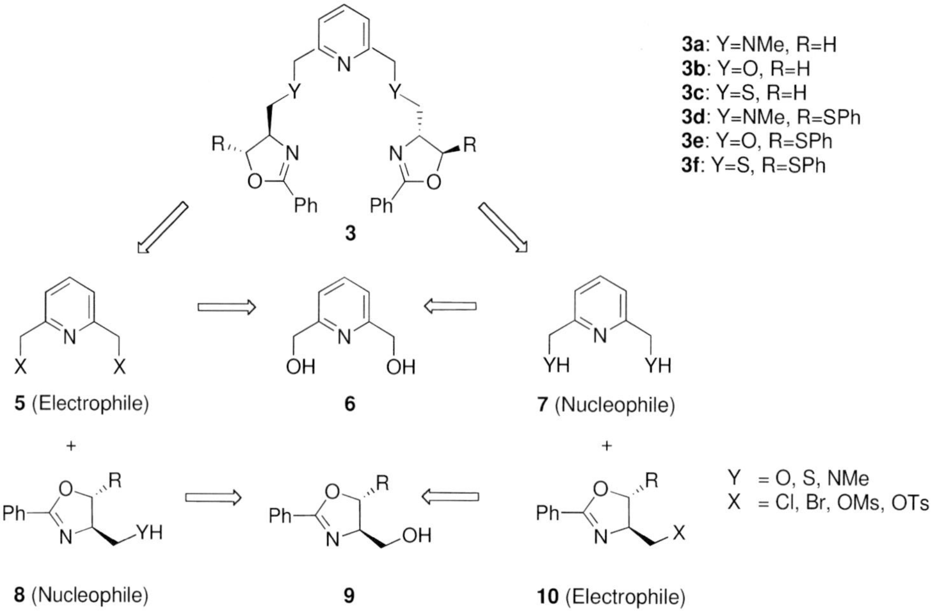

3a: Y=NMe, R=H
3b: Y=O, R=H
3c: Y=S, R=H
3d: Y=NMe, R=SPh
3e: Y=O, R=SPh
3f: Y=S, R=SPh
3
5 (Electrophile)
6
7 (Nucleophile)
Y = O, S, NMe
X = Cl, Br, OMs, OTs
8 (Nucleophile)
9
10 (Electrophile)

Scheme 18.1 Modular strategy for the synthesis of pentadentate bis(oxazoline) ligands **3**.

3a
3b
3c
Zn(ClO4)2
• (ClO4)2
7a: Y = NMe, 60%
7b: Y = O, 86%
7c: Y = S, 80%
Δg [a.u.]
Wavelength [nm]
7a
7b
7c
Δ2-7a
Δ2-7b
Δ2-7c

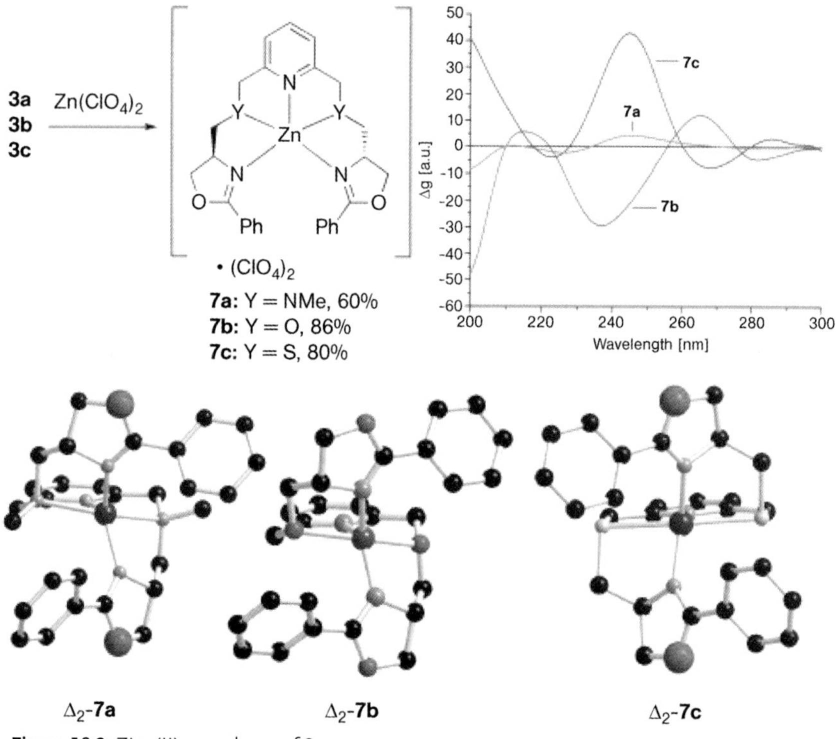

Figure 18.3 Zinc(II)-complexes of **3a–c**

of **3a** exists in solution. This finding was quite general for other metal complexes as well: based on CD- and whenever possible on NMR-analysis, metal complexes of **3b, c** always gave rise to single species in solution, but complexes of **3a** showed dynamic behavior, suggesting that rapidly converting species existed. The main difference between the structures **7a–c** is the $C_{benzylic}$-Y (C–N: 1.47 Å; C–O: 1.44 Å; C–S: 1.81 Å) and Zn–S bond lengths (Zn–N: 2.29 Å; Zn–O: 2.27 Å; Zn–S: 2.53 Å), which must account for the complete reversal of stereoselectivity from Λ_2 to Δ_2.

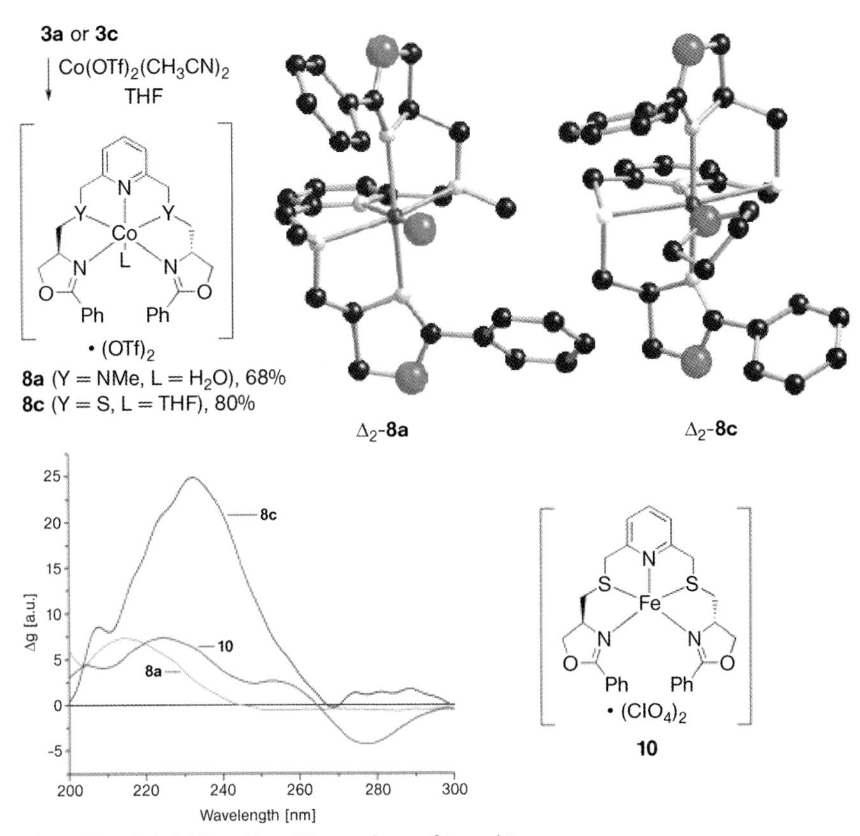

Figure 18.4 Cobalt(II) and iron(II) complexes of **3a** and **3c**.

Octahedral complexes that were envisioned to mimic bleomycin could also be obtained with **3a** and **3c**. Cobalt complexes **8a** and **8c** [19] were both formed exclusively in their Δ_2-configurations in the solid state (Figure 18.4). It is noteworthy that it became evident that the configurational switch observed in the zinc complexes **7** upon changing from ligand **3a** to **3c** does not occur in cobalt complexes **8**. Comparison of **8a** and **8c** in solution by CD-spectroscopy was problematic given the different intensities and maxima observed. Also, analysis of the CD-spectrum of the iron(II) complex **10** prepared from $Fe(ClO_4)_2$ and **3c** was ambiguous: mass spectroscopy indicated the formation of an $Fe(ClO_4)_2$•**3c** complex as the only iron species detected, but the weak CD-pattern observed did not allow a clear comparison with other complexes and might be indicative of more than one species present in solution. Unfortunately, we were unable to obtain suitable crystals to characterize the complex in the solid state by an X-ray structure.

In contrast, the iron(II) complex **11b** prepared from $Fe(ClO_4)_2$ and **3b** showed a distinct CD-spectrum (Figure 18.5) that was completely different from that of any complex previously obtained with **3**, pointing toward a complex geometry being unrelated to octahedral or trigonal bipyramidal. In this case X-ray analysis was possible, and, indeed, the iron center was encompassed by the ligand in a pentagonal bipyramidal way with two additional water molecules in axial positions to give rise to an overall heptacoordinated complex. The chirality of the ligand induces only one type of structure, i.e. *P*-helicity around the metal, which is assumed to be also retained in solution based on the strong CD-signal intensity observed. The driving force responsible for adopting this unusual geometry is likely to be an attractive interaction

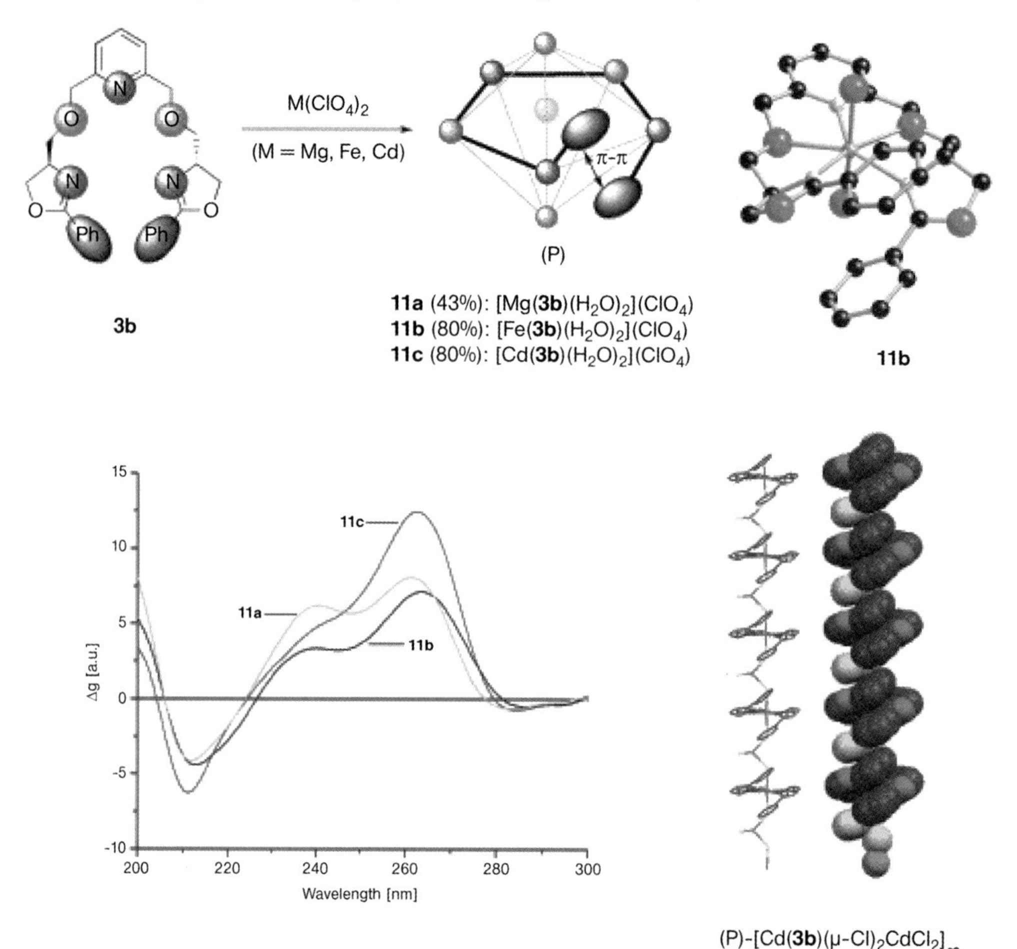

Figure 18.5 Pentagonal bipyramidal complexes with **3b**.

through π-stacking of the phenyl groups attached to the oxazoline moieties, being separated by approximately 340 Å in **11b**, a distance similar to that found in graphite (335 Å). The same coordination geometry was also found for complexes of **3b** formed with magnesium or cadmium perchlorate [20]. With cadmium chloride and bromide, one-dimensional organic-inorganic hybrid polymers could be constructed in an analogous way, having in the solid state a $(Cd{-}X)_n$ chain in which every second cadmium atom is coordinated by **3b** in a *P*-helical way [21].

Iron, cobalt and ruthenium complexes derived from **3a–f** showed only low activity for oxidation of sulfides, ethylbenzene or cyclohexane, with turnover rates in the range of other neutral pentacoordinating ligands known, but not competitive with nature's systems [22]. It is interesting to note, however, that the pentagonal bipyramidal iron complexes (entry 2,3) gave a higher selectivity for the formation of cyclohexanol with respect to subsequent oxidation to cyclohexanone than the octahedral complex presumably formed with **3f** (entry 3). The selectivity observed for the sulfide oxidation was disappointing: the sulfoxides were obtained in racemic form in all cases. Representative results are shown in Table 18.1

Table 18.1 Oxidation of cyclohexane and phenylmethylsulfide

Entry	Metal salt (1 mol%)	Ligand (1 mol%)	Substrate	Oxidant	Yield (%)
1[a)]	$Fe(ClO_4)_3$	(*ent*)-**3d**			4.8/0.6[b)]
2[a)]		(*ent*)-**3e**	cyclohexane	TBHP	7.7/1.1[b)]
3[a)]		(*ent*)-**3f**			6.2/4.2[b)]
4[c),d)]	$Fe(ClO_4)_3$	(*ent*)-**3f**			39
5[c),d)]	$Co(ClO_4)_2$	(*ent*)-**3f**			8
6[c),d)]	$[Ru(C_6H_6)Cl_2]$	(*ent*)-**3f**	PhSMe	H_2O_2	64
7[c),e)]		(*ent*)-**3d**			13
8[c),e)]	$VO(acac)_2$	(*ent*)-**3e**			99
9[c),e)]		(*ent*)-**3f**			91

a) Cyclohexane/*tert*-butylhydroperoxide (TBHP)/catalyst = 1000/100/1, 1h, r.t.
b) Yield cyclohexanol/cyclohexanone (GC based on TBHP, bromobenzene as internal standard).
c) $PhSMe/H_2O_2$/catalyst = 100/500/1, 0 °C.
d) 22 h.
e) 1.5 h.

We next synthesized ligands corresponding to **3a–3f** in which the phenyl group in the oxazoline moiety was replaced by a 2-hydroxyphenyl substituent. Molecular modeling suggested (Figure 18.6) that the hydroxyl function of this group would be in proximity (~2.5 Å) to the metal center being coordinated in a octahedral geometry by **3**, thus stabilization of oxo-ligands on the metal center by hydrogen bonding should become feasible.

Following the strategy outlined in Scheme 18.2 we were able to synthesize **3g–h** and (*ent*)-**3i–k**, however, **3h** could not be obtained as a pure stereoisomer since the precursor **8h** was prone to racemization. Further substituents on the oxazoline (R^1, $R^2 \neq H$) were found to circumvent this problem, giving rise to **3i–k** in enantiomerically pure form.

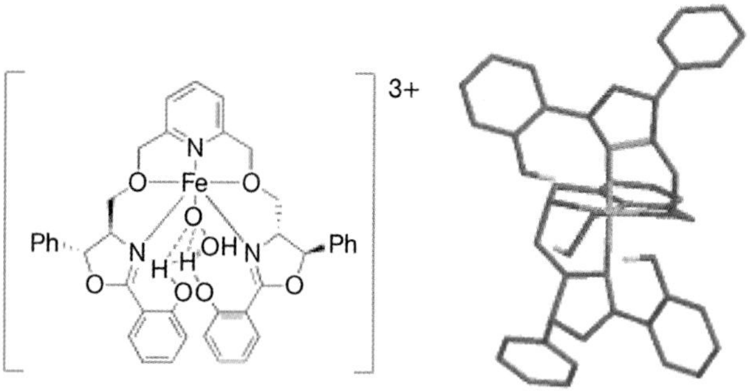

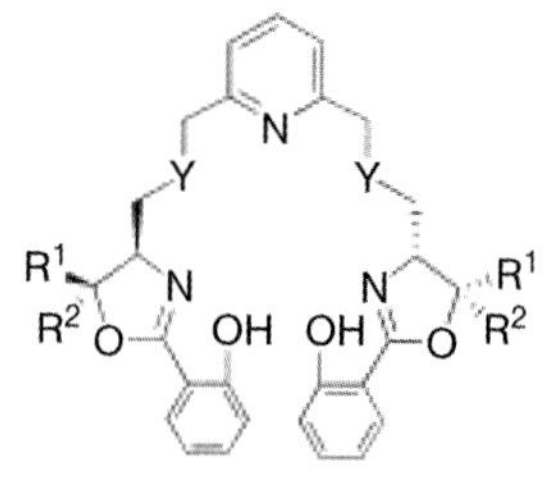

3g: Y = O, $R^1 = R^2 = H$
3h: Y = S, $R^1 = R^2 = H$
3i: Y = O, $R^1 = R^2 = Me$
3j: Y = S, $R^1 = R^2 = Me$
3k: Y = S, R^1 = MeSPh, $R^2 = H$

Figure 18.6 Molecular modeling (PM3) of an iron(III)-peroxo complex with ligand **3** containing a 2-hydroxyphenyl moiety; ligands **3g–h**, (*ent*)-**3i–k**, synthesized according to Scheme 18.1.

Zinc(II) complexes were obtained from **3g**, (*rac*)-**3h**, and (*ent*)-**3i–j**, but only the complex of (*rac*)-**3h** could be characterized by X-ray analysis. In complete analogy to Δ_2-**7c** (*cf.* Figure 18.3), (*S,S*)-**3h** exclusively formed the distorted trigonal bipyramidal complex Δ_2-**7h**. Notably, the oxygen atoms of the phenolic group are pointed toward the zinc center (Zn–O = 2.60 Å), suggesting an electrostatic interaction between these atoms.

Complexation of **3g–3k** with a variety of ruthenium(II)-, iron(II)- or iron(III)-salts revealed that these metals prefer to coordinate with the oxazoline and phenol moieties rather than with the central pyridine core; that is, the envisioned secondary binding site has become the primary one in these cases. This was concluded from NMR-, CD-, and IR-spectroscopy as well as from mass-spectrometry. Furthermore, an X-ray structure could be obtained from the reaction of $Fe(ClO_4)_2$ and *meso*-**3h** (Figure 18.8) that confirmed that mode of coordination and at the same time the potential for oxidation of such assemblies, since one of the sulfur bridges of the ligand was converted to its sulfoxide.

Based on this result, we screened only the oxygen-bridged ligands **3g** and **3i** in oxidation reactions. Unfortunately, the catalytic activity of iron complexes derived from iron(II)-salts and these ligands remained low. Somewhat encouraging was the finding that complexes $FeX_2\cdot$**3g** were more active in the oxidation of

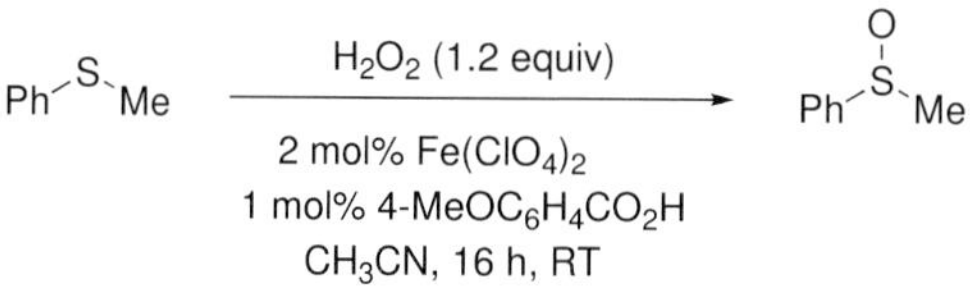

H_2O_2 (1.2 equiv); 2 mol% $Fe(ClO_4)_2$; 1 mol% 4-$MeOC_6H_4CO_2H$; CH_3CN, 16 h, RT

no ligand : 10-26%
2 mol% **3g**: 28-57%; 15% *ee*

Scheme 18.2 Iron-catalyzed oxidation of phenylmethylsulfide with hydrogen peroxide in the presence and in absence of **3g**.

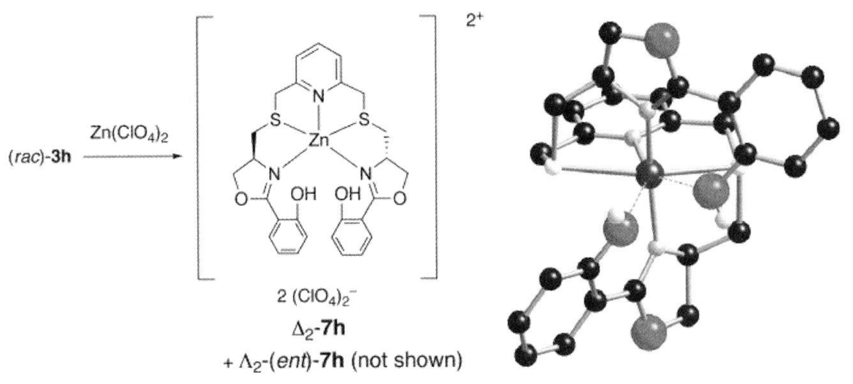

Figure 18.7 Synthesis of zinc(II)-complexes Δ_2-**7h** and Λ_2-(*ent*)-**7h**; 2 $(ClO_4)^-$ in the X-ray structure of Δ_2-**7h** (right) omitted for clarity.

(*meso*)-**3h**
$Fe(ClO_4)_2$
12
ClO_4^-

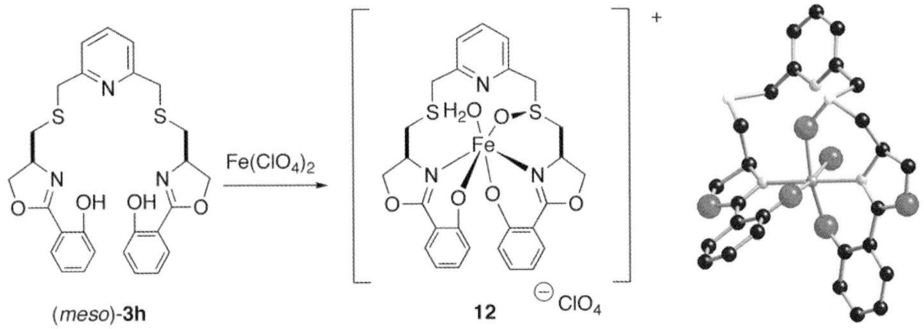

Figure 18.8 Reaction of (*meso*)-**3h** with $Fe(ClO_4)_2$ to give **12**; $(ClO_4)^-$ in the Xray structure of **12** (right) omitted for clarity.

phenylmethylsulfide with hydrogen peroxide than the iron salts in the absence of ligand (Scheme 18.2), when adopting reaction conditions reported by Bolm and coworkers [23].

In conclusion, novel pentacoordinating bis(oxazoline) ligands could be developed that form trigonal bipyramidal (zinc), octahedral (Co, Ru), and pentagonal bipyramidal (Fe, Mg, Cd) complexes with complete transfer of chirality from the ligand backbone to the metal complex. However, only modest activity for the oxidation of sulfides or alkanes was observed with these complexes, and this could not be enhanced by introducing secondary binding sites into the ligand core.

Acknowledgments

This work was supported by the Deutsche Forschungsgemeinschaft (SPP 1118) and the Fonds der Chemischen Industrie (fellowship for A.K. and Sachbeihilfe).

References

1 Sono, M., Roach, M.P., Coulter, E.D. and Dawson, J.H. (1996) *Chemical Reviews*, **96**, 2841.

2 Que, L. Jr. and Ho, R.Y.N. (1996) *Chemical Reviews*, **96**, 2607.

3 Collman, J.P. and Fu, L. (1999) *Accounts of Chemical Research*, **32**, 455.

4 Denisov, I.G., Makris, T.M., Sligar, S.G. and Schlichting, I. (2005) *Chemical Reviews*, **105**, 2253.

5 Messerschmidt, A., Huber, R., Poulos, T. and Wieghardt, K. (2001) *Handbook of Metalloproteins*, Vol. 1, J. Wiley & Sons, Chichester, UK.

6 Beinert, H., Holm, R.H. and Munck, E. (1997) *Science*, **277**, 653.

7 Bian, S. and Cowan, J.A. (1999) *Coordination Chemistry Reviews*, **190–192**, 1049.

8 Rees, D.C. and Howard, J.B. (2003) *Science*, **300**, 929.

9 Solomon, E.I., Brunold, T.C., Davis, M.I., Kemsley, J.N., Lee, S..-K., Lehnert, N., Neese, F., Skulan, A.J., Yang, Y..-S. and Zhou, J. (2000) *Chemical Reviews*, **100**, 235.

10 Solomon, E.I., Decker, A. and Lehnert, N. (2003) *PNAS*, **100**, 3589.

11 Stubbe, J., Kozarich, J.W., Wu, W. and Vanderwall, D.E. (1996) *Accounts of Chemical Research*, **29**, 322.

12 Lubben, M., M.A. Wilkinson, E.C., Feringa, B. and Que, L.J. (1995) *Angewandte Chemie*, **107**, 1610.

13 Newkome, G.R., Gupta, V.K., Fronczek, F.R. and Pappalardo, S. (1984) *Inorganic Chemistry*, **23**, 2400.

14 Bernauer, K., Pousaz, P., Porret, J. and Jeanguenat, J. (1988) *Helvetica Chimica Acta*, **71**, 1339.

15 Rasappan, R., Laventine, D. and Reiser, O. (2007) *Coordination Chemistry Reviews*, **252**, 702.

16 Borovik, A.S. (2005) *Accounts of Chemical Research*, **38**, 54.

17 Seitz, M., Kaiser, A., Tereshchenko, A., Geiger, C., Uematsu, Y. and Reiser, O. (2006) *Tetrahedron*, **62**, 9973.

18 Seitz, M., Stempfhuber, S., Zabel, M., Schutz, M. and Reiser, O. (2005) *Angewandte Chemie – International Edition*, **44**, 242.

19 Seitz, M., Kaiser, A., Powell, D.R., Borovik, A.S. and Reiser, O. (2004) *Advanced Synthesis Catalysis*, **346**, 737.

20 Seitz, M., Kaiser, A., Stempfhuber, S., Zabel, M. and Reiser, O. (2005) *Inorganic Chemistry*, **44**, 4630.

21 Seitz, M., Kaiser, A., Stempfhuber, S., Zabel, M. and Reiser, O. (2004) *Journal of the American Chemical Society*, **126**, 11426.

22 *Review*: Jackson, T.A. and Que, L. Jr. (2006) Structural and functional models for oxygen - activating nonheme iron enzymes *in Concepts and Models in Bioinorganic Chemistry*, (eds H-.B. Kraatz and N. Metzler-Nolte), VCH, Weinheim, p. 259

23 (a) Legros J. and Bolm C. (2004) *Angewandte Chemie – International Edition*, **43**, 4225; (b) Legros J. and Bolm C. (2005) *Chemistry-A European Journal*, **11**, 1086.

19
Flavin Photocatalysts with Substrate Binding Sites

Harald Schmaderer, Jiri Svoboda, and Burkhard König

19.1
Introduction

Photochemical activation of inert substrates is desirable whenever chemical storage of light energy is attempted. The best example of this is the highly efficient and sophisticated process of photosynthesis found in Nature. To mimic photosynthesis in a technical process is one of the engaging challenges in chemistry, molecular biology, and physics. Recent heterogeneous approaches addressed the photocatalytic reduction of carbon dioxide with silicates [1], semiconductors [2] or metal oxides and hydrogen [3]. Homogeneous photocatalysts may allow a more rational optimization, because of their defined structure. Examples of molecular photocatalysts are cyclodextrin-stabilized palladium clusters used for the reduction of hydrogen carbonate [4], the photohydrogenation of alkynes [5] or the photooxidation of benzyl alcohol [6].

The initial key step of photoredox catalysis is the light-induced transfer of an electron. Such processes have been intensively studied with the help of covalently and noncovalently connected electron donor-acceptor dyads [7]. As expected from Marcus theory [8], the efficiency of the electron transfer was shown to be strongly dependent on the distance and the orientation of the reaction partners. An efficient and selective photocatalyst should therefore reversibly bind the reaction substrate, rather than undergo a diffusion-controlled reaction, to ensure optimal interaction with the chromophore of the catalytic system. Examples of such templated photochemistry [9] showed high selectivity; chiral templates even allow controlling the absolute stereochemistry of a reaction [10]. Scheme 19.1 shows the general structure of a template guiding a photochemical reaction. The shield restrains the orientation of the photoactive reactants or participates in the reaction, if it is a sensitizer. A recent review has summarized the achievements in the field of photochemical reactions with topological control [11]. In this article we focus on photocatalysts with a substrate binding site bearing flavin as the chromophore.

Flavin adenine dinucleotide (FAD) or flavin mononucleotide (FMN) are prominent redox co-factors in many enzymes. Their redox properties, UV absorption and

Activating Unreactive Substrates: The Role of Secondary Interactions.
Edited by Carsten Bolm and F. Ekkehardt Hahn

ISBN: 978-3-527-31823-0

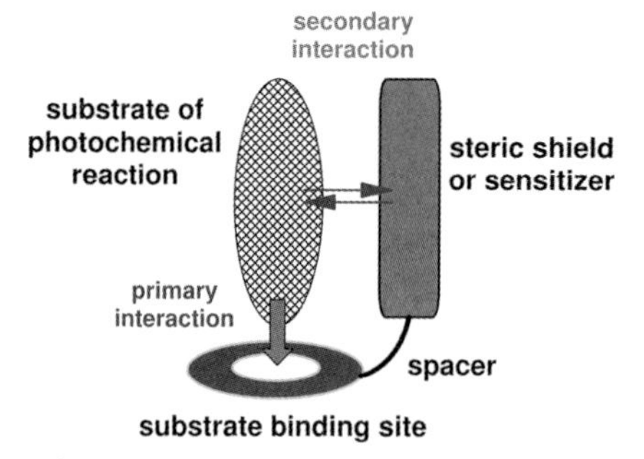

Scheme 19.1 General structure of a template controlling and enhancing photochemical homogeneous reactions.

reactivity change with substitution, noncovalent interactions, such as hydrogen bonds, and the nature of the surrounding protein [12]. Numerous flavoenzyme models which try to simulate a particular feature of the protein have been studied [13]. Nearly all of them investigate changes of the chromophore's redox potential, but the use of modified flavins in chemical catalysis is less common [14]. Scheme 19.2 shows the typical redox and protonation states of flavin [15]. The oxidized form of flavin is reduced via a direct two-electron transition to the flavohydroquinone anion. On the other hand, after one-electron reduction to the semiquinone radical or radical anion, it can accept a second electron to reach the fully reduced form. The different states are easily distinguished by UV/Vis-spectroscopy.

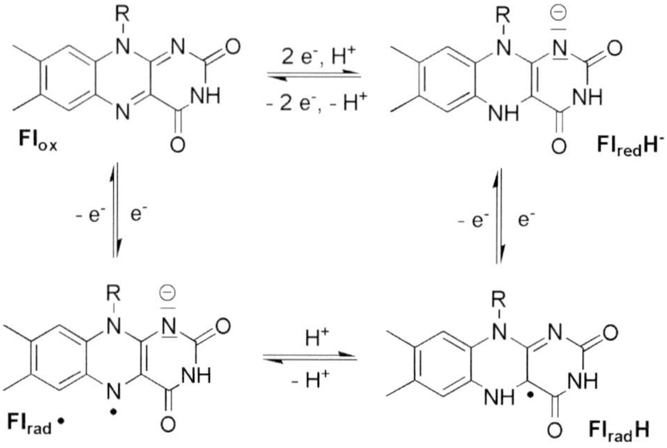

Scheme 19.2 Typical redox and protonation states of flavins.

In principle, both halves of the flavin redox cycle can be utilized for photocatalytic conversions (Scheme 19.3). If substrates are to be reduced (right side), a sacrificial electron donor is added to regenerate the reduced form of flavin. Typical electron donors are EDTA or triethylamine, which reduce flavins efficiently upon irradiation by visible light [13]. For the oxidation of substrates (left side) in most cases oxygen serves as terminal oxidant, regenerating the oxidized flavin. The excited states of reduced and oxidized flavin provide sufficient redox energy [16], as estimated by the

Rehm-Weller equation [17], to convert even substrates with low chemical reactivity, which recommends its use in photocatalysis.

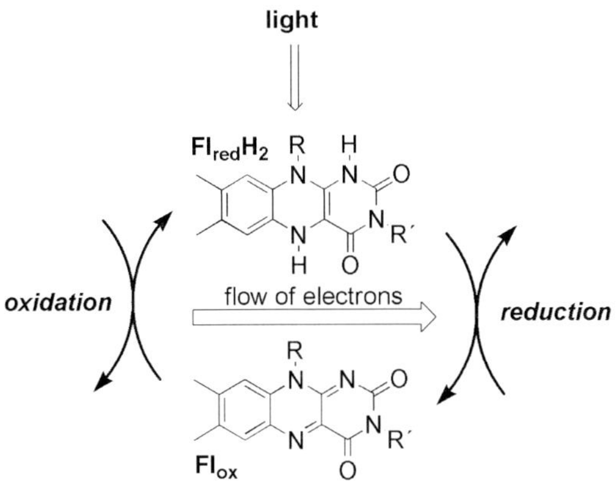

light

$Fl_{red}H_2$

flow of electrons

oxidation

reduction

Fl_{ox}

Scheme 19.3 Photocatalysis with flavins.

We begin our survey of templated flavin photocatalysis with examples of photoreductions and continue with photooxidations.

19.2
Templated Flavin Photoreductions

Flavin derivatives have been used as the sensitizing chromophore for the reductive cycloreversion of pyrimidine photocycloadducts in DNA strands. These cyclobutanes occur in Nature as a result of environmental damage to DNA exposed to UV-light. A bis-pyrimidine cyclobutane is formed, for example, between two adjacent thymine residues in DNA-strands, thus destroying the genetic information and leading to cell death or skin cancer [18]. The DNA lesions are selectively recognized by the bacterial enzyme DNA photolyase and repaired by photoinduced electron transfer using a noncovalently bound reduced flavin as electron donor. To mimic and understand this repair mechanism, artificial DNA-repair systems were prepared [19]. Covalent constructs of flavins and synthetic bis-pyrimidine cyclobutanes proved the principle of photo repair. Carell and coworkers incorporated flavin as an artificial amino acid into oligopeptides via a modified Fmoc peptide synthesis protocol [20]. These peptides were able to repair short oligonucleotide sequences containing bis-pyrimidine cyclobutanes, such as 5′-CGCGT-U = U-TGCGC-3′. Irradiation in the presence of EDTA led to fully reduced flavin species, which are also the active compounds in Nature's photolyase. The reduced flavin then cycloreverts the cyclobutane and thus repairs the DNA lesions (Scheme 19.4). Dimerized oligopeptides were synthesized to mimic helix-loop-helix proteins and to maximize the DNA-binding properties.

In 2004 Wiest et al. described an even more simplified photolyase model which reversibly coordinates bis-pyrimidine cyclobutanes with millimolar affinity in both

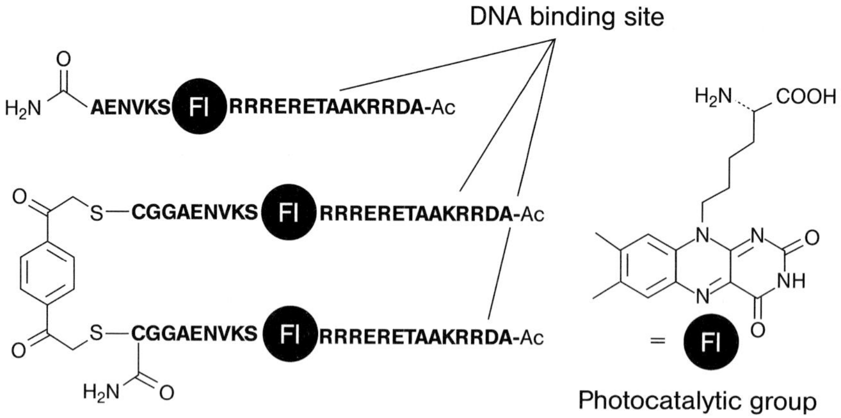

Scheme 19.4 Artificial flavin-containing peptides for DNA repair.

protic and nonprotic solvents [21]. Irradiation by visible light cyclorevert the dimeric compounds into monomeric pyrimidines: the excited flavin chromophore is reduced (1), transfers an electron onto the nucleobase cyclobutane dimer (2), which after cycloreversion (3) returns the electron to flavin (4) to close the catalytic cycle. The substrate binding site is essential to achieve an efficient conversion. However, the reaction ceases at about 75% conversion. The monomeric heterocyclic products of the reaction compete with their imide groups for coordination to the metal complex, similarly to enzymatic product inhibition, and block further binding of bis-pyrimidine cyclobutanes (Scheme 19.5).

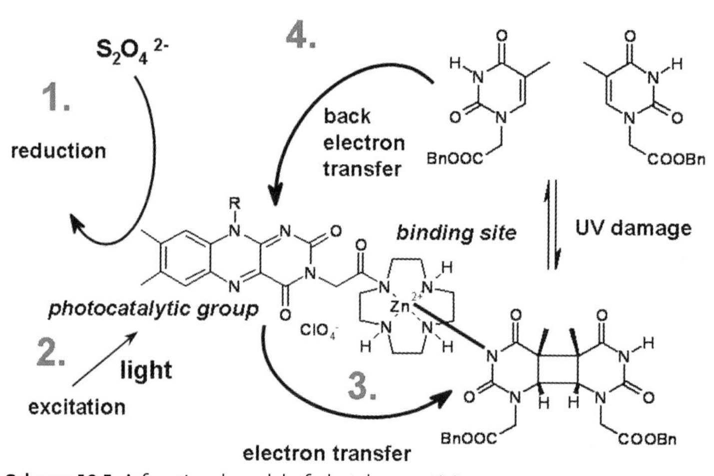

Scheme 19.5 A functional model of photolyase activity.

If the product of a flavin photoreduction is itself a catalytically active species, modulation of a catalytic reaction by light becomes possible. This was realized with riboflavin tetraacetate and copper(II) ions as substrate of the photoreduction in the presence of amines as electron donor [22]. No additional substrate binding sites on flavin are necessary for an efficient photoreduction to copper(I), as the metal ions coordinate to heteroatoms of the flavin [23]. The generated copper(I) ions serve as catalyst for a subsequent azide–alkyne cycloaddition (Huisgen reaction, Scheme 19.6). It was shown that the light quantity correlates with the amount of copper(I) and the rate of the cycloaddition. The system is an example of signal amplification by regulated catalysis: one photon induces the synthesis of 15 triazoles by catalyzed cycloaddition.

Strongly oxidising

Strongly reducing

NEt_3

$+ 2 H^+ + 2 e^-$

hν | λ > 420 nm

$2 Cu^{2+}$

$+ 2 e^-$

$- 2 H^+$

$2 Cu^+$

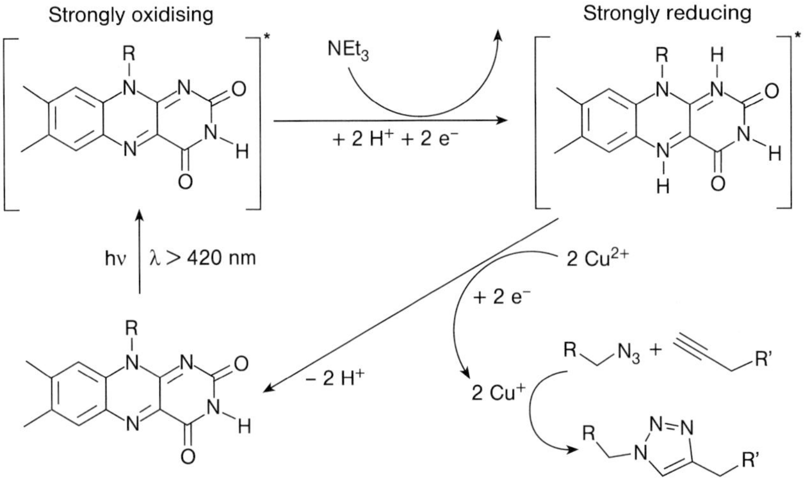

Scheme 19.6 Flavin photoreduction of copper(II) to copper(I) and subsequent copper(I)-catalyzed cycloaddition.

19.3 Templated Flavin Photooxidations

The use of flavin as photooxidant has been described in many examples [24]. However, observed selectivity and stability are not satisfactory in many cases. Therefore, optimization by the addition of a substrate binding site was attempted. Azamacrocyclic complexes, such as zinc(II) cyclen, are Lewis acids and coordinate Lewis-basic functional groups even in polar protic solvents. A hybrid compound of zinc(II)cyclen and flavin was therefore prepared and its properties in the oxidation of 4-methoxybenzyl alcohol were tested (Scheme 19.7) [25]. The coordination of the substrate's hydroxyl group by the metal complex brings it into close proximity to the flavin. After light irradiation (1), a photoinduced electron transfer from the alcohol to the flavin occurs (2). Reoxidation (3) of the flavin by oxygen dissolved in the solution regenerates the oxidized form of the flavin photocatalyst. The reaction proceeds in acetonitrile and aqueous solutions with catalytic amounts of the flavin sensitizer (10 mol%) leading to 90% conversion after two hours of irradiation.

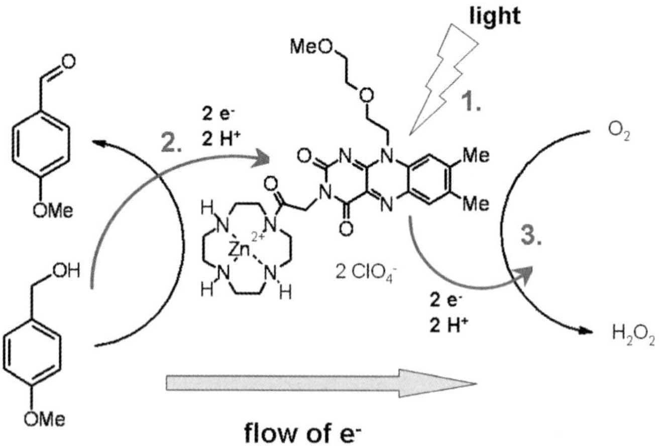

Scheme 19.7 Schematic representation of the catalytic oxidation of 4-methoxybenzyl alcohol by a flavin unit.

Thiourea is well known for its ability to reversibly form hydrogen bonds with a variety of functional groups. This has been widely used in the design of supramolecular aggregates and organocatalysts. Thiourea derivatives of flavin were therefore selected as a potential photocatalyst lead structure [26]. Their preparation uses highly reactive flavin isothiocyanates, which were derived from flavin amines, accessible by modified Kuhn synthesis. Scheme 19.8 shows two of the compounds from a larger series that was prepared.

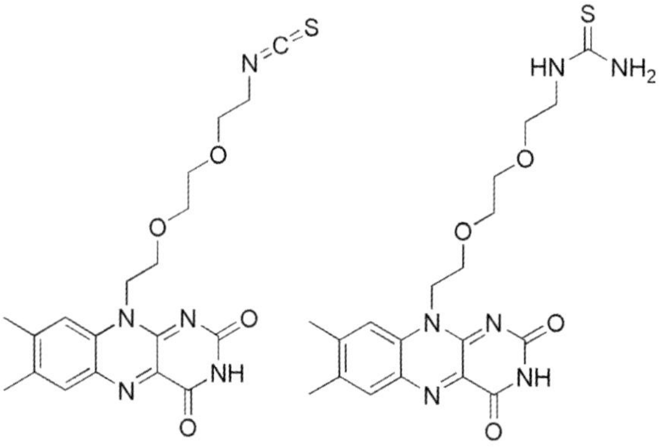

Scheme 19.8 Flavin isothiocyanate and thiourea derivatives.

Thiourea derivatives of flavin showed high activity in the catalytic photooxidation of benzyl alcohol. In a very clean reaction complete conversion of the alcohol was achieved within one hour of irradiation by a light-emitting diode (440 nm, 5 W) in air, with 10 mol% of flavin derivative as photocatalyst. The photostability of the catalyst is good and recycling of up to five times possible. With a catalyst loading of 0.1%, high turnover numbers of up to 580 were achieved. The quantum yield of the intermolecular reaction is in the order of $\Phi \approx 0.02$. Values for comparison of the efficiency are available for the enzyme photolyase ($\Phi = 0.7 - 0.9$) [19] and artificial photolyase models ($\Phi = 0.005 - 0.11$) [13j, 27] cleaving pyrimidine cyclobutanes intramolecularly. A series of control experiments were performed to reveal the role of thiourea in enhancing the oxidation. Surprisingly, mixtures of thiourea and flavins show a similar rate-enhancing effect on the alcohol oxidation to that of covalent flavin-thiourea hybrid compounds, which disproves the idea of thiourea acting as a binding site. A comparison of the redox potentials of flavin, thiourea and the benzyl alcohol substrate indicates the possible role of thiourea as an electron transfer mediator between the alcohol and flavin (Scheme 19.9).

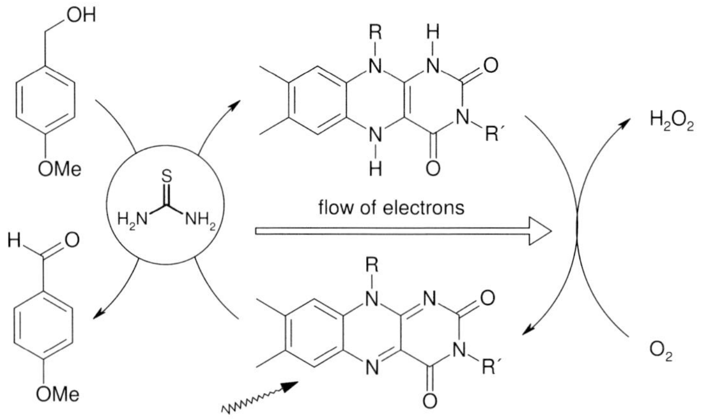

Scheme 19.9 Thiourea assisted flavin photooxidation of 4-methoxybenzyl alcohol.

19.4 Summary and Outlook

Flavin derivatives have been successfully used as photocatalysts for reductions and oxidations. The introduction of binding sites typically enhances the selectivity and efficiency of the reactions compared with diffusion-controlled processes.

The photostability of flavins remains a concern in the development of efficient catalysts. However, fine tuning of the reaction conditions may allow us to overcome the problem. Even if all physical parameters, such as redox potential, excitation energies and lifetimes of excited states are available, the coupling of the physical processes of chromophore excitation and electron transfer with a chemical reaction is difficult and still has to relay on experimental trials. Flavin-mediated photocatalytic reactions leading to nucleophilic products, such as the reduction of carbonyl compounds to alcohols, are still a challenge because of the facile covalent addition of the products to flavin, destroying the chromophore. Reactions at interfaces and new techniques of noncovalent immobilization of catalyst in ionic liquids or fluorous phases may provide solutions to this problem and pave the way for more frequent use of flavins as photoactive groups to activate less reactive molecules in photocatalysis.

Acknowledgments

We thank the DFG (SPP 1118 and GRK 640), the Fonds der chemischen Industrie and the University of Regensburg for supporting our work.

References

1 Lin, W., Han, H. and Frei, H. (2004) *The Journal of Physical Chemistry. B*, **108**, 18269–18273.

2 Pathak, P., Meziani, M.J., Li, Y., Cureton, L.T. and Sun, Y.-P. (2004) *Chemical Communications*, 1234–1235.

3 (a) Kohno, Y., Tanaka, T., Funabiki, T. and Yoshida, S. (2000) *Physical Chemistry Chemical Physics*, **2**, 2635–2639; (b) Kohono, Y., Ishikawa, H., Tanaka, T., Funabiki, T. and Yoshida, S. (2001) *Physical Chemistry Chemical Physics*, **3**, 1108–1113.

4 Mandler, D. and Willner, I. (1987) *Journal of the American Chemical Society*, **109**, 7884–7885.

5 Rau, S., Schäfer, B., Gleich, D., Anders, E., Rudolph, M., Friedrich, M., Görls, H., Henry, W. and Vos, J.G. (2006) *Angewandte Chemie – International Edition*, **45**, 6215–6218. (2006) *Angewandte Chemie*, **118**, 6361–6364.

6 Fukuzumi, S., Imahori, H., Okamoto, K., Yamada, H., Fujitsuka, M., Ito, G. and Guldi, D.M. (2002) *Journal of Physical Chemistry A*, **106**, 1903–1908.

7 For reviews on photoinduced electron transfer in noncovalently bonded assemblies, see: (a) Ward, M.D., (1997) *Chemical Society Reviews*, **26**, 365–375; (b) Willner, I., Kaganer, E., Joselevich, E., Dürr, H., David, E., Günter, M.J. and Johnston, M.R. (1998) *Coordination Chemistry Reviews*, **171**, 261–285; (d) Hayashi, T. and Ogoshi, H. (1997) *Chemical Society Reviews*, **26**, 355–364; (a) Recent examples of photoinduced electron transfer in noncovalently bonded assemblies. Braun, M., Atalick, S., Guldi, D.M., Lanig, H., Brettreich, M., Burghardt, S., Hatzimarinaki, H., Ravanelli, E., Prato, M., van Eldik, R. and Hirsch, A. (2003) *Chemistry - A European Journal*, **9**, 3867–3875; (b) Yagi, S., Ezoe, M., Yonekura, I., Takagishi, T. and Nakazumi, H. (2003) *Journal of the American Chemical Society*, **125**, 4068–4069; (c) Nelissen, H.F.M., Kercher, M., De Cola, L., Freiters, M.C. and Nolte, R.J.M. (2002) *Chemistry - A European Journal*, **8**, 5407–5414; (d) Ballardini, R., Balzani, V., Clemente-León, M., Credi, A., Gandolfi, M.T., Ishow, E., Perkins, J., Stoddart, J.F., Tseng, H.-R. and Wenger, S. (2002) *Journal of the American Chemical Society*, **124**, 12786–12795; (e) Kercher, M., König, B., Zieg, H. and De Cola, L. (2002) *Journal of the American Chemical Society*, **124**, 11541–11551; (f) Lang, K., Král, V., Kapusta, P., Kubát, P. and Vašek, P. (2002) *Tetrahedron Letters*, **43**, 4919–4922; (g) Kojima, T., Sakamoto, T., Matsuda, Y., Ohkubo, K. and Fukuzumi, S. (2003) *Angewandte Chemie*, **115**, 5101–5104; (2003) *Angewandte Chemie – International Edition* **42** 4951–4954; (h) Potvin, P.G., Luyen, P.U. and Bräckow, J. (2003) *Journal of the American Chemical Society*, **125**, 4894–4906.

8 Marcus, R.A. (1993) *Angewandte Chemie – International Edition*, **32**, 1111–1121; (1993) *Angewandte Chemie*, **105**, 1161–1172.

9 The template either controls the reaction geometry of the photoactive reactants or participates in the reaction, if it bears as chromophore as sensitizer.

10 Bauer, A., Westkämper, F., Grimme, S. and Bach, T. (2005) *Nature*, **436**, 1139–1140.

11 Svoboda, J. and König, B. (2006) *Chemical Reviews*, **106**, 5413–5430.

12 (a) Müller, F. (ed.), (1991) *Chemistry and Biochemistry of Flavoenzymes*, CRC, Boca Raton. (b) Fritz, B.J., Kasai, S. and Matsui, K. (1987) *Photochemistry and Photobiology*, **45**, 113–117; (c) Bowd, A., Byrom, P., Hudson, J.B. and Turnbull, J.H. (1968) *Photochemistry and Photobiology*, **8**, 1–10.

13 Selected examples: (a) Jordan, B.J., Cooke, G., Garety, J.F., Pollier, M.A., Kryvokhyzha, N., Bayir, A., Rabani, G. and Rotello, V.M. (2007) *Chemical Communications*, 1248–1250; (b) Caroll, J.B., Jordan, B.J., Xu, H., Erdogan, B., Lee, L., Cheng, L., Tiernan, C., Cooke, G. and Rotello, V.M. (2005) *Organic Letters*, **7**, 2551–2554; (c) Gray, M., Goodmann, A.J., Carroll, J.B., Bardon, K., Markey, M., Cooke, G. and Rotello, V.M. (2004) *Organic Letters*, **6**, 385–388; (d) Butterfield, S.M., Goodman, C.M., Rotello, V.M. and Waters, M.L. (2004) *Angewandte Chemie – International Edition* **43**, 724–727; (2004) *Angewandte Chemie*, **116**, 742–745; (e) Legrand, Y.-M., Gray, M., Cooke, G. and Rotello, V.M. (2003) *Journal of the American Chemical Society*, **125**, 15789–15795; (f) Cooke, G. (2003) *Angewandte Chemie – International Edition*, **42**, 4860–4870; (2003) *Angewandte Chemie*, **115**, 5008–5018; (g) Guo, F., Chang, B.H. and Rizzo, C.J. (2002) *Bioorganic & Medicinal Chemistry Letters*, **12**, 151–154; (h) Behrens, C., Ober, M. and Carell, T. (2002) *European Journal of Organic Chemistry*, 3281–3289; (i) König, B., Pelka, M., Reichenbach-Klinke, R., Schelter, J. and Daub, J. (2001) *European Journal of Organic Chemistry*, 2297–2303; (j) Butenandt, J., Epple, R., Wallenborn, E.-U., Eker, A.P.M., Gramlich, V. and Carell, T. (2000) *Chemistry - A European Journal*, **6**, 62–72; (k) Rotello, V.M. (1999) *Current Opinion in Chemical Biology*, **3**, 747–751; (l) Deans, R. and Rotello, V.M. (1997) *The Journal of Organic Chemistry*, **62**, 4528–4529; (m) Breinlinger, E., Niemz, A. and Rotello, V.M. (1995) *Journal of the American Chemical Society*, **117**, 5379–5380.

14 (a) Lindén, A.A., Johansson, M., Hermanns, N. and Bäckvall, J.-E. (2006) *The Journal of Organic Chemistry*, **71**, 3849–3853; (b) Lindén, A.A., Hermanns, N., Ott, S., Krüger, L. and Bäckvall, J.-E. (2005) *Chemistry - A European Journal*, **11**, 112–119; (c) Imada, Y., Iida, H., Murahashi, S.-I. and Naota, T. (2005) *Angewandte Chemie – International Edition*, **44**, 1704–1706; (2005) *Angewandte Chemie*, **117**, 1732–1734; (d) Cibulka, R., Vasold, R. and König, B. (2004) *Chemistry - A European Journal*, **10**, 6223–6231; (e) Imada, Y., Iida, H., Ono, S. and Murahashi, S.-I. (2003) *Journal of the American Chemical Society*, **125**, 2868–2869; (f) Moonen, M.J.H., Fraaije, M.W., Rietjens, I.M.C.M., Laane, C. and van Berkel, W.J.H. (2002) *Advanced Synthesis and Catalysis*, **344**, 1023–1035; (g) Murahashi, S.-I., Ono, S. and Imada, Y. (2002) Angewandte Chemie – International Edition, **41**, 2366–2368; (2002) *Angewandte Chemie*, **114**, 2472–2474; (h) Murahashi, S.-I., Oda, T. and Masui, Y. (1989) *Journal of the American Chemical Society*, **111**, 5002–5003; (i) Shinkai, S., Ishikawa, Y.-i. and Manabe, O. (1982) *Chemistry Letters*, **11**, 809–812.

15 Niemz, A. and Rotello, V.M. (1999) *Accounts of Chemical Research*, **32**, 44–52.

16 Julliard, M. and Chanon, M. (1983) *Chemical Reviews*, **83**, 425–506.

17 (a) Rehm, D. and Weller, A. (1969) *Berichte der Deutschen Chemischen Gesellschaft*, **73**, 834–839; (b) Scandola, F., Balzani, V. and Schuster, G.B. (1981) *Journal of the American Chemical Society*, **103**, 2519–2523.

18 (a) Lindahl, T. (1993) *Nature*, **362**, 709–715. (b) Taylor, J.-S. (1994) *Accounts of Chemical Research*, **27**, 76–82; (c) Begley, T.P. *Comprehensive Natural Products Chemistry C*, (1999) (ed. D. Poulter), Elsevier, **5**, p. 371; (d) Heelis, P.F., Hartman, R.F. and Rose, S.D. (1995) *Chemical Society Reviews*, **24**, 289–297; (e) Cadet, J. and Vigny, P. (1990) *Bioorganic Photochemistry Vol. I:*

Photochemistry and Nucleic Acids, (ed. H. Morrison), Wiley & Sons, New York.

19 (a) Sancar, A. (2003) *Chemical Reviews*, **103**, 2203–2238;(b) Sancar, A. (1992) *Advances in Electron Transfer Chemistry*, Vol. 2 (ed. P.S. Mariano), JAI Press, New York. p 215; (c) Begley, T.P. (1994) *Accounts of Chemical Research*, **27**, 394–401;

20 Carell, T. and Butenandt, J. (1997) *Angewandte Chemie – International Edition*, **36**, 1461–1464; (1997) *Angewandte Chemie*, **109**, 1590–1593.

21 Wiest, O., Harrison, C.B., Saettel, N.J., Cibulka, R., Sax, M. and König, B. (2004) *The Journal of Organic Chemistry*, **69**, 8183–8185.

22 Ritter, S.C. and König, B. (2006) *Chemical Communications*, 4694–4696.

23 (a) Beekman, B., Drijfhout, J.W., Bloemhoff, W., Ronday, H.K., Tak, P.P. and te Koppele, J.M. (1996) *FEBS Letters*, **390**, 221–225; (b) Hirata, J., Ariese, F., Gooijer, C. and Irth, H. (2003) *Analytica Chimica Acta*, **478**, 1–10; (c) Zhu, L., Lynch, V.M. and Anslyn, E.V. (2004) *Tetrahedron*, **60**, 7267–7275; (d) Shinkai, S., Kameoka, K., Ueda, K. and Manabe, O. (1987) *Journal of the American Chemical Society*, **109**, 923–924; (e) Shinkai, S., Nakao, H., Ueda, K., Manabe, O. and Ohnishi, M. (1986) *Bulletin of the Chemical Society of Japan*, **59**, 1632–1634.

24 (a) Fukuzumi, S., Yasui, K., Suenobu, T., Ohkubo, K., Fujitsuka, M. and Ito, O. (2001) *Journal of Physical Chemistry A*, **105**, 10501–10510; (b) Yasuda, M., Nakai, T., Kawahito, Y. and Shiragami, T. (2003) *Bulletin of the Chemical Society of Japan*, **76**, 601–605; (c) Naya, S., Miyama, H., Yasu, K., Takayasu, T. and Nitta, M. (2003) *Tetrahedron*, **59**, 1811–1821; (d) Fukuzumi, S. and Kuroda, S. (1999) *Research on Chemical Intermediates*, **25**, 789–811; (e) Del Giacco, T., Ranchella, M., Rol, C. and Sebastiani, G. (2000) *Journal of Physical Organic Chemistry*, **13**, 745–751.

25 Cibulka, R., Vasold, R. and König, B. (2004) *Chemistry - A European Journal*, **10**, 6223–6231.

26 Svoboda, J., Schmaderer, H. and König, B. (2007) *Chemistry - A European Journal*, in print.

27 (a) Friedel, M.G., Cichon, M.K., Arnold, S. and Carrel, T. (2005) *Organic and Biomolecular Chemistry*, **3**, 1937–1941; (b) Epple, R. and Carell, T. (1999) *Journal of the American Chemical Society*, **121**, 7318–7329; (c) Epple, R., Wallenborn, E.-U. and Carell, T. (1997) *Journal of the American Chemical Society*, **119**, 7440–7451.

20
New Catalytic Cu-, Pd- and Stoichiometric Mg-, Zn-Mediated Bond Activations

Tobias Thaler, Hongjun Ren, Nina Gommermann, Giuliano C. Clososki, Christoph J. Rohbogner, Stefan H. Wunderlich, and Paul Knochel

20.1
Introduction

Activations of C–H and C–X (X: I, Br, Cl, F) bonds in organometallic chemistry are achieved by metallic species, which are either used catalytically or stoichiometrically. These two principal pathways are depicted in Figure 20.1.

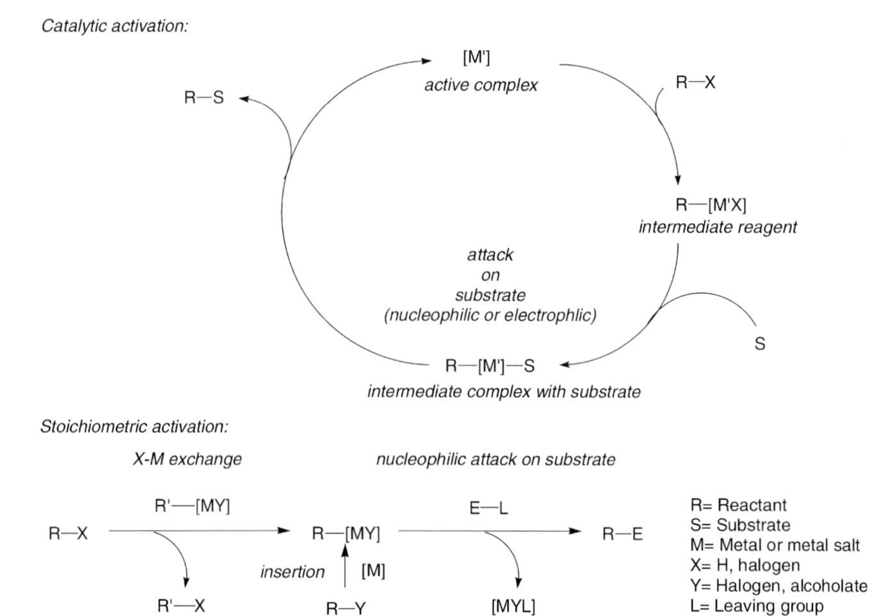

Figure 20.1 Catalytic and stoichiometric metal-mediated bond activation.

Activating Unreactive Substrates: The Role of Secondary Interactions.
Edited by Carsten Bolm and F. Ekkehardt Hahn

ISBN: 978-3-527-31823-0

In both pathways, the reactive species R-[M′]/R-[MY] are obtained by the interaction of the metallic reagent with the reactant R–X/R–Y. The catalytic activation is especially well suited for domino reactions [1]. The catalytic use of an active metal species [M′] also facilitates enantioselective reactions. These catalytic activations mostly involve the use of expensive and/or toxic transition metals, and it is therefore important to minimize the amount of catalyst used. In the stoichiometric activation pathway, [M] and [R′–MY] have to be used in stoichiometric amounts, are quantitatively consumed during the reaction, and therefore entail large amounts of chemical waste. That is why inexpensive and readily available metal species have to be used. The present chapter will give an insight into the current state-of-the-art of organometallic C–H and C–X bond activations by highlighting recent research results in this field. Thus, we will describe two examples of efficient catalytic transition metal-mediated bond activations, namely (1) a Pd-catalyzed C–H activation reaction which allows the formation of condensed polycyclic *N*-heterocycles from easily accessible starting materials [2] and (2) a Cu(I)-catalyzed activation of terminal alkynes in a one-pot three-component synthesis of propargylamines [3]. We will also report on the stoichiometric use of (1) secondary alkylmagnesium reagents in the halogen-magnesium exchange reaction [4] and (2) novel magnesium and zinc amides for the performance of selective deprotonation reactions [5, 6].

20.2
Catalytic Activation

20.2.1
C–H Bond Activation for the Preparation of Condensed Polycyclic Alkaloids

The C(sp^3)–H bond is generally reluctant to undergo C–H activation [7]. Because of its high activation energy very few examples have been reported in the literature involving its use for the preparation of polyheterocycles [8]. Recently, we have reported a chemoselective activation of the benzylic C–H bonds of *N*-arylpyrroles for the preparation of condensed polycyclic *N*-heterocycles [2]. C–H activation was achieved using a $Pd(OAc)_2/(p\text{-Tol})_3P$ catalyst system in the presence of stoichiometric amounts of Cs_2CO_3 at elevated temperatures (Figure 20.2).

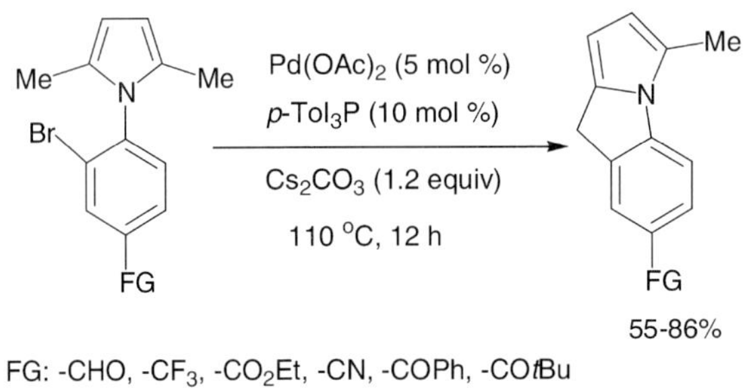

Figure 20.2 Pd-catalyzed C-H activation of *N*-arylpyrroles.

When this reaction was performed with unsymmetrically substituted arylpyrroles bearing a methyl group on one side and a larger aliphatic residue on the other (**1**), a remarkable selectivity was observed in all cases with only the methyl group being activated. An attempt to apply the reaction to 2,5-diethylpyrrole (**2**) showed that no cyclization took place at all (Figure 20.3). This may be explained by steric effects, since the formation of the intermediate palladacycle implies coplanarity of the aryl and pyrrole moieties. The rotation around the aryl-pyrrole linkage should therefore be as little obstructed as possible.

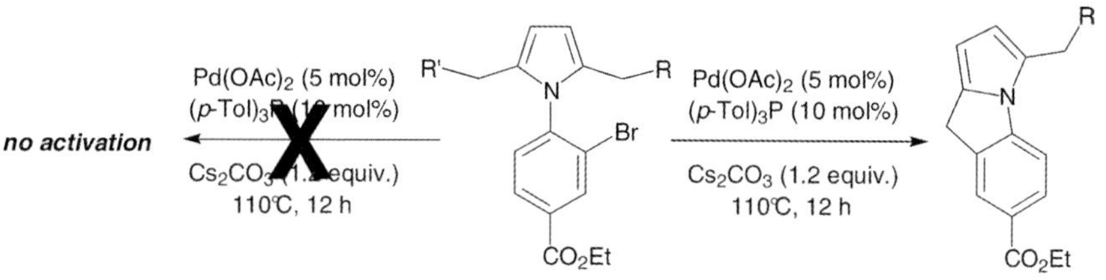
no activation
Pd(OAc)2 (5 mol%)
(p-Tol)3P (10 mol%)
Cs2CO3 (1.2 equiv.)
110℃, 12 h
R'
R
N
Br
CO2Et
Pd(OAc)2 (5 mol%)
(p-Tol)3P (10 mol%)
Cs2CO3 (1.2 equiv.)
110℃, 12 h
R
N
CO2Et

1: R = Me; R'= H
2: R = Me; R' = Me

75%

Figure 20.3 Substitution pattern for C-H activations.

In the case of 2-phenyl-5-methyl arylpyrrole (**3**), it was found that the rate of the $C(sp^2)$-activation of the phenyl group was much faster, thus resulting in the formation of the thermodynamically less favored seven-membered ring (Figure 20.4).

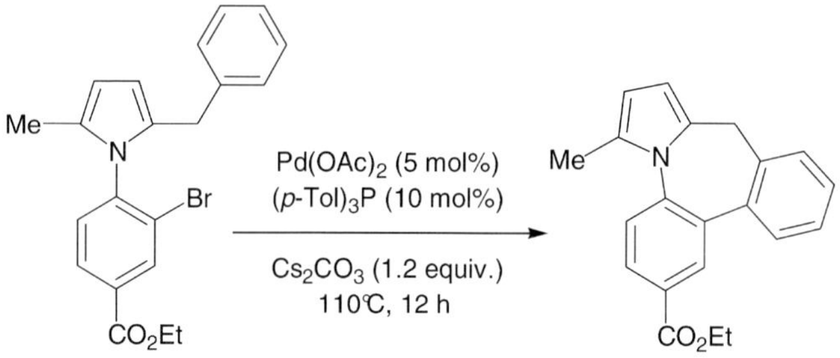
Me
N
Br
CO2Et
Pd(OAc)2 (5 mol%)
(p-Tol)3P (10 mol%)
Cs2CO3 (1.2 equiv.)
110℃, 12 h
Me
N
CO2Et

3

62%

Figure 20.4 Seven-membered ring formation due to faster $C(sp^2)$-H activation.

A successive $C(sp^2){-}H$ and $C(sp^3){-}H$ activation was achieved for dibrominated 2-phenyl-5-methyl arylpyrrole derivatives (**4**), which resulted in the formation of complex condensed polycyclic structures such as **5**. By using shorter reaction times (110 °C, 12 h) the tetracyclic heterocycle **6** could be isolated in 62% yield. Heating the reaction mixture for 24 h provided the pentacyclic product **6** in 61% yield (Figure 20.5).

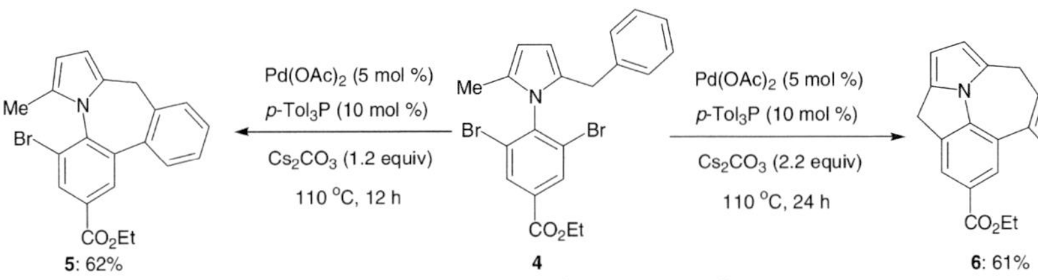

Me
Br
CO_2Et
5: 62%

$Pd(OAc)_2$ (5 mol %)
p-Tol_3P (10 mol %)
Cs_2CO_3 (1.2 equiv)
110 °C, 12 h

Me
N
Br
Br
CO_2Et
4

$Pd(OAc)_2$ (5 mol %)
p-Tol_3P (10 mol %)
Cs_2CO_3 (2.2 equiv)
110 °C, 24 h

N
CO_2Et
6: 61%

Figure 20.5 Single C(sp^2)-H and successive C(sp^2)-H and C(sp^3)-H reaction.

Condensed heterocycles of this type have so far been difficult to prepare by conventional methods. The tentative mechanism shown in Figure 20.6. describes the successive steps of the domino reaction. The process is initiated by an oxidative insertion of Pd^0 into the dibromoarylpyrrole (7). The formation of the

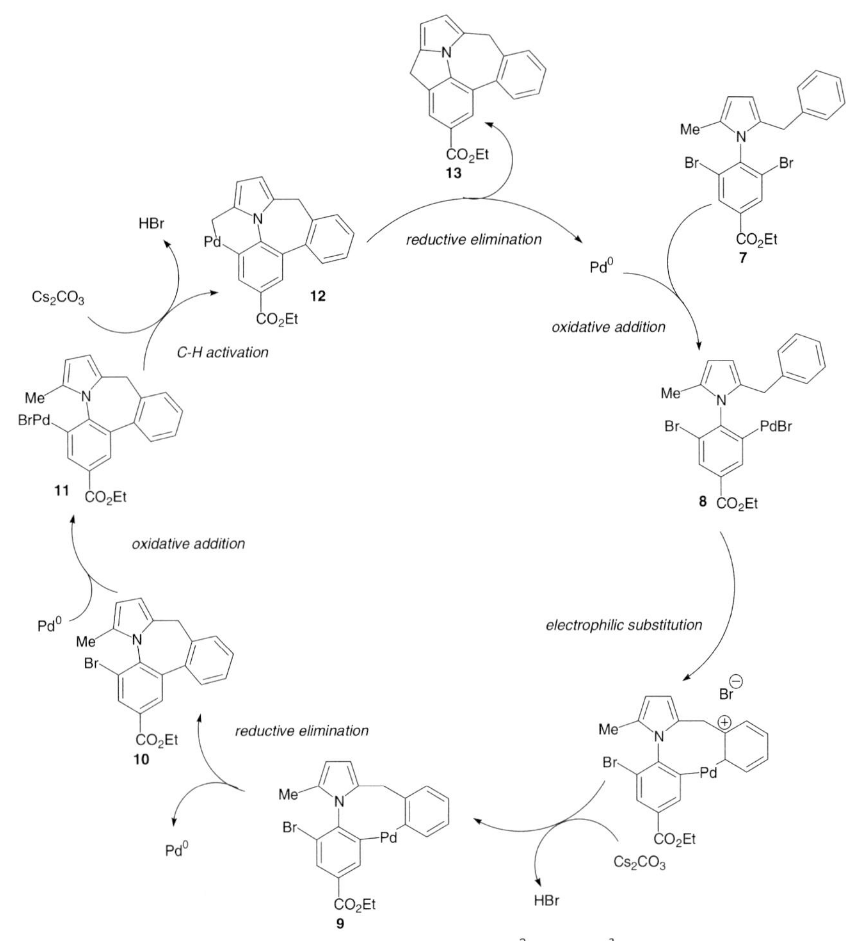

Figure 20.6 Proposed mechanism for the successive C(sp^2)-H, C(sp^3)-H activation reaction.

first carbon-carbon bond may take place by an electrophilic substitution resulting in an intermediate eight-membered palladacycle (**9**) which undergoes a reductive elimination affording the tetracyclic product **10** and regenerates the active Pd^0 catalyst. Then, oxidative insertion into the second C–Br bond takes place. Simultaneous C–H activation and HBr elimination lead to the second palladacycle (**12**), resulting in the pentacyclic product **13** after reductive elimination.

The scope of the reaction was also extended to a successive one-pot reaction involving a C(sp^3)-H activation followed by a Suzuki cross-coupling (Figure 20.7). Thus, the reaction of the functionalized *N*-arylpyrrole **14** with the boronic acid **15** in the presence of $Pd(OAc)_2$ (10 mol%) and *p*-Tol_3P (20 mol%) provides the pentacyclic heterocycle **16** in 59% yield. This reaction underlines the potential of the $Pd(OAc)_2$/(*p*-Tol)$_3$P catalyst system for initiating complex domino reactions [9].

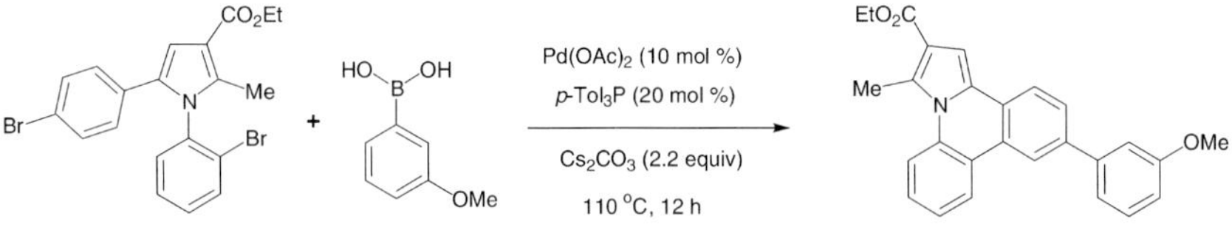

Figure 20.7 Successive C-H activation and Suzuki cross-coupling reaction.

20.2.2
Activation of Terminal Alkynes in a One-pot Three-component Enantioselective Synthesis of Propargylamines

An activation of terminal alkynes can be performed by the use of π-philic Cu salts which are known to interact with C–C triple bonds, thus offering an easy access to alkynylcopper intermediates [10]. Based on this C–H activation by copper(I) salts, we have developed an asymmetric three-component synthesis of propargylamines (**20**) [3].

This reaction has several major advantages: (1) the starting materials, an aldehyde (**17**), an amine (**18**) and a terminal alkyne (**19**) are readily available; (2) it can be carried out in one pot; (3) it is conducted with only catalytic amounts of a Cu(I) salt allowing for both an optimum atom economy [11] and enantioselective catalysis by the application of a suitable chiral ligand (Figure 20.8). The reaction was found to proceed faster without P,N-ligand, underlining the crucial role of the catalytic CuBr as key mediator. By investigating the mechanism of the reaction, it was found that the enantioselective reaction displayed a strong positive nonlinear effect [12]. Thus, an enantiomerically enriched ligand of only 10% *ee* still furnished the corresponding propargylamines with up to 68% *ee*. This suggested that a dimeric Cu/QUINAP complex acted as the catalytically active species, which was in good agreement with

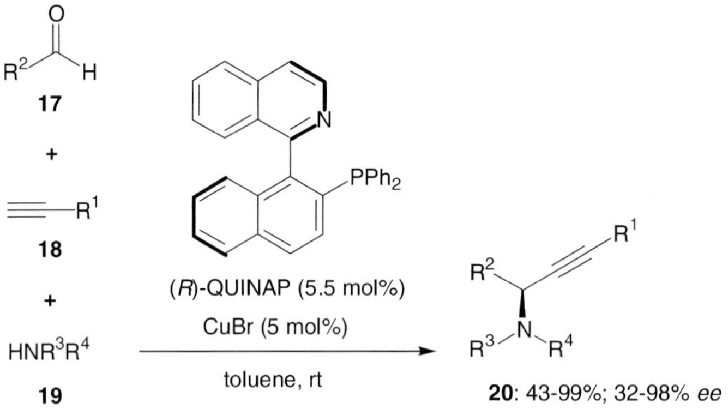

R^2CHO **17** + ≡–R^1 **18** + HNR^3R^4 **19**

(*R*)-QUINAP (5.5 mol%)
CuBr (5 mol%)
toluene, rt

PPh$_2$

20: 43-99%; 32-98% *ee*

Figure 20.8 Enantioselective three-component synthesis of propargylamines.

the crystal structure of the [CuBr{(*R*)-QUINAP}]$_2$ complex [3]. Here, the heterochiral complex [Cu_2Br_2{(*R*)/(*S*)-QUINAP}] reacts at a much slower rate than the corresponding homochiral [Cu_2Br_2{(*R*)/(*R*)-QUINAP}]. These insights led to the tentative mechanistic pathway depicted in Figure 20.9.

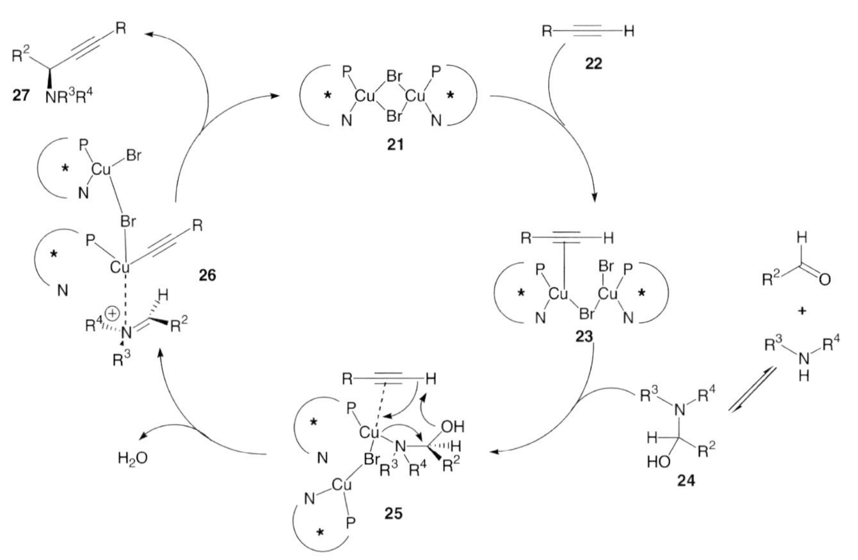

Figure 20.9 Proposed mechanism of the enantioselective three-component synthesis of propargylamines.

The dimeric chiral copper complex **21** coordinates the terminal alkyne **22**, resulting in a side-on complex (**23**). Complexation with the intermediate aminal **24** provides intermediate complex **25**. Deprotonation of the coordinated alkyne and elimination of H_2O leads to an end-on copper acetylide (**26**) with a coordinated iminium ion. The attack of the acetylide on the iminium ion in the coordination sphere furnishes the chiral propargylamine (**27**) and regenerates the active Cu(I)/QUINAP complex (**21**). This reaction has a high synthetic value, as the resulting chiral propargylamines are useful building blocks for the preparation of biologically relevant molecules [13]. A broad spectrum of possibilities for further modification of the propargyl unit extends the application scope of this method. Thus, the chiral propargylic amines such as **28** can either be reduced to secondary amines (**29**) [3] or converted to heterocyclic compounds such as α-aminoalkylpyrimidines (**30**) [14] and α-aminoalkyl-1,2,3-triazoles (**31**) [15] (Figure 20.10).

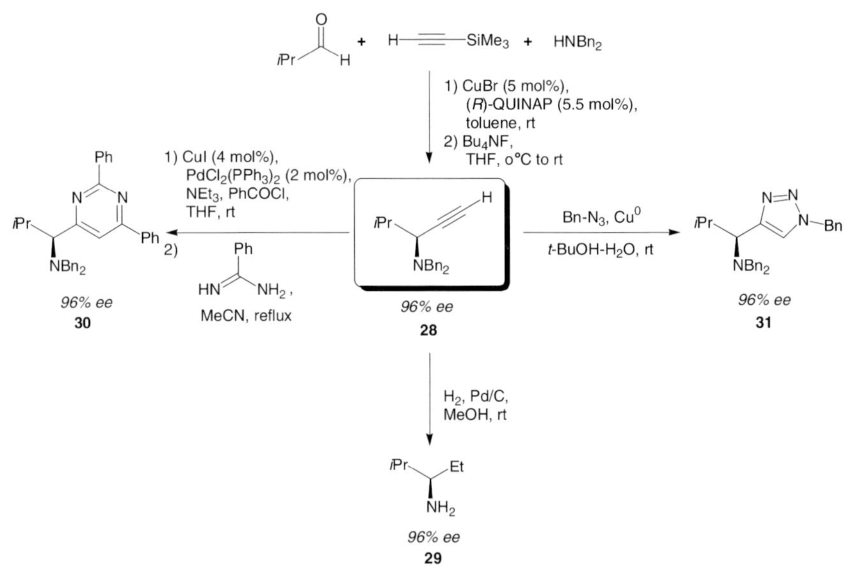

Figure 20.10 Further conversion of the chiral propargylamine **28**.

The three-component synthesis was also successfully applied as key step in the synthesis of (*S*)-(+)-coniine (**34**), a highly toxic alkaloid which occurs in the Schierling mushroom that was used to poison Socrates in 399 BC. Its synthesis is depicted in Figure 20.11. The chiral propargylamine key intermediate **32** was obtained in 88% yield and 90% *ee* from commercially available starting materials. The terminal propargylamine **32** was deprotonated with *n*BuLi and alkylated with ethylene oxide in the presence of $BF_3 \cdot OEt_2$ under mild conditions (−78 °C, 2 h). Silylation with TIPSCl in the presence of DMAP (5 mol%) and imidazole (1.5 equiv.) in DMF (0 °C to rt, 12 h) led to the TIPS-protected derivative **33** in 70% yield. Hydrogenation (1 atm) of **33** in methanol in the presence of Pd/C (10 mol%) resulted in the hydrogenolysis of the benzyl groups and reduction of the triple bond. Desilylation with Bu_4NF and subsequent submission to an intramolecular Mitsunobu reaction (DEAD (1.1 equiv.), PPh_3 (1.1 equiv.), THF, −10 °C to rt, 12 h) eventually yielded (*S*)-(+)-coniine (**34**) with 90% *ee* [16].

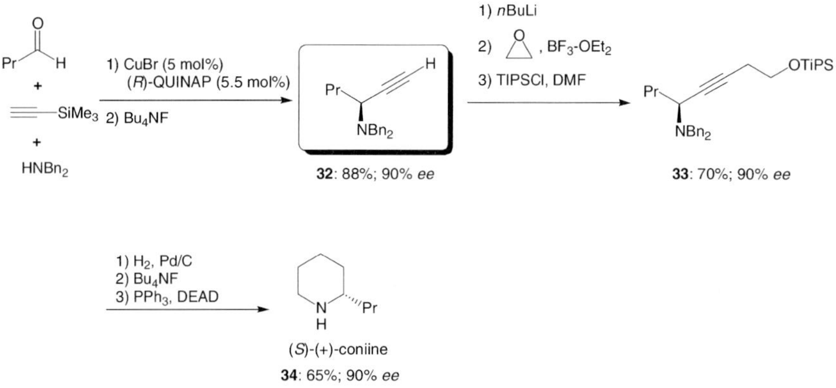

Pr–CHO + ≡–$SiMe_3$ + $HNBn_2$

1) CuBr (5 mol%) (*R*)-QUINAP (5.5 mol%) 2) Bu_4NF

32: 88%; 90% *ee*

1) *n*BuLi 2) oxetane, BF_3-OEt_2 3) TIPSCl, DMF

33: 70%; 90% *ee*

1) H_2, Pd/C 2) Bu_4NF 3) PPh_3, DEAD

(*S*)-(+)-coniine

34: 65%; 90% *ee*

Figure 20.11 Synthesis of (*S*)-(+)-coniine (**34**).

20.3
Stoichiometric Activation

20.3.1
The Halogen-Magnesium Exchange

The halogen-magnesium exchange has become a well-established reaction with a broad scope of applications in modern organic synthesis [4a]. We have demonstrated that the iodine-magnesium exchange proceeded with an excellent tolerance of sensitive functionalities [17, 18]. It represents an attractive alternative to the conventional preparation of Grignard reagents via insertion of Mg into a C-halogen bond, and it complements the halogen-lithium exchange, which, because of the high reactivity of the organolithium reagents, is only compatible with a limited range of functional groups (Figure 20.12) [22a].

The rate of exchange is dramatically increased using *i*PrMgCl-LiCl (**35**), a highly active *ate* complex [19], allowing for the metalation of the much less activated aryl and heteroaryl bromides (Figure 20.13) [4c]. Sensitive functional groups, such as cyano and ester groups were still tolerated.

The scope of the halogen-magnesium exchange has recently been extended to the functionalization of both unprotected aromatic and heteroaromatic carboxylic acids (**37**) [20] and uracils (**36**) [21] without the need of protecting the carboxylic acid or amide functionalities (Figure 20.14).

The exchange reagents required, such as *i*PrMgCl-LiCl, (**35**) are commercially available (Aldrich, Chemetall). The inorganic Mg salts and *i*PrBr/*i*PrI which accumulate as by products of the reaction can easily be disposed of. A broad range of aromatic C–I and C–Br bonds are efficiently converted into reactive nucleophilic C–MgX bonds. Because of the high tolerance of functional groups, the halogen-magnesium exchange offers an expedient access to polyfunctionalized Grignard reagents, which today are useful nucleophilic intermediates in modern organic synthesis.

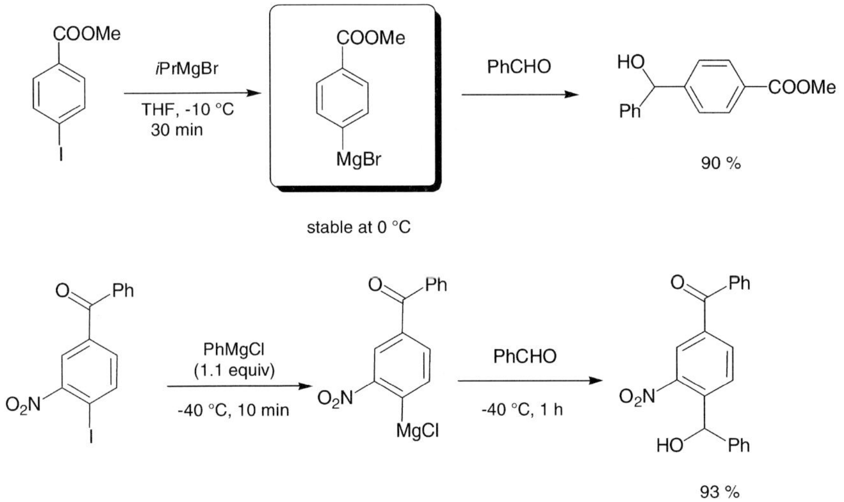

COOMe — *i*PrMgBr, THF, -10 °C, 30 min — MgBr (stable at 0 °C) — PhCHO — 90 %

PhMgCl (1.1 equiv), -40 °C, 10 min — MgCl — PhCHO, -40 °C, 1 h — 93 %

Figure 20.12 Functional group tolerance in the iodine-magnesium exchange.

$Li^{\oplus}(THF)_4$

35

***i*PrMgCl-LiCl** (1.05 equiv.), THF, -7 °C — PhCHO, 0 °C — 81 %

***i*PrMgCl-LiCl** (1.05 equiv.), THF, -7 °C — PhCOCl, CuCN·2 LiCl — 87 %

***i*PrMgCl-LiCl** (1.05 equiv.), THF, -7 °C — PhCOCl, CuCN·2 LiCl — 88 %

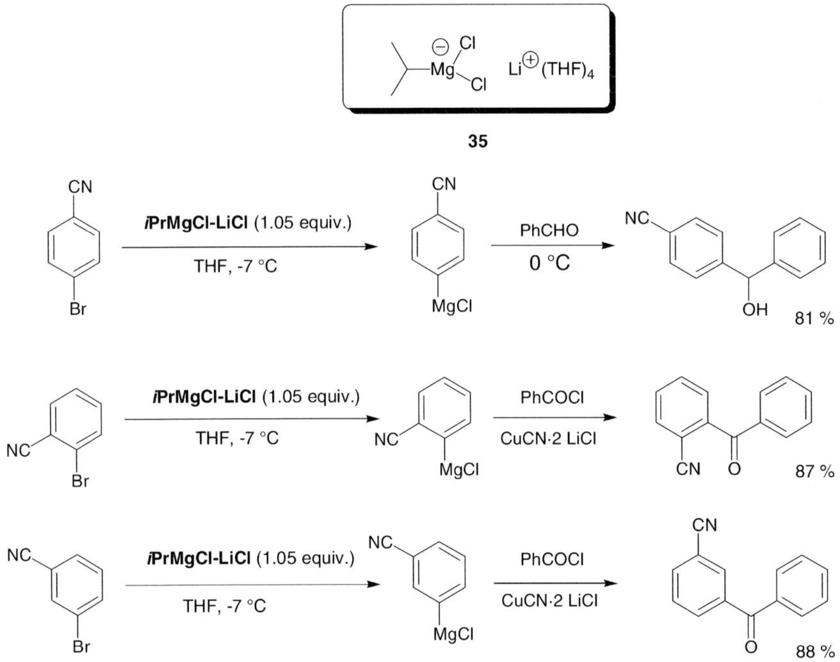

Figure 20.13 The bromine-magnesium exchange.

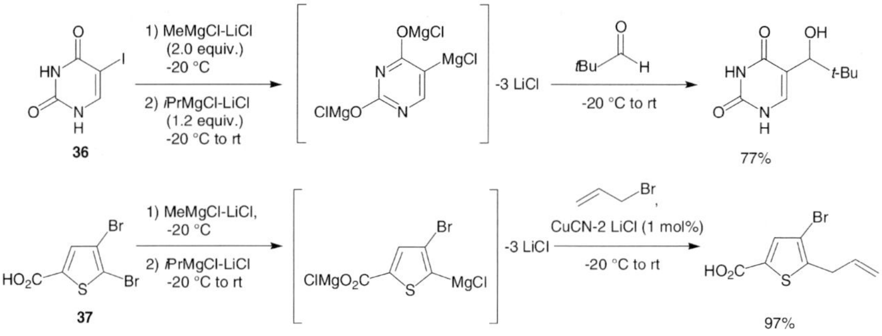

Figure 20.14 Functionalization of unprotected uracils and heteroaromatic carboxylic acids.

20.3.2
Selective Deprotonation Reactions with Magnesium and Zinc Amides

So far, mostly alkyllithium reagents (RLi) and lithium amides (R_2NLi) have found applications in organic synthesis as reagents for directed metalation [22]. However, because of their strong reactivity and nucleophilicity, these bases have only a moderate functional group tolerance. Furthermore, the low stability of lithium reagents in THF at room temperature is a major drawback, as it makes the storage of these solutions difficult. In order to avoid these problems, alternative bases have been developed such as magnesium amides [23] (**40**, **41** and **42**) and amidozincates [24] (**43**) (Figure 20.15).

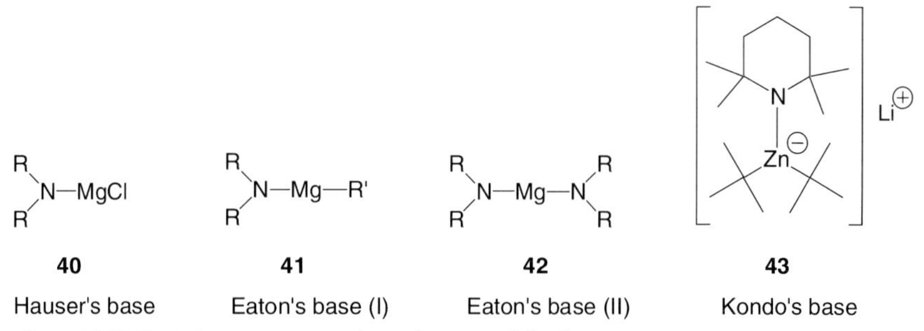

Figure 20.15 Typical magnesium and zinc bases used for the metalation of aryl and heteroaryl compounds.

These bases, however, have some synthetic disadvantages: (1) the magnesium amides only have very restricted solubility, and (2) large excesses (2–12 equiv.) often must be used in order to achieve high conversions, which further complicates the reaction. The amidozincate **43** may even require up to 3.5–4 equiv of an electrophile in subsequent quenching. Therefore, these bases cannot be readily used in large-scale processes. Recently, we have found that the addition of LiCl as stoichiometric additive greatly enhanced the activity and solubility of many organometallic species. We have hence also examined its use in the preparation of reactive and soluble magnesium amide bases. Deprotonation of 2,2,6,6-tetramethylpiperidine (TMPH; **44**) with *i*PrMgCl-LiCl (**35**) in THF furnished a stable and highly soluble mixed lithium/magnesium amide (**45**) which demonstrated an improved kinetic basicity and regioselectivity for the magnesiation of aryl and heteroaryl compounds (Figure 20.16) [5a].

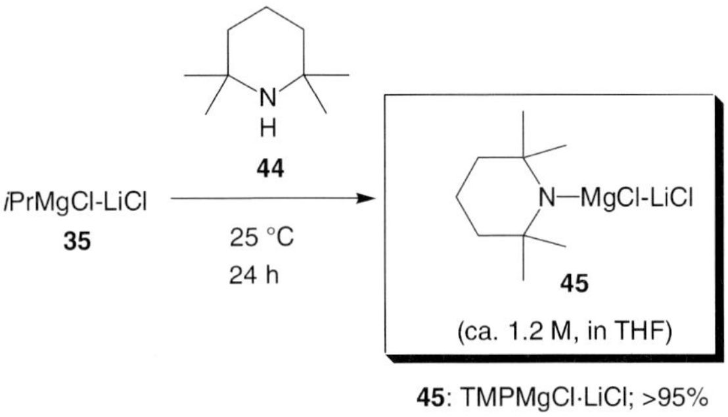

45: TMPMgCl·LiCl; >95%

Figure 20.16 Preparation of TMPMgCl-LiCl.

The sterically demanding TMPMgCl-LiCl (**45**) was able to selectively deprotonate pyrimidines (**46**) [5a] and could even be used in directed multiple magnesiation reactions of functionalized benzenes (**47–50**) (Figure 20.17) [25]. Remarkable regioselectivity was observed in the aforementioned reactions. The increased solubility of **45** toward the conventional magnesium bases (**40**, **41**, **42**) suggests that the

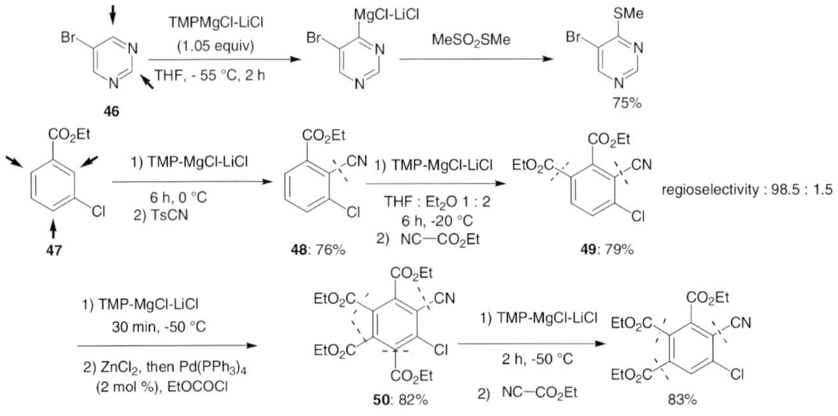

Figure 20.17 Regioselective deprotonation of pyrimidines and directed multiple deprotonation of arenes.

presence of LiCl prevents the formation of oligomeric aggregates of magnesium amides. By its *ate* character [19] a well-balanced compromise between reactivity and selectivity is accomplished allowing for the formation of new metalated aryl and heteroaryl species, which are not accessible by conventional methods.

Although TMPMgCl-LiCl (**45**) could be applied to the deprotonation of a broad range of aromatic and heteroaromatic substrates, moderately activated arenes, such as *tert*-butyl benzoate (**53**), gave unsatisfactory yields. We therefore developed a novel mixed lithium/magnesium bisamide $(TMP)_2MgCl$-2 LiCl (**52**) which shows a considerably increased reactivity. It can either be prepared by reacting two equivalents of TMPLi (**51**) with $MgCl_2$ in THF or more conveniently by simply mixing TMPMgCl-LiCl (**45**) with TMPLi (**51**) (Figure 20.18) [5b].

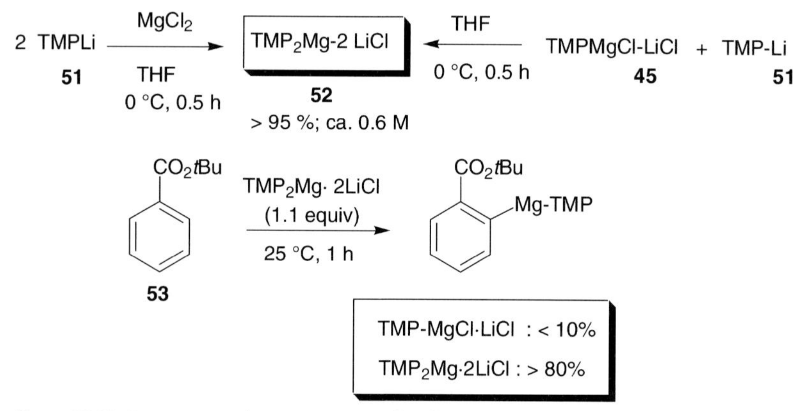

Figure 20.18 Preparation of $(TMP)_2Mg$ -2 LiCl and its application in the deprotonation of less-activated arenes.

Electron-rich aromatic substrates are especially hard to deprotonate. However, the new base $(TMP)_2Mg$-2 LiCl (**52**) is capable of performing such magnesiations. Thus, it was successfully used in the synthesis of a natural product found in the essential oil of *Pelargonium sidoides* DC (**62**): the electron-rich dimethyl-1,3-benzodioxan-4-one ring (**54**) was converted, using $(TMP)_2MgCl$-2 LiCl (**52**), to the corresponding magnesium reagent within 10 min at −40 °C. Transmetalation with $ZnCl_2$, Pd-catalyzed cross-coupling with (*E*)-1-hexenyl iodide (**55**), hydrogenation of the double bond and subsequent deprotection furnished the substituted salicylic acid derivative **56** with 67% overall yield (Figure 20.19) [5b].

This base, despite its high reactivity, also displays good tolerance towards sensitive functionalities, such as ester, keto, carbonate (OBoc; Boc = *tert*-butoxycarbonyl) or phosphorodiamidate groups. The *ortho*-directed metalation [22, 26] is a well-known method for the functionalization of aryl and heteroaryl compounds. Here, the chelation of the DMG (directed metalation group) is used to control the regioselectivity of the reaction (Figure 20.20).

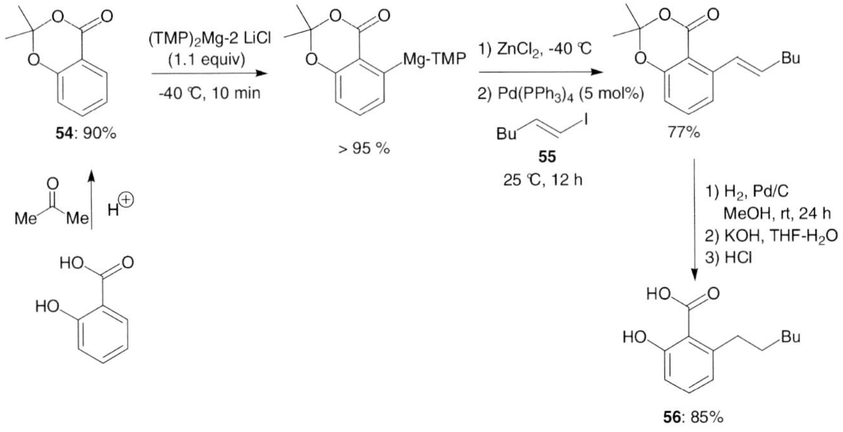

54: 90%
(TMP)$_2$Mg-2 LiCl (1.1 equiv) -40 °C, 10 min
Mg-TMP > 95 %
1) $ZnCl_2$, -40 °C
2) $Pd(PPh_3)_4$ (5 mol%)
55
25 °C, 12 h
77%
1) H_2, Pd/C MeOH, rt, 24 h
2) KOH, THF-H_2O
3) HCl
56: 85%

Figure 20.19 Natural product synthesis involving (TMP)$_2$Mg-2 LiCl as key reagent.

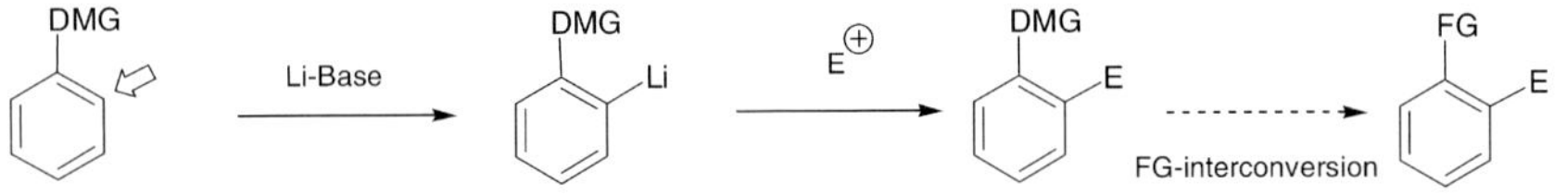

DMG — Li-Base → DMG, Li — $E^{\oplus}$ → DMG, E — FG-interconversion → FG, E

Figure 20.20 *ortho*-Direction in deprotonation reactions.

Polar DMGs may furthermore transfer electron density to the metal base and thus increase their metalating power [22]. Interestingly, the Boc and the $OP(O)(NMe_2)_2$ DMGs display complementary directing effects in deprotonation reactions with (TMP)$_2$MgCl-2 LiCl (Figure 20.21). When **57** was reacted with (TMP)$_2$MgCl-2 LiCl (**52**), metalation took place selectively at the 2-position in between the ester and the

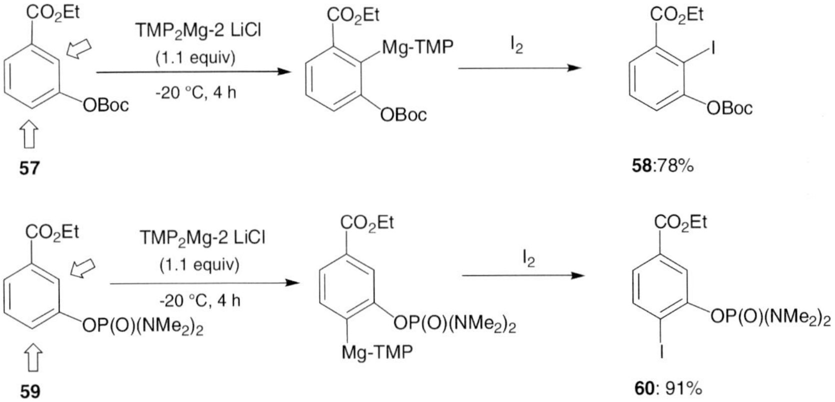

57: CO_2Et, OBoc — TMP_2Mg-2 LiCl (1.1 equiv), -20 °C, 4 h → Mg-TMP — I_2 → 58: 78%

59: CO_2Et, $OP(O)(NMe_2)_2$ — TMP_2Mg-2 LiCl (1.1 equiv), -20 °C, 4 h → Mg-TMP — I_2 → 60: 91%

Figure 20.21 Complementary directing effects of the OBoc and $OP(O)(NMe_2)_2$ group.

OBoc functionality. The $OP(O)(NMe_2)_2$ group in **59** by contrast directed the metalation to the 4-position of the aromatic ring, leading, after iodolysis, to the *para*-substituted ethyl iodobenzoate **60** in 91% yield.

In further experiments, the directing ability of the $OP(O)(NMe_2)_2$ group was found to generally overrule the directing power of most functional groups, even including esters. Therefore, we wondered whether this finding could be used for the establishment of a new general method for simple *meta*-, *para*- and even *meta*-, *meta'* as well as *para*-, *meta*-functionalization, as these substitution patterns are not easily accessible by conventional synthetic methods. In Figure 20.22 this approach is demonstrated.

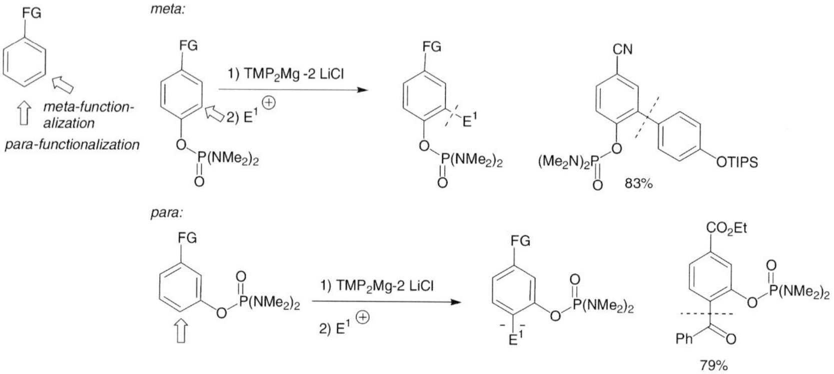

Figure 20.22 Novel synthetic strategies for *meta*- and *para*-functionalization.

Thus, the superior directing power of the $OP(O)(NMe_2)_2$ group is used to introduce functionalities in the *meta*- or *para*-position of the aromatic ring. The reaction proceeds with excellent regioselectivity [27]. The $OP(O)(NMe_2)_2$ group is replaceable by other functionalities via cleavage to the respective phenol and subsequent conversion to a triflate or nonaflate (**61**), which can be replaced by nucleophiles in, for example, cross-coupling reactions (**62**) (Figure 20.23).

Subsequent deprotonation triggered by the $OP(O)(NMe_2)_2$ group in the 4-position of the aromatic ring led to *bis-meta*-functionalized products (**63** and **64**). In this way, unsymmetrical substitution patterns could also be efficiently created, as shown in Figure 20.24.

The novel mixed lithium/magnesium amide bases in combination with the highly powerful $OP(O)(NMe_2)_2$ allows new highly regioselective functionalization of aromatics and heteroaromatics. Compared to the already established lithium bases, these bases have a higher tolerance towards functionalities. Functional groups, such as ester, keto and nitrile groups are tolerated, and relatively sensitive heteroaromatics, such as pyrimidines, can also be subjected to deprotonation with these bases. However, more sensitive functionalities, such as aldehydes and nitro groups, are not tolerated. Also, extremely sensitive classes of heterocycles, as for example 1,3,4-oxadiazoles (**65**), provide unstable magnesiated intermediates which are prone to ring opening (Figure 20.25) [28].

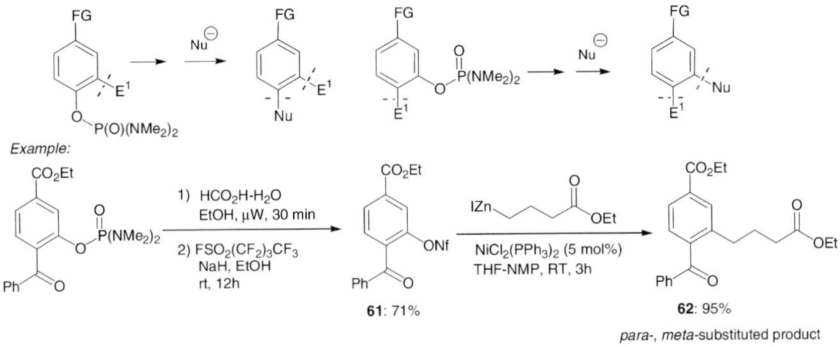

Figure 20.23 Conversion of the OP(O)$(NMe_2)_2$ group leading to the *para-*, *meta-*substituted product.

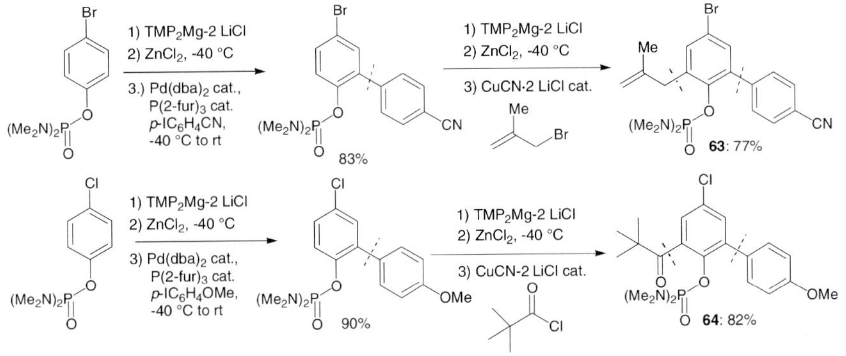

Figure 20.24 Unsymmetrical *meta-*, *meta'*-substitution by subsequent deprotonation.

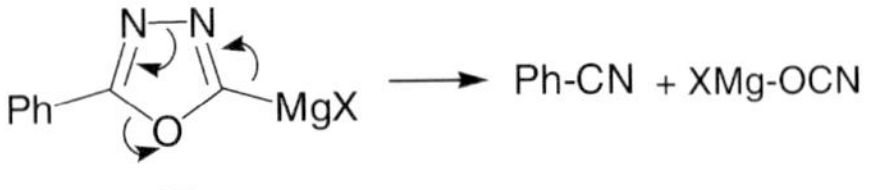

Figure 20.25 Ring opening of magnesiated 1,3,4-oxazoles.

We therefore sought to develop a new metallic base which would show selectivity and activity comparable to that of the magnesium bases and, at the same time, allow higher functional group tolerance. By considering organozinc reagents compared to their magnesium and lithium counterparts, we have prepared a new zinc amide base. TMPMgCl-LiCl (**45**) was reacted with $ZnCl_2$ (0.5 equiv.), providing $(TMP)_2Zn{\cdot}2\ MgCl_2\ {\cdot}2\ LiCl$ (**66**) (Figure 20.26) [6].

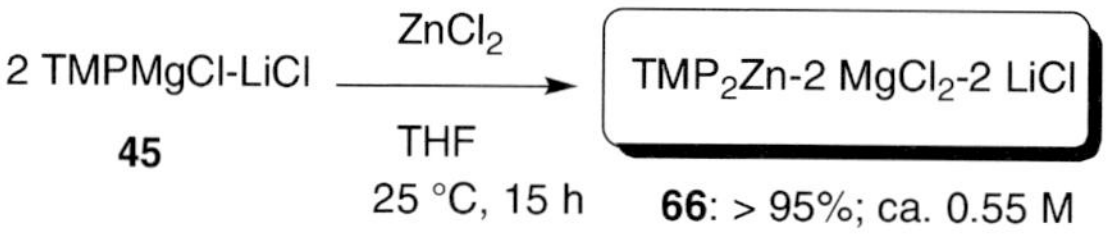

2 TMPMgCl-LiCl **45** → ($ZnCl_2$, THF, 25 °C, 15 h) TMP_2Zn-2 $MgCl_2$-2 LiCl **66**: > 95%; ca. 0.55 M

Figure 20.26 Preparation of TMP_2Zn-2 $MgCl_2$-2 LiCl

Unlike Kondo's amidozincate (**44**) [24] (Figure 20.27) the high reactivity of this base does not rely on its *ate* nature but on the combination with a Lewis acid ($MgCl_2$, LiCl).

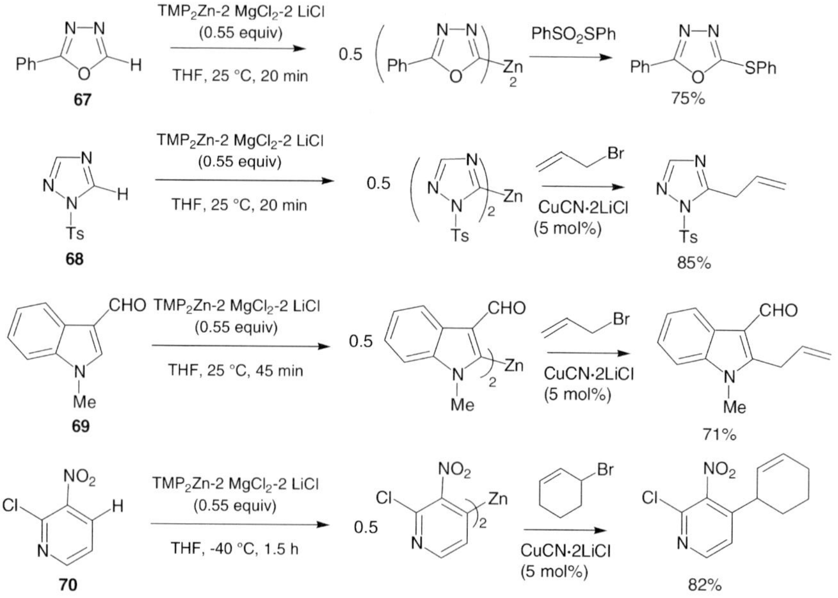

Figure 20.27 Tolerance of very sensitive functionalities during deprotonation with $(TMP)_2Zn$-2 $MgCl_2$-2 LiCl.

It could be used for the deprotonation of aromatics and heteroaromatics bearing sensitive functional groups, even including aldehyde (**69**) and nitro (**70**) functionalities. Also, 2-phenyl-1,3,4-oxadiazole (**67**) and *N*-tosyl-1,2,4-triazole (**68**) were efficiently converted without ring opening (Figure 20.27). The novel zinc base $(TMP)_2Zn$-2 $MgCl_2$-2 LiCl (**66**) thus proved to be a very valuable complement to the mixed lithium/magnesium amides, even extending the scope for selective deprotonations to highly sensitive aryl and heteroaryl compounds. Using the methods presented in this section, acidic aromatic and heteroaromatic C—H bonds can be regio- and chemoselectively activated for functionalization with various electrophiles.

20.4 Summary

In the first part we presented a Pd-catalyzed activation of *N*-arylpyrroles. The readily available $Pd(OAc)_2/(p\text{-}Tol)_3P$ system was found to be a suitable mediator for C−H activation (cyclization and C−H activation) cross-coupling reaction sequences. The reactions proceeded chemoselectively with a preference of $C(sp^2)$-H over $C(sp^3)$-H activations, resulting in the formation of less favored seven-membered rings. The fact that C(sp)−H activations are mediated by only catalytic amounts of Cu(I) salts was used to establish an enantioselective three-component synthesis of propragylamines, which are valuable building blocks for the synthesis of chiral natural products and heterocycles. In the second part we focused on the stoichiometric activation of $C(sp^2)$−X and $C(sp^2)$−H bonds. Both reactions were shown to proceed with a high tolerance of functional groups. Very sensitive functionalities, such as aldehyde and nitro groups, can be tolerated by using the highly chemoselective base $(TMP)_2Zn$-2 $MgCl_2$-2 LiCl (**66**). Deprotonation with the magnesium bases TMP MgCl-LiCl (**45**) and $(TMP)_2Mg$ -2 LiCl (**52**) proceeds with remarkable regioselectivity and thus offers easy access to unsymmetrically substituted polyfunctionalized arenes and heteroarenes. The zinc and magnesium salts which are obtained as waste products are nontoxic and can easily be disposed of in an environmentally friendly way.

References

1 (a) Tietze, L.F. (1996) *Chemical Reviews*, **96**, 115; (b) de Meijere, A., Zerschwitz, P.v., Nuske, H. and Stulgies, B. (2002) *The Journal of Organic Chemistry*, **129**, 653; (c) Ikeda, S.-I. (2000) *Accounts of Chemical Research*, **33**, 511.

2 (a) Ren, H. and Knochel, P. (2006) *Angewandte Chemie – International Edition*, **45**, 3462; (b) Ren, H., Li, Z. and Knochel, P. (2007) *Chemistry, an Asian Journal*, **2**, 416.

3 (a) Gommermann, N., Koradin, C., Polborn, K. and Knochel, P. (2003) *Angewandte Chemie – International Edition*, **31**, 1023; (b) Koradin, C., Gommermann, N., Polborn, K. and Knochel, P. (2003) *Chemistry - A European Journal*, **9**, 2797; (c) Koradin, C., Polborn, K. and Knochel, P. (2002) *Angewandte Chemie – International Edition*, **41**, 2535.

4 (a) Knochel, P., Dohle, W., Gommermann, N., Kneisel, F.F., Kopp, F., Korn, T., Sapountzis, I. and Vu, V.A. (2003) *Angewandte Chemie – International Edition*, **42**, 4302; (b) Boymond, L., Rottländer, M., Cahiez, G. and Knochel, P. (1998) *Angewandte Chemie – International Edition*, **110**, 1801; (c) Krasovskiy, A. and Knochel, P. (2004) *Angewandte Chemie – International Edition*, **43**, 3333; (d) Krasovskiy, A., Straub, B.F. and Knochel, P. (2006) *Angewandte Chemie – International Edition*, **45**, 159.

5 (a) Krasovskiy, A., Krasovskaya, V. and Knochel, P. (2006) *Angewandte Chemie – International Edition*, **45**, 2958; (b) Clososki, G.C., Rohbogner, C.J. and Knochel, P. (2007) *Angewandte Chemie – International Edition*, **46**, 7681.

6 Wunderlich, S.H. and Knochel, P. (2007) *Angewandte Chemie – International Edition*, **46**, 7685.

7 (a) Kakiuchi, F. and Murai, S. (1999) Activation of C-H Bonds: Catalytic

Reactions. *Topics in Organometallic Chemistry*, **3**, 47; (b) Dyker, G. (1999) *Angewandte Chemie – International Edition*, **38**, 1698; (c) Shilov, A.E. and Shul'pin, G.B. (1997) *Chemical Reviews*, **97**, 2879; (d) Ritleng, V., Sirlin, C. and Pfeffer, M. (2002) *Chemical Reviews*, **102**, 1731; (e) Jia, C., Kitamura, T. and Fujiwara, Y. (2001) *Accounts of Chemical Research*, **34**, 633; (f) Li, C.-J. (2002) *Accounts of Chemical Research*, **35**, 533; (g) Ma, S. and Gu, Z. (2005) *Angewandte Chemie – International Edition*, **44**, 7512; (h) Dyker, G. (ed.), (2005) *Handbook of C-H transformations: Application in Organic Synthesis*, **Vols. 1 and 2**, Wiley-VCH, Weinheim.

8 (a) Dyker, G. (1992) *Angewandte Chemie (International Edition in English)*, **31**, 1023; (b) Dyker, G. (1993) *The Journal of Organic Chemistry*, **58**, 6426; (c) Dyker, G. (1994) *Chemische Berichte*, **127**, 739; (d) Catellani, M., Motti, E. and Ghelli, S. (2000) *Chemical Communications*, 2003; (e) Baudoin, O., Herrbach, A. and Guérrite, F. (2003) *Angewandte Chemie – International Edition*, **42**, 5736; (f) Desai, L.V., Hull, K.L. and Sandford, M.S. (2004) *Journal of the American Chemical Society*, **126**, 9542.

9 (a) Tietze, L.F. and Rackelmann, N. (2004) *Pure and Applied Chemistry*, **76**, 1967; (b) de Meijere, A., Zezschwitz, P. and Bräse, S. (2005) *Accounts of Chemical Research*, **38**, 413; (c) Padwa, A. (2003) *Pure and Applied Chemistry*, **75**, 47; (d) Breit, B. (2000) *Chemistry - A European Journal*, **6**, 1519; (e) Ikeda, S. (2000) *Accounts of Chemical Research*, **33**, 511.

10 Hefner, J.G., Zizelman, P.M., Durfee, L.D. and Lewandos, G.S. (1984) *The Journal of Organic Chemistry*, **260**, 369.

11 (a) Trost, B.M. (1995) *Angewandte Chemie – International Edition*, **34**, 259; (b) Trost, B.M. (1991) *Science*, **254**, 1471.

12 (a) Kagan, H.B. (2007) "Non-linear Effects in Asymmetric Catalysis", in *Asymmetric Synthesis – The Essentials*, (eds. M. Christmann and S. Bräse), Wiley-VCH, Weinheim; (b) Kagan, H.B. (2001) *Advanced Synthesis and Catalysis*, **343**, 227; (c) Girard, C. and Kagan, H.B. (1998) *Angewandte Chemie – International Edition*, **37**, 2923; (d) Bolm, C. (1996) *Advances in Asymmetric Synthesis*, 9.

13 (a) Huffmann, M.A., Yasuda, N., DeCamp, A.E. and Grabowski, E.J.J. (1995) *The Journal of Organic Chemistry*, **60**, 1590; (b) Konishi, M., Ohkuma, H., Tsuno, T., Oki, T., VanDuyne, G.D. and Clardy, J. (1990) *Journal of the American Chemical Society*, **112**, 3715.

14 Dube, H., Gommermann, N. and Knochel, P. (2004) *Synthesis*, 2015

15 Gommermann, N., Gehrig, A. and Knochel, P. (2005) *Synlett* 2796.

16 (a) Gommermann, N. and Knochel, P. (2006) *Chemistry - A European Journal*, **12**, 4380; (b) Gommermann, N. and Knochel, P. (2004) *Chemical Communications*, 2324.

17 Jensen, A.E., Dohle, W., Sapountzis, I., Lindsay, D.M., Vu, V.A. and Knochel, P. (2002) *Synthesis*, 565.

18 Sapountzis, I. and Knochel, P. (2002) *Angewandte Chemie – International Edition*, **41**, 1610.

19 (a) Wittig, G., Meyer, F.J. and Lange, G. (1951) *Justus Liebigs Annalen der Chemie*, **571**, 167; (b) Wittig, G. (1958) *Angewandte Chemie*, **70**, 65; (c) Mulvey, R.E., Mongin, F., Uchiyama, M. and Kondo, Y. (2007) *Angewandte Chemie – International Edition*, **46**, 3802.

20 Kopp, F., Wunderlich, S. and Knochel, P. (2007) *Chemical Communications*, 2075.

21 Kopp, F. and Knochel, P. (2007) *Organic Letters*, **9**, 1639.

22 (a) Schlosser, M. (2005) *Angewandte Chemie – International Edition*, **44**, 376; (b) Turck, A., Ple, N., Mongin, F. and Quéguiner, G. (2001) *Tetrahedron*, **57**, 4489; (c) Mongin, F. and Quéguiner, G. (2001) *Tetrahedron*, **57**, 4059; (d) Schlosse, M. (2001) *European Journal of Organic Chemistry* 3975; (e) Hodgson, D.M., Brady, C.D. and Kindon, N.D. (2005) *Organic Letters*, **7**, 2305; (f) Plaquevent, J.-C., Perrard, T. and Cahard, D. (2002) *Chemistry - A European Journal*, **8**, 3300; (g) Chang, C.-C. and Ameerunisha, M.S. (1999)

Coordination Chemistry Reviews, **189**, 199; (h) Clayden, J. (2002) *Organolithiums: Selectivity for Synthesis* (eds J.E. Baldwin and R.M. Williams) Elsevier; (i) Schlosser, M., Zohar, E. and Marek, I. (2004) "The Preparation of Organolithium Reagents and Intermediates": F. Leroux, in: *Chemistry of Organolithium Compounds*, (eds. Z. Rappoport and I. Marek), Wiley, New York. chap. 1 p. 435; (j) Henderson, K.W. and Kerr, W.J. (2001) *Chemistry - A European Journal*, **7**, 3430; (k) Henderson, K.W., Kerr, W.J. and Moir, J.H. (2002) *Tetrahedron*, **58**, 4573; (l) Whisler, M.C., MacNeil, S., Snieckus, V. and Beak, P. (2004) *Angewandte Chemie – International Edition*, **43**, 2206; (m) Quéguiner, G., Marsais, F., Snieckus, V. and Epsztajn, J. (1991) *Advances in Heterocyclic Chemistry*, **52**, 187; (n) Veith, M., Wieczorek, S., Frics, K. and Huch, V. (2000) *Zeitschrift Fur Anorganische Und Allgemeine Chemie*, **626**, 1237.

23 (a) Zhang, M.-X. and Eaton, P.E. (2002) *Angewandte Chemie – International Edition*, **41**, 2169; (b) Kondo, Y., Akihiro, Y. and Sakamoto, T. (1996) *Journal of the Chemical Society-Perkin Transactions 1*, 2331; (c) Eaton, P.E., Lee, C.H. and Xiong, Y. (1989) *Journal of the American Chemical Society*, **111**, 8016; (d) Eaton, P.E., Zhang, M.-X., Komiya, N., Yang, C.-G., Steele, I. and Gilardi, R. (2003) *Synlett*, 1275; (e) Eaton, P.E. and Martin, R.M. (1988) *The Journal of Organic Chemistry*, **53**, 2728; (f) Shilai, M., Kondo, Y. and Sakamoto, T. (2001) *Journal of the Chemical Society-Perkin Transactions 1*, 442.

24 (a) Kondo, Y., Shilai, M., Uchiyama, M. and Sakamoto, T. (1999) *Journal of the American Chemical Society*, **121**, 3539, (b) Imahori, T., Uchiyama, M., Sakamoto, T. and Kondo, Y. (2001) *Chemical Communications*, **23**, 2450.

25 Lin, W., Baron, O. and Knochel, P. (2006) *Organic Letters*, **8**, 5673.

26 (a) Snieckus, V. (1990) *Chemical Reviews*, **90**, 879; (b) Metallinos, C., Nerdinger, S. and Snieckus, V. (1999) *Organic Letters*, **1**, 1183; (c) Chauder, B., Green, L. and Snieckus, V. (1999) *Pure and Applied Chemistry*, **71**, 1521.

27 Rohbogner, C.J., Clososki, G.C. and Knochel, P. (2000) unpublished results.

28 (a) Micetich, R.G. (1970) *Canadian Journal of Chemistry*, **48**, 2006; (b) Meyers, A.I. and Knaus, G.N. (1973) *Journal of the American Chemical Society*, **95**, 3408; (c) Knaus, G.N. and Myers, A.I. (1974) *The Journal of Organic Chemistry*, **39**, 1189; (d) Miller, R.A., Smith, M.R. and Marcune, B. (2005) *The Journal of Organic Chemistry*, **70**, 9074; (e) Hilf, C., Bosold, F., Harms, K., Marsch, M. and Boche, G. (1997) *Chemische Berichte-Recueil*, **130**, 1213.

21
From Cobalt(II)-activated Molecular Oxygen to Hydroxymethyl-substituted Tetrahydrofurans

Bárbara Menéndez Pérez, Dominik Schuch, and Jens Hartung

21.1
Introduction [1]

Molecular oxygen in its triplet electronic state has a marked affinity toward complexes of cobalt(II) [2–6]. Dioxygen binding occurs in a fast and often reversible manner [7] to afford adducts that are able to convert, for example, phenols into quinones [8–10], hydrocarbons into products of autoxidation [11], multiple-substituted olefins into epoxides [12–16], and monosubstituted olefins into secondary alcohols [17]. The diverse mechanisms of this chemistry have in common that at least one of the selectivity-determining steps proceeds via free radical intermediates [11].

Although significant progress has been made within the last decades, the issue of selectivity control in aerobic cobalt(II)-catalyzed oxidations is subject to ongoing investigations. It has been solved, for example, for the CAB-process that is used in order to convert aliphatic substituents of aromatic compounds into α-aryl carbonyl functionalities [18]. The state of the art of oxidative alkenol ring closure (Scheme 21.1), on the other hand, still reflects the knowledge that was provided with the original report in 1990 [19]. Its application in natural product synthesis is so far restricted to the oxidation of 1-substituted pent-4-en-1-ols using 20 mol% [bis(1-morpholinocarbamoyl-4,4-dimethyl-1,3-pentanedionato)]cobalt(II) (not shown) as catalyst for the activation of at least one molar equivalent of *tert*-butyl hydroperoxide *and* molecular oxygen [19–21]. The reaction generally furnishes *trans*-2,5-disubstituted tetrahydrofurans in synthetically useful yields and selectivities.

Substituted tetrahydrofurans are frequently used as heterocyclic building blocks in natural product synthesis [22, 23]. Unfortunately, terminal unsubstituted alkenols (e.g. **1**) are poor substrates in terms of reactivity and stereoselectivity in peroxide-based tetrahydrofuran syntheses [24, 25]. In view of its outstanding 2,5-trans selectivity, the cobalt method thus has the potential to significantly broaden the scope of oxidative alkenol ring closures. Major improvements to be addressed in order to verify this aim require the advent of a catalyst that is able to provide high chemo- and stereoselectivities in oxidations of the above-mentioned critical substrates using dioxygen as sole terminal oxidant. The system has to be applicable

Activating Unreactive Substrates: The Role of Secondary Interactions.
Edited by Carsten Bolm and F. Ekkehardt Hahn

ISBN: 978-3-527-31823-0

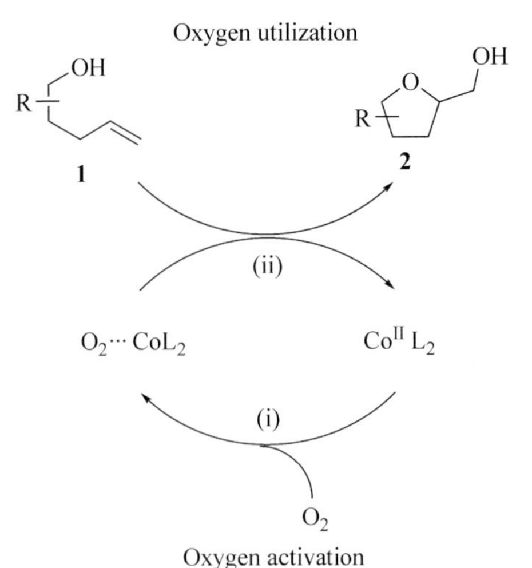

Scheme 21.1 Schematic presentation of (i) oxygen activation and (ii) utilization in the synthesis of substituted 2-(hydroxymethyl) tetrahydrofuran **2** via cobalt(II)-catalyzed aerobic oxidation of δ,ε-unsaturated alcohol **1** (L = monoanionic auxiliary).

for the conversion of a broad range of substrates, including cyclic and open-chain alkenols having alkyl- or aryl-substituents attached in other positions than the hydroxyl-substituted carbon atom. In this sense, alkenol **1a** as potential substrate poses a challenge: the scope of the method would be demonstrated if the oxidative aerobic ring closure furnishes tetrasubstituted tetrahydrofuran **2a** in a highly selective manner (Scheme 21.2). Its synthesis would open new perspectives for the preparation of potentially bioactive derivatives of the antiallergic natural product magnosalicin [26, 27]. In view of this background, a project has been pursued during the past few years to transform the cobalt method into a versatile procedure for oxidatively cyclizing alkenols into functionalized tetrahydrofurans. Major achievements of the study are summarized in the following progress report, which closes with brief reflections on scope, limitation, and stereochemical guidelines.

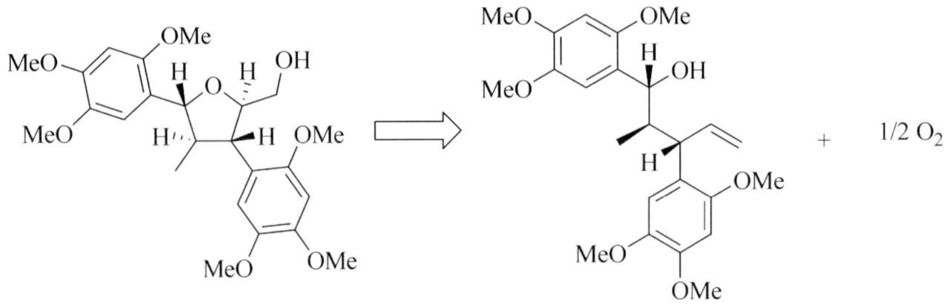

Scheme 21.2 Retrosynthetic strategy for stereoselectively preparing a derivative of natural product magnosalicin via aerobic ring closure of δ,ε-unsaturated alcohol **1a**.

21.2
Thermochemical Considerations

Aerobic monooxygenation of organic substrates requires splitting of the dioxygen molecule [28]. In principle, both O-atoms are transferable to an acceptor. In most instances, however, one of the oxygens is reduced to furnish water while the second one is consumed in substrate oxygenation [28, 29]. In nature, the reducing equivalents are delivered in a sequence of consecutive electron and proton transfer steps. In organic synthesis, this process is considered to occur via direct H-atom translocation from compounds with weak C,H connectivities to the oxidant. The term 'weak' here relates to bond dissociation energies (BDEs) in the range of approximately 360–380 kJ mol^{-1}, which for instance corresponds to the 2-C,H BDE of 2-propanol (Table 21.1, entry 5) [30, 31]. The stoichiometry of the reaction of 4-pentenol **1b** with O_2 in the presence of 2-propanol [32] requires the formation of acetone and H_2O besides 2-(hydroxymethyl)tetrahydrofuran (**2b**) (Scheme 21.3, top). Thermochemical considerations (Table 21.1) and results from density functional theory calculations (B3LYP/6-31 + G**) [33] indicate that this transformation is associated with a significant thermochemical driving force [$\Delta_R H(298.15) = -353$ kJ mol^{-1}, $\Delta_R G(298.15) = -356$ kJ mol^{-1}, $\Delta_R S(298.15) = +10$ J mol^{-1} K^{-1}]. It is by far more exergonic than the hypothetical alternative (Scheme 21.3, bottom), i.e. the entropically disfavored conversion of 2 molecules of alkenol **1b** with $^3O_2$ into 2 molecules of tetrahydrofuran **2b** [$\Delta_R H(298.15) = -363$ kJ mol^{-1}, $\Delta_R G(298.15) = -306$ kJ mol^{-1}, $\Delta_R S(298.15) = -191$ J mol^{-1} K^{-1}].

Table 21.1 Tabulation of bond dissociation energies (BDEs) [kJ mol^{-1}] of connectivities associated with chemical modifications in aerobic alkenol oxidations [30, 31].

Entry	Connectivity	Type	BDE [kJ mol^{-1}]
1	•O–O• (in $^3O_2$)		498.4 ± 0.2[a)]
2	C–O	σ	355 ± 6[b)]
3	C–O	σ	400 ± 4[c)]
4	C=O	π	323[d)]
5	C–H	σ	381 ± 4[e)]
6	O–H	σ	499.2 ± 0.2[f)]
7	O–H	σ	445[g)]
8	O–H	σ	442 ± 2[c)]
9	O–H	σ	438 ± 3[h)]

a) Ref. [31];
b) in diethyl ether;
c) in 2-propanol;
d) calculated [34];
e) 2-H in 2-propanol;
f) in H_2O;
g) in acetic acid;
h) in methanol.

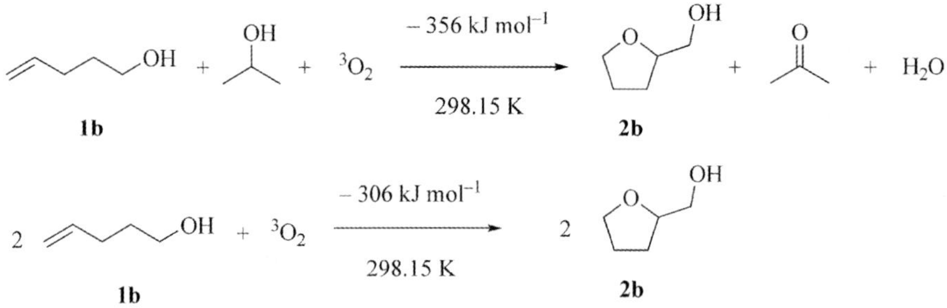

1b + 2-propanol + $^3O_2$ —(−356 kJ mol^{-1}, 298.15 K)→ 2b + acetone + H_2O

2 1b + $^3O_2$ —(−306 kJ mol^{-1}, 298.15 K)→ 2 2b

Scheme 21.3 Free reaction energy changes (B3LYP/6-31+G**//B3LYP/6-31+G**) in two alternative routes to 2-(hydroxymethyl) tetrahydrofuran (**2b**) via aerobic oxygenation of pent-4-en-1-ol (**1b**) {**1b** (linear conformer): $E = -271.77300$, E + ZPVE $= -271.631299$, $H = -271.622636$, $G = -271.663311$; (R)-**2b** ($_1T^5$ – conformer, O–C–C–O = 60.1°; O–H···O = 106.3°; H···O = 2.398 Å): $E = -347.011984$, E + ZPVE $= -346.861786$, $H = -346.853762$, $G = -346.893508$; 2-propanol: $E = -194.3806888$, E + ZPVE $= -194.272866$, $H = -194.266457$, $G = -194.300280$; acetone: $E = -193.1745092$, E + ZPVE $= -193.090962$, $H = -193.084632$, $G = -193.119178$; H_2O: $E = -76.4340477$, E + ZPVE $= -76.412761$, $H = -76.408981$, $G = -76.430410$; $^3O_2$: $E = -150.327577$, E + ZPVE $= -150.323838$, $H = -150.323838$, $G = -150.343818$, $\langle S^2 \rangle = 2.0889$; $\nu = 1641\ cm^{-1}$; E, H, G, ZPVE are given in Hartree (1 Hartree = 2625.50 kJ mol^{-1}; H- and G-values refer to 298.15 K; data for $^3O_2$ were computed with the unrestricted wave function}.

21.3 Cobalt(II)-Diketonate Complexes

Based on the affinity of β-diketonates to coordinate to cobalt(II) [35–37], auxiliaries **3–5** with an inductively electron-withdrawing substituent (R = CF_3 [38]) and with inductively *and* mesomerically interacting groups (R = C_6H_5 [39] and 3,5-$(CF_3)_2C_6H_3$) were prepared via α-acylation of (+)-camphor (Table 21.2). The camphor skeleton was considered to positively affect the solubility of cobalt complexes **6–8** in organic media and, more importantly, to guide stereoselection in the course of substrate oxygenation,

Table 21.2 Synthesis of (+)-camphor-derived β-diketonato(−1) complexes of cobalt(II)

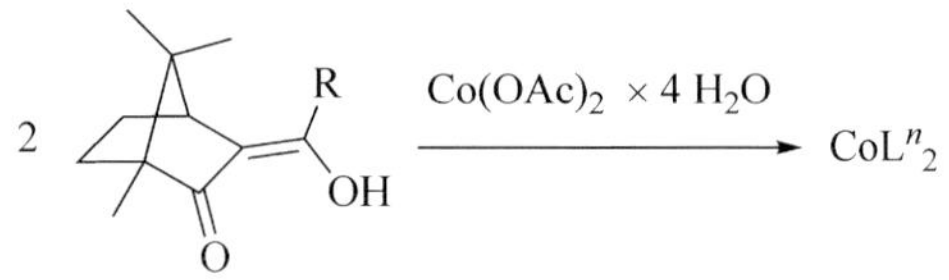

2 HL^n : **3–5** —($Co(OAc)_2 \times 4\ H_2O$)→ CoL^n_2 **6–8**

Entry	3–5	R	Yield of 6–8 [%]
1[a]	3 (HL^1)	CF_3	**6**: quant.
2[b]	4 (HL^2)	C_6H_5	**7**: 60
3[b]	5 (HL^3)	3,5-$(CF_3)_2C_6H_3$	**8**: quant.

[a] Reaction in EtOH at 20 °C;
[b] reaction in EtOH/aq. NaOH at 20 °C.

i.e. tetrahydrofuran formation. Diketonate complexes **6–8** were obtained as red (**6**) or light brown (**7**) to orange (**8**) crystalline solids that are readily soluble in slightly to strongly polar organic solvents, e.g., EtOH, 2-propanol, THF, Et_2O, acetone, and CH_2Cl_2. They are almost insoluble in H_2O or aliphatic hydrocarbons.

Bis[3-trifluoroacetyl-(+)-camphorato(−1)]cobalt(II) (**6**) is a paramagnetic compound ($\mu_{eff} = 1.5\,\mu_B$, 298 K) that crystallizes from aqueous THF as hydrated mono-tetrahydrofuran adduct [40]. The cobalt atom is located in a distorted octahedral coordination sphere. One diketonate ligand $(L^1)^-$ is coordinated in meridional position, while the second is bound with one O-atom in apical and one in equatorial position. The H_2O molecule occupies the second apex leaving the fourth equatorial site for tetrahydrofuran coordination. If this compound served as a reagent for $^3O_2$ activation, its integrity would probably not prevail in solution. Since activation is considered to occur via binding to cobalt(II), dissociation of at least one of the ligands would be necessary in order to allow dioxygen activation. Preliminary computational data provide evidence that tetrahydrofuran loss of the adduct is energetically favored compared with dehydration [41].

The compound referred to as precatalyst **6** in this project was prepared upon treatment of 2 molar equivalents of auxiliary **3** with one equivalent of $Co(OAc)_2 \times 4\ H_2O$ in EtOH. There is cumulative evidence that the material that is obtained after drying of the product is the hydrated acetic acid adduct of bis[3-trifluoroacetyl-(+)-camphorato(−1)]cobalt(II) (**6**). Compounds **7** and **8** were characterized via combustion analysis, UV/Vis- and IR-spectroscopy. In view of the existing information, neither of the complexes **7** and **8** has additional solvent molecules attached to the cobalt(II) ion.

If stirred in a solution of 2-propanol at 20 °C, 1.00 mmol of bis[3-trifluoroacetyl-(+)-camphorato(−1)]cobalt(II) (**6**) (see above) ($c_o = 0.10$ M) absorb 1.06 mmol of O_2 within 30 h. The event of oxygen uptake is associated with a change in color from red (**6**) via brown to green. It is unclear at the moment whether these changes originate exclusively from oxygen binding to the transition metal or are indicative of an oxidative conversion of the solvent, or a combination of both. Removal of the volatiles from this solution affords a green solid, which shows, apart from minor shifts and changes, two dominating new IR-absorptions at 1205 and 1147 cm^{-1}. Combustion analysis (C 43.17%, H 5.32%) of the sample points to a slight increase in oxygen content compared to the data obtained for **6** (C 52.26, H 5.49). Its UV/Vis spectrum (2-propanol) shows a broad band at 270 nm and shoulders at 361 and 600 nm. The spectral information of this material closely matches values obtained for a compound that is formed upon treatment of complex **6** with TBHP in 2-propanol.

21.4 Reactivity

Cobalt(II)-complexes **6–8** are able to catalyze the formation of *trans*-(2-hydroxymethyl)-5-(*tert*-butyl)tetrahydrofuran **2c** if added to a solution of *tert*-butylpentenol **1c** in 2-propanol in a stationary O_2 atmosphere (Table 21.3). The product is a racemate, as determined by ^{31}P-NMR of a derived chiral dioxaphospholane [42].

The stereoisomer *cis*-**2c** was not detected either in the crude reaction mixture or in the purified sample (GC). $Co(OAc)_2 \times 4\,H_2O$ was found to be completely (GC) ineffective to catalyze formation of **2c** from **1c** under such conditions. Purging of O_2 through solutions of, e.g., **6** and **1c** in 2-propanol at 60 °C furnished complete turnover of the alkenol, however, without providing a satisfactory mass balance (not shown). The yield of cyclization product **2c** gradually decreases along the sequence of applied cobalt(II) complexes **6** ($R = CF_3$) > **8** [$R = 3,5\text{-}(CF_3)_2C_6H_3$] > **7** ($R = C_6H_5$) (Table 21.3, entries 1–3]. The reaction may be induced by microwave irradiation instead of conductive heating, thus leading to shorter reaction times (e.g. Table 21.3, entry 4).

Table 21.3 Solvent effect in aerobic cobalt(II)-catalyzed oxidations of alkenol **1c**.

1c → (O_2 / Co(II) cat., *solvent*) → **2c**

Entry	Co(II) [mol%]	Solvent[a]	*T* [°C]	*t* [h]	Conversion [%][b]	Yield [%][b] (*cis*:*trans*)[b]
1	**6** [10]	*i*PrOH	60	2.5	91	63 (<1 : 99)[c]
2	**7** [10]	*i*PrOH	60	3.0	14	4 (<1 : 99)[c]
3	**8** [10]	*i*PrOH	60	3.0	87	42 (<1 : 99)[c]
4	**6** [10]	*i*PrOH	60	1.5[d,e]	91	49 (<1 : 99)[c]
5	**6** [20]	EtOH	60	15[e]	quant.	50 (<1 : 99)[c]
6	**6** [20]	*t*BuOH	60	15[e]	quant.	32 (<1 : 99)[c]
7	**6** [20]	cC_5H_9OH	60	15[e]	quant.	31 (<1 : 99)[c]
8	**6** [20]	CF_3CH_2OH	60	15[e]	48	25 (8 : 92)
9	**6** [20]	CH_2Cl_2	20	48[e]	85	28 (6 : 94)

[a] 8 mL of p.a. grade solvent;
[b] quantitative GC measurements; *er* = 50 : 50 for *trans*-**2c**;
[c] *cis*-**2c** not detected (GC);
[d] microwave heating;
[e] MS 4 Å added.

Control experiments indicated that complex concentrations below 0.01 M provide a slower but more chemoselective turnover of substrate **1c** (not shown). A precatalyst concentration of 0.0125 M (10 mol%) poses a balance between selectivity and time-yield factor. Since catalyst deactivation gradually occurs, a complete conversion of alkenol **1c** was not attainable even in extended runs (>48 h) at cobalt concentrations well below 0.006 M.

The UV/Vis spectrum of the spent catalyst points to formation of a cobalt(III) residue of unknown constitution. UV/Vis absorptions of this material ($\lambda = 271$, 308, 363, 607 nm) and strong IR bands at 1203 and 1135 cm^{-1} are similar but not identical to spectral information of a product originating from the reaction between **6** and either O_2 or TBHP (see above). Attempts to regain catalytic activity for oxidative bishomoallylic alcohol cyclization upon treatment of the cobalt residue with typical Co(III)/Co(II) reductants [43], e.g. formate or hypophosphite, failed for unknown reasons. Addition of formate, hypophosphite, or L-ascorbate to a solution of **6** and **1c** in 2-propanol in a typical run completely inhibited the formation of 2-hydroxymethyl (*tert*-butyl)tetrahydrofuran (**2c**).

A gradual increase in the yield of **2c** is seen along the series of solvents CH_2Cl_2 $CF_3CH_2OH < cC_5H_9OH \sim tBuOH < EtOH < iPrOH$ (Table 21.3, entries 1, 5–9). Experiments performed in hot 2-propanol occurred in a highly effective manner using catalyst concentrations below 0.025 M (20 mol%) of **6** in the absence of molecular sieves (4 Å).

Addition of *tert*-butyl hydroperoxide (TBHP), either as 70% (w/w) aqueous solution or as anhydrous 5.5 M solution in nonane did not improve the efficiency of the aerobic oxidative alkenol cyclization (Table 21.4, entries 1 and 2). Attempts to prepare a methylcyclohexane-derived hydroperoxide *in situ* via cobalt(II)-mediated autoxidation [11] prior to aerobic oxidative cyclizations were ineffective for increasing the yield of target compound **2c** (Table 21.4, entry 3).

Table 21.4 Peroxide effect in aerobic cobalt(II)-catalyzed oxidations of alkenol **1c**.

O_2 / **6** (20 %)
additive
*i*PrOH /MS 4 Å

1c → **2c**

er = 50 : 50

Entry	Additive	*T* [°C]	*t* [h]	Conversion[a]	Yield [%][a] (*cis:trans*)[a]
1	TBHP[b]	50	5	82	56 (<1 : 99)[c]
2	TBHP[d]	50	5	quant.	62 (<1 : 99)[c]
3	$cC_6H_{11}CH_3$[e]	60	15	quant.	36 (<1 : 99)[c]

[a] Quantitative GC measurements;
[b] 1.0 equiv. of a 70% (*w*/*w*) aqueous solution;
[c] *cis*-**2c** was not detected (GC);
[d] 1.0 equiv. of a 5.5 M anhydrous solution in nonane;
[e] methylcyclohexane/2-propanol 50/50 (*v*/*v*).

An almost quantitative mass balance was obtained for aerobic oxidations of 1-(*tert*-butyl)-4-pentenol **1c** and 1-phenyl derivative **1d** (97–99%) using either cobalt(II) complex **6** (Table 21.5, entries 1 and 3), or an *in situ* formulation of the reagent (Table 21.5, entries 2 and 4). In all instances, 1 major and 5 minor products were formed. Apart from target compounds **2c**,**2d** and 2-methyl-5-phenyltetrahydrofuran (**9d**), none of the identified products has so far been reported as an additional component in aerobic cobalt(II)-catalyzed alkenol oxidations [19, 21, 24]. Products originating from a formal C,C-π-bond cleavage (**12–14**) are subject to ongoing

Table 21.5 Mass balancing in aerobic cobalt(II)-catalyzed oxidations of 1-substituted pent-4-en-1-ols **1c** and **1d**.

O_2 / **6** (10 %), *i*PrOH / 60 °C: **1** → **2** + **9** + **10** + **11** + **12** + **13** + **14** + **15**

Entry	R	1, 2, 9–15	*t* [h]	Conv. [%][a]	Yield [%][a]							
					2	9	10	11	12	13	14	15
1[b]	*t*Bu	c	2.5	91	63	–[c]	6	1[d]	11	4	3	–[c]
2[e]	*t*Bu	c	2.5	98	60	2	5	7[f]	15	6	2	–[c]
3[b]	Ph	d	3.0	98	61	7	4	–[c]	6	–[c]	17	2
4[e]	Ph	d	4.0	99	59	9	3	–[c]	6	–[c]	17	2

[a] Quantitative GC measurements;
[b] 10 mol% of **6**;
[c] not detected;
[d] $dr = 62:38$;
[e] catalyst generated *in situ* from 10 mol% of $Co(OAc)_2 \times 4\ H_2O$ and 20 mol% of **3**;
[f] $dr = 55:45$.

Table 21.6 Quantitative analysis of H_2O and acetone formed in aerobic cobalt(II)-catalyzed oxidations in 2-propanol.

O_2 / **6** (10 %), *i*PrOH / 60 °C

Entry	1	pO_2 [bar]	Conv. of **1** [%][a]	Yield [%]		
				2[a]	Acetone[b]	H_2O[c]
1	**1c**[d]	3.0	80 ± 1	**2c**: 50 ± 2	60 ± 6	78 ± 8
2	**1c**	1.5	82	**2c**: 53	53	71
3	**1d**[d]	3.0	66 ± 4	**2d**: 25 ± 2	47 ± 5	64 ± 10
4	**1d**	1.5	71	**2d**: 20	46	67

[a] Determined via GC;
[b] HPLC analysis of derived 2,4-dinitrophenylhydrazone [45];
[c] Karl Fischer titration [46];
[d] mean value ± standard deviation as determined from 5 independent runs.

investigations. Autoxidation products (e.g. **15d**) were formed to a minor extent, although cobalt(II) compounds usually very effectively catalyze this type of reaction [44].

Formation of acetone and water as additional products from the oxidative cyclization of substrates **1c** and **1d** was verified and quantified by suitable analytical techniques (Table 21.6) [45, 46]. The yields of water and acetone slightly diverged. They were consistently higher than the yields of tetrahydrofurans **2c** and **2d** (Table 21.6). One of the reasons for this observation may originate from above-mentioned side reactions. Autoxidations for instance, increase the percentage of reaction water without necessarily affecting the acetone balance.

21.5 Stereoselectivity Survey

The selectivity of tetrahydrofuran formation from 1-phenylpentenol **1d** inversely correlates with catalyst concentration (Table 21.7). Cobalt complex concentrations below 0.008 M (6.5 mol%) afford slow and incomplete turnover of **1d** and therefore are not recommended for synthetic purposes. Tetrahydrofuran **2d** is formed as racemate as determined by ^{31}P-NMR of a derived chiral dioxaphospholane [42]. The yield of 2-methyl-5-phenyltetrahydrofuran (**9d**) gradually increases with the concentration of the cobalt(II) reagent and with the reaction temperature (Figure 21.1). Heterocycle **9d** is formed with identical diastereoselectivity to that for the hydroxyl-substituted derivative **2d**.

Table 21.7 Affecting selectivity in oxidative cyclizations of 1-phenylpentenol **1d** (15 h, 60 °C, see Figure 21.1).

Entry	**6** [mol L^{-1}] (mol%)	Conversion	Yield **2d** [%][a)]	Yield **9d** [%][b)]
1	0.025 (20)	quant.	48	24
2	0.009 (7.5)	quant.	59	11
3	0.008 (6.5)	86	56	–[c)]

a) *cis:trans* < 1 : 99 (GC), *er* = 50 : 50 for *trans*-**2d**;
b) *cis:trans* < 1 : 99 (GC), *er* not determined for *trans*-**9d**;
c) not detected (GC).

Exposure of a solution of 2-phenylpent-4-en-1-ol (**1e**) for 5 h to O_2 in a solution of 2-propanol (60 °C) and 0.013 M (10 mol%) of bis[3-trifluoroacetyl-(+)-camphorato(−1)] cobalt(II) (**6**) afforded 67% of 2-hydroxymethyl-4-phenyltetrahydrofuran (**2e**) (*cis*:*trans* = 78 : 22) and 14% of 2-methyl derivative **9e** (*cis*:*trans* = 71 : 29) (Figure 21.2). Both diastereoisomers of **2e** are formed as racemates. Apart from a minor scatter of the data, diastereoselectivity of product formation remains unchanged throughout

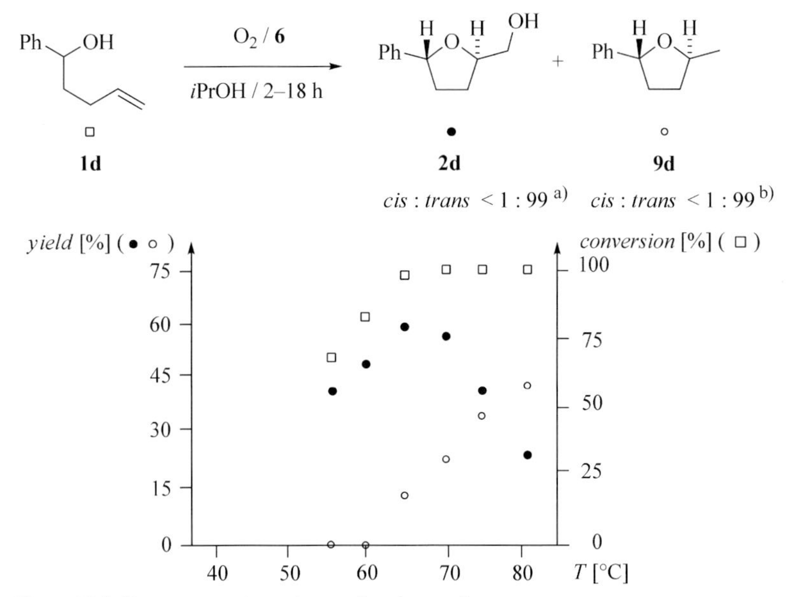

Figure 21.1 Temperature dependence of major product formation in Co(II)-catalyzed oxidations of 1-phenylpentenol **1d**. Experiments at 55–70 °C were performed with 10 mol% of **6**, experiments at 75 and 82 °C with 20 mol%. [a)]*er* = 50 : 50 for *trans*-**2d**; [b)]*er* not determined for *trans*-**9d**.

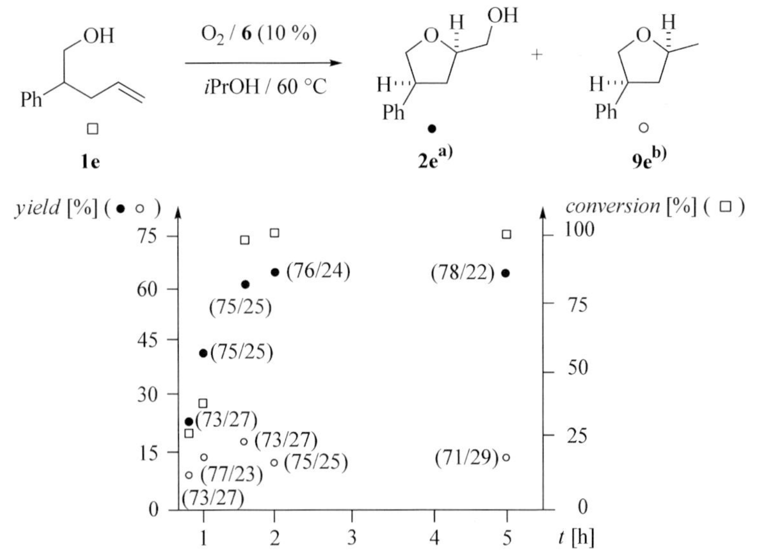

Figure 21.2 Monitoring product formation in aerobic oxidation of 2-phenylpent-4-en-1-ol (**1e**). [a)]*er* = 50 : 50 for *cis*-**2e** and for *trans*-**2e**; [b)]*er* not determined for *cis*-**9e** and *trans*-**9e**. Figures in brackets refer to *cis/trans*-ratios of **2e** (•) and **9e** (○).

the conversion of alkenol **1e**. An improvement in product selectivity is attainable upon lowering the catalyst concentration to 0.008 M (6.5 mol%), which provides 76% of **2e** (*cis:trans* = 78 : 22) and 2% of 2-methyl derivative **9e.**

3-Phenyl-4-pentenol **1f** provides 2-hydroxymethyl-3-phenyltetrahydrofuran **2f** (*cis:trans* = 2 : 98) if treated in an oxygen atmosphere in the presence of cobalt(II) reagent **6** (c_0 = 0.009 M; Table 21.8, entry 3). An increase in the catalyst concentration is associated with a more efficient but less selective conversion of substrate **1f** (Table 21.8, entries 1–2).

Table 21.8 Selectivity dependence of aerobic oxidations of 3-phenylpent-4-en-1-ol (**1f**) from cobalt(II)-complex concentrations.

O_2 / **6**, *i*PrOH / 60 °C, 15 h: **1f** → **2f** + **9f**

2f: *cis* : *trans* = 2 : 98 [a)]

9f: *cis* : *trans* < 1 : 99 [b)]

Entry	6 [mol L^{-1}] (mol%)	Conversion	Yield 2f [%]	Yield 9f [%]
1	0.025 (20)	quant.	44	16
2	0.013 (10)	98	63	5
3	0.009 (7.5)	91	59	–[c)]

a) *er* = 50 : 50 for *trans*-**2f**;
b) *cis*-**9f** not detected (GC); *er* not determined for *trans*-**9f**;
c) not detected (GC).

21.6
A Derivative of Magnosalicin

1- and 3-substituted δ,ε-unsaturated alcohols provide *trans*-disubstituted tetrahydrofurans as major products. A substituent attached at position 2 directs the aerobic alkenol cyclization cis-diastereoselectively. In view of this information, a novel synthesis of β-functionalized tetrasubstituted tetrahydrofuran **2a** was devised. The product is difficult to obtain in peroxide-based oxygenations, using for instance vanadium(V) complexes as catalysts [25]. Treatment of diastereomerically pure racemic substrate **1a** with cobalt(II) complex **6** (10 mol%, c_0 = 0.005 M) and molecular oxygen in 2-propanol for 4 h at 60 °C and afterwards with the same amount of the cobalt reagent (10 mol%) for a further 15 h at 75 °C affords the all-trans-configured target compound **2a** (72%) as a single diastereoisomer (^{1}H-NMR). The compound is a valuable reagent for the preparation of derivatives of magnosalicin in upcoming studies, since the natural product has recently attracted attention as an antiallergic constituent of *Magnolia salicifolia* [26].

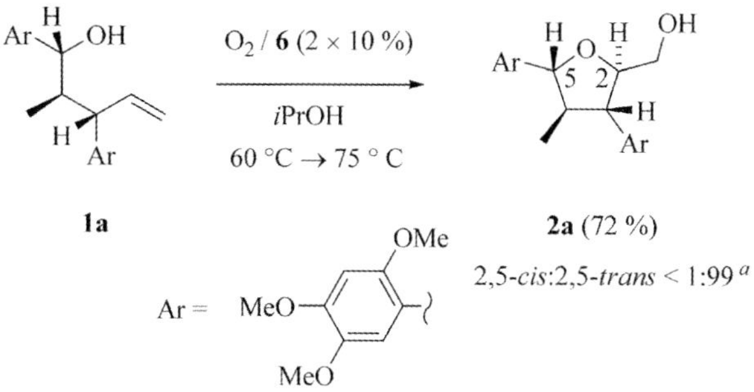

Ar = 2,4,5-trimethoxyphenyl

1a → O_2 / **6** (2 × 10 %), *i*PrOH, 60 °C → 75 °C → **2a** (72 %), 2,5-*cis*:2,5-*trans* < 1:99 [a]

Scheme 21.4 Stereoselective synthesis of a derivative of magnosalicin. [a] GC and NHR.

21.7 Expanding the Scope

The reagent combination of molecular oxygen and bis[3-trifluoroacetyl-(+)-camphorato(−1)]cobalt(II) (**6**) in hot isopropanol selectively oxidizes diastereomers of 2-allylcyclohexanol **1g** and **1h** to furnish 8-substituted oxabicyclo[4.3.0]nonanes **2h** (61%) and **2g** (68%) (Scheme 21.5). Considering diastereoselection in the synthesis of monocyclic tetrahydrofurans, a 1,8-*cis*- and 6,8-*trans*-arrangement, as seen in the major stereoisomer of **2h**, constitutes a stereochemically matching configuration. Formation of diastereomerically pure **2g** from *cis*-configured alkenol **1g**, however, does not seem to follow this general guideline. A 1,8-*trans*-6,8-*trans*- arrangement poses in this sense a stereochemically mismatching couple of configurations, however, without leading to poorer stereoselection in the formation of **2h**. The two examples show that an extended set of stereochemical guidelines will be necessary for predicting major products in the formation of alicyclic-fused tetrahydrofurans.

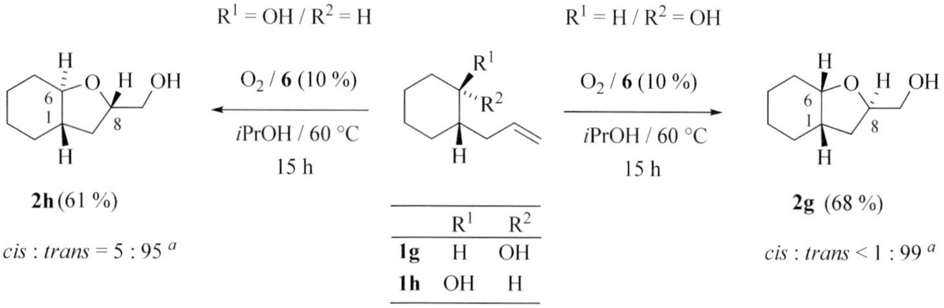

R^1 = OH / R^2 = H: O_2 / **6** (10 %), *i*PrOH / 60 °C, 15 h → **2h** (61 %), *cis* : *trans* = 5 : 95 [a]

R^1 = H / R^2 = OH: O_2 / **6** (10 %), *i*PrOH / 60 °C, 15 h → **2g** (68 %), *cis* : *trans* < 1 : 99 [a]

	R^1	R^2
1g	H	OH
1h	OH	H

Scheme 21.5 Stereoselective synthesis of oxabicyclo[4.3.0] nonanes. [a]*cis/trans*-ratios refer to relative configurations of atoms attached at positions 6 and 8.

Bishomoallylic alcohols that have two alkyl substituents located in terminal position of the olefinic π-bond occur as a widespread structural motif in, for instance, terpene-derived natural products. The feasibility of 2,5-*trans*-selective oxidative cyclizations of 1-substituted alkenols having this substitution pattern certainly would be regarded as significant progress, since peroxide-based oxidations favor the cis mode of ring closure in this instance. The substrate to explore this approach, i.e. 1-phenyl-5-methylhex-4-en-1-ol (**1i**), however, undergoes almost no conversion if treated with cobalt(II) complex **6** in 2-propanol at 60 °C. A change of solvent to 2-butanol allows one to conduct the reaction at an elevated temperature (98 °C), leading to 2,5-disubstituted tetrahydrofuran **2i**, diol **16**, and hydroxyketone **17** (Scheme 21.6, top). Heterocycle **2i** shows a cis/trans ratio of 75 : 25. Formation of diol **16** and hydroxyketone **17** in significant amounts points to alternative reaction channels, i.e. olefin hydratation and autoxidation, that gain importance as the reaction temperature rises [47].

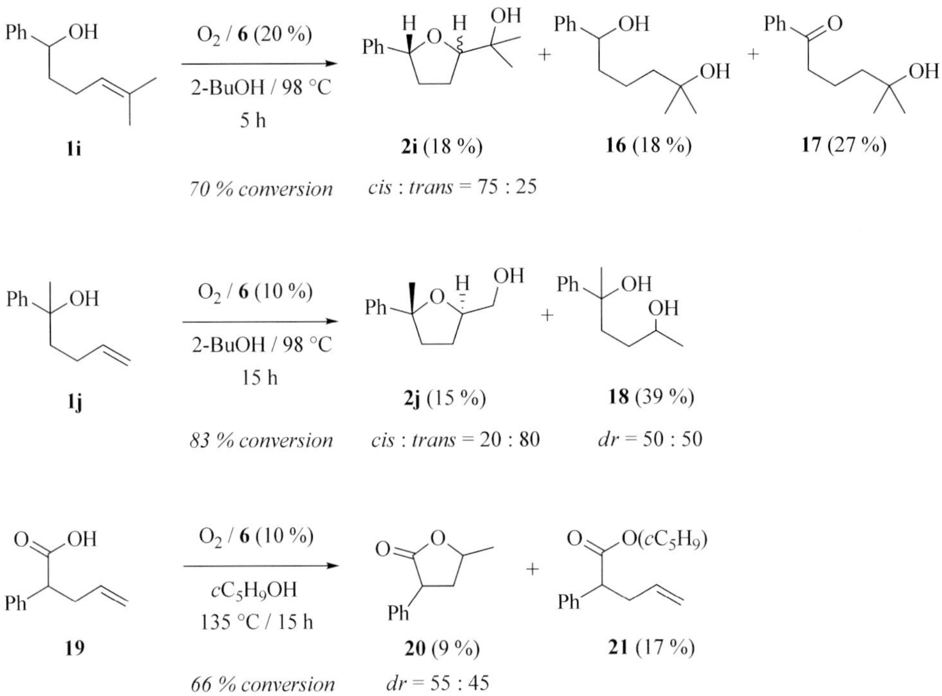

Ph
OH
O_2 / **6** (20 %)
2-BuOH / 98 °C
5 h
1i
H
H
O
OH
Ph
+
Ph
OH
OH
+
Ph
O
OH
2i (18 %)
16 (18 %)
17 (27 %)
70 % conversion
cis : *trans* = 75 : 25
Ph
OH
O_2 / **6** (10 %)
2-BuOH / 98 °C
15 h
1j
H
OH
O
Ph
+
Ph
OH
OH
2j (15 %)
18 (39 %)
83 % conversion
cis : *trans* = 20 : 80
dr = 50 : 50
O
OH
Ph
19
O_2 / **6** (10 %)
cC_5H_9OH
135 °C / 15 h
O
O
Ph
+
O
$O(cC_5H_9)$
Ph
20 (9 %)
21 (17 %)
66 % conversion
dr = 55 : 45

Scheme 21.6 Formation of oxyfunctionalized products from other substrates than monosubstituted pent-4-en-1-ols.

Tertiary δ,ε-unsaturated alcohol **1j** requires elevated temperatures (98 °C) in order to undergo significant transformation with the reagent combination of O_2 and cobalt complex **6**. Target compound **2j**, however, is formed in modest yields under these circumstances. Stereoisomer *trans*-**2j** is obtained as major cyclization product. The reason for the reluctance of substrate **1j** to undergo oxidative ring closure at 60 °C in 2-propanol, however, is the subject of ongoing investigations.

The aerobic oxidation of γ,δ-unsaturated acid **19** in the presence of cobalt(II) complex **6** affords lactone **20** and cyclopentylester **21** in low yields, if heated in an oxygen atmosphere in a solution of cyclopentanol (135 °C). No significant transformation occurs at lower temperatures. Both products are obtained in poor yields. This finding is reminiscent of the propensity of carboxylic acids in general to inhibit aerobic oxidations, which allows other processes to gain importance, such as the esterification observed in the synthesis of **21** [17].

21.8 Concluding Remarks

Almost two decades after the original report on the subject, several important synthetic questions dealing with the aerobic cobalt(II)-catalyzed oxidative alkenol cyclization have been answered. The procedure has been developed into a versatile method for the synthesis of, for example, 2,3-*trans*- (96 *de*), 2,4-*cis*- (~60% *de*), and 2,5-*trans*-disubstituted tetrahydrofurans (>99 *de*) from the corresponding monosubstituted pent-4-en-1-ols. TBHP is no longer required as a co-oxidant in order to obtain synthetically useful yields and selectivities. The scope of the reaction is extendable to the formation of tetrasubstituted tetrahydrofurans and bicyclic compounds with notable diastereoselectivity (>90–99%). The formation of cyclic ethers from ω,ω-dimethyl-substituted bishomoallylic alcohols or tertiary substrates affords results that are more difficult to interpret in terms of a general reactivity selectivity scheme.

An attempt to correlate the existing selectivity data with one of the known mechanisms for oxidative bishomoallylic alcohol cyclization shows that none of the pathways is able to fit all of the data satisfactorily (Figure 21.3). Some (e.g. **2a**, **2c–h**, **2j**) are similar to selectivities that are attainable in alkenol oxidations using high-valent metal oxo compounds (metal-oxo route) [48, 49]. Others are diagnostic of the metal peroxy-mechanism (e.g. **2i**) [24, 25, 49], while some selectivities almost coincide with values reported for 5-*exo*-trig-ring closures of underlying alkenoxyl radicals (e.g. **2e–f**, **2h**) [50, 51]. If relative configuration of products served as a guideline for classifying a reaction mechanism, it seemed that either more than one of the known mechanisms operated or a new pathway applied in aerobic cobalt-catalyzed alkenol cyclizations. Support for the former argument arises from the fact that subtle modifications in reaction parameters, such as temperature, precatalyst concentration and oxygen pressure are associated with notable changes in product selectivity (e.g. Figure 21.1). Further evidence for the existence of competing reaction channels originates from experiments performed in the temperature range 98–135 °C (Scheme 21.6). Diols that are formed under these conditions in notable amounts are typical products of olefin hydratation. A derived hydroxyketone (i.e. **17**) is considered to originate from autoxidation. Both pathways are almost irrelevant if oxidations are performed with, for example 1-substituted alkenols **1c** and **1d**, at 60 °C (2-propanol). The exact role of the solvents 2-butanol and cyclopentanol, apart from providing media that allow

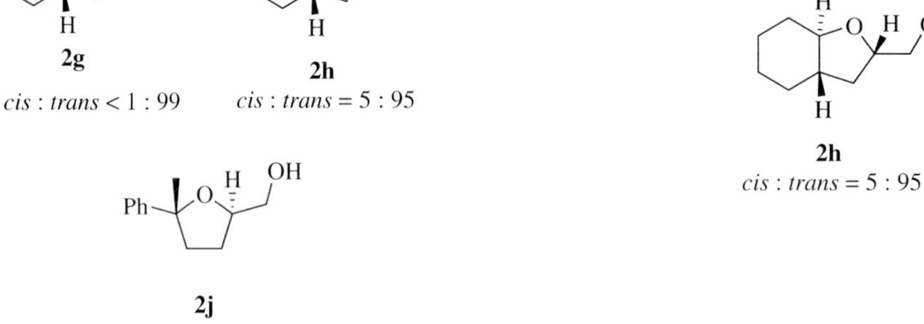

Figure 21.3 Correlating selectivities in aerobic cobalt(II)-catalyzed oxidations to diagnostic stereochemical information originating from alternative oxidative alkenol cyclizations [R = $C(CH_3)_3$, C_6H_5, Ar = 2,4,5-$(H_3CO)_3C_6H_2$] [24, 25, 47, 49, 50].

aerobic oxidations to be conducted at elevated temperatures, certainly has to be clarified in future experiments.

The common intermediate that has the potential to unify the reaction pathways outlined in the selectivity survey into a single mechanistic scheme could be a cobalt (III) hydroperoxy complex derived from precursor **6**, for example. Intermediates of this type have been discussed over the years in order to explain results from cobalt(II)-catalyzed oxidations [17]. The chemistry of these compounds, however, is at present too little developed to serve as a basis for an in-depth discussion at this stage of the investigation.

Acknowledgments

Generous financial support was provided by the Deutsche Forschungsgemeinschaft (Schwerpunktprogramm 1118, Sekundäre Wechselwirkungen, Projekt Ha1705/9–1) and the Fonds der Chemischen Industrie (scholarship for D.S.). We thank Dipl.-Chem. Georg Stapf for helpful advice on enantiomer analysis via ^{31}P-NMR, Dipl.-Chem. Michèle Graf for measuring magnetic properties of cobalt(II) complex **6**, Nathalie Laponche and Markus Weyland for technical assistance in quantitative acetone and water analysis, and Dr. Simone Drees for performing preliminary investigations associated with aerobic cobalt(II)-catalyzed oxidations.

References

1 The following abbreviations were used: BDE = bond dissociation energy; CAB = cobalt/acetate/bromide; *de* = diastereomeric excess; *dr* = diastereomeric ratio; *er* = enantiomeric ratio; ms = molecular sieve; T = twist conformer of tetrahydrofuran; TBHP = *tert*-butyl hydroperoxide; ZPVE = zero-point vibrational energy.

2 Werner, A. and Mylius, A.Z.Z. (1898) *Zeitschrift fur Anorganische und Allgemeine Chemie*, **16**, 245–267.

3 Tsumaki, T. (1938) *Bulletin of the Chemical Society of Japan*, **13**, 252–260.

4 Samándi, L.I. (2003) in *Advances in Catalytic Activation of Dioxygen by Metal Complexes*, (ed. L.I. Simándi), Kluwer Academic Publishers, Dordrecht, pp. 265–328.

5 Drago, R.S. and Corden, B.B. (1980) *Accounts of Chemical Research*, **13**, 353–360.

6 Henrici-Oliv,é G. and Oliv,é S. (1974) *Angewandte Chemie*, **86**, 1–12.Henrici-Oliv,é G. and Oliv,é S. (1974) *Angewandte Chemie – International Edition*, **13**, 29–38.

7 Zhang, M., van Eldik, R., Espenson, J.H. and Bakac, A. (1994) *Inorganic Chemistry*, **33**, 130–133.

8 Nishinaga, A., Tomita, H. and Matsuura, T. (1980) *Tetrahedron Letters*, **21**, 4853–4854.

9 Zombeck, A., Drago, R.S., Corden, B. and Gaul, J.H. (1981) *Journal of the American Chemical Society*, **103**, 7530–7585.

10 Vogt, L.H., Jr. Wirth, J.G. and Finkbeiner, H.L. (1969) *The Journal of Organic Chemistry*, **34**, 273–277.

11 Freemann, F. (1986) in *Organic Synthesis by Oxidations with Metal Compounds*, (eds W.J. Mijs and De Jonge C.R.H.), Plenum Press, New York, pp. 315–371.

12 Takai, T., Hata, E., Yorozu, K. and Mukaiyama, T. (1992) *Chemistry Letters*, 2077–2080.

13 Tang, Q., Zhang, Q., Wu, H. and Wang, Y. (2005) *Journal of Molecular Catalysis A*, **230**, 384–397.

14 Hunter, R., Turner, P. and Rimmer, S. (2000) *Synthetic Communications*, **30**, 4461–4466.

15 Punniyamurthy, T., Bhatis, B., Reddy, M.M., Maikap, G.C. and Iqbal, J. (1997) *Tetrahedron*, **53**, 7649–7670.

16 Mukaiyama, T., Yorozu, K., Takai, T. and Yamada, T. (1993) *Chemistry Letters*, 439–442.

17 Hamilton, D.E., Drago, R.S. and Zombeck, A. (1987) *Journal of the American Chemical Society*, **109**, 374–379.

18 Partenheimer, W. (2006) *Advanced Synthesis and Catalysis*, **348**, 559–568.

19 Inoki, S. and Mukaiyama, T. (1990) *Chemistry Letters*, 67–70.

20 Wang, Z.-M., Tian, S.-K. and Shi, M. (1999) *Tetrahedron Letters*, **40**, 977–980.

21 Wang, Z.-M., Tian, S.-K. and Shi, M. (1999) *Tetrahedron: Asymmetry*, **10**, 667–670.

22 Gribble, G.W. (1998) *Accounts of Chemical Research*, **31**, 141–152.

23 Faul, M.M. and Huff, B.E. (2000) *Chemical Reviews*, **100**, 2407–2473.

24 Hartung, J. and Greb, M. (2002) *Journal of Organometallic Chemistry*, **661**, 67–84.

25 Hartung, J., Drees, S., Greb, M., Schmidt, P., Svoboda, I., Fuess, H., Murso, A. and Stalke, D. (2003) *European Journal of Organic Chemistry*, 2388–2408.

26 Mori, K., Komatsu, M., Kido, M. and Nakagawa, K. (1986) *Tetrahedron*, **42**, 523–528.

27 Greb, M., Hartung, J., Köhler, F., Špehar, K., Kluge, R. and Csuk, R. (2004) *European Journal of Organic Chemistry*, 3799–3812.

28 Sheldon, R.A. (1993) *Topics in Current Chemistry*, **164**, 23–43.

29 Kaim, W. and Schwederski, B. (1991) *Bioanorganische Chemie*, Teubner, Stuttgart, pp. 87–134.

30 Luo, Y.-R. (2003) *Handbook of Bond Dissociation Energies in Organic Compounds*, CRC Press, Boca Raton.

31 Brix, P. and Herzberg, G. (1953) *Journal of Chemical Physics*, **21**, 2240.

32 Mukaiyama, T. and Yamada, T. (1995) *Bulletin of the Chemical Society of Japan*, **68**, 17–35.

33 Frisch, M.J., Trucks, G.W., Schlegel, H.B., Scuseria, G.E., Robb, M.A., Cheeseman, J.R., Zakrzewski, V.G., Montgomery, J.A., Jr. Stratmann, R.E., Burant, J.C., Dapprich, S., Millam, J.M., Daniels, A.D., Kudin, K.N., Strain, M.C., Farkas, O., Tomasi, J., Barone, V., Cossi, M., Cammi, R., Mennucci, B., Pomelli, C., Adamo, C., Clifford, S., Ochterski, J., Petersson, G.A., Ayala, P.Y., Cui, Q., Morokuma, K., Malick, D.K., Rabuck, A.D., Raghavachari, K., Foresman, J.B., Cioslowski, J., Ortiz, J.V., Baboul, A.G., Stefanov, B.B., Liu, G., Liashenko, A., Piskorz, P., Komaromi, I., Gomperts, R., Martin, R.L., Fox, D.J., Keith, T., Al-Laham, M.A., Peng, C.Y., Nanayakkara, A., Gonzalez, C., Challacombe, M., Gill, P.M.W., Johnson, B., Chen, W., Wong, M.W., Andres, J.L., Gonzalez, C., Head-Gordon, M., Replogle, E.S. and Pople J.A. (1998) *Gaussian 98, Revision A.7*, Gaussian, Inc., Pittsburgh PA.

34 Schmidt, M.W., Truong, P.N. and Gordon, M.S. (1987) *Journal of the American Chemical Society*, **109**, 5217–5227.

35 Holm, R.H. and Cotton, F.A. (1960) *Journal of Inorganic and Nuclear Chemistry*, **15**, 63–66.

36 Lintwedt, R.L. and Fatta, A.M. (1968) *Inorganic Chemistry*, **7**, 2489–2495.

37 Kato, K., Yamada, T., Takai, T., Inoki, S. and Isayama, S. (1990) *Bulletin of the Chemical Society of Japan*, **63**, 179–186.

38 Hsu, E.C. and Holzwarth, G. (1973) *Journal of the American Chemical Society*, **95**, 6902–6906.

39 Drees, S. (2002) *Dissertation*, Universtität Würzburg.

40 Menéndez, B., Schuch, D. and Hartung, J. (2008) *Organic Biomolecular Chemistry*, **6**, 3532–3541.

41 Hartung, J. and Fink, K. *unpublished results*.

42 Brunel, J.-M., Pardigon, O., Maffei, M. and Buono, G. (1992) *Tetrahedron: Asymmetry*, **3**, 1243–1246.

43 Holleman, A.F. and Wiberg, E. (1995) *Lehrbuch der Anorganischen Chemie*, de Gruyter, 101. Auflage, Berlin, pp. 1548–1574.

44 Sheldon, R.A. and Kochi, J.K. (1981) *Metal-Catalyzed Oxidations of Organic Compounds*, Academic Press, New York.

45 Lipari, F. and Swarin, S.J. (1982) *Journal of Chromatography*, **247**, 297–306.

46 Wieland, G. (1985) *Wasserbestimmung durch Karl-Fischer-Titration*, GIT-Verlag, Darmstadt.

47 Zombeck, A., Hamilton, D.E. and Drago, R.S. (1982) *Journal of the American Chemical Society*, **104**, 6782–6784.

48 Tang, S. and Kennedy, R.M. (1992) *Tetrahedron Letters*, **33**, 5299–5302.

49 Lempers, H.E.B., Ripollès i Garcia, A. and Sheldon, R.A. (1998) *The Journal of Organic Chemistry*, **63**, 1408–1413.

50 Hartung, J., Gottwald, T. and Špehar, K. (2002) *Synthesis*, 1469–1498.

51 Hartung, J., Kopf, T.M., Kneuer, R. and Schmidt, P. (2001) *Comptes Rendus Academic de Sciences Paris Chimie/Chemistry*, 649–666.

22
Regiodivergent Epoxide Opening

Andreas Gansäuer, Florian Keller, Chun-An Fan, and Peter Karbaum

22.1
Epoxide Opening via Nucleophilic Substitution: Limitations Arising from the S_N2-mechanism

Epoxides are amongst the most versatile compounds in organic chemistry. This is because they are not only important synthetic endpoints [1] but also key intermediates for further manipulations [2]. Because of the high ring strain of oxiranes (about 27 kcal mol^{-1}) they are 'spring loaded' for a number of interesting ring-opening reactions. In the past, this field has been heavily dominated by nucleophilic substitution reactions [3]. The usefulness of epoxides in S_N2 reactions has been expanded even further by the development of transition metal-catalyzed desymmetrizations of *meso*-epoxides [3] and kinetic resolutions [3] mainly of monosubstituted and 1,1-disubstituted epoxides. In this manner a wide range of important 1,2-difunctionalized compounds have been obtained with high enantio- and complete diastereoselectivity through the back-side approach of the nucleophile. Two of the many excellent examples of catalytic epoxide openings are shown in Scheme 22.1.

Despite the resounding success of these catalytic processes, important limitations are caused precisely by the mechanistic key aspect of all S_N2 reactions. The displacement of the leaving group is sensitive to both the substitution pattern of the carbon atom involved and the bulk of the nucleophile employed. Hence a number of potentially useful reactions are kinetically hindered or even impossible.

Two simple examples underline these restrictions: First, intermolecular epoxide openings via S_N2 usually do not occur at the higher-substituted carbon and second reactions of epoxides with sterically demanding substituents are frequently too slow to be preparatively useful.

Another more subtle limitation has emerged more recently. It has, to date, not been possible to observe the high regioselectivities obtained in the enantioselective opening of *meso*-epoxides in opening reactions of other epoxides. Surprisingly, this is even so for the most similar case, *cis*-1,2-disubstituted epoxides with two sterically and electronically unbiased groups [4]. The two examples shown in Scheme 22.2

Activating Unreactive Substrates: The Role of Secondary Interactions.
Edited by Carsten Bolm and F. Ekkehardt Hahn

ISBN: 978-3-527-31823-0

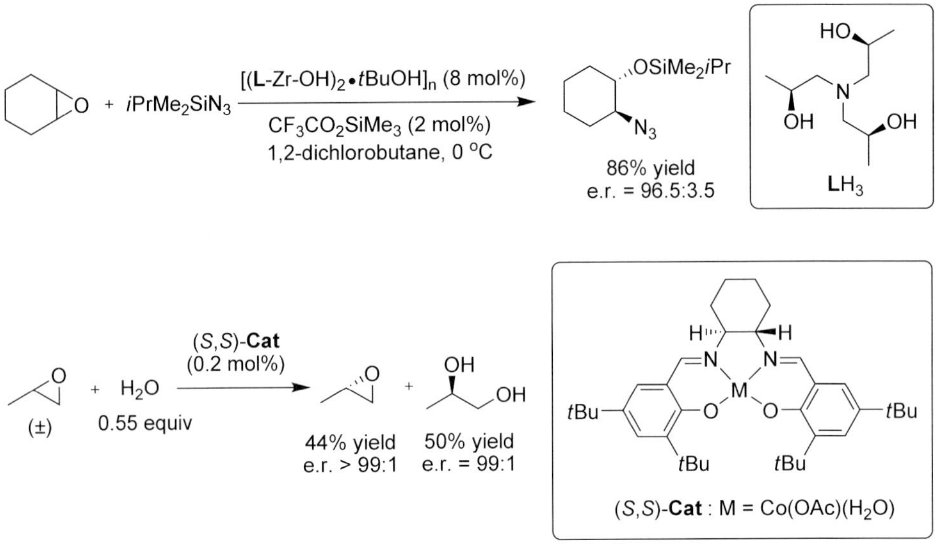

Scheme 22.1 Catalyzed epoxide openings via S_N2.

highlight the problems. It seems hard to combine high regioselectivity of epoxide opening with high reactivity of both enantiomers of the substrate even with today's most successful catalysts. Hence, either rather unselective reactions or kinetic resolutions are observed.

Nevertheless, such reactions are in principle extremely attractive for applications in the synthesis of complex molecules and also very interesting conceptually. Here, we outline these synthetic perspectives and propose the name 'regiodivergent epoxide opening' (REO) for such transformations if ring opening constitutes the stereo-differentiating event. In a more general context, for racemic substrates such processes have been termed 'parallel resolutions' or 'divergent reactions of racemic mixtures' [5]. A mathematical treatment has been provided [5c, f].

A complementary reaction of vinyloxiranes with dialkylzinc reagents has been reported by Pineschi and Feringa [6]. Reductive elimination from an allylcopper complex constitutes the stereodifferentiating step and not the ring opening that we concentrate on here.

22.2 Regiodivergent Epoxide Opening (REO): Mechanistic Implications, Synthetic Potential, and Aspects of Catalyst Design

The analysis of a typical REO proceeding via a classical S_N2-mechanism reaction is depicted in Scheme 22.3. REOs are, however, by no means confined to nucleophilic substitutions.

Before entering the mechanistic discussion, we note that kinetic resolutions and openings of *meso*-epoxides constitute subclasses of REO. In the former case only one

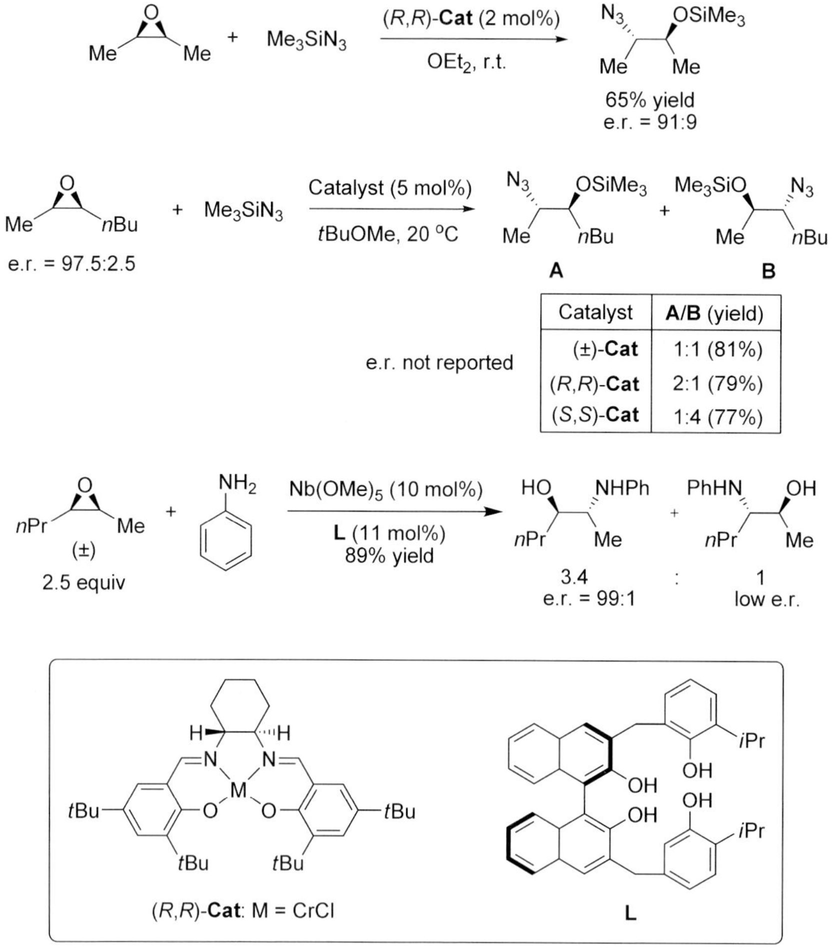

Catalyst	**A/B** (yield)
(±)-**Cat**	1:1 (81%)
(*R*,*R*)-**Cat**	2:1 (79%)
(*S*,*S*)-**Cat**	1:4 (77%)

Scheme 22.2 Catalytic regioselective epoxide opening vs *meso*-epoxide opening of an enantiomerically enriched substrate.

of the substrates reacts and in the latter case the identity of the substituents (R′ = R) reduces the analysis to the reaction of only one substrate.

For the reaction of a racemic substrate, efficient REOs result in parallel resolutions. Both enantiomers are turned over, preferentially with similar rates. The products are formed via the major pathway of one substrate and the minor pathway of the other substrate and therefore with high enantiomeric purity.

The use of enantiomerically enriched substrates in REOs leads to extremely attractive double asymmetric processes not possible with *meso*-epoxides. The major product is obtained through the dominant opening of the major enantiomer and the minor pathway of the minor enantiomer and hence with much higher optical purity than the substrate. This scenario is especially appealing when the synthesis of

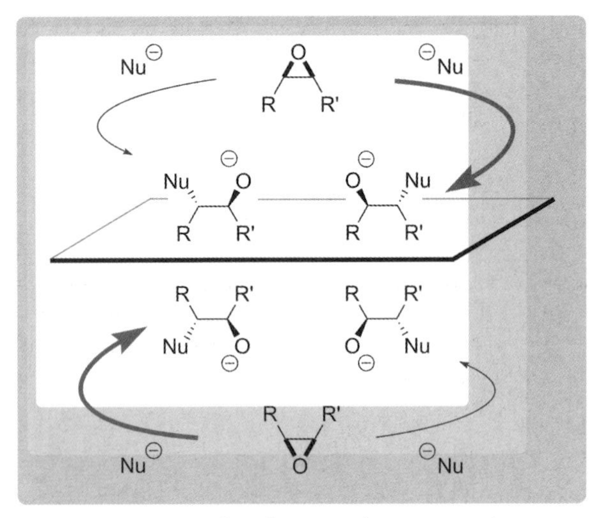

Scheme 22.3 Concept of catalytic 'regiodivergent epoxide opening' (REO).

compounds with very high enantiomeric excess is required. The minor products are formed with low enantioselectivity.

Even in the case of enantiomerically pure substrates the method is quite valuable. Semisynthetic modifications of epoxide-containing natural products seem especially well suited in this respect. By employing either enantiomer of the catalyst it should be possible to obtain two regioisomeric functionalized derivatives of the natural products with high selectivity, respectively.

In contrast to the opening of *meso*-epoxides, the design of efficient REO catalysts must address two issues and is hence more complex. Both enantiomers of the substrate must bind with similar efficiency to the catalyst and must then be opened with high and opposite regioselectivity.

To serve this dual purpose an efficient means for the differentiation of the heterotopic C–O bonds must be provided that is at the same time relatively insensitive to the steric demand of the epoxides' substituents. As already shown in Scheme 22.2, this seems hard to realize in S_N2 reactions.

We therefore decided to initiate a program toward the development of efficient REOs that does not rely on the nucleophilic substitution reactions but rather on the reductive epoxide opening via electron transfer from titanocene(III) complexes [7]. The mechanistic key features of this reaction will be described briefly in order to be able to develop our concept of catalyst design.

22.3
Reductive Epoxide Opening via Electron Transfer from Titanocene(III) Reagents

Catalytic reductive epoxide opening via electron transfer from titanocenes has emerged as an attractive alternative to the classical nucleophilic substitution strategies [8].

It combines the traditional advantages of radical chemistry [9] with regioselectivity of ring opening opposite to the nucleophilic substitutions. This has led to a number of interesting and unusual applications in the synthesis of complex molecules [10]. To date, other low-valent metal complexes have performed less satisfactory [11].

22.3.1
Mechanism of Reductive Epoxide Opening: Predetermined for REO!

Based on synthetic experience, reductive epoxide opening suggests itself as highly attractive for the development of REOs for three major reasons. First, the course of ring opening can be controlled by the catalyst, as we have demonstrated in our enantioselective opening of *meso*-epoxides with Kagan's Zn-reduced **1**[12] to yield **3** from **2**, as shown in Scheme 22.4 [13].

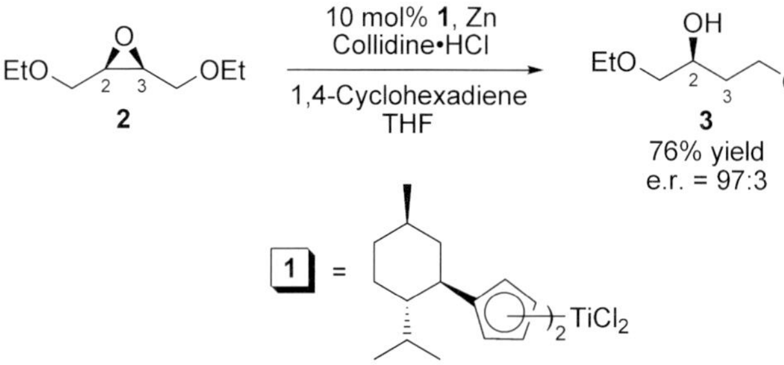

EtO OEt **2** — 10 mol% **1**, Zn, Collidine•HCl, 1,4-Cyclohexadiene, THF → EtO OH OEt **3**, 76% yield, e.r. = 97:3

1 = $TiCl_2$

Scheme 22.4 Enantioselective *meso*-epoxide opening via ET.

Second, the ring-opening event, i.e. the electron transfer from the titanocene catalyst, is decoupled from the trapping of the generated radical. Thus, the problems encountered through the limitations of the S_N2 reactions do not necessarily apply for reductive epoxide opening. Third, compound **1**, to date the most efficient catalyst in our enantioselective opening of *meso*-epoxides does not induce kinetic resolutions [14].

A study of the mechanism of the reductive epoxide opening that we carried out in cooperation with the groups of Kim Daasbjerg in Aarhus, Denmark, and Stefan Grimme in Münster, Germany, provided the detailed insights that convinced us to pursue the titanocene-catalyzed REO [15].

The electrochemical studies from Aarhus established that for Zn-reduced solutions of titanocenes with sterically demanding substituents, monomeric Ti(III) complexes are predominant in solution. These species also constitute the active species for epoxide opening and not dimers as in the case of Zn-Cp_2TiCl_2. The rate of the overall reduction is independent of the presence of a radical trap, such as 1,4-cyclohexadiene. Thus, ring opening constitutes the rate-controlling step of the overall process and is under kinetic control exactly as desired for our REO.

On the basis of these results the reaction mechanism was investigated on a molecular level by computational chemistry in Münster. For Cp_2TiCl and its dimer the activation and reaction energies of ring opening and the structures of all pertinent intermediates, transition states and products were determined by DFT calculations.

Moreover, the complexation mode of a number of epoxides by various titanocenes including Kagan's Zn-reduced **1** was established.

A combination of two points suggests that Zn-**1** fulfills the requirements for an efficient REO catalyst in a rather promising manner and certainly much better than the above-mentioned catalyst for the S_N2 reactions.

First, for a number of epoxides, Zn-**1** binds both enantiomers equally well. Second, even though the catalyst's reactive site is rather loose, steric interactions between the ligands and the epoxides' substituents constitute the main reason for the regioselectivity of ring opening. Surprisingly, this is not only the case for the generation of radicals from substrates with identical substitution patterns, such as *cis*-1,2-disubstituted epoxides including *meso*-epoxides, but also for the reactions of mono-, 1,1- and trisubstituted oxiranes.

These steric interactions are most easily understood when analyzing the opening of *meso*-epoxide **2** as shown schematically in Figure 22.1. From, the discussion it will become clear that the same arguments apply for the REOs.

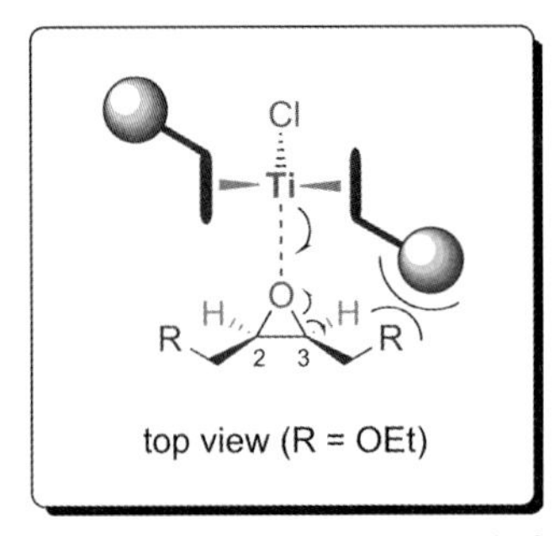

Figure 22.1 Mechanistic rationale for the enantioselective titanocene-catalyzed opening of *meso*-epoxides.

Breaking of the left-hand C-O bond in Zn-**1*****2**, the first intermediate of ring opening, results in an increased and energetically unfavorable interaction of the left-hand $CH_2OCH_2CH_3$ group with the ligand in the transition state. In this manner (*R*)-**3** would be formed after reduction.

Breaking of the right-hand C-O leads to a reduction of these steric interactions. As a result, formation of (*S*)–**3** is preferred, exactly as observed experimentally. Clearly, this mechanism for controlling the regioselectivity of ring opening should also be operating with *cis*-1,2-disubstituted epoxides containing non-identical but sterically and electronically unbiased substituents. Combined with the fact that Zn-**1** does not induce kinetic resolutions, this strongly suggests that REOs with such substrates proceed efficiently. The experimental proof for this assumption is presented next.

22.4
Synthetic Realization of Titanocene-catalyzed REO

We chose epoxides *rac*-**4** and (4*S*,5*R*)-**4** as substrates for our initial investigation with **1**. Both **1** and *ent*-**1** gave higher yields of readily separated **5** and **6** (84–95%) than

Cp_2TiCl_2 (77%, **5** : **6** = 63 : 37) as shown in Scheme 22.5 [16]. All results are summarized in Table 22.1.

OH CO2tBu
10 mol% 1 or ent-1
Mn, Collidine•HCl, THF
1,4-Cyclohexadiene
rac-4
(4S,5R)-4: e.r. = 91:9

Scheme 22.5 The first efficient titanocene-catalyzed REO.

Table 22.1 Regiodivergent epoxide opening of *rac*-**4** and (4*S*,5*R*)-**4**.

Entry	Substrate	Catalyst	6/%	e.r.[a]	7/%	e.r.[a]
1	*rac*-4	1	51	88.5 : 11.5	44	5 : 95
2	*rac*-4	*ent*-1	45	9.5 : 90.5	42	96.5 : 3.5
3	(4*S*,5*R*)-4	1	13	46 : 54	71	99.5 : 0.5
4	(4*S*,5*R*)-4	*ent*-1	76	97 : 3	10	25 : 75

[a] (*R*):(*S*) by comparison to authentic samples and from substrates.

As hoped for, a rather effective parallel resolution was observed with *rac*-**4**. It should be noted that both enantiomers of **1** give similar ratios of **5** to **6**. Hence, no chelation by the ester is involved and both substituents must indeed be considered unbiased. This is, of course, important for the generality of our process. As expected, the minor isomers are formed with higher selectivity than the major products.

Gratifyingly, the desired double asymmetric process with the enantiomerically enriched substrate **4** was also realized. With catalyst *ent*-**1**, **5** was obtained in 76% yield and an e.r. of 97 : 3. The minor product **6** was obtained in only 10% yield and with low enantioselectivity. With **1**, the opposite regioisomer of epoxide opening **6** was obtained as the major product in 71% yield and, as above, with exceptionally high enantiomeric purity (99.5 : 0.5). Again, the minor product was formed in low yield and low enantioselectivity.

It should be noted that opening of the epoxide shown in Scheme 22.6 with Jacobsen's system resulted in very low enantioselectivity of ring opening as summarized.

A mechanistic rationale for the outcome of the titanocene-catalyzed REOs of *rac*-**4** and (4*S*,5*R*)-**4** is shown in Scheme 22.7.

According to our mechanistic study, the depicted titanocene epoxide complexes constitute the first intermediate of ring opening. The selectivity of the parallel resolution in the case of *rac*-**4** and of the double asymmetric reactions with (4*S*,5*R*)-**4** is imposed by the high regioselectivity of ring opening in the complexes of both enantiomers of the epoxide with the enantiomerically pure titanocene. In this manner, (4*S*,5*R*)-**4** is opened via the dominant opening of the major enantiomer

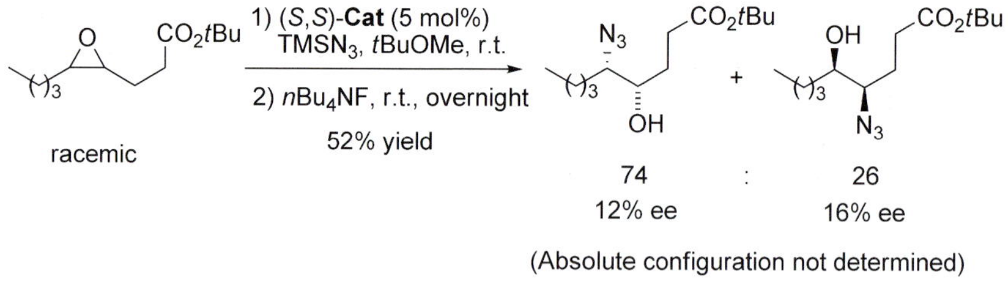

racemic
1) (S,S)-Cat (5 mol%)
TMSN3, tBuOMe, r.t.
2) nBu4NF, r.t., overnight
52% yield
74 : 26
12% ee
16% ee
(Absolute configuration not determined)
(S,S)-Cat : M = CrCl

Scheme 22.6 Attempted REO via nucleophilic substitution.

minor
major
major
minor

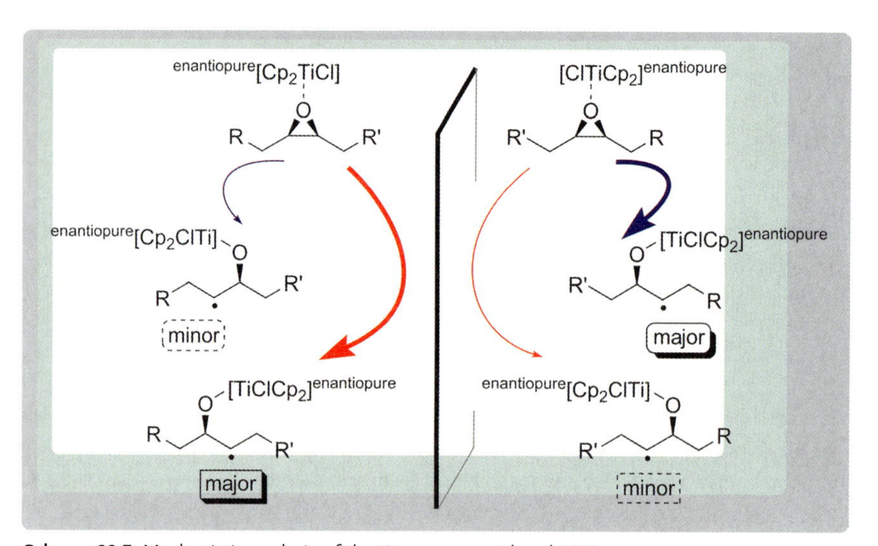

Scheme 22.7 Mechanistic analysis of the titanocene-catalyzed REO.

and the minor pathway of the minor enantiomer. As a result, the main product is formed in much higher selectivity than the substrate.

The key difference between this and the openings via S_N2 is due to the decoupling of ring opening and radical trapping. At first glance this loss of the diastereospecificity of the ring opening may be regarded as a severe disadvantage. However, we have already demonstrated that radical trapping can be highly diastereoselective with properly adjusted titanocene complexes [13c, 17]. Our recently introduced modular titanocene synthesis [18] should increase the potential of this approach even further, as the catalysts can be tailored according the specific demands of the situation.

Our reaction is general, and ester substitution is not necessary, as summarized in Scheme 22.8 and Table 22.2 (entries 5–8).

(*R*,*S*)-**7** → 10 mol% **1** or *ent*-**1**, Mn; Collidine•HCl, THF; 1,4-Cyclohexadiene → **8** + **9**

7a: R = *n*Bu, R' = CH_2CO_2tBu, (5*R*,4*S*):(5*S*,4*R*) = 94:6
7b: R = *n*Hexyl, R' = CH_2CO_2tBu, (5*R*,4*S*):(5*S*,4*R*) = 94:6
7c: R = *n*Bu, R' = Me, (4*R*,3*S*):(4*S*,3*R*) = 94:6
7d : R = *n*Pentyl, R' = Me, (4*R*,3*S*):(4*S*,3*R*) = 87:13

Scheme 22.8 REO for the synthesis of chiral alcohols.

Table 22.2 REO of **7**.

Entry	Substrate	Catalyst	Product	Yield/%	8/9[a]	8/e.r.[b]	9/e.r.[b]
1	7a	1	8a/9a	77	86 : 14	99 : 1	66 : 34
2	7a	*ent*-1	8a/9a	88	18 : 82	35 : 65	99 : 1
3	7b	1	8b/9b	73	82 : 18	99 : 1	43 : 47
4	7b	*ent*-1	8b/9b	76	12 : 88	45 : 55	99 : 1
5	7c	1	8c/9c	82	84 : 16	0.5 : 99.5	34 : 66
6	7c	*ent*-1	8c/9c	86	4 : 96	40 : 60	0.5 : 99.5
7	7d	1	8d/9d	89	83 : 17	0.5 : 99.5	70 : 30
8	7d	*ent*-1	8d/9d	88	12 : 88	50 : 50	0.5 : 99.5

a) Determined by ^1H NMR analysis of the crude mixture; **8**/**9** with Cp_2TiCl_2 was about 40 : 60 in all cases.
b) (*R*):(*S*) from configuration of substrates.
c) Zn used instead of Mn; **8**/**9** not separated.

The hydroxyesters (entries 1–4) are important intermediates for the synthesis of γ- and δ-lactones [19]. By treatment of **9b** with TsOH (*R*)-4-dodecanolide [20], a defensive secretion of beetles, was obtained (92%) with the highest enantiomeric purity reported. From **8b** (*R*)-5-dodecanolide, a component of the odor of cheddar [21], was obtained (96% yield). Hence, both natural products were prepared by the action of *ent*-**1** or **1** on the same precursor **7b**! Reactions with glycidol ethers were also carried out (Scheme 22.9).

In these substrates, chelation of titanium constitutes an additional potential control element for the regioselectivity of ring opening. This has been demonstrated to be very important [7b] in the reactions of the 'Sharpless' epoxides [1a]. The products of ring opening, derivatives of 1,2- and 1,3-diols, are important intermediates for the synthesis of complex molecules. Our results are summarized in Table 22.3.

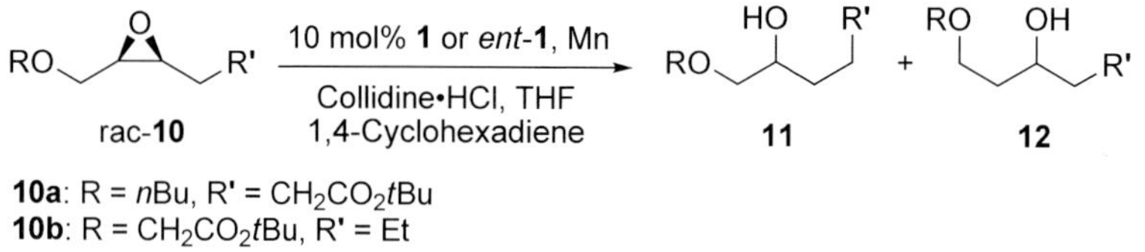

RO
O
R'
rac-10
10 mol% 1 or ent-1, Mn
Collidine•HCl, THF
1,4-Cyclohexadiene
HO
R'
RO
11
+
RO
OH
R'
12
10a: R = nBu, R' = CH2CO2tBu
10b: R = CH2CO2tBu, R' = Et

Scheme 22.9 REO for the synthesis of derivatives of 1,2- and 1,3-diols.

Table 22.3 REO of **10a** and **10b**.

Entry[a)]	Substrate	Catalyst	Product	Yield/%	11/12[a)]	11/e.r.[b)]	12/e.r.[b)]
1	**10a**	**1**	**11a/12a**	85	54 : 46	10 : 90	96 : 4
2	**10a**	*ent*-**1**	**11a/12a**	81	52 : 48	91 : 9	3 : 97
3	**10b**	**1**	**11b/12b**	77	66 : 34	19 : 81	99 : 1
4	**10b**	*ent*-**1**	**11b/12b**	74	80 : 20	65 : 35	10 : 90

a)Determined by ^{1}H NMR analysis of the purified products. For Cp_2TiCl_2: **10a** 72 : 28; **10b** 90 : 10.
b)Absolute configuration not determined. Measured from products and separated lactones.

For simple alkylethers (entries 1 and 2), the possible chelation does not affect the reagent-controlled course of the reaction. Thus, derivatives of both 1,2- and 1,3-diols become available in high enantiomeric purity from racemic substrates, similarly to the reactions of *rac*-**4**. However, with two chelating groups in one substituent a partially substrate-controlled reaction results, displaying matched and mismatched cases of selectivity of ring opening (entries 3, 4).

However, alkyl ethers, such as **10a**, **11a** and **12a**, are not easily cleaved to the corresponding alcohols, and thus more suitable protecting groups are required for synthetic applications. We therefore investigated a number of silyl ethers and found that the TBS-protected 'Sharpless' epoxides **13** constitute good substrates for the titanocene-catalyzed epoxide opening, as summarized in Scheme 22.10 and Table 22.4.

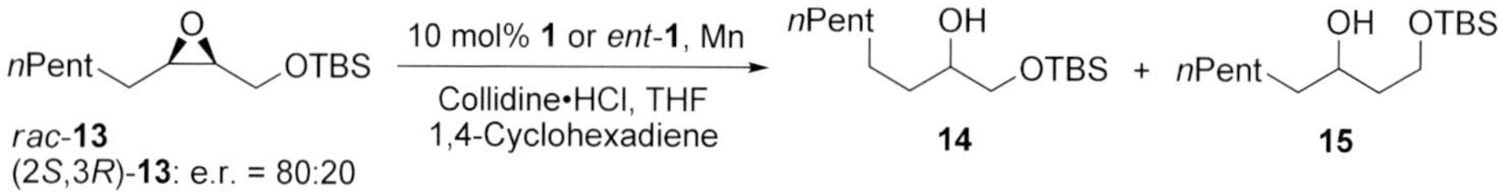

nPent
O
OTBS
rac-13
(2S,3R)-13: e.r. = 80:20
10 mol% 1 or ent-1, Mn
Collidine•HCl, THF
1,4-Cyclohexadiene
nPent
OH
OTBS
14
+
nPent
OH
OTBS
15

Scheme 22.10 First steps toward efficient REOs of 'Sharpless' epoxides via protecting-group tuning.

Table 22.4 REO of *rac*-**13** and (2*S*,3*R*)-**13**.

Entry	Substrate	Catalyst	Yield of 14 [%]	Yield of 15 [%]	14 : 15
1	*rac*-**13**	Cp_2TiCl_2	76	7	92 : 8
2	*rac*-**13**	**1**	55	31	64 : 36
3	*rac*-**13**	*ent*-**1**	52	33	61 : 39
4	(2*S*,3*R*)-**13**	**1**	30	65	32 : 68
5	(2*S*,3*R*)-**13**	*ent*-**1**	86	4	96 : 4

We note that the high regioselectivity of ring opening of **13** observed with Cp_2TiCl_2 is reduced to about 60 : 40 with both **1** and *ent*-**1**. With (2*S*,3*R*)-**14** a matched and a mismatched case of the regioselectivity of ring opening was observed. While these results are not yet fully satisfactory, they clearly indicate that the regioselectivity of the ring opening of 'Sharpless' epoxides can be controlled by the choice of an appropriate protecting group and catalyst.

In summary, we have devised the first efficient REO through our titanocene-catalyzed reductive epoxide opening. By decoupling ring opening from radical trapping, the typical disadvantages of S_N2 reactions are avoided. Our method combines a high regioselectivity of ring opening with the well-established advantages of radical chemistry and should hence be of interest for many applications in the synthesis of complex molecules.

Acknowledgements

We are grateful to the *Deutsche Forschungsgemeinschaft (Gerhard Hess-Programm, Schwerpunktprogramm 1181)*, the *Alexander von Humboldt-Stiftung (Forschungsstipendium to C.-A. F.)*, the *Studienstiftung des deutschen Volkes, (Stipendium to P. K.)* and the *Fonds der Chemischen Industrie* (*Sachbeihilfen*) for continuing financial support.

References and Notes

1 (a) Johnson, R.A. and Sharpless, K.B. (2000) in *Catalytic Asymmetric Synthesis*, (ed. I. Ojima), Wiley-VCH, New York, pp. 231–280; (b) Lane, B.S. and Burgess, K. (2003) *Chemical Reviews*, **103**, 2457–2474; (c) Yang, D. (2004) *Accounts of Chemical Research*, **37**, 497–505; (d) Shi, Y. (2004) *Accounts of Chemical Research*, **37**, 488–496; (e) McGarrigle, E.M. and Gilheany, D.G. (2005) *Chemical Reviews*, **105**, 1564–1602; (f) Xia, Q.-H., Ge, H.-Q., Ye, C.-P., Liu, Z.-M. and Su, K.-X. (2005) *Chemical Reviews*, **105**, 1603–1662.

2 Yudin, A.K. (ed.), (2006) *Aziridines and Epoxides in Organic Synthesis*, Wiley-VCH, Weinheim, Germany.

3 (a) Paterson, I. and Berrisford, D.J. (1992) *Angewandte Chemie*, **104**, 1204–1205; (1992) *Angewandte Chemie – International Edition*, **31**, 1179–1180; (b) Jacobsen, E.N. and Wu, M.H. (1999) in *Comprehensive Asymmetric Catalysis*, Vol. III, (eds. E.N. Jacobsen, A. Pfaltz and H. Yamamoto), Springer, Heidelberg, Chapter 35. (c) Jacobsen, E.N. (2000) *Accounts of Chemical Research*, **33**, 421–431; (d) Nielsen, L.C. and Jacobsen, E.N. (2006) in *Aziridines and Epoxides in Organic Synthesis* (ed. A.K. Yudin), Wiley-VCH, Weinheim, Germany, pp. 229–269; (e) Schneider, C. (2006) *Synthesis*, 3919–3944.

4 (a) Brandes, B.D. and Jacobsen, E.N. (2001) *Synlett*, 1013–1015; (b) Arai, K., Lucarini, S., Salter, M.M., Ohta, K., Yamashita, Y. and Kobayashi, S. (2007) *Journal of the American Chemical Society*, **129**, 8103–8111.

5 (a) Vedejs, E. and Chen, X. (1997) *Journal of the American Chemical Society*, **119**, 2584–2585; (b) Eames, J. (2000) *Angewandte Chemie*, **112**, 913–916; (2000) *Angewandte Chemie – International Edition*, **39**, 885–888; (c) Kagan, H.B. (2001) *Tetrahedron*, **57**, 2449–2468; (d) Dehli, J.R.

and Gotor, V. (2002) *Chemical Society Reviews*, **31**, 365–370; (e) Vedejs, E. and Jure, M. (2005) *Angewandte Chemie*, **117**, 4040–4069; (2005) *Angewandte Chemie – International Edition*, 44 3974–4001; (f) Kagan, H.B. (1996) *Croatica Chemica Acta*, **69**, 669–680; (g) Bolm, C. and Schlingloff, G. (1995) *Chemical Communications*, 1247–1248.

6 (a) Bertozzi, F., Crotti, P., Macchia, F., Pineschi, M. and Feringa, B.L. (2001) *Angewandte Chemie*, **113**, 956–958; (2001) *Angewandte Chemie – International Edition*, **40**, 930–932; (b) Review Pineschi, M. (2004) *New Journal of Chemistry*, **28**, 657–665.

7 (a) Nugent, W.A. and RajanBabu, T.V. (1988) *Journal of the American Chemical Society*, **110**, 8561–8562; (b) RajanBabu, T.V. and Nugent, W.A. (1989) *Journal of the American Chemical Society*, **111**, 4525–4527; (c) RajanBabu, T.V., Nugent, W.A. and Beattie, M.S. (1990) *Journal of the American Chemical Society*, **112**, 6408–6409; (d) RajanBabu, T.V. and Nugent, W.A. (1994) *Journal of the American Chemical Society*, **116**, 986–997; Reviews. (e) Gansäuer, A. and Bluhm, H. (2000) *Chemical Reviews*, **100**, 2771–2788; (f) Gansäuer, A. and Narayan, S. (2002) *Advanced Synthesis and Catalysis*, **344**, 465–475; (g) Gansäuer, A., Lauterbach, T. and Narayan, S. (2003) *Angewandte Chemie*, **115**, 5714–5431; (2003) *Angewandte Chemie – International Edition*, **42**, 5556–5573.

8 (a) Gansäuer, A., Pierobon, M. and Bluhm, H. (1998) *Angewandte Chemie*, **110**, 107–109; (1998) *Angewandte Chemie – International Edition*, **37**, 101–103; (b) Gansäuer, A., Bluhm, H. and Pierobon, M. (1998) *Journal of the American Chemical Society*, **120**, 12849–12859; (c) Barrero, A.F., Rosales, A., Cuerva, J.M. and Oltra, J.E. (2003) *Organic Letters*, **5**, 1935–1938; (d) Justicia, J., Rosales, A., Buñuel, E., Oller-López, J.L., Valdivia, M., Haïdour, A., Oltra, J.E., Barrero, A.F., Cárdenas, D.J. and Cuerva, J.M. (2004) *Chemistry - A European Journal*, **10**, 1778–1788.

9 (a) Giese, B. (1986) *Radicals in Organic Synthesis: Formation of Carbon-Carbon Bonds*, Pergamon Press, Oxford. (b) Curran, D.P., Porter, N.A. and Giese, B. (1996) *Stereochemistry of Radical Reactions*, VCH, Weinheim. (c) Renaud, P. and Sibi, M.P. (eds) (2001), Radicals in Organic Synthesis Weinheim, Vol. 1–2. (d) Zard, S.Z. (2003) *Radical Reactions in Organic Synthesis;*, Oxford University, Oxford.

10 (a) Gansäuer, A., Rinker, B., Pierobon, M., Grimme, S., Gerenkamp, M. and Mück-Lichtenfeld, C. (2003) *Angewandte Chemie*, **115**, 3815–3818; (2003) *Angewandte Chemie – International Edition*, **42**, 3687–3690; (b) Gansäuer, A., Rinker, B., Ndene-Schiffer, N., Pierobon, M., Grimme, S., Gerenkamp, M. and Mück-Lichtenfeld, C. (2004) *European Journal of Organic Chemistry*, 2337–2351; (c) Gansäuer, A., Lauterbach, T. and Geich-Gimbel, D. (2004) *Chemistry - A European Journal*, **10**, 4983–4990; (d) Friedrich, J., Dolg, M., Gansäuer, A., Geich-Gimbel, D. and Lauterbach, T. (2005) *Journal of the American Chemical Society*, **127**, 7071–7077; (e) Justicia, J., Oller-Lopez, J.L., Campaña, A., Oltra, J.E., Cuerva, J.M., Buñuel, E. and Cárdenas, D.J. (2005) *Journal of the American Chemical Society*, **127**, 14911–14921.

11 Gansäuer, A. and Rinker, B. (2002) *Tetrahedron*, **58**, 7017–7026.

12 (a) Cesarotti, E., Kagan, H.B., Goddard, R. and Krüger, C. (1978) *Journal of Organometallic Chemistry*, **162**, 297–309; (b) Structure Gansäuer, A., Bluhm, H., Pierobon, M. and Keller, M. 2001 *Organometallics*, **20**, 914–919; (c) Practical synthesis. Gansäuer, A., Narayan, S., Schiffer-Ndene, N., Bluhm, H., Oltra, J.E., Cuerva, J.M., Rosales, A. and Nieger, M. (2006) *Journal of Organometallic Chemistry*, **691**, 2327–2331.

13 (a) Gansäuer, A., Lauterbach, T., Bluhm, H. and Noltemeyer, M. (1999) *Angewandte Chemie*, **111**, 3112–3114; (1999)

Angewandte Chemie – International Edition, **38**, 2909–2910; (b) Gansäuer, A., Bluhm, H. and Lauterbach, T. (2001) *Advanced Synthesis and Catalysis*, **343**, 785–787; (c) Gansäuer, A., Bluhm, H., Rinker, B., Narayan, S., Schick, M., Lauterbach, T. and Pierobon, M. (2003) *Chemistry - A European Journal*, **9**, 531–542.

14 (a) Daasbjerg, K., Svith, H., Grimme, S., Gerenkamp, M., Mück-Lichtenfeld, C., Gansäuer, A., Barchuk, A. and Keller, F. (2006) *Angewandte Chemie*, **118**, 2095–2098; (2006) *Angewandte Chemie – International Edition*, **45**, 2041–2044; (b) Gansäuer, A., Barchuk, A., Keller, F., Schmitt, M., Grimme, S., Gerenkamp, M., Mück-Lichtenfeld, C., Daasbjerg, K. and Svith, H. (2007) *Journal of the American Chemical Society*, **129**, 1359–1371.

15 Gansäuer, A., Barchuk, A. and Fielenbach, D. (2004) *Synthesis*, 2567–2573.

16 Gansäuer, A., Fan, C.-A., Keller, F. and Keil, J. (2007) *Journal of the American Chemical Society*, **129**, 3484–3485.

17 Gansäuer, A., Rinker, B., Barchuk, A. and Nieger, M. (2004) *Organometallics*, **23**, 1168–1171.

18 Gansäuer, A., Franke, D., Lauterbach, T. and Nieger, M. (2005) *Journal of the American Chemical Society*, **127**, 11622–11623.

19 Hanessian, S. (1983) *Total Synthesis of Natural Products: The Chiron Approach;* Pergamon Press, New York. Chapter 9.

20 For synthetic approaches to (*R*)-4-dodecanolide see: (a) Sabitha, G., Reddy, E.V., Yadagiri, K. and Yadav, J.S. (2006) *Synthesis*, 3270–3274 and references cited therein.

21 Milo, C. and Reineccius, G.A. (1997) *Journal of Agricultural and Food Chemistry*, **45**, 3590–3594.

23
Supramolecular Containers: Host-guest Chemistry and Reactivity

Markus Albrecht

23.1
Introduction

Chemical reactions proceed within a well-defined space formed by neighboring molecules and often lead to selective transformations. Examples range from simple solvent effects, providing a specific environment, to enzymatically directed reactions, which take place at reaction centers on the surface of the enzyme or in a cavity buried within. Even (stereo)selective reactions highly depend on the structure of the cleft.

Based on this, it is a challenge for supramolecular chemists to develop simple approaches to the generation of molecular clefts or containers with an interior space to take up reactands, to stabilize unusual species (or intermediates) and to promote or catalyze chemical reactions [1]. The stepwise covalent preparation of molecular containers, as was described, for example, by Cram, Sherman, Warmuth and others, turned out to be a highly challenging synthetic task and provides only small amounts of material [2]. Alternatively, the noncovalent assembly of simple molecular building blocks represents an easy way to obtain large amounts of complex supramolecular architectures [3].

For example, container molecules are obtained by hydrogen bonding between appropriate moieties, as was exemplified early on by Rebek's molecular 'tennis and soft balls' [4] and as has recently become very popular with resorcinol arene capsules [5].

Coordination chemistry, on the other hand, has the advantage that the complex units represent very stable entities, and thus robust metallosupramolecular containers are obtained. Therefore oligotopic organic ligands with a specific symmetry have to be prepared in order to form oligonuclear coordination clusters. Fujita and Stang introduced oligopyridine derivatives in combination with palladium(II) or platinum (II) ions to get impressive supramolecular containers with up to multinanometer dimensions. The hydrophobic interior of the coordination compounds enables rich host-guest chemistry to take place in water. Reactive intermediates can be stabilized and unusual selectivities can be enforced in chemical reactions [6]. Raymond *et al.* developed a supramolecular container based on naphthyl-bridged dicatechol amides, which encapsulates positively charged guest species and acts as an active catalyst in, for example, aza-Cope rearrangements [7].

Activating Unreactive Substrates: The Role of Secondary Interactions.
Edited by Carsten Bolm and F. Ekkehardt Hahn

ISBN: 978-3-527-31823-0

In this chapter, approaches to the formation of supramolecular cavities (containers or clefts) are described. Initially the self-assembly of tetrahedral M_4L_4 complexes is shown and the modification as well as the host-guest chemistry of such complexes are described. Later, dinuclear complexes with amino acid spacers, which provide a cleft-type cavity for the binding of alcoholate guests, are discussed. Finally, a Diels-Alder reaction with unusual exo/endo selectivity is presented.

23.2
M_4L_4Tetrahedra

A tetrahedron is a relatively simple geometric body, which can contain a large internal void to encapsulate some guest species. There are two different approaches to the formation of metallosupramolecular tetrahedra (Figure 23.1). Very often, linear ditopic ligands, which bridge the edges of an M_4L_6 tetrahedron, are used [8]. Alternatively, triangular tritopic ligands can be used to span the faces and to form an M_4L_4 complex. Although the former approach is used frequently, there are only rare examples of the latter [9].

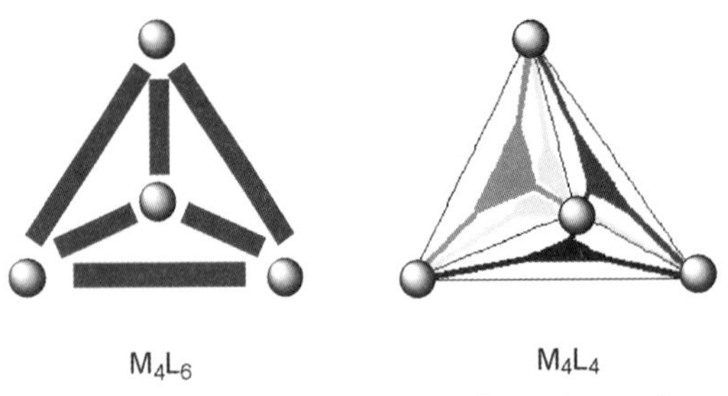

Figure 23.1 Representations of two different forms of metallosupramolecular tetrahedra: six linear ligands bridging the edges and four triangular ligands bridging the faces.

Based on our experience with linear dicatechol ligands and with the self-assembly of triple-stranded helicates, we started a program to prepare triangular triscatechol ligands for M_4L_4 tetrahedra [10]. The primary interest was to control the assembly of the supramolecular aggregates in order to finally use them for host-guest chemistry and maybe in the future for reaction control. Although the use of rigid triangular ligands would seem to be straightforward, we were also interested in introducing flexible ligand systems and controlling their coordination behavior by conformational preferences.

23.2.1
Flexible Triangular Ligands

Triangular triscatechol ligands with flexible spacers are well known from the literature as potent metal chelators. Prominent examples are the naturally occurring enterobactin [11] or the synthetic analog TRENCAM (Figure 23.2) [12]. The latter forms mononuclear complexes with a series of metal ions. The complexation is supported by intramolecular hydrogen bonding between the amide NH and internal catecholate oxygen atoms (A) [13]. Substitution of the amide by an imine linkage changes the situation dramatically. In a mononuclear complex the imine nitrogen and the catecholate oxygen lone pairs are pointing towards each other, leading to strong repulsion. This can be compensated by a conformational change (B), which prevents the formation of a mononuclear complex. Here, the ligand spreads out to form a 'triangle' and thus is predisposed for the assembly of an M_4L_4 tetrahedron. As an alternative, the repulsion can be compensated by binding a templating cation M′ to the catecholate oxygen as well as to the nitrogen lone pair (C) [14].

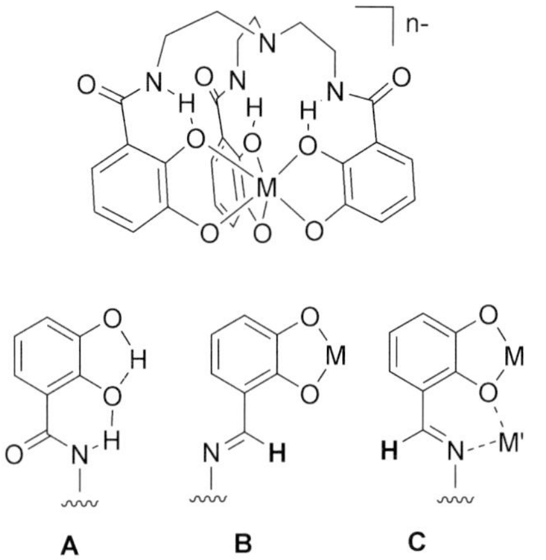

Figure 23.2 A mononuclear complex of the TRENCAM ligand (top) and comparison of the preferred conformations in catecholamides (A) and catecholimines without (B) or with (C) templating cations.

The flexible imine ligand $\mathbf{L}^1$-H_6 is easily prepared by imine condensation of tren (tris(aminoethyl amine)) with 2,3-dihydroxybenzaldehyde (Scheme 23.1) [15]. Reaction of this ligand with a source of titanium(IV) ions and sodium carbonate results in the mononuclear complex $Na[Na(\mathbf{L}^1)Ti]$ in which the repulsion of the lone pairs is compensated by encapsulation of a sodium cation in the cavity formed upon complexation of the ligand to titanium(IV). Substitution of sodium carbonate by potassium carbonate does not lead to a well-defined complex but rather to a mixture of oligomers. Addition of sodium salts transforms the mixture into the mononuclear compound again.

Sodium cations are of adequate size to stabilize the mononuclear complex. Potassium is too big, and thus the mixture of oligomers is observed. This is supported by an X-ray structure analysis of crystals obtained from the complex mixture with potassium cations. The coordination compound $K_2[K_2(\mathbf{L}^1)_2Ti_3O_2]$ possesses two of the cavities formed by ligands $\mathbf{L}^1$. Both are 'filled' with potassium cations. However, the bigger size of potassium compared to sodium does not allow the coordination of all three catecholate binding sites of the ligand to only one titanium(IV) ion. Therefore a tristitanium dioxo cluster is coordinated by one $\{(\mathbf{L}^1)K\}$-unit from the top and a second one from the bottom.

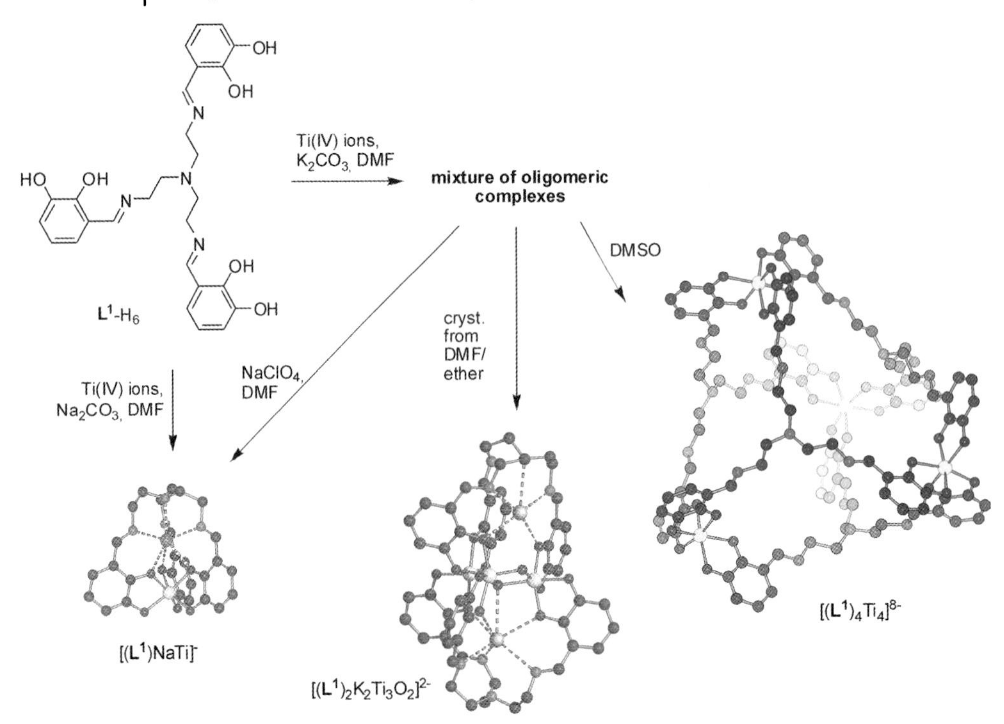

Scheme 23.1 Formation of different kinds of titanium(IV) complexes with ligand **L**¹. The outcome of the coordination studies depends on templating cations, crystallization conditions, and solvents.

To obtain an M_4L_4 tetrahedron with $\mathbf{L}^1$, the binding of potassium cations to the ligand has to be prevented. Therefore the mixture of oligomers can be dissolved in DMSO (or DMSO-[D_6]), which is a strongly coordinating solvent and is able to remove (fully dissolve) the potassium cations. NMR spectroscopy reveals that within a couple of hours only a highly symmetric species is present in this solvent. ESI FT-ICR MS shows this species to be the M_4L_4 tetrahedron [16].

The situation changes if ligand $\mathbf{L}^1$ is substituted by the chiral derivative $\mathbf{L}^2$ (Scheme 23.2). Because of steric interaction between the substituents at the spacer, the ligand is not able to adopt the triangular conformation which is essential for the formation of a tetrahedron. However, it is ideally preorientated for the formation of mononuclear gallium(III) (or titanium(IV)) complexes with an encapsulated sodium ion. In the absence of Na^+, this preorganization is still valid, supporting the formation of mononuclear compounds. These take up protons from the solvent (methanol or residual water) and form the triply protonated species [($\mathbf{L}^2$)H_3Ga]. In order to reduce the accumulation of positive charge in the cavity, the compound adopts the enaminone/quinomethine structure, which can be observed by X-ray structure analysis. The presence of this mesomeric form as a minimum structure is also supported by computational studies [17].

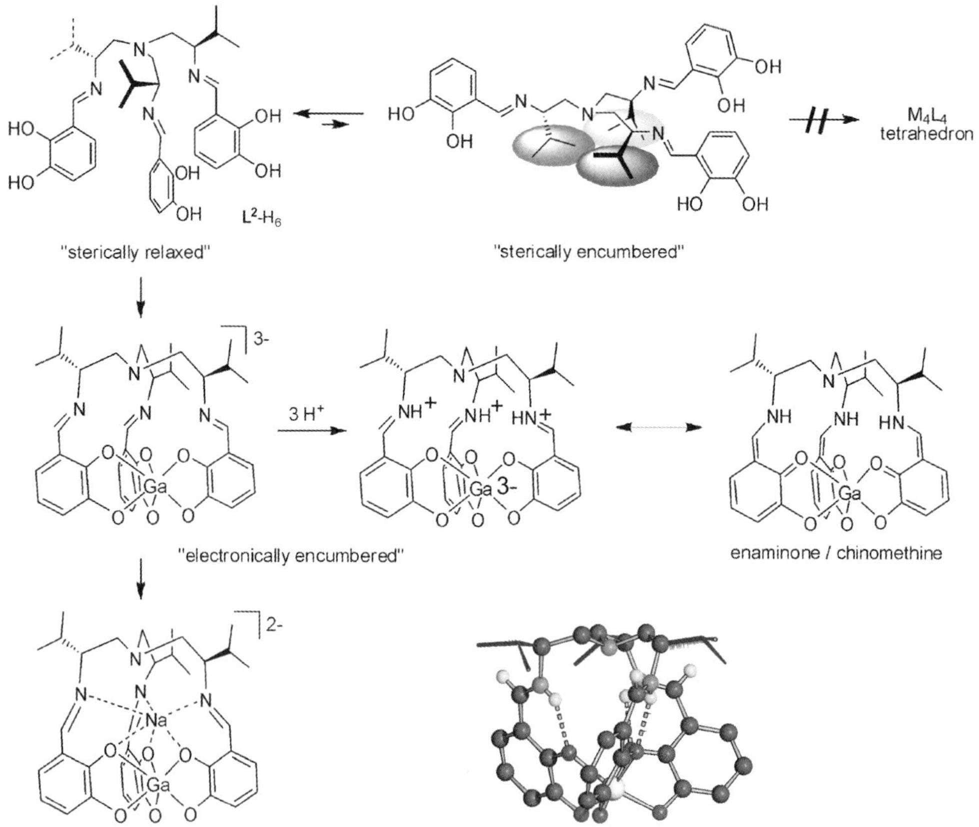

Scheme 23.2 Coordination studies with ligand $\mathbf{L}^2$ and gallium(III) ions. Because of the steric hindrance in the spacer, the ligand can only form mononuclear complexes. To avoid repulsion between electron pairs either Na^+ or three protons are bound in the cavity. In the latter situation an enaminone/chinomethine structure is adopted to reduce the positive charge within the cavity. The inset shows the X-ray structure of $[(\mathbf{L}^2\text{-}H_3)Ga]$.

The investigations with flexible ligands show that the electronic control of conformations at these compounds allows the formation of metallosupramolecular M_4L_4 tetrahedra. However, subtle steric changes (like introduction of substituents) can counteract the electronic effects and lead to unexpected species.

23.2.2
Rigid Triangular Ligands

The preorientation of rigid ligands leads to an easier formation of molecular tetrahedra. Imine-type ligands $\mathbf{L}^3\text{-}H_6$ and $\mathbf{L}^4\text{-}H_6$ are prepared from the trisaniline derivatives and 2,3-dihydroxybenzaldehyde.

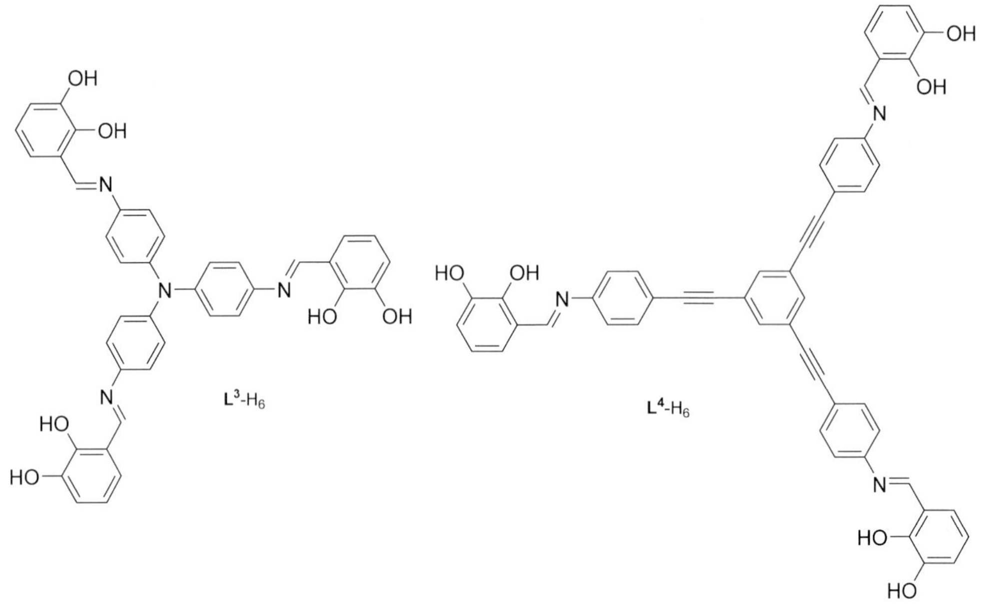

$\mathbf{L}^3$-H_6

$\mathbf{L}^4$-H_6

Figure 23.3 Rigid triangular ligands $\mathbf{L}^3$-H_6 and $\mathbf{L}^4$-H_6.

Reaction of the ligands with titanium(IV) ions in the presence of alkali metal carbonates results in the quantitative self-assembly of the corresponding metallo-supramolecular tetrahedra (Figure 23.3).

X-ray quality crystals were obtained for $K_8[(\mathbf{L}^3)_4Ti_4]$. The analysis reveals that the complex possesses the structure of a molecular tetrahedron with the titanium(IV) ions located on the corners and the ligands spanning the faces (Figure 23.4).

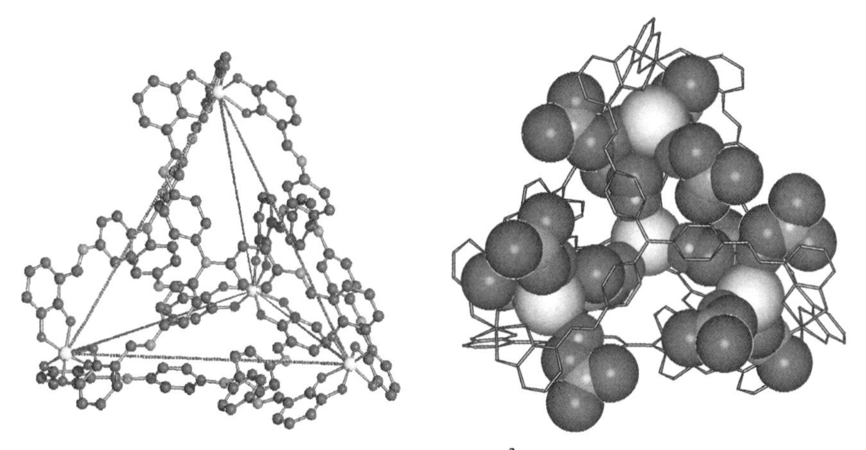

Figure 23.4 Results of the X-ray structural analysis of $K_8[(\mathbf{L}^3)_4Ti_4]$. Left: The metallosupramolecular tetrahedron $[(\mathbf{L}^3)_4Ti_4]^{8-}$. Right: The tetrahedron is indicated as lines while the internal potassium cations and DMF molecules are shown as space-filling models.

The metal–metal distances are approximately 17 Å. Thus a huge internal cavity is formed which in the crystal is filled by four of the potassium counter cations. Those are located close to the corners of the tetrahedron and coordinate to the internal oxygen atoms of the triscatecholate titanium(IV) complex units. Three molecules of DMF are additionally coordinated to each of the potassium ions [18].

The outer size of the molecular tetrahedron can be measured by the distance of the terminal hydrogen atoms of the catecholate ligands. In ligand $\mathbf{L}^3$ this distance is approximately 20 Å, while the larger ligand $\mathbf{L}^4$ shows a distance of about 28 Å. Despite the difference in this H $\cdots$ H separation, there seems to be no significant difference in the volume of the cavity of $[(\mathbf{L}^3)_4Ti_4]^{8-}$ compared to $[(\mathbf{L}^4)_4Ti_4]^{8-}$. Model considerations indicate that the longer spacers lead to a stronger twisting at the corners of the tetrahedron resulting in a reduction of the inner space. However, $[(\mathbf{L}^4)_4Ti_4]^{8-}$ does possess bigger pores because of the slim alkynyl groups, and this should enable a fast guest exchange between the interior of the container and the surrounding medium (Figure 23.5).

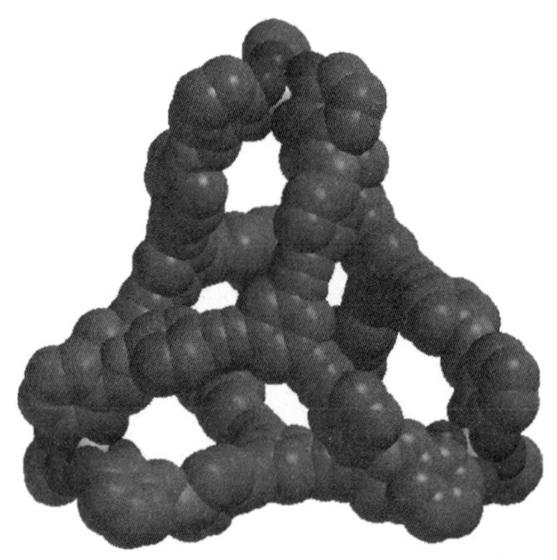

Figure 23.5 Modeled structure of $[(\mathbf{L}^4)_4Ti_4]^{8-}$ (force field) showing the twisting of the complex units at the corners and large pores in the walls.

Preliminary studies on host-guest chemistry were performed with $[(\mathbf{L}^3)_4Ti_4]^{8-}$. The idea was to substitute alkali metal cations in the interior of the cavity by appropriate organic cations. Modeling studies suggest that it should be possible to substitute all four cations which are observed in the interior in the solid state by anilinium cations.

Figure 23.6 depicts an NMR spectroscopic study in DMSO-[D_6]. Spectrum (1) shows the signals of $Li_8[(\mathbf{L}^3)_4Ti_4]$ (a): one for the imine protons, two for the aromatic spacers and three for the catecholate units (total of 6 signals). Addition of one equivalent of 4-bromoanilinium chloride (c) leads to broad signals. This is because of the loss of the symmetry of the tetrahedron by incorporation of only one organic cation. The system should be highly dynamic, resulting in the broad resonances. The spectrum starts to structure again upon addition of a total of four equivalents of 4-bromoanilinium chloride (3) [19].

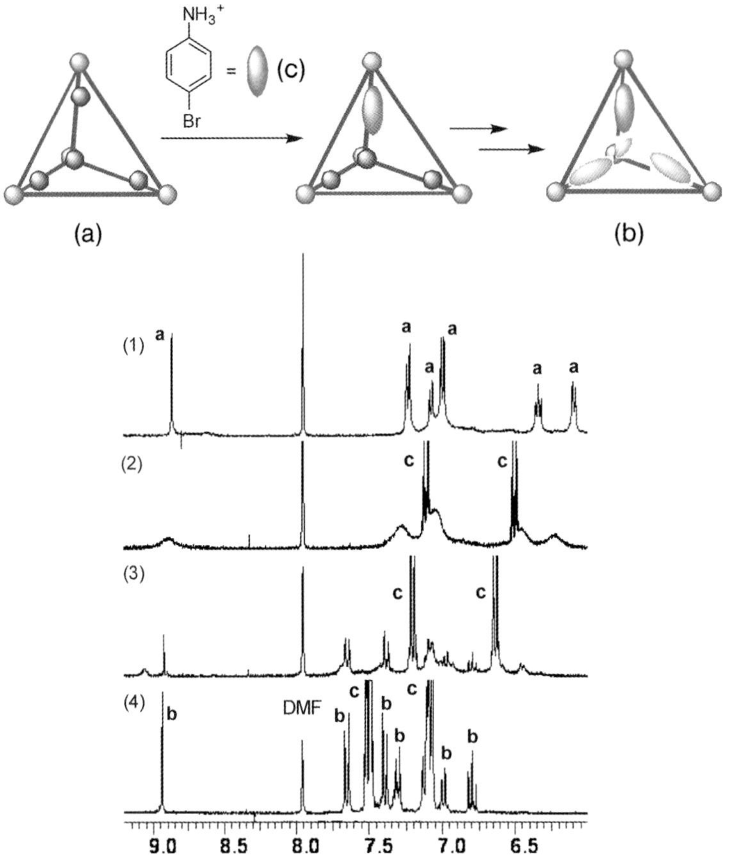

Figure 23.6 NMR titration experiment in DMSO-[D_6]: (1) Spectrum of $Li_8[(\mathbf{L}^3)_4Ti_4]$ (a); (2) addition of 1 equiv. of 4-bromoanilinium chloride (c); (3) addition of 4 equiv. of 4-bromoanilinium chloride; (4) addition of 10 equiv. of 4-bromoanilinium chloride. In spectrum (4) only signals of the tetrahedron (b) incorporating 4-bromoanilinium are observed.

Addition of ten equivalents of 4-bromoanilinium chloride leads to a new set of signals for the tetrahedron. This is because of the substitution of all the alkali metal cations by anilinium. Control experiments with related dinuclear titanium(IV) complexes show that the shift in the NMR spectra are due to incorporation of the organic cation and not due to an 'outside' binding by electrostatic and $\pi - \pi$ attraction.

The example described reveals that it is possible to incorporate organic guests in the interior of the supramolecular tetrahedra. A future challenge will be to perform reactions or catalysis in the container. However, for this it is necessary to have more robust capsules which are stable in water, and it would be an advantage to assemble the containers in enantiomerically pure form.

In order to increase the stability of the ligands, the imine linkages were substituted by ethylene bridges. Those were introduced by triple Wittig reaction followed by catalytic hydrogenation of the resulting double bonds. The ligands **L**^{5}-H_6 and **L**^{6}-H_6 were prepared in this fashion (Figure 23.7).

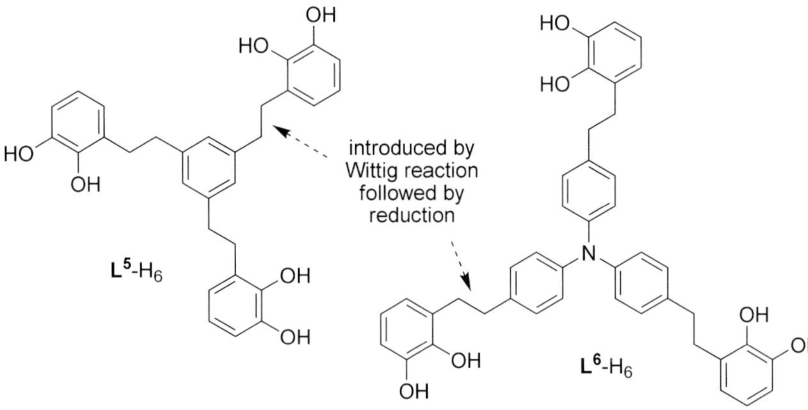

Figure 23.7 Triangular ligands with ethylene linkages instead of imines.

Both triangular alkyl-bridged ligands **L**^{5}-H_6 and **L**^{6}-H_6 form tetranuclear titanium(IV) tetrahedra in the presence of alkali metal carbonate. The complexes are characterized by NMR and ESI FT-ICR mass spectrometry. In addition they are very soluble in water, where they show long term stability [20].

Chiral information is introduced into the tetrahedron by attaching chiral amide substituents at the three termini of the ligand **L**^{7}-H_6. Upon metal coordination, the amide NHs form hydrogen bonds to the external catecholate oxygen atoms bringing the sterocenter close to the metal and inducing the helicity at the complex units.

Ligand **L**^{7}-H_6 forms tetrahedral tetranuclear complexes $[(\mathbf{L}^7)_4Ti_4]^{8-}$ which can be characterized by ESI FT-ICR MS. NMR indicates that only one enantiomerically pure diastereoisomer is formed (Figure 23.8). Modeling based on the known structures of the corresponding achiral tetrahedron and on related mononuclear catecholate complexes with the phenylethyl amide substituent shows that most probably the Λ-configuration is induced at the complex units by the S-configured phenylethyl amide substituent [21].

Our investigations into the formation of molecular tetrahedra show that we can prepare well-defined supramolecular structures with huge voids. The complexes exhibit interesting host-guest behavior toward organic cations. They are prepared in a water-stable form and can be obtained enantiomerically pure. These results lay the foundations for the use of the systems for supramolecular reactions and catalysis in water. These topics need to be investigated in the future. However, an example in which a cleft-type dinuclear complex mimics enzyme behavior and controls chemical reactivity is given in the next section of this chapter.

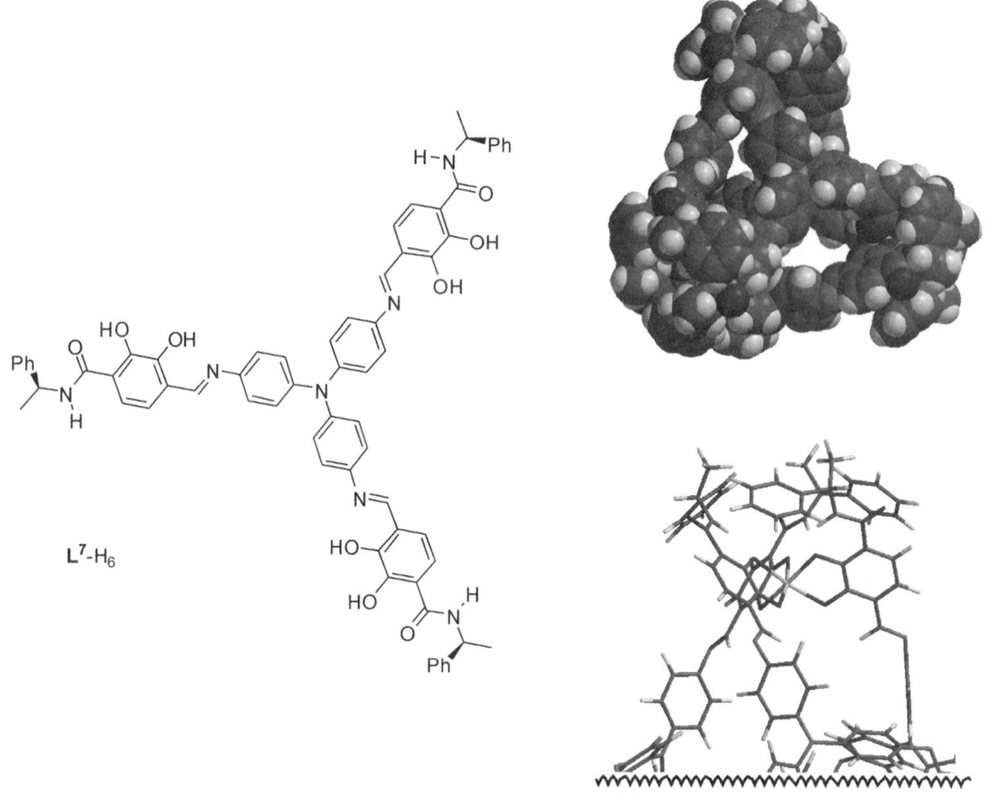

Figure 23.8 The enantiomerically pure triangular ligand $\mathbf{L}^7$-H_6 and a computer model of its titanium(IV) complex $[(\mathbf{L}^7)_4Ti_4]^{8-}$. One titanium complex 'corner' of the tetrahedron is also shown (bottom right).

23.3 Amino Acid-bridged Dinuclear Titanium(IV) Complexes as Metalloenzyme Mimicry

Enzymes provide reactive centers within clefts in order to control reactivity and selectivity in chemical reactions. Metalloenzymes possess metal ions which are very often involved in the reactions, but in addition they can act as an anchor to fix the substrate in the cavity. In a subsequent reaction, selectivities are influenced by the amino acids in the vicinity of the cavity [22].

In order to mimic the action of metalloenzymes, amino acid-bridged dicatechol ligands $\mathbf{L}^8$-H_4 were prepared [23], and these were used to form double-stranded dinuclear titanium(IV) complexes with two of the dicatecholates bridging the metal centers and two additional alcoholates fixed between them (Figure 23.9) [24]. Although this system is completely artificial, it has some features which resemble nature's metalloproteins: (i) the distance of the metal centers is approximately 5.6 Å, as in dinuclear metalloenzymes, (ii) alcoholate is fixed as a substrate, and (iii) the

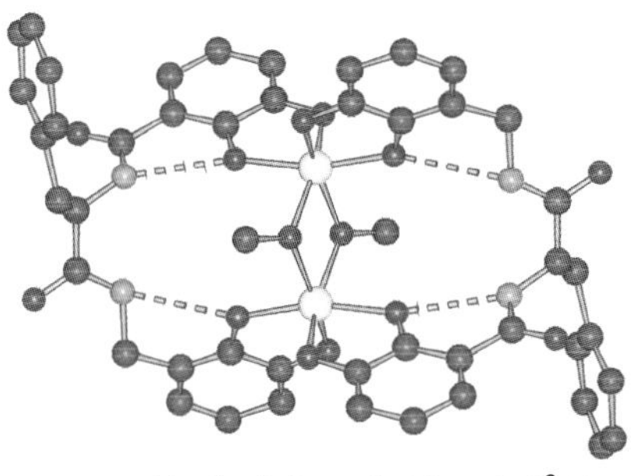

Figure 23.9 Amino acid-bridged dicatechol ligands $\mathbf{L^8}$-H_4 as precursors for dinuclear double-stranded titanium(IV) complexes with two bridging alcoholate coligands. The X-ray structure of the complex with phenylalanine in the spacer and methanolate is shown.

cavity is formed by amino acid residues. However, to mimic the function of the enzymes, reactions should be performed at the substrate and should be controlled by the binding pocket.

Therefore some requirements had to be fulfilled:

1. Because of the directionality of the amino acid spacers (parallel or antiparallel) and the stereochemistry at the metal centers ($\Lambda\Lambda$, $\Delta\Delta$, $\Lambda\Delta$), up to seven isomeric dinuclear complexes can be formed. All isomers are observed in the early stage of the complexation (kinetic control). However, after some days at room temperature the thermodynamically favored species is formed as the major one and can be obtained by precipitation. The structure of this isomer was derived from NMR and X-ray investigations along with a conformational analysis (following Ramachandrans method) and PM3 calculations. The major isomer is the $\Lambda\Delta$ isomer with parallel orientation of the ligands, the N-terminal catecholates binding to the Δ-configured metal center and the C-terminal to the Δ-configured metal [25].

2. The alcoholate coligands are usually derived from the solvent. However, in order to perform reactions in the periphery of the complex, the coligands should be introduced separately and the complex should be soluble in 'innocent' solvents which do not take part in the reaction. This is achieved by formation of the coordination compounds in acetonitrile or *tert*-butanol (which because of its size is not incorporated in the complex) in the presence of equimolar amounts of an appropriate alcohol and with tetrabutylammonium hydroxide as base. The product is obtained by precipitation as the thermodynamically favored isomer [26].

3. Following this protocol, a dinuclear titanium(IV) complex with acryl glycolate as substrate has been prepared. This complex has been introduced in a Diels-Alder reaction with cyclopentadiene and the organic product could be isolated by hydrolysis of the complex. In this way the bicyclic Diels-Alder product was isolated as a mixture of the exo and endo isomers in a 7:1 ratio. In contrast to this, the

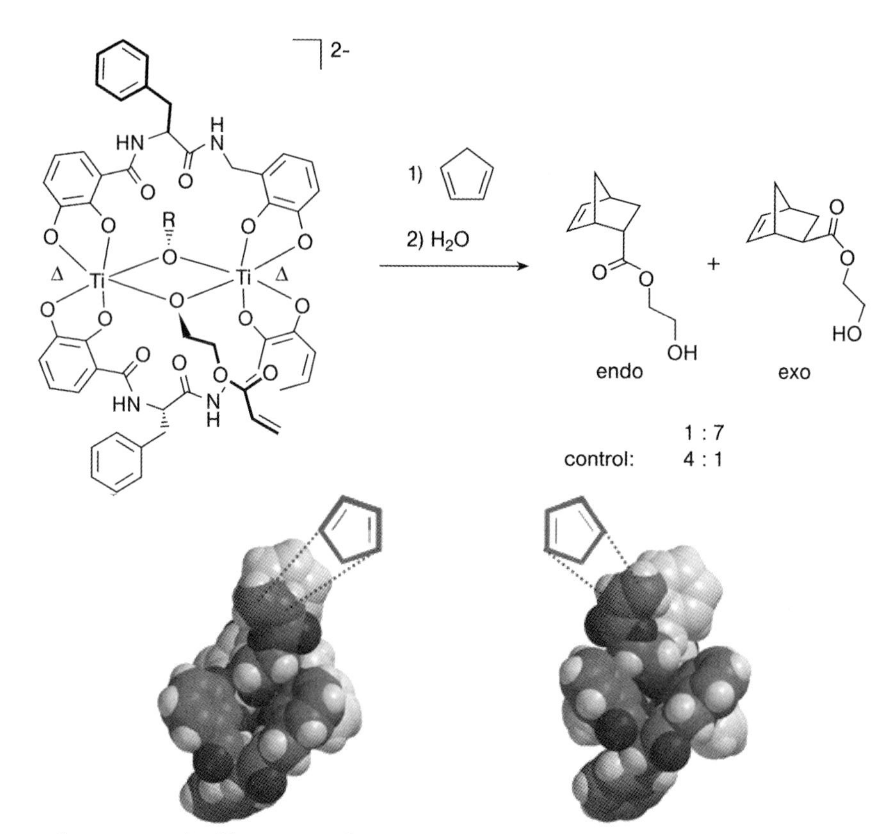

Scheme 23.3 Diels-Alder reaction of acrylglycolate with cyclopentadiene in the cleft of the dinuclear phenylalanine bridged titanium(IV) complex. The model on the bottom rationalizes the unusual diastereoselectivity of the reaction due to the approach of cyclopentadiene with the small CH_2 pointing into the cleft.

expected endo selectivity (1:4 exo/endo) is observed if the cycloaddition is performed as a thermal reaction at the uncomplexed acryl glycolate.

The unusual diastereoselectivity of the Diels-Alder reaction is due to the fixation of the acryl glycolate in the cavity of the coordination compound, which allows a cycloaddition only for an orientation of the cyclopentadiene with the smaller CH_2 unit directed into the cleft. Despite the chiral information at the amino acid residues, no asymmetric induction by the chiral amino acids can be observed in this process (Scheme 23.3) [27].

23.4 Conclusions

Supramolecular chemistry allows the formation of complex structures in simple self-assembly processes with relatively readily available organic building blocks. The challenge is to form aggregates which provide cavities for host-guest chemistry or reactions. In this chapter, an approach toward the preparation of metallosupramolecular tetrahedra is presented which affords stable and sometimes enantiomerically pure complexes in high yields. Appropriate cationic organic guest species are able to substitute inorganic cations in the interior of the compounds. Mechanistic investigations show that flexible as well as rigid ligand systems can be used, but that the coordination chemistry with the rigid systems is more predictable. Unforeseen reaction pathways can prevent the formation of containers with flexible ligands but could lead to new kinds of mononuclear coordination compounds.

Although no chemical reactions have yet been performed in the interior of the tetrahedra, an amino acid cleft-type dinuclear complex shows unusual exo selectivity in a Diels-Alder reaction which proceeds within the cleft.

The results presented represent approaches toward molecular cavities which bind guest species by secondary interactions (coordinatively or electrostatic/hydrophobic) and are able to alter reactivity. We are just at the starting point of this kind of chemistry, but in the future more sophisticated systems can be expected which will act as enzyme mimics or as supramolecular catalysts.

Acknowledgment

This work was supported by the DFG (SPP 1118). I thank all my coworkers and collaborators who participated in the various parts of the project, and I specially want to mention Dr. Roland Fröhlich, who performed all the structural analyses.

References

1 Lützen, A. (2005) *Angewandte Chemie*, **117**, 1022 (2005) *Angewandte Chemie – International Edition*, **44**, 1000.

2 (a) Sherman, J.C. and Cram, D.J. (1989) *Journal of the American Chemical Society*, **111**, 4527; (b) Helgeson, R.C., Paek, K.,

Knobler, C.B., Maverick, E.F. and Cram, D.J. (1996) *Journal of the American Chemical Society*, **118**, 5590; (c) Beno, B.R., Sheu, C., Houk, K.N., Warmuth, R. and Cram, D.J. (1998) *Chemical Communications*, 301; (d) Warmuth, R. (1998) *Chemical Communications*, 60; (e) Warmuth, R. (2001) *European Journal of Organic Chemistry*, 423.

3 Saalfrank, R.W. and Bernt, I. (1998) *Current Opinion in Solid-State & Materials Science*, **3**, 407.

4 (a) Wyler, R., de Mendoza, J. and Rebek, J. (1993) *Angewandte Chemie*, **105**, 1820; (1993) *Angewandte Chemie (International Edition in English)*, **32**, 1699; (b) Grotzfeld, R.M., Branda, N. and Rebek, J. (1996) *Science*, **271**, 487.

5 (a) Atwood, J.L., Barbour, L.J. and Jerga, A. (2002) *Proceedings of the National Academy of Sciences of the United States of America*, **99**, 4837; (b) Gerkensmeier, T., Iwanek, W., Agena, C., Fröhlich, R., Kotila, S., Nather, C. and Mattay, J. (1999) *European Journal of Organic Chemistry*, 2257.

6 (a) Fujita, M., (1998) *Chemical Society Reviews*, **27**, 417; (b) Fujita, M., Umemoto, K., Yoshizawa, M., Fujita, N., Kusukawa, T. and Biradha, K. (2001) *Chemical Communications*, 509; (c) Yoshizawa, M., Kusukawa, T., Fujita, M. and Yamaguchi, K. (2000) *Journal of the American Chemical Society*, **122**, 6311; (d) Ito, H., Kusukawa, T. and Fujita, M. (2000) *Chemistry Letters*, 598; (e) Yoshizawa, M., Takeyama, Y., Okano, T. and Fujita, M. (2003) *Journal of the American Chemical Society*, **125**, 3243; (f) Olenyuk, B., Fechtenkötter, A. and Stang, P.J. (1998) *Journal of The Chemical Society, Dalton Transactions*, 1707.

7 (a) Ziegler, M., Brumaghim, J. and Raymond, K.N. (2000) *Angewandte Chemie*, **112**, 4285. (2000) *Angewandte Chemie – International Edition*, **39**, 4119; (b) Brumaghim, J.L., Michels, M., Pagliero, D. and Raymond, K.N. (2004) *European Journal of Organic Chemistry*, 5115;. (c) Brumaghim, J.L., Michels, M. and Raymond, K.N. (2004) *European Journal of Organic Chemistry*, 4552; (d) Leung, D.H., Fiedler, D., Bergman, R.G. and Raymond, K.N. (2004) *Angewandte Chemie*, **116**, 981. (2004) *Angewandte Chemie – International Edition*, **43**, 963; (e) Fiedler, D., Bergman, R.G. and Raymond, K.N. (2004) *Angewandte Chemie*, **116**, 6916. (2004) *Angewandte Chemie – International Edition*, **43**, 6748.

8 Saalfrank, R.W., Stark, A., Peters, K. and Von Schnering, H.G. (1988) *Angewandte Chemie*, **100**, 878. (1988) *Angewandte Chemie – International Edition*, **27**, 851; (b) Saalfrank, R.W., Stark, A., Bremer, M. and Hummel, H.-U. (1990) *Angewandte Chemie*, **102**, 292. (1990) *Angewandte Chemie – International Edition*, **29**, 311; (c) Saalfrank, R.W., Hörner, B., Stalke, D. and Salbeck, J. (1993) *Angewandte Chemie*, **105**, 1223. (1993) *Angewandte Chemie – International Edition*, **32**, 1179.

9 (a) Amoroso, A.J., Jefferey, J.C., Jones, P.L., McCleverty, J.A., Thornton, P. and Ward, M.D. (1995) *Angewandte Chemie*, **107**, 1577. (1995) *Angewandte Chemie – International Edition*, **34**, 1443; (b) Paul, R.L., Amoroso, A.J., Jones, P.L., Couchman, S.M., Reeves, Z.R., Rees, L.H., Jefferey, J.C., McCleverty, J.A. and Ward, M.D. (1999) *Dalton Transactions*, 1563; (c) Brückner, C., Powers, R.E. and Raymond, K.N. (1998) *Angewandte Chemie*, **110**, 1937.(1998) *Angewandte Chemie – International Edition*, **37**, 1837; (d) Caulder, D., Brückner, C., Powers, R.E., König, S., Parac, T.N., Leary, J.A. and Raymond, K.N. (2001) *Journal of the American Chemical Society*, **123**, 8923; (e) Saalfrank, R.W., Glaser, H., Demleitner, B., Hampel, F., Chowdhry, M.M., Schünemann, V., Trautwein, A.X., Vaughan, G.B.M., Yeh, R., Davis, A.V. and Raymond, K.N. (2001) *Chemistry - A European Journal*, **8**, 493.

10 (a) Albrecht, M., Janser, I. and Fröhlich, R. (2005) *Chemical Communications*, 157; (b) Albrecht, M., (2007) *Bulletin of the Chemical Society of Japan*, **80**, 797.

11 Dunford, H.D., Dolphin, D., Raymond, K.N. and Sieker, L. (1981) *The Biological Chemistry of Iron*, D. Reidel Publishing Co., Dordrecht, pp 85–105.

12 Rodgers, S.J., Lee, C.W., Ng, C.Y. and Raymond, K.N. (1987) *Inorganic Chemistry*, **26**, 1622.

13 Karpishin, T.B., Stack, T.D.P. and Raymond, K.N. (1993) *Journal of the American Chemical Society*, **115**, 182.

14 Albrecht, M., Janser, I., Kamptmann, S., Weis, P., Wibbeling, B. and Fröhlich, R. (2004) *Dalton Transactions*, 37.

15 (a) Aguiari, A., Bullita, E., Casellato, U., Guerriero, P., Tamburini, S. and Vigato, P.A. (1992) *Inorganica Chimica Acta*, **202**, 157; (b) Albrecht, M., Janser, I. and Fröhlich, R. (2004) *Synthesis*, 1977.

16 Albrecht, M., Janser, I., Runsink, J., Raabe, G., Weis, P. and Fröhlich, R. (2004) *Angewandte Chemie*, **116**, 6832. (2004) *Angewandte Chemie – International Edition*, **43**, 6662.

17 Albrecht, M., Burk, S., Stoffel, R., Lüchow, A., Fröhlich, R., Kogej, M. and Christoph, A. Schalley (2007) *European Journal of Inorganic Chemistry*, 1361.

18 Albrecht, M., Janser, I., Meyer, S., Weis, P. and Fröhlich, R. (2003) *Chemical Communications*, 2854.

19 Albrecht, M., Janser, I., Burk, S. and Weis, P. (2006) *Dalton Transactions*, 2875.

20 Albrecht, M., Burk, S., Weis, P., Schalley, C.A. and Kogej, M. (2007) *Synthesis*, **23**, 3736.

21 Albrecht, M., Burk, S. and Weis, P. (2008) *Synthesis*, **18**, 2963.

22 Lippard, S.J. and Berg, J.M. (1994) *Principles of bioinorganic chemistry*, University Science Books, Mill Valley.

23 (a) Albrecht, M., Napp, M. and Schneider, M. (2001) *Synthesis*, 468; (b) Albrecht, M., Spieß, O. and Schneider, M. (2002) *Synthesis*, 126.

24 Albrecht, M., Napp, M., Schneider, M., Weis, P. and Fröhlich, R. (2001) *Chemical Communications*, 409.

25 Albrecht, M., Napp, M., Schneider, M., Weis, P. and Fröhlich, R. (2001) *Chemistry - A European Journal*, **7**, 3966.

26 (a) Albrecht, M., Nolting, R. and Weis, P. (2005) *Synthesis*, 1125; (b) Albrecht, M., Stortz, P. and Nolting, R. (2003) *Synthesis*, 1307.

27 (a) Albrecht, M. (2007) *Naturwissenschaften*, in press. (b) Clapham, G. and Shipman, M. (1999) *Tetrahedron Letters*, **40**, 5639.

24
Self-assembly of Dinuclear Helical Metallosupramolecular Coordination Compounds

Ulf Kiehne, Jens Bunzen, and Arne Lützen

24.1
Introduction

The idea to create defined cavities that can encapsulate molecules and – even more excitingly – can be used to modify compounds in enzyme-like transformations, where the special confined conditions within these cavities allow reactions to take place much faster than in bulk solution or even in a completely new way due to the specific steric environment with its unique arrangement of functional groups, fascinates chemists around the world and has led to considerable efforts and achievements already.

Container molecules like cryptophanes and (hemi-)carcerands, for example, have been demonstrated to encapsulate small molecules and even allow the formation and study of reactive compounds like cyclobutadiene, benzyne, cycloheptatetraene, or *anti*-Bredt-bridgehead olefins within their interior [1]. However, despite all the admirable progress that has been made in covalent organic synthesis [2], the enormous effort associated with it severely limits the generation of such complicated artificial devices. Thus, noncovalent strategies using reversible self-assembly processes where complex structures are formed by 'weak' intermolecular interactions like hydrogen bonds or metal coordination do represent an attractive alternative [3]. The high degree of convergence usually significantly lowers the preparative efforts necessary for the formation of sophisticated architectures compared to the work of preparing covalently assembled analogs (if these can be prepared at all). Nevertheless, depending on the surrounding conditions, the assembly is highly precise, because these equilibrium-controlled processes are self-controlling and self-repairing leading to thermodynamically stable aggregates, which, however, often show interesting dynamic behavior. In addition, cooperative effects are often observed that cause fast assembly (after an initial rate-determining nucleation), whereby several bonds can be made more easily than a single one.

Activating Unreactive Substrates: The Role of Secondary Interactions.
Edited by Carsten Bolm and F. Ekkehardt Hahn

ISBN: 978-3-527-31823-0

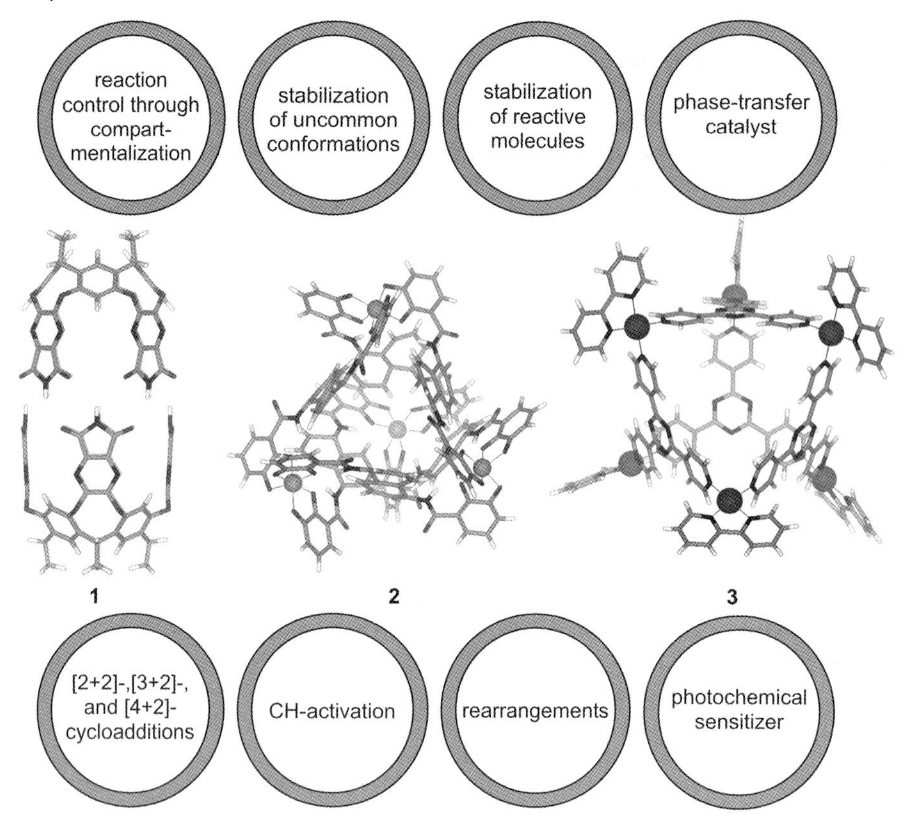

Figure 24.1 Computer model of J. Rebek, Jr.'s cylindrical capsule **1** and X-ray crystal structure analyses of K. N. Raymond's tetranuclear gallium complex (Δ,Δ,Δ,Δ)-**2**, and M. Fujita's hexanuclear metal coordination compound **3** and some of the successfully established applications of these aggregates. Long alkyl chains of the lower rim of the resocin[4]arenes in **1**, counter ions, solvent molecules, and the guest molecule in **2** and **3** have been omitted. **3** is shown with 2,2-bipyridine ligands. However, most of the studies were performed with ethylene diamine as chelating ligand instead.

Following this approach, capsular aggregates could also be obtained [4], and Figure 24.1 shows three prominent examples from the groups of J. Rebek, Jr. (**1**), K. N. Raymond (**2**), and M. Fujita (**3**), that have been demonstrated not only to stabilize reactive species or molecules in uncommon conformations through encapsulation, but also to act as molecular reaction vessels for encapsulated substrates [5].

Of these, only the tetranuclear tetrahedron of K. N. Raymond is chiral. However, since it is composed only of achiral components it can only be obtained in racemic form. Although Raymond could nicely show that these can be resolved into the respective enantiomers [6], we wondered whether capsular or macrocyclic supramolecular metal coordination compounds could also be formed in a stereoselective manner in order to get access to chiral confined spaces of defined stereochemistry.

24.2
The Concept of Diastereoselective Self-assembly of Dinuclear Helicates

Helices are ubiquitous types of structures in nature that are often used to make sure that a given system can fulfil its complex function, for example DNA, peptides, or proteins. The helicity, however, is usually induced by a sophisticated network of noncovalent interactions, which – aside from its beauty – makes this structural motif very attractive for supramolecular chemists. Helicates – oligonuclear helical metal coordination compounds that are built up by at least two metal centers and one ligand strand – for instance, represent simple model systems that can be used to study the underlying noncovalent interactions, questions about how the chirality can be controlled during the assembly, but also the challenge of introducing functionality into supramolecular aggregates. Thus, these supramolecular coordination compounds have been subject of intense research for some time now [7].

Enantiomerically pure helicates can be obtained when chiral elements are introduced either at the termini or in the spacer of the ligand [7–18]. In particular, the latter strategy offers the opportunity to obtain supramolecular structures that bear chiral cavities with inward-directed functionalities which could be used for further purposes, such as molecular recognition or supramolecular reactivity – although very few such helicates have so far been designed. However, in order to avoid problems arising from the orientation of the ligands in these helicates, which would result in an almost uncontrollable number of possible stereoisomers, the use of dissymmetric ligand units is especially advantageous.

Our first contribution to this field was the design and synthesis of bis-(2,2′-bipyridyl)-substituted 2,2′-dihydroxy-1,1′-binaphthyl (BINOL) **4** that could be demonstrated to undergo diastereoselective self-assembly to enantiomerically pure D_n-symmetric double- ($M = Ag^+$, Cu^+) or triple-stranded ($M = Zn^{2+}$, Fe^{2+}) helicates (Scheme 24.1) [19].

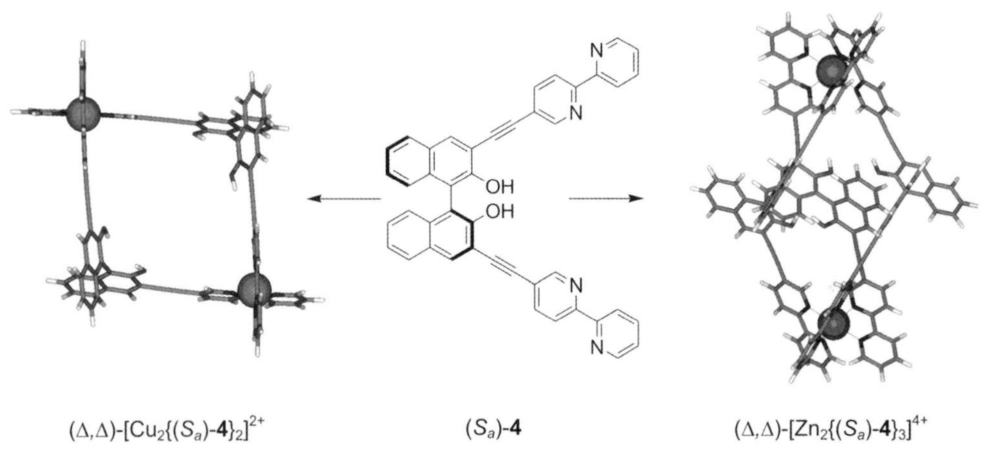

Scheme 24.1 Diastereoselective self-assembly of dinuclear D_2- or D_3-symmetric helicates (computer models confirmed by MS, NMR and X-ray diffraction studies) from bis(bipyridyl) substituted BINOL **4**.

These aggregates bear chiral cavities of considerable size with inward-directed hydroxyl groups that could be shown to be useful for the recognition of monosaccharide derivatives [20]. Therefore, we wondered whether this was a single finding or whether further bis(chelating) ligands of this type could be created that also undergo diastereoselective self-assembly to dinuclear coordination compounds, which would prove this concept to be more general [21]. Therefore, **4** was regarded as a kind of lead structure to explore the possibilities of employing other building blocks. In order to do this, one could vary the metal binding site, the central chiral element, and the spacer in between these. Figure 24.2 shows a summary of the different approaches.

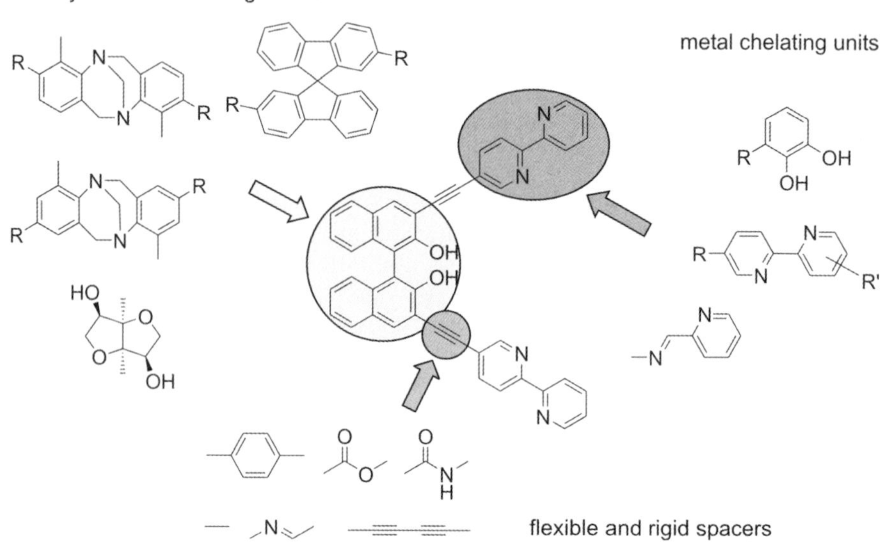

Figure 24.2 Molecular building blocks for the synthesis of future generations of ligands.

24.3
Synthesis of Building Blocks for the Covalent Assembly of Bis(chelating) Ligands

24.3.1
Synthesis of Dissymmetric Elements

The central building block has to fulfil the following requirements to be useful for our design: (1) it has to be C_2-symmetric, as mentioned before, (2) it has to be fairly rigid to avoid a conformational collapse in the sense that the two metal binding sites could bind to the same metal center, (3) it has to carry suitable functional groups to attach the chelating binding site (or the spacer), and (4) it should have a concave or V-shaped form in order to create a cavity. These considerations led us to identify three other promising classes of compounds as well as BINOL: 2,2′-disubstituted 9,9′-spirobifluorenes, 2,8- or 3,9-disubstituted derivatives of Tröger's base, both being more rigid than the BINOL which can adopt different conformations due to (slight) rotation around the aryl-aryl-bond, and D-isomannide, a commercially available compound derived from D-mannitol, as an example of a nonaromatic compound.

24.3.2
Synthesis and Resolution of 9,9′-Spirobifluorenes

Although 9,9′-spirobifluorene had been synthesized as early as 1930 by M. Gomberg [22], it has been only recently that these compounds have found tremendous interest because of their exciting properties and applications in materials science [23]. Nevertheless, the first enantiomerically pure spirobifluorenes were prepared by V. Prelog in 1969 [24]. However, his protocols were quite tedious as he admitted himself. A better procedure was introduced by F. Toda in 1988, when he was able to isolate the (+)-enantiomer of 2,2′-dihydroxy-9,9′-spirobifluorene (**5**) upon clathrate formation with a tartaric acid derivative [25]. We took this approach a step further and succeeded in the isolation of both enantiomers of **5** and the assignment of their absolute stereochemistry [26]. The resolved material was then further elaborated to get a number of optically pure difunctionalized building blocks (Scheme 24.2).

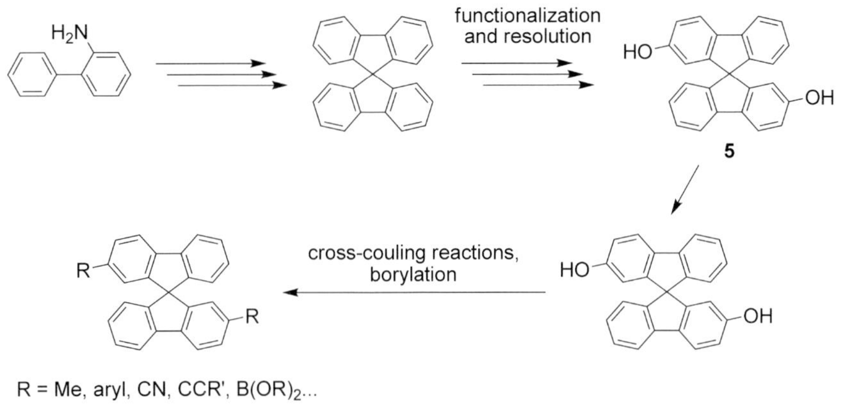

H_2N — functionalization and resolution — HO — OH — **5** — R — R — cross-couling reactions, borylation — HO — OH

R = Me, aryl, CN, CCR′, $B(OR)_2$...

Scheme 24.2 Synthesis of optically pure 2,2′-difunctionalized 9,9′-spirobifluorenes.

24.3.3
Synthesis and Resolution of Tröger's Base Derivatives

Tröger's base derivatives have experienced a true renaissance recently [27], after K. Wärnmark reported on a facile route to dihalogenated derivatives [28] like **6** and **7** that allowed the formation of a variety of new and valuable precursors for larger architectures with extended V-shaped cores through various cross-coupling methodologies (Scheme 24.3) [29, 30].

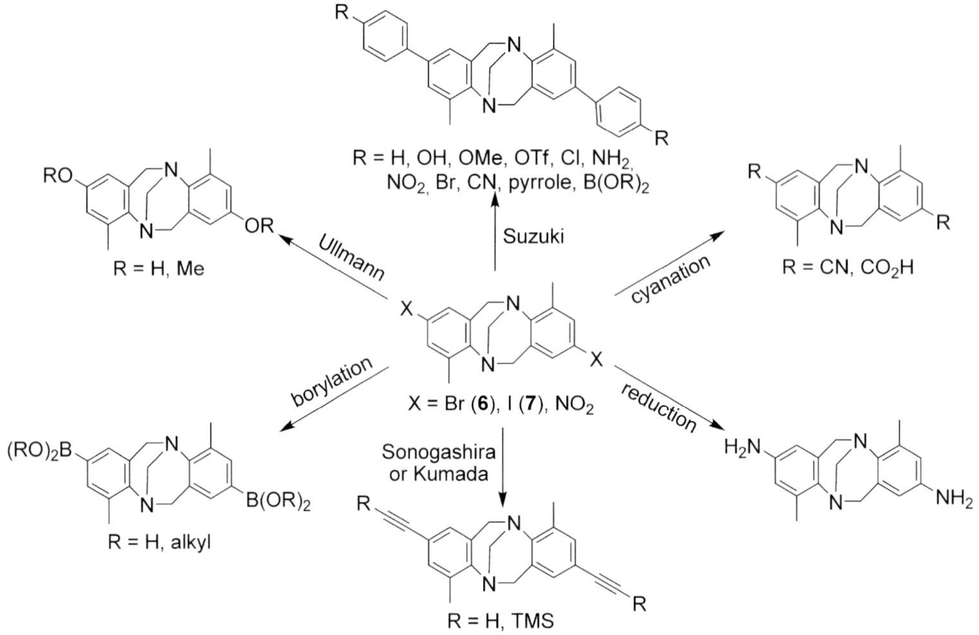

Scheme 24.3 Synthesis of 2,8-disubstituted Tröger's base derivatives.

We also succeeded in the resolution of two of these racemic derivatives of Tröger's base – the 2,8-diboronic ester **8** and the 3,9-dibromo substituted derivative **9** – by HPLC on a stationary Whelk-01 phase on a semi-preparative scale, giving rise to both enantiomers in pure form. These are valuable precursors for a variety of further applications. Their absolute configurations were determined by comparison of their calculated CD and UV-Vis spectra with the experimental ones and independently confirmed by X-ray diffraction analysis (Figure 24.3) [[29]c].

24.3.4 Synthesis of 2,2′-Bipyridines

The last 15 years witnessed a great deal of effort directed toward the use of chiral ligand structures to efficiently control the stereochemistry of metal centers of coordination compounds in a diastereoselective manner [31]. This is especially true for helicates [7], as pointed out before. In particular, chiral *N*-donor ligands like 2,2′-bipyridines [8], terpyridines [9], quaterpyridine [10], oxazolines [11], pyridylmethanimines [12] and also *P*-donor ligands [13] leading to cationic complexes

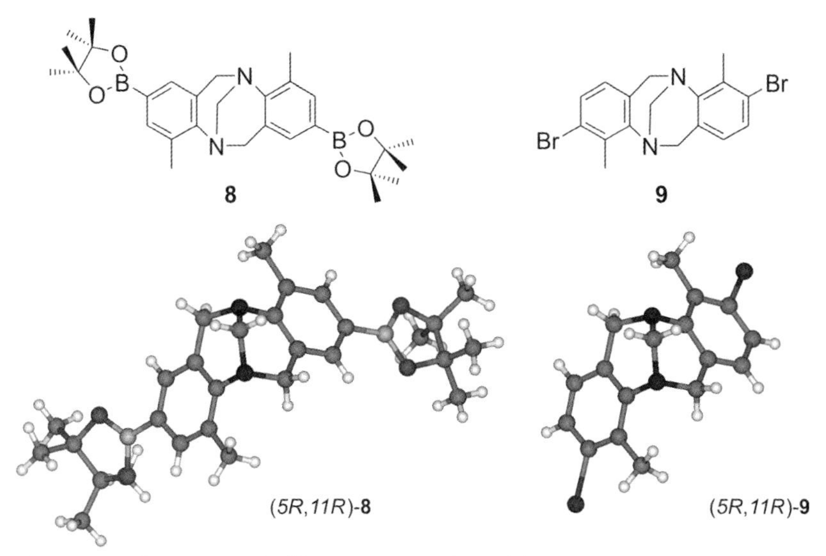

Figure 24.3 Diboronic ester **8**, dibromo compound **9** and the X-ray crystal structure analyses of their (5R,11R)-configured enantiomers.

with suitable late transition metal or lanthanide ions were shown to be very effective in this sense. Helicates, however, have not only been obtained from *N*-, or *P*-donor ligands but also from *O*-donor ligands like catechols and hard metal ions like titanium(IV) or gallium(III) ions [7c, g, h]. These helicates are anionic and hence have different properties from those of the cationic aggregates self-assembled from ligands bearing *N*-heterocyclic metal chelation units and late transition metal ions. After the first examples of this kind, in particular those published by K. N. Raymond [14], M. Albrecht has extensively studied a variety of helical catecholato-complexes, reported in numerous publications [7c, g, h, 15], but some examples were published recently by F. E. Hahn [16]. So far, however, there have been very few reports on the use of chiral ligands to control the stereochemistry of the newly formed stereogenic metal centers in these types of helicates, and the stereogenic information was introduced in the outer periphery of the bis(catechol) ligand structure [15c, 17] rather than in the core structural element [18].

In our study we focused on three different types of chelating ligand units: catechols, pyridylmethanimines, and 2,2′-bipyridines. Especially the last of these certainly belong to the most widely used ligand structure motifs in coordination chemistry [32], but the efficient synthesis of functionalized species is still challenging. Thus, we had to find ways to prepare symmetrically difunctionalized derivatives and mono- or differently disubstituted 2,2′-bipyridines first. This was achieved by the development of a nickel-catalyzed homocoupling, and a modified Negishi cross-coupling procedure, respectively (Scheme 24.4) [15, 33].

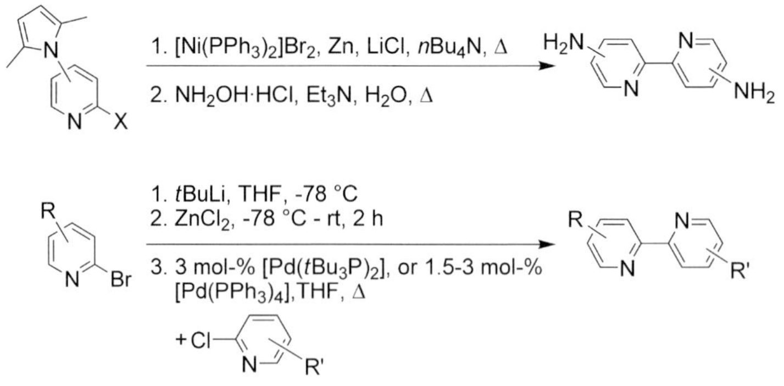

1. $[Ni(PPh_3)_2]Br_2$, Zn, LiCl, $n$$Bu_4N$, Δ

2. $NH_2OH{\cdot}HCl$, Et_3N, H_2O, Δ

1. tBuLi, THF, -78 °C
2. $ZnCl_2$, -78 °C - rt, 2 h
3. 3 mol-% $[Pd(tBu_3P)_2]$, or 1.5-3 mol-% $[Pd(PPh_3)_4]$, THF, Δ

R = H, Me, OMe, pyrrole, $CCSiMe_3$ R' = Me, Ph, CF_3, $CCSiMe_3$, (4-MeOPh), (3,5-Me_2Ph), CN, Br, CO_2R, SPh, OMe, pyrrole, alkyl

Scheme 24.4 Synthesis of functionalized 2,2′-bipyridines via nickel-catalyzed homocoupling or modified Negishi cross-coupling protocol.

24.4
Synthesis of Bis(chelating) Ligands and Their Dinuclear Metal Complexes

With all necessary building blocks in hand we turned to the synthesis of a series of ligands and studied their coordination behavior toward different transition metal ions. Before we present the most important properties of these, we describe an example of our strategy in more detail for a ligand based on D-isomannide in order to demonstrate which analytical techniques were used and what kind of conclusions can be drawn from these analyses.

24.4.1
D-Isomannide-based Ligand and Its Complexes

Ligand **10** was prepared by a simple two-fold ester formation of D-isomannide upon reaction with 5-(2,2′-bipyridyl)acid chloride as shown in Figure 24.4. **10** was then used for the complexation studies with copper(I) and silver(I), which usually prefer a tetrahedral coordination by two bipyridine ligands, iron(II), which prefers an octahedral coordination by three ligands, and zinc(II), that can form both tetrahedral bis- and octahedral tris(bipyridine) complexes because of its 'chameleon' character [34]. The resulting coordination compounds were examined by NMR- and CD-spectroscopy as well as mass spectrometry (Figures 24.4 and 24.5) [35].

Upon addition of the metal salts to **10**, characteristic color changes occurred indicating the expected formation of the respective bipyridine complexes of Ag^+ (yellow), Cu^+ (brown-red), and Fe^{2+} (red), whereas the solution of the respective Zn^{2+}-aggregate remained almost colorless. This could also be proven by analysis of the ^1H NMR spectra that show only one set of signals for the metal complexes that are significantly shifted compared to the corresponding signals of the free **10** (like Figure 24.4, iii). This observation refers to the formation of discrete

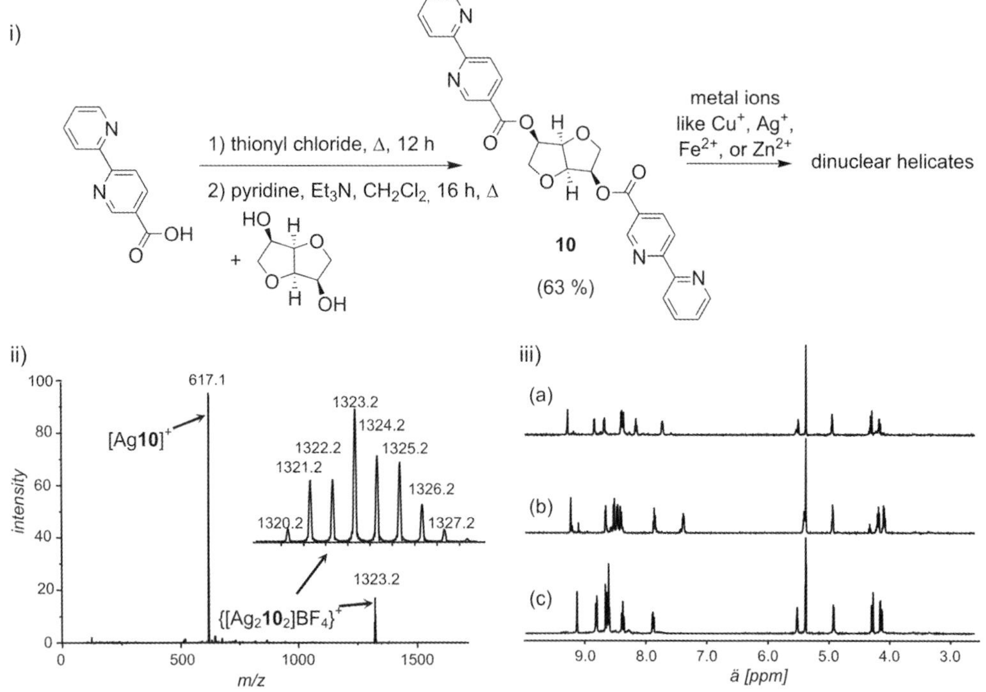

Figure 24.4 (i) Synthesis of ligand **10**; (ii) Positive ESI MS of a solution of $[Ag_2\mathbf{10}_2](BF_4)_2$ in CH_2Cl_2/CH_3CN; (iii) ^1H NMR spectra (CD_2Cl_2/CD_3CN 3:1, 298 K) of (a) **10** + $[Ag(CH_3CN)_2]BF_4$ (1:1), (b) **10**, and (c) **10** + $[Zn(BF_4)_2]\cdot 6.5\ H_2O$ (1:1).

dinuclear helicates and excludes the formation of oligomeric or polymeric species. This could also be proven by ESI MS experiments because only signals arising from the intact helicates and fragments resulting from them could be detected (as in Figure 24.4, ii). Interestingly, only double-stranded dinuclear zinc(II) helicates were found to be formed in this case.

However, the analysis of the NMR spectra allows us to draw some more conclusions concerning the stereoselectivity of the helicate formation: the number of signals of the free ligands is equal to the number of signals of their silver(I) and zinc(II) complexes (as well as of their copper(I) and iron(II) complexes which are not shown in Figure 24.4). This is only possible if the ligands retain their symmetry in the aggregates. This, however, is only feasible when the newly formed chiral metal centers have the same configuration (Δ,Δ or Λ,Λ), thus excluding the formation of Δ,Λ- or Λ,Δ-configured '*meso*'-helicates. Since Δ,Δ- or Λ,Λ-configured helicates of these chiral ligands are diastereomers, however, it is at least very unlikely, that all of the proton signals of the Δ,Δ- or Λ,Λ-configured helicates are isochronic by coincidence. Therefore, one can conclude that the self-assembly processes are indeed completely diastereoselective, yielding enantiomerically pure D_2- or D_3-symmetrical dinuclear helicates with equally configured metal centers.

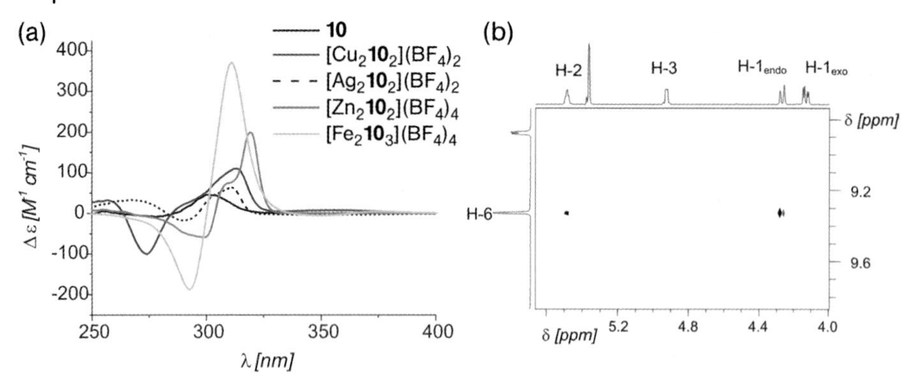

CD-bands and ROE-contacts correspond to (Λ,Λ)-configurated complexes

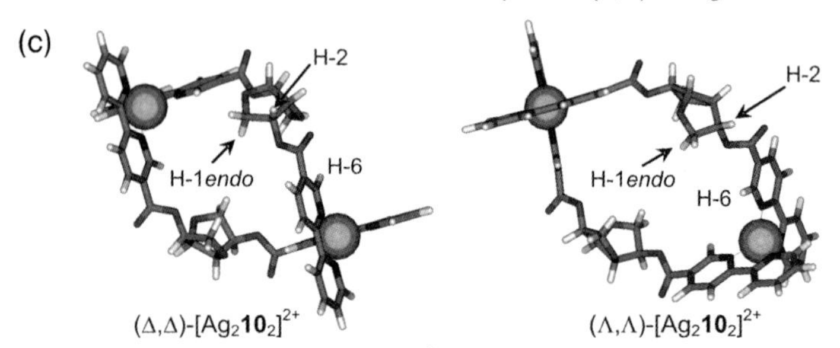

Figure 24.5 (a) CD-spectra of 10^{-5} M solution of the dinuclear helicates of **10** in dichloromethane and (b) part of the 2D-ROESY spectrum of $[Ag_2\mathbf{10}_2](BF_4)_2$ (CD_2Cl_2/CD_3CN 3:1, 298 K). (c) Computer models of (Δ,Δ)-$[Ag_2\mathbf{10}_2]^{2+}$ and (Λ,Λ)-$[Ag_2\mathbf{10}_2]^{2+}$.

The question resulting from this is, which stereoisomer is formed? Unfortunately, we have not yet been able to obtain crystals suitable for X-ray analysis of our aggregates. However, to elucidate the stereochemistry, ROESY NMR experiments can also be used because they allow an assignment of the relative orientation of the metal chelating unit and the D-isomannide core. As Figure 24.5 shows, for $[Ag_2\mathbf{10}_2](BF_4)_2$ clear evidence points toward a (Λ,Λ)-configuration: ROE contacts of bipyridine proton H-6 to H-1*endo* and H-2 of the isomannide bicyclic structure are not possible in the case of (Δ,Δ)-configured helicates because the distances are too large, although they can be observed. Similar results were obtained for $[Zn_2\mathbf{10}_2](BF_4)_4$.

CD spectroscopy is a powerful chirooptical spectroscopic method and has been widely used in recent years to elucidate the stereochemistry of chiral bipyridine transition metal complexes [36]. Assignments of the absolute configuration of such helical systems by local Δ- or Λ-descriptors for each metal ion of the complex can be derived from the sign of lc (ligand-centered) π-π^*-transitions around 300–340 nm in most cases [7d, e, g, 9, 36–38].

Figure 24.5 shows the CD spectra of **10** and its metal complexes in the lc region: in general, intensities of the aggregates are much higher than those of the free ligand **10**. All complexes show a CD activity in the relevant lc region. The observed absorptions can be assigned to two coupled exciton π-π^*-transitions of two chromophoric bipyridine units which are coordinated to the same metal ion. All absorptions for the double-stranded complexes show a positive sign, which generally is characteristic for a *P* helicity. If these transitions are attributed to π-π^*-transitions and exciton coupling is assumed, the metal centers bear a local (Λ,Λ)-configuration (according to 'skew line' convention) which could be observed for similar examples [8c, 10a, 36]. These results are in full agreement with the conclusions that could be drawn from the interpretation of the ^{1}H NMR ROESY spectra.

Thus, dissymmetric D-isomannide derivative **10**, which bears four stereogenic centers, forms enantiomerically pure D_2- or D_3-symmetrical (Λ,Λ)-configured dinuclear double- and triple-stranded complexes of helical shape with metal ions such as Ag^+, Cu^+, Zn^{2+}, and Fe^{2+} in a diastereoselective manner. This demonstrates that the concept of diastereoselective self-assembly of dinuclear helicates can also be applied to nonaromatic dissymmetric core skeletons and rather flexible linkers like ester groups. In fact, this is only the second example of D-isomannide being successfully incorporated into metallosupramolecular aggregates [39].

24.4.2 9,9′-Spirobifluorene-based Ligand and Its Complexes

Interestingly, enantiomerically pure ligand **11** bearing a 9,9′-spirobifluorene core behaves differently in this kind of self-assembly processes [40]. Again, single species were formed in a diastereoselective manner upon coordination to copper(I) or silver(I) ions, and these could be identified as double-stranded dinuclear copper(I) and silver(I) complexes (Figure 24.6). However, the metal centers in these helicates do not have the same configuration. Thus, they can be regarded as '*meso*'-helicates, although these C_2-symmetric enantiomerically pure aggregates are, of course, still chiral. This can easily be deduced from the analysis of the ^{1}H NMR spectra, since the number of signals observed for the isolated helicates is exactly doubled compared to the ligand itself which expresses the lower symmetry of a C_2-symmetric (Δ,Λ)-configured metallosupramolecular aggregate compared to the corresponding D_2-symmetrical (Δ,Δ)- or (Λ,Λ)-configured diastereomers obtained from dissymmetrical ligands.

24.4.3 Tröger's Base Derivatives-based Ligands and Their Complexes

We then turned our attention to ligands derived from Tröger's bases. Interestingly, when we started this work there had been no other examples of the incorporation of this scaffold into metallosupramolecular aggregates, and it has only been recently that another study was reported by C. A. Mirkin [41]. When we tested racemic 2,8-bis(catechol)-functionalized Tröger's base derivatives (*rac*)-**12**(H_4) and (*rac*)-**13**(H_4) for their ability to form triple-stranded dinuclear helicates upon coordination to titanium(IV) ions in the presence of alkali metal carbonates, however, we found out that the shape of these ligands is just not well preorganized to form the

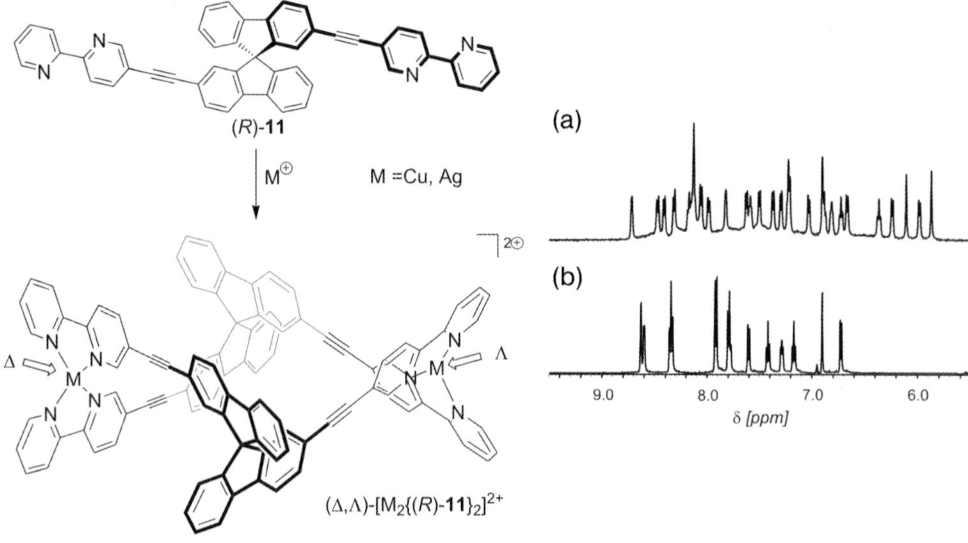

Figure 24.6 Diastereoselective self-assembly of "*meso*"-helicates from ligand **11**. ^{1}H NMR spectra (CD_2Cl_2/CD_3CN, 298 K) shown on the right were taken from (Δ,Λ)-$[Ag_2\{(R)\text{-}\mathbf{11}\}_2](BF_4)_2$ (a) and ligand (*R*)-**11** (b).

desired helicates. Nevertheless, the dinuclear titanium complexes $M_4[Ti_2\mathbf{12}_3]$ and $M_4[Ti_2\mathbf{13}_3]$ (M = Li, Na, K) are observed as the exclusive products of these experiments (Scheme 24.5), but the strain which is introduced into the ligand structure compensates the energetic differences between the individual diastereomers and

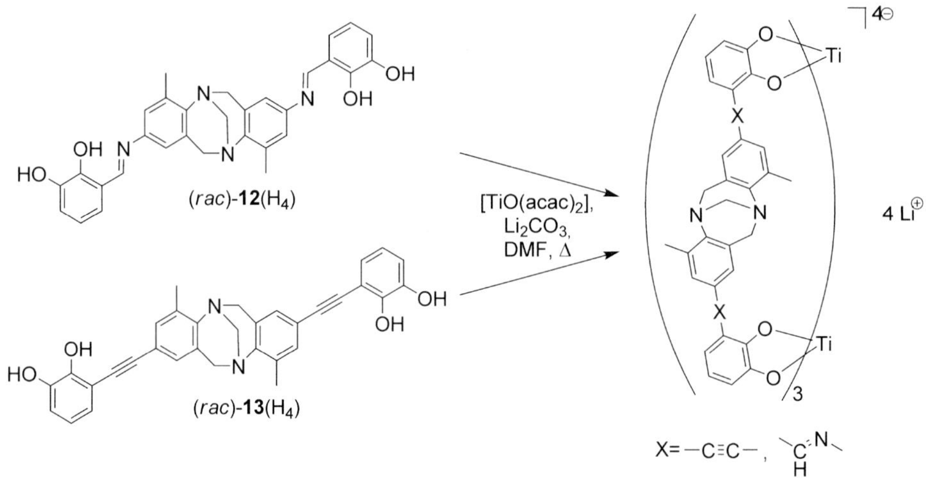

Scheme 24.5 Formation of triple-stranded dinuclear titanium(IV) complexes from racemic 2,8-functionalised Tröger's base-derived ligands (*rac*)-**12**(H_4) and (*rac*)-**13**(H_4).

results in a complex mixture of stereoisomers [42]. This can easily be seen in the NMR spectra recorded for the complexation studies of (*rac*)-**12** (Figure 24.7).

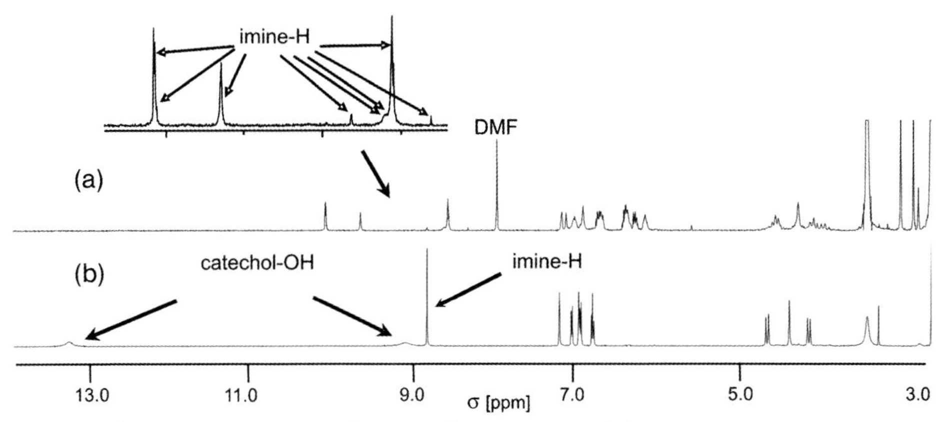

Figure 24.7 ^{1}H NMR spectra (DMSO-d_6, 298 K) of (a) $Li_4[Ti_2\mathbf{12}_3]$ and (b) (*rac*)-**12**.

Please note that the use of racemic ligands not only allows for the formation of homoleptic complexes with equally configured ligand strands but also heteroleptic complexes that incorporate two ligands of the same configuration and one of the other. Thus, a total of six different stereoisomers (together with their respective enantiomers which cannot be distinguished by normal NMR spectroscopy, of course) can be formed in these experiments. For these one would expect a maximum of 12 different imine proton signals ($1 + 1 + 2 + 2 + 2 + 4 = 12$) which can almost all be detected in the spectra of the titanium complexes. Hence, the self-assembly process is clearly not diastereoselective but rather gives rise to a complex mixture of all possible metallosupramolecular aggregates.

Similar results were obtained with the corresponding bis(bipyridyl) ligands (*rac*)-**14** - (*rac*)-**18** and also bis(pyridylmethanimine) ligand (*rac*)-**19** all based on the 2,8-difunctionalized Tröger's base core (Figure 24.8) when we tried to prepare triple-stranded zinc(II) helicates, although these were obtained in a completely diastereoselective manner from BINOL-based ligand **4** in its MOM-protected form [19]. As in the case of the titanium complexes described above, the desired dinuclear triple-stranded zinc(II) complexes could be detected by ESI MS, but unfortunately none of the ligands (*rac*)-**14** - (*rac*)-**19** assembles stereoselectively [43].

Nevertheless, all of these ligands based on the 2,8-difunctionalized Tröger's base core are obviously well preorganized to undergo self-assembly to dinuclear double-stranded complexes with copper(I) or silver(I) complexes because these were found to be formed in a completely diastereoselective manner [29b, 43]. Thus, the configuration of the newly formed chiral metal centres is completely controlled by the configuration of the Tröger's base, even when ligands with elongated spacer units such as **17**, **18** or even **16** are used that set the stereogenic centers of the Tröger's base and the metal centers up to 14.3 Å apart from each other. Figure 24.9 shows computer models of the complexes, which are in agreement with the mass spectrometric and (*rac*)-**14** (*rac*)-**15** (*rac*)-**16** (*rac*)-**17** (*rac*)-**18** (*rac*)-**19**

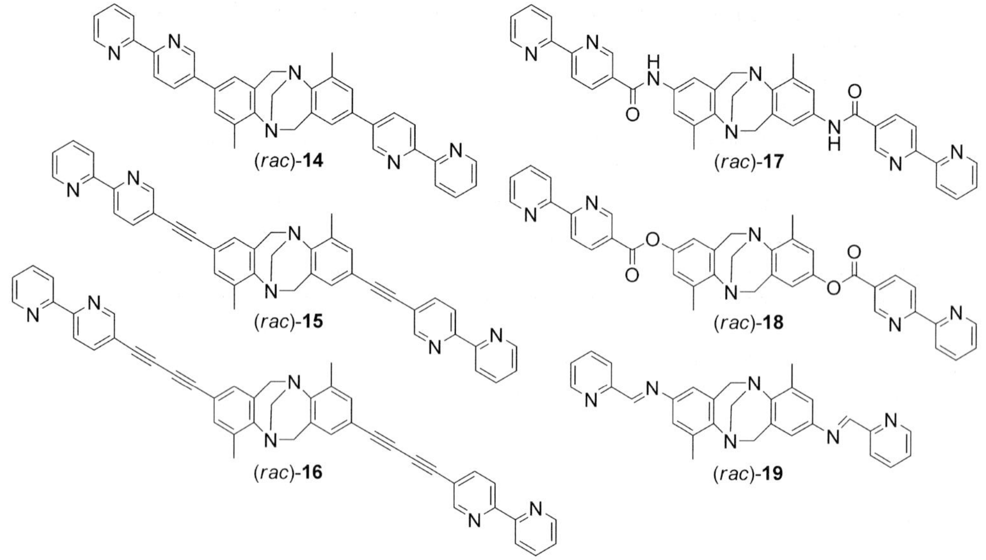

Figure 24.8 *N*-donor ligands (*rac*)-**14** – (*rac*)-**19** with a 2,8-functionalized Tröger's base core.

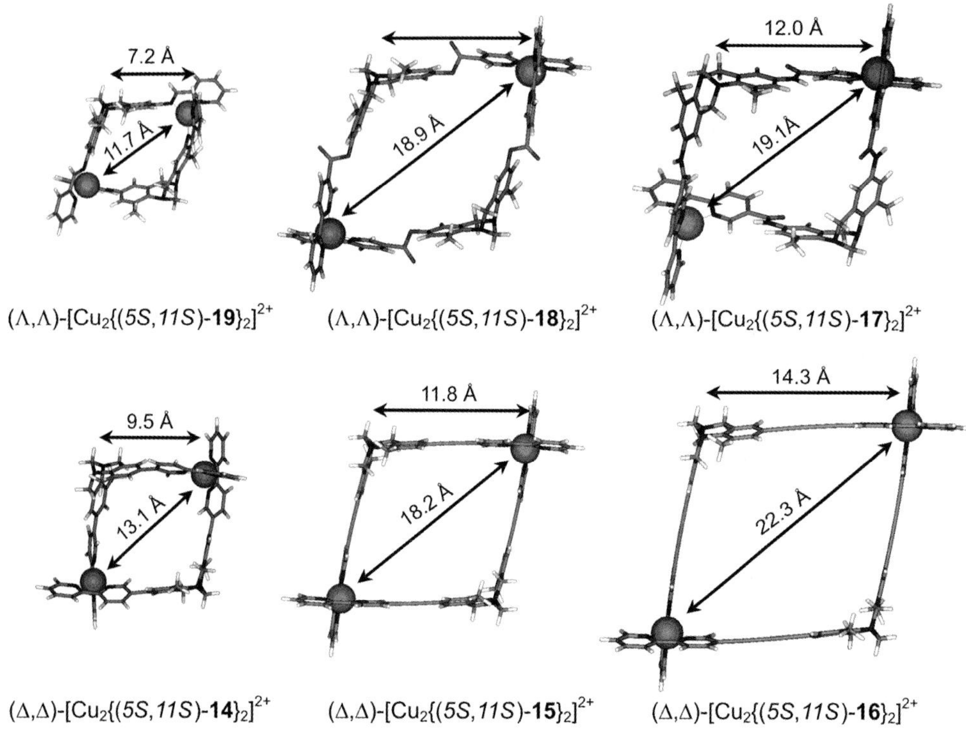

Figure 24.9 Computer models of double-stranded copper(I) helicates obtained upon diastereoselective self-assembly of Tröger's base-derived ligands (*rac*)-**14** to (*rac*)-**19** upon coordination to copper(I) ions (only the enantiomer formed from the (*5S,11S*)-enantiomers are shown).

NMR spectroscopic data. Thus, macrocyclic aggregates with tailored chiral cavities can be formed selectively with a metal–metal distance of up to 22.3 Å.

How sensitive these processes are to subtle changes, however, could be proved by preparing ligand (*rac*)-**20**, which is very similar to (*rac*)-**15** but has a different substitution pattern of the Tröger's base core. Changing the positions of the attached bipyridylethynyl groups from the 2,8- to the 3,9-positions has a huge impact on the diastereoselectivity of the self-assembly processes. Because of the larger opening angle defining the extent of the V-shaped character of the ligand, the 3,9-disubstituted building blocks are better preorganized to form triple-stranded helicates, which could be proved for ligand **20** which undergoes diastereoselective self-assembly with zinc (II) and iron(II) ions to the respective dinuclear helicates (Scheme 24.6). Interestingly, however, the angle is still not too big to prevent diastereoselective formation of double-stranded helicates upon coordination to silver(I) or copper(I) ions, as could be proved by NMR spectroscopic and MS studies [44].

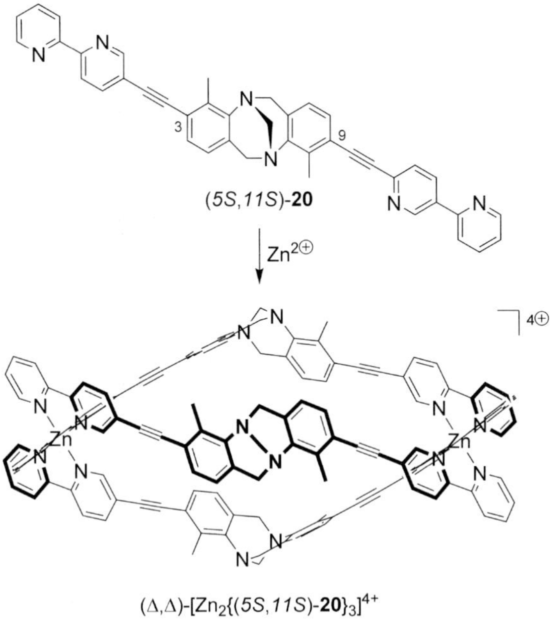

(5*S*,11*S*)-**20**

Zn^{2+}

(Δ,Δ)-$[Zn_2\{(5S,11S)\text{-}\mathbf{20}\}_3]^{4+}$

Scheme 24.6 Diastereoselective self-assembly of racemic triple-stranded dinuclear zinc(II) helicates from ligand (*rac*)-**20**. (Only the (Δ,Δ)-configurated complex, which is formed from the (5*S*,11*S*)-enantiomer, is shown).

24.5 Conclusions

The concept of diastereoselective self-assembly of dinuclear helicates from dissymmetrical concave bis(chelating) ligand strands has been proved to be very general based on the results we obtained with a large number of different ligands that we have studied. Whereas double-stranded copper(I) or silver(I) helicates were found to be formed from all of the *N*-donor ligands under invesigation, the formation of triple-stranded zinc(II) helicates was found to be more sensitive to a suitable conformational preorganization, which is obviously necessary for the stereoselectivity of these processes. This approach is useful for the formation of a large variety of macrocyclic and capsular metallosupramolecular aggregates with cavities tailored to afford size and functionalities potentially useful for further purposes like molecular recognition and supramolecular reactivity.

Acknowledgments

This work was supported by the DFG (SPP 1118). We are grateful to all of our colleagues and collaborators who participated in the various parts of these projects. We would also like to thank the DFG for funding some very exciting and fruitful workshops on the issue of 'secondary interactions' as well as the regular symposia within the priority programme 1118.

References

1 Some recent reviews: (a) Jasat, A. and Sherman, J.C. (1999) *Chemical Reviews*, **99**, 931; (b) Warmuth, R. and Marvel, M.A. (2001) *European Journal of Organic Chemistry*, 423; (c) Warmuth, R. and Yoon, J. (2001) *Accounts of Chemical Research*, **34**, 95. Two new examples; (d) Roach, P. and Warmuth, R. (2003) *Angewandte Chemie*, **115**, 3147; Roach, P. and Warmuth, R. (2003) *Angewandte Chemie – International Edition*, **42**, 3039; (e) Kedelhué, J.-L., Langenwalter, K.J. and Warmuth, R. (2003) *Journal of the American Chemical Society*, **125**, 973.

2 (a) Nicolaou, K.C. and Sorensen, E.J. (1996) *Classics in Total Synthesis*, Wiley-VCH, Weinheim; (b) Nicolaou, K.C. and Snyder, S.A. (2003) *Classics in Total Synthesis II*, Wiley-VCH, Weinheim.

3 Some reviews: (a) Lehn, J.-M. (1990) *Angewandte Chemie*, **102**, 1347; Lehn, J.-M. (1990) *Angewandte Chemie – International Edition*, **29**, 1304; (b) Lindsey, J.S. (1991) *New Journal of Chemistry*, **15**, 153; (c) Lawrence, D.S., Jiang, T. and Levett, M. (1995) *Chemical Reviews*, **95**, 2229; (d) Philp, D. and Stoddart, J.F. (1996) *Angewandte Chemie*, **108**, 1242; Philp, D. and Stoddart, J.F. (1996) *Angewandte Chemie – International Edition*, **35**, 1154; (e) Fyfe, M.C.T. and Stoddart, J.F. (1997) *Accounts of Chemical Research*, **30**, 393; (f) Saalfrank, R.W. and Bernt, I. (1998) *Current Opinion in Solid State & Materials Science*, **3**, 407; (g) Greig, L.M. and Philp, D. (2001) *Chemical Society Reviews*, **30**, 287.

4 Some recent reviews: (a) Holliday, B.J. and Mirkin, C.A. (2001) *Angewandte Chemie*, **113**, 2076; Holliday, B.J. and Mirkin, C.A. (2001) *Angewandte Chemie – International Edition*, **40**, 2022; (b) Seidel, S.R. and Stang, P.J. (2002) *Accounts of Chemical Research*, **35**, 972; (c) Hof, F., Craig, S.L., Nuckolls, C. and Rebek, J., Jr (2002) *Angewandte Chemie*, **114**, 1556; Hof, F., Craig, S.L., Nuckolls, C. and Rebek, J., Jr (2002) *Angewandte Chemie – International Edition*, **41**, 1488; (d) Würthner, F., You, C.-C. and Saha-Möller, C.R. (2004)

Chemical Society Reviews, **33**, 133; (e) Fiedler, D., Leung, D.H., Bergman, R.G. and Raymond, K.N. (2005) *Accounts of Chemical Research*, **38**, 351; (f) Fujita, M. (2005) *Accounts of Chemical Research*, **38**, 369.

5 Short reviews: (a) Lützen, A. (2005) *Angewandte Chemie*, **117**, 1022; Lützen, A. (2005) *Angewandte Chemie – International Edition*, **44**, 1000; (b) Schmuck, C. (2007) *Angewandte Chemie*, **119**, 5932; Schmuck, C. (2007) *Angewandte Chemie – International Edition*, **46**, 5830.

6 (a) Terpin, A.J., Ziegler, M., Johnson, D.W. and Raymond, K.N. (2001) *Angewandte Chemie*, **113**, 161; Terpin, A.J., Ziegler, M., Johnson, D.W. and Raymond, K.N. (2001) *Angewandte Chemie – International Edition*, **40**, 157; (b) Ziegler, M., Davis, A.V., Johnson, D.W. and Raymond, K.N. (2003) *Angewandte Chemie*, **115**, 689; Ziegler, M., Davis, A.V., Johnson, D.W. and Raymond, K.N. (2003) *Angewandte Chemie – International Edition*, **42**, 665.

7 Reviews on helicates: (a) Constable, E.C. (1992) *Tetrahedron*, **48**, 10013; (b) Piguet, C., Bernardinelli, G. and Hopfgartner, G. (1997) *Chemical Reviews*, **97**, 2005; (c) Albrecht, M. (1998) *Chemical Society Reviews*, **27**, 281; (d) von Zelewsky, A. (1999) *Coordination Chemistry Reviews*, **190–192**, 811; (e) Knof, U. and von Zelewsky, A. (1999) *Angewandte Chemie*, **111**, 312; Knof, U. and von Zelewsky, A. (1999) *Angewandte Chemie – International Edition*, **38**, 302; (f) von Zelewsky, A. and Mamula, O. (2000) *Journal of the Chemical Society, Dalton Transactions*, 219; (g) Albrecht, M. (2000) *Chemistry - A European Journal*, **6**, 3485; (h) Albrecht, M. (2001) *Chemical Reviews*, **101**, 3457; (i) Bünzli, J.-C.G. and Piguet, C. (2002) *Chemical Reviews*, **102**, 1897; (j) Mamula, O. and von Zelewsky, A. (2003) *Coordination Chemistry Reviews*, **242**, 87.

8 Examples of the diastereoselective formation of helicates with 2,2′-bipyridines: (a) Zarges, W., Hall, J., Lehn, J.-M. and Bolm, C. (1991) *Helvetica Chimica Acta*, **74**, 1843; (b) Fletcher, N.C., Keene, F.R., Viebrock, H. and von Zelewsky, A. (1997) *Inorganic Chemistry*, **36**, 1113; (c) Mamula, O., von Zelewsky, A. and Bernardinelli, G. (1998) *Angewandte Chemie*, **110**, 302; Mamula, O., von Zelewsky, A. and Bernardinelli, G. (1998) *Angewandte Chemie – International Edition*, **37**, 289; (d) Murner, H., von Zelewesky, A. and Hopfgartner, G. (1998) *Inorganica Chimica Acta*, **271**, 36; (e) Mamula, O., von Zelewsky, A., Bark, T. and Bernardinelli, G. (1999) *Angewandte Chemie*, **111**, 3129; Mamula, O., von Zelewsky, A., Bark, T. and Bernardinelli, G. (1999) *Angewandte Chemie – International Edition*, **38**, 2945; (f) Mamula, O., Monlien, F.J., Porquet, A., Hopfgartner, G., Merbach, A.E. and von Zelewsky, A. (2001) *Chemistry - A European Journal*, **7**, 533; (g) Mamula, O., von Zelewsky, A., Brodard, P., Schläpfer, C.-W., Bernardinelli, G. and Stoeckli-Evans, H. (2005) *Chemistry - A European Journal*, **11**, 3049; (h) Woods, C.R., Benaglia, M., Cozzi, F. and Siegel, J.S. (1996) *Angewandte Chemie*, **108**, 1977; Woods, C.R., Benaglia, M., Cozzi, F. and Siegel, J.S. (1996) *Angewandte Chemie – International Edition*, **35**, 1830; (i) Annunziata, R., Benaglia, M., Cinquini, M., Cozzi, F., Woods, C.R. and Siegel, J.S. (2001) *European Journal of Organic Chemistry*, 173; (j) Baum, G., Constable, E.C., Fenske, D., Housecroft, C.E. and Kulke, T. (1999) *Chemical Communications*, 195; (k) Prabaharan, R., Fletcher, N.C. and Nieuwenhuyzen, M. (2002) *Journal of The Chemical Society-Dalton Transactions*, 602; (l) Prabaharan, R. and Fletcher, N.C. (2003) *Inorganica Chimica Acta*, **355**, 449; (m) Telfer, S.G., Tajima, N. and Kuroda, R. (2004) *Journal of the American Chemical Society*, **126**, 1408.

9 Examples for the diastereoselective formation of helicates with terpyridines: (a) Constable, E.C., Kulke, T., Neuburger,

M. and Zehnder, M. (1997) *Chemical Communications*, 489 (b) Baum, G., Constable, E.C., Fenske, D., Housecroft, C.E. and Kulke, T. (1998) *Chemical Communications*, 2659; (c) Baum, G., Constable, E.C., Fenske, D., Housecroft, C.E., Kulke, T., Neuburger, M. and Zehnder, M. (2000) *Journal of The Chemical Society-Dalton Transactions*, 945.

10 Examples for the diastereoselective formation of helicates with quaterpyridines: (a) Baum, G., Constable, E.C., Fenske, D. and Kulke, T. (1997) *Chemical Communications*, 2043 (b) Constable, E.C., Kulke, T., Baum, G. and Fenske, D. (1998) *Inorganic Chemistry Communications*, **1**, 80; (c) Baum, G., Constable, E.C., Fenske, D., Housecroft, C.E. and Kulke, T. (1999) *Chemistry - A European Journal*, **5**, 1862.

11 Examples of the diastereoselective formation of helicates with oxazolines: (a) Provent, C., Rivara-Minten, E., Hewage, S., Brunner, G. and Williams, A.F. (1999) *Chemistry - A European Journal*, **5**, 3487 (b) Gelalcha, F.G., Schulz, M., Kluge, R. and Sieler, J. (2002) *Journal of The Chemical Society-Dalton Transactions*, 2517.

12 Examples of the diastereoselective formation of helicates with pyridylmethanimines: (a) van Stein, G.C., van Koten, G., Vrieze, K., Brevard, C. and Spek, A.L. (1984) *Journal of the American Chemical Society*, **106**, 4486; (b) Masood, M.A., Enemark, E.J. and Stack, T.D.P. (1998) *Angewandte Chemie*, **110**, 973; Masood, M.A., Enemark, E.J. and Stack, T.D.P. (1998) *Angewandte Chemie – International Edition*, **37**, 928; (c) Amendola, V., Fabbrizzi, L., Mangano, C., Pallavicini, P., Roboli, E. and Zema, M. (2000) *Inorganic Chemistry*, **39**, 5803; (d) Hamblin, J., Childs, L.J., Alcock, N.W. and Hannon, M.J. (2002) *Journal of The Chemical Society-Dalton Transactions*, 164 See also Ref. [8m].

13 Examples of the diastereoselective formation of helicates with *P*-donor ligands: (a) Airey, A.L., Swiegers, G.F., Willis, A.C. and Wild, S.B. (1997) *Inorganic Chemistry*, **36**, 1588; (b) Cook, V.C., Willis, A.C., Zank, J. and Wild, S.B. (2002) *Inorganic Chemistry*, **41**, 1897; (c) Bowyer, P.K., Cook, V.C., Gharib-Naseri, N., Gugger, P.A., Rae, A.D., Swiegers, G.F., Willis, A.C., Zank, J. and Wild, S.B. (2002) *Proceedings of the National Academy of Sciences of the United States of America*, **99**, 4877.

14 (a) Scarrow, R.C., White, D.L. and Raymond, K.N. (1985) *Journal of the American Chemical Society*, **107**, 6540; (b) Hou, Z., Sunderland, C.J., Nishio, T. and Raymond, K.N. (1996) *Journal of the American Chemical Society*, **107**, 5148; (c) Kersting, B., Meyer, M., Powers, R.E. and Raymond, K.N. (1996) *Journal of the American Chemical Society*, **118**, 7221.

15 (a) Albrecht, M., Janser, I., Houjou, H. and Fröhlich, R. (2004) *Chemistry - A European Journal*, **10**, 2839; (b) Albrecht, M., Janser, I., Kamptmann, S., Weis, P., Wibbeling, B. and Fröhlich, R. (2004) *Journal of The Chemical Society-Dalton Transactions*, 37; (c) Albrecht, M., Janser, I., Fleischhauer, J., Wang, Y., Raabe, G. and Fröhlich, R. (2004) *Mendeleev Communications*, 250; (d) Albrecht, M., Janser, I., Lützen, A., Hapke, M., Fröhlich, R. and Weis, P. (2005) *Chemistry - A European Journal*, **11**, 5742.

16 (a) Hahn, F.E., Isfort, C.S. and Pape, T. (2004) *Angewandte Chemie*, **116**, 4911; Hahn, F.E., Isfort, C.S. and Pape, T. (2004) *Angewandte Chemie – International Edition*, **43**, 4807; (b) Isfort, C.S., Kreickmann, T., Pape, T., Fröhlich, R. and Hahn, F.E. (2007) *Chemistry - A European Journal*, **13**, 2344.

17 Albrecht, M. (1996) *Synlett*, 565.

18 (a) Enemark, E.J. and Stack, T.D.P. (1995) *Angewandte Chemie*, **107**, 1082; Enemark, E.J. and Stack, T.D.P. (1995) *Angewandte Chemie – International Edition*, **34**, 996; (b) Enemark, E.J. and Stack, T.D.P. (1998) *Angewandte Chemie*, **110**, 977; Enemark, E.J. and Stack, T.D.P. (1998) *Angewandte Chemie – International Edition*, **37**, 932.

19 Lützen, A., Hapke, M., Griep-Raming, J., Haase, D. and Saak, W. (2002) *Angewandte Chemie*, **114**, 2190; Lützen, A., Hapke, M., Griep-Raming, J., Haase, D. and Saak, W. (2002) *Angewandte Chemie – International Edition*, **41**, 2086.

20 Hapke, M., Thiemann, F. and Lützen, A. (2004) unpublished results. Preliminary results have been presented on the International Conference on Supramolecular Science & Technology (ICSS&T) 2004, Prague (Czech Republic), 5-9. September 2004, *Chemicke Listy* 98 S47.

21 Schalley, C.A., Lützen, A. and Albrecht, M. (2004) *Chemistry - A European Journal*, **10**, 1072.

22 Clarkson, R.G. and Gomberg, M. (1930) *Journal of the American Chemical Society*, **52**, 2881.

23 A CAS inquiry from October 28th, 2007 revealed that more than 1800 derivatives of 9,9′-spirobifluorene have been described in more than 740 publications so far. However, especially the last couple of years (2002-2007) have witnessed a tremendous development of this class of compounds, when 320 patents were filed and in total more than 570 publications appeared. Recent review: Saragi, T.P.I., Spehr, T., Siebert, A., Fuhrmann-Lieker, T. and Salbeck, J. (2007) *Chemical Reviews*, **107**, 1011.

24 Haas, G. and Prelog, V. (1969) *Helvetica Chimica Acta*, **52**, 1202.

25 Toda, F. and Tanaka, K. (1988) *The Journal of Organic Chemistry*, **53**, 3607.

26 Thiemann, F., Piehler, T., Saak, W., Haase, D. and Lützen, A. (2005) *European Journal of Organic Chemistry*, 1991.

27 (a) Valik, M., Strongin, R.M. and Kral, V. (2005) *Supramolecular Chemistry*, **17**, 347; (b) Dolenský, B., Elguero, J., Král, V., Pardo, C. and Valík, M. (2007) *Advances in Heterocyclic Chemistry*, **93**, 1.

28 (a) Jensen, J. and Wärnmark, K. (2001) *Synthesis*, 1873; (b) Hansson, A., Jensen, J., Wendt, O.F. and Wärnmark, K. (2003) *European Journal of Organic*, 3179.

29 (a) Kiehne, U. and Lützen, A. (2004) *Synthesis*, 1687; (b) Kiehne, U., Weilandt, T. and Lützen, A. (2007) *Organic Letters*, **9**, 1283; (c) Kiehne, U., Bruhn, T., Schnakenburg, G., Fröhlich, R., Bringmann, G. and Lützen, A. (2008) *Chemistry - A European Journal*, **14**, 4246. Kiehne, U. and Lützen, A. *unpublished results*.

30 (a) Jensen, J., Strozyk, M. and Wärnmark, K. (2002) *Synthesis*, 2761; (b) Hof, F., Schar, M., Scofield, D.M., Fischer, F., Diederich, F. and Sergeyev, S. (2005) *Helvetica Chimica Acta*, **88**, 2333; (c) Solano, C., Svensson, D., Olomi, Z., Jensen, J., Wendt, O.F. and Wärnmark, K. (2005) *European Journal of Organic*, 3510; (d) Bew, S.P., Legentil, L., Scholier, V. and Sharma, S.V. (2007) *Chemical Communications*, 389.

31 von Zelewsky, A. (1995) *Stereochemistry of Coordination Compounds*, John Wiley & Sons, Chichester.

32 (a) Kaes, C., Katz, A. and Hosseini, M.W. (2000) *Chemical Reviews*, **100**, 3553; (b) Newkome, G.R., Patri, A.K., Holder, E. and Schubert, U.S. (2002) *European Journal of Organic Chemistry*, 235.

33 (a) Lützen, A. and Hapke, M. (2002) *European Journal of Organic Chemistry*, 2292; (b) Lützen, A., Hapke, M. and Meyer, S. (2002) *Synthesis*, 2289; (c) Lützen, A., Hapke, M., Staats, H. and Bunzen, J. (2003) *European Journal of Organic Chemistry*, 3948; (d) Kiehne, U., Bunzen, J., Staats, H. and Lützen, A. (2007) *Synthesis*, 1061; (f) Hapke, M., Staats, H., Wallmann, I. and Lützen, A. (2007) *Synthesis*, 2711; (g) Haphe, M., Brauds, L. and Lützen, A. (2008) *Chemical Society Reviews, in press*.

34 Sigel, H. and Martin, R.B. (1994) *Chemical Society Reviews*, **23**, 83.

35 Kiehne, U. and Lützen, A. (2007) *Organic Letters*, **9**, 5333.

36 Ziegler, M. and von Zelewsky, A. (1998) *Coordination Chemistry Reviews*, **177**, 257.

37 Telfer, S.G., Kuroda, R. and Sato, T. (2003) *Chemical Communications*, 1064. See also Ref. [8m].

38 Amendola, V., Fabbrizzi, L., Mangano, C., Pallavicini, P., Roboli, E. and Zema, M. (2000) *Inorganic Chemistry*, **39**, 5803.

39 Grosshans, P., Jouaiti, A., Bulach, V., Planeix, J.-M., Hosseini, M.W. and Nicoud, J.-F. (2003) *Chemical Communications*, 1336.

40 Piehler, T. and Lützen, A. *unpublished results.*

41 Khoshbin, M.S., Ovchinnikov, M.V., Mirkin, C.A., Golen, J.A. and Rheingold, A.L. (2006) *Inorganic Chemistry*, **45**, 2603.

42 Kiehne, U. and Lützen, A. (2007) *European Journal of Organic Chemistry*, **34**, 5703.

43 Kiehne, U., Weilandt, T. and Lützen, (2008) *European Journal of Organic Chemistry*, **12**, 2056.

44 Kiehne, U. and Lützen, A. *unpublished results.*

Index

Activating Unreactive Substrates: The Role of Secondary Interactions.
Edited by Carsten Bolm and F. Ekkehardt Hahn

ISBN: 978-3-527-31823-0